Solid-State Devices and Integrated Circuits

Solid-State Devices and Integrated Circuits

William D. Cooper
Henry B. Weisbecker

RESTON PUBLISHING COMPANY, INC., Reston, Virginia
A Prentice-Hall Company

Library of Congress Cataloging in Publication Data

Cooper, William David.
Solid-state devices and integrated circuits.

Includes bibliographical references and index.
1. Semiconductors. 2. Integrated circuits.
I. Weisbecker, Henry. II. Title.
TK7871.85.C623 621.3815'2 82-307
ISBN 0-8359-7045-0 AACR2

10 9 8 7 6 5 4 3 2 1

Printed in the United States of America.

Contents

CHAPTER 6 – POWER AMPLIFIERS 216

CHAPTER 7 – RECTIFIERS AND FILTERS 241

CHAPTER 8 – POWER SUPPLY REGULATORS 283

Preface

Few phenomena of our twentieth century can equal in importance the growth and development of the electronics industry. Starting with such modest beginnings as the discovery of the Edison effect (the finding that a heated filament could emit electrons) and the transmission by Marconi of a wireless signal across the Atlantic, the industry today affects every aspect of our existence. In making this tremendous progress, the industry went through many generations of equipment, so many that it seems hard to believe that the entire process covers less than a century. The first giant step was the development by DeForest of the vacuum tube, a device that enabled the industry to create many of its electronic wonders. Such devices as AM and FM radios, television, computers, radar, sonar, servomechanisms, microwave relays, etc., all came into existence during the time when the vacuum tube was the only available active device. However, the tube had three fundamental defects—the need for energy to heat the filaments, the large size, and the limited life—which caused electronic equipment to remain largely bulky, inefficient, and expensive.

The development of the transistor by Shockley and his associates during the early 1950's solved all three problems at once. A filament was no longer required so energy consumption dropped radically, the size was drastically reduced, and reliability was greatly enhanced.

This was to be only an early step in the phenomenal development of the industry. It was discovered that not only transistors but also such associated elements as resistors, capacitors, and wires could be etched right into the semiconductor material, resulting in the integrated circuit. This meant that, instead of constructing a circuit of many transistors and their associated components, a single unit could be built that would perform a given function—such as an amplifier, for example.

But this was not the end. Further progress resulted in large-scale integration, which enabled many circuits to be placed on a very small chip of semiconductor material. This brought with it not only smaller size and greater efficiency but also vastly greater speed of operation.

And now we have microprocessors. These are not only large-scale-integrated circuits. Not only do they encompass the components for performing electronic functions, but also they can be programmed, so that an entire computer is contained in a chip. The result of all this is that what once required a roomful of equipment is today a unit comparable in size to a coin. Costwise, what can be put into a $10 calculator used to cost, perhaps, a million old dollars.

This book, being written at the time it is, no longer even mentions vacuum tubes. It does treat transistors and other discrete devices because the integrated versions cannot be understood without an understanding of the discrete devices. But future books may not even mention these, as the function of the engineer gradually changes from being a tinkerer with tubes, resistors, capacitors, transistors, diodes, etc., to either a designer of chips or a user of these marvelous devices, only concerned with their applications.

This book is a step along the way, but by no means the last word in this rapidly developing industry.

Henry B. Weisbecker

About the Authors

Henry B. Weisbecker graduated with a bachelor's degree from Pratt Institute, Brooklyn, New York, earned a master's degree from New York University, and received his doctorate from the Technical University of Munich, Germany. Originally, working as a radio engineer, he later participated in the development of color television, worked in the field of servomechanisms, and developed infrared equipment and facsimile systems. He then became an Associate Professor of Electrical Engineering at the New Jersey Institute of Technology in Newark. Dr. Weisbecker has written two books and numerous articles in his field. He has been a Senior Member of the IEEE and is now a Consultant in Old Bridge, New Jersey.

Solid-State Devices and Integrated Circuits

Chapter 1

Introduction to Semiconductor Electronics

1-1 PROPERTIES OF SEMICONDUCTOR CRYSTALS

1-1.1 Silicon and Germanium Atoms

Silicon and germanium are crystalline solids used in the manufacture of solid-state devices. For early transistors, germanium (Ge) was the more important of the two materials, but because of superior temperature characteristics silicon (Si) is now the major semiconductor material.

A *silicon atom* consists of a nucleus and 14 orbiting electrons. The nucleus is comprised of 14 protons (the positive charge) and 14 neutrons. An isolated silicon atom is electrically neutral; that is, it contains equal numbers of positive charges (protons) and negative charges (electrons).

The electrons are in orbit around the nucleus. Figure 1-1(a) shows the Bohr model of the silicon atom. The two inner orbits or *shells* contain their maximum number of electrons: 2 electrons in the first shell, closest to the nucleus, and 8 electrons in the second shell. The remaining 4 electrons only partially fill the eight available positions in the third, or outermost, shell. The 10 inner electrons are tightly bound to the nucleus; the 4 electrons in the partially filled outer shell are loosely tied to the nucleus.

Figure 1-1(b) shows the Bohr model of a *germanium atom*. The positive charge of the atom is in the nucleus. There are 32 electrons in the atom; 28 of these fill the three shells closest to the nucleus, the remaining four electrons partially fill the fourth shell. The four electrons in the outer shells of the silicon and germanium atoms are called *valence* electrons. They contribute to electrical conduction in semiconductor materials and so are of special interest in semiconductor physics. Orbiting electrons possess kinetic energy, since each electron is a mass in motion. Each electron has a certain energy value and operates at a distinct *energy level.* This is a function of (1) the distance between the nucleus and the orbiting electron, and (2) the momentum of the electron.

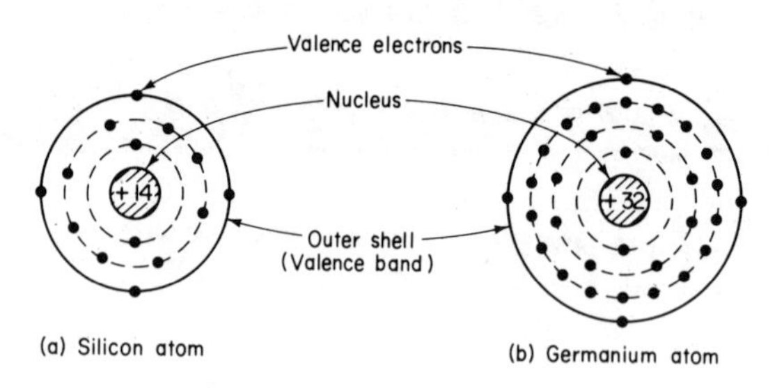

Fig. 1–1 Bohr models of silicon and germanium atoms. The silicon atom has 14 orbiting electrons and the germanium atom 32.

Electrons orbiting in the same shell operate at the same energy level; those in shells close to the nucleus at a higher level than those in shells farther away. Hence, it requires less energy to remove an electron from the outer shell (a valence electron) than from an inner shell, and the inner electrons are more tightly bound to the nucleus. The outer shell, containing the valence electrons, is called the *valence shell* or *valence band.*

The Bohr model of these atoms can be simplifed by showing a *central core*, which consists of the nucleus and the tightly bound inner electrons, and the *valence shell*, which contains the four loosely bound valence electrons (Fig. 1–2). These simplified models of the silicon and germanium atoms are identical, since each atom consists of a central core and four valence electrons.

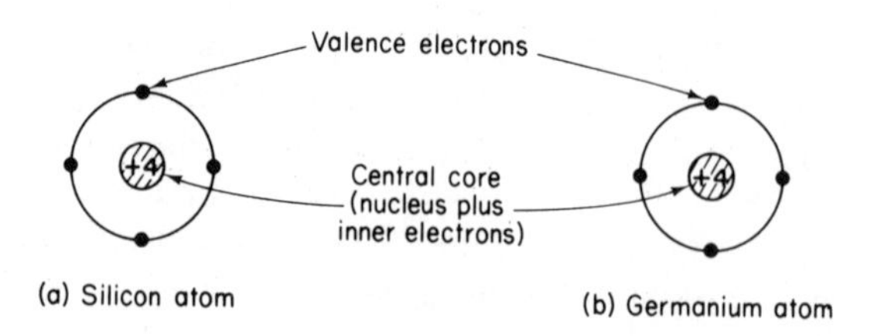

Fig. 1–2 Simplified models of silicon and germanium atoms. Each core consists of the nucleus plus the inner electrons and in both cases has a positive charge of 4.

1–1.2 Single Crystals

A silicon or germanium *crystal* consists of many atoms, bound together in a regular and distinctive pattern called a *lattice structure*. The mechanism by which these atoms are bound together in a fixed spatial relation to one another is called *covalent bonding*. Figure 1–3 is a two-dimensional representation of a silicon or germanium lattice structure, which shows covalent bonding of the atoms. The central atom in the structure contains eight electrons in its valence shell—four valence electrons belong to the central atom itself and four additional valence electrons are supplied by four neighbors of the central atom. The valence shell of this central atom, with eight electrons in covalent bonding, is in a *stable* state. Expanding by adding more silicon or germanium atoms to the structure, every atom in the crystal, sharing its valence electrons with its neighbors, has a stable valence shell of eight electrons.

When a piece of silicon or germanium contains several crystals, it is a *polycrystalline* material. The best solid-state devices, however, are manufactured from *single crystals*.

1–1.3 Conductivity of the Crystal

In a perfect crystal lattice, all valence electrons are part of the covalent bonding system, and there are no excess electrons available as possible electrical charge carriers. A perfect and stable lattice structure behaves as a perfect insulator. This occurs at 0°K.

However, the covalent bonds of a pure (*intrinsic*) silicon or germanium crystal can be broken by the application of heat (thermal agitation). When a covalent bond is broken, a valence electron is dislodged from its natural position in the lattice structure and becomes a *free* electron. Free electrons, generated by thermal energy, are responsible for the conductive properties of the intrinsic crystal.

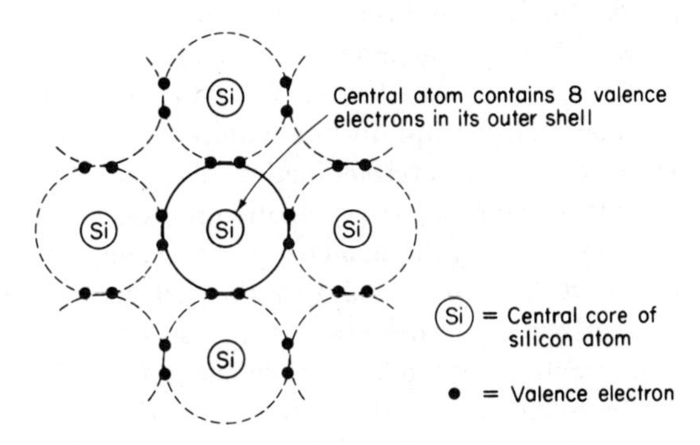

Fig. 1–3 Sharing of valence electrons results in covalent bonding.

When the temperature of the crystal is raised, some covalent bonds are broken, liberating some valence electrons. These free valence electrons act as electric charge carriers and the crystal exhibits some conductivity.

In other materials, such as copper (Cu) or aluminum (Al), the valence electrons are not all used in the covalent bonding system of the crystal lattice. In fact, in a bar of pure copper one free unbound electron exists for every copper atom, so that even at 0° K, many free electrons are available as charge carriers. Again, if the temperature is raised, covalent bonds in the crystal structure are broken, liberating additional valence electrons and causing the conductivity of the copper bar to increase. The very large number of available free electrons in copper makes this material an excellent *conductor*.

In silicon and germanium crystals, the number of free electrons at room temperature (T = 25°C) is relatively small and they behave neither as good insulators nor as good conductors. Hence, we refer to silicon and germanium as *semiconductors*.

1–1.4 Electrons and Holes

Heating a semiconductor crystal causes valence electrons to break away from their parent atoms to become free electrons. The departure of a valence electron creates an *electron deficiency* in the valence shell of the atom, and the atom assumes a *positive charge* equal in magnitude to the charge of an electron. This electron deficiency is called a *hole*. The hole, defined as a vacancy in the atomic structure, gives the atom its positive charge, and it is therefore convenient to consider the hole as a positively charged particle. The application of thermal energy creates one hole for every valence electron set free, so that an equal number of positive and negative charges is generated. This process is known as thermal generation of *electron-hole pairs*.

The thermally generated electrons move through the crystal in a random and unorganized manner called *diffusion*. When a free electron collides with a hole, it is captured by the hole, and the normal stable state of eight electrons in the valence band of that particular atom is reestablished. The hole disappears and the captured electron becomes a bound electron again. This action is called *recombination*. To maintain the equilibrium within the crystal, a new electron-hole pair is generated every time a recombination takes place.

The electron involved in the recombination process is not necessarily one of the free electrons generated by thermal energy. For example, an atom that has "lost" an electron (and "gained" a hole) by thermal generation can acquire another electron from one of its neighboring atoms. When this happens, the covalent bond of the neighboring atom is broken and a new hole is created. This hole, in turn, can be filled by an electron from one of its neighboring atoms, thereby creating another hole. It is evident that the movement of an electron "filling" a hole is accompanied by the creation of another hole in a different location. Hence, the holes move through the crystal as freely and randomly as the electrons.

1–2 DOPED SEMICONDUCTORS

1–2.1 n-type Semiconductors

It was shown that pure or intrinsic semiconductor material contains equal numbers of electrons and holes, and that the conductivity of an intrinsic semiconductor crystal is low and strongly dependent on temperature. But the conductivity can be greatly increased, and closely controlled, by introducing a small number of *impurity atoms* into the semiconductor crystal through a process called *doping*.

Consider the silicon crystal of Fig. 1–4, where one of the silicon atoms in the crystal lattice has been replaced by a phosphorus (P) atom. A phosphorus atom is *pentavalent*, i.e., it has five electrons in its valence shell.

When a phosphorus atom occupies the position of a silicon atom in the crystal lattice, four of its valence electrons form covalent bonds with adjacent silicon atoms. The fifth valence electron cannot take part in a covalent bond and is only lightly attached to the parent atom. This fifth (unbound) electron is a *free electron* in the sense that it can easily move to other parts of the crystal. Since the phosphorus atom *contributes* a free electron for conduction, we call phosphorus a *donor impurity*. Antimony (Sb) and arsenic (As) are two other pentavalent materials used in doped semiconductor crystals.

By adding the correct concentration of pentavalent impurity atoms to the silicon crystal, the conductivity of the doped material can be controlled very closely. In a representative semiconductor device, the concentration of donor

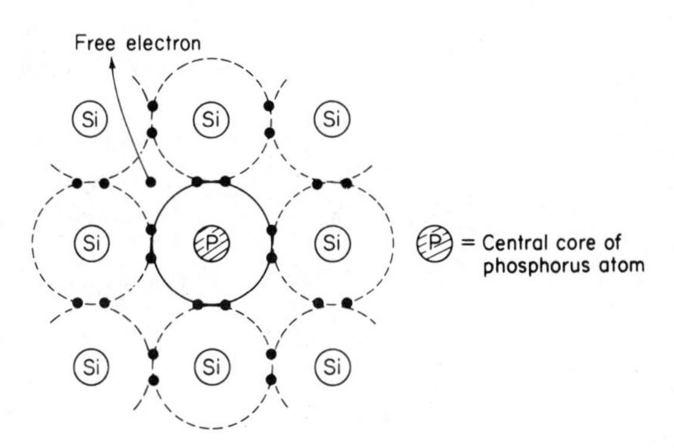

Fig. 1–4 The pentavalent phosphorus atom produces a free electron, available for conduction.

atoms in the silicon crystal is typically 1 part in 10^6, a slight amount of impurity. The addition of the impurity leaves the metallurgical properties of the crystal practically unchanged, but the electrical characteristics are greatly altered.

There is a difference between free electrons contributed by donor atoms, and free electrons generated by thermal agitation (ionization of the silicon atoms). Heating provides a number of free electrons *plus* an equal number of holes. Addition of donor impurity atoms creates *only* free electrons. The doped crystal therefore contains an *excess of free electrons* and is called an *n-type* semiconductor. Hence, both electrons and holes are present in the *n*-type crystal, and both contribute to the conduction process. Since the electrons in the *n*-type semiconductor predominate, they are called the *majority carriers*; the holes are the *minority carriers*.

1-2.2 p-type Semiconductors

Figure 1-5 shows that one of the silicon atoms in the crystal lattice has been replaced by an aluminum (Al) atom. The aluminum atom is *trivalent*, i.e., it has only three electrons in its valence shell. These are used to form covalent bonds with adjacent silicon atoms. There is no valence electron available to form the fourth bond in the regular lattice structure, and a *hole* therefore exists in the lattice. We consider this hole a *positive charge*, because it can attract and capture an electron from one of the adjacent silicon atoms.

Since the aluminum atom can capture an electron, we call aluminum an *acceptor impurity*. Other elements used as acceptor impurities are boron (B), gallium (Ga), and indium (In).

Concentration of acceptor atoms in the silicon structure is again about 1 part in 10^6.

Trivalent impurity atoms introduced into the silicon crystal create an *excess of holes* (*positively charged* particles), and this doped crystal is therefore called a *p-type semiconductor*. Both holes and electrons take part in the

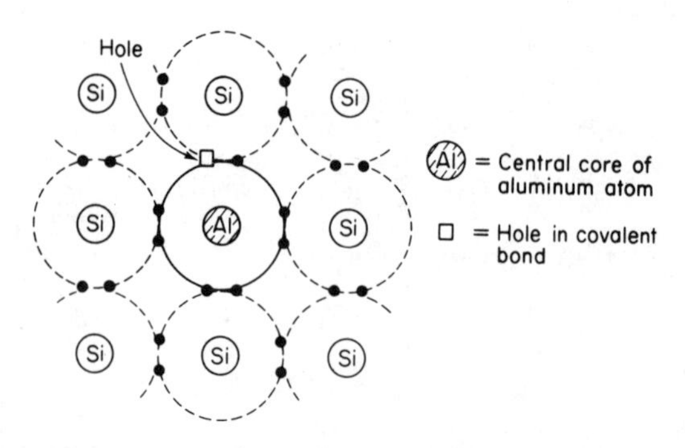

Fig. 1-5 The trivalent aluminum atom produces a hole in the covalent bond.

conduction process, but holes are the *majority carriers* and electrons are the *minority carriers.*

The concentration of charge carriers created by doping is virtually independent of temperature, while the normal ionization of the silicon crystal is highly dependent on temperature.

1-2.3 Conduction in the n-type Semiconductor

The n-type semiconductor crystal of Fig. 1-6 is provided with contacts and a battery is connected across it. The switch is initially open. At room temperature ($T = 25°C$), the crystal contains a number of thermally generated *electron-hole pairs*, plus a number of *free electrons* contributed by the donor impurities. These positive and negative charges are in a state of thermal diffusion, moving in a random manner.

When the switch is closed, an electric field is produced in the crystal. This field causes a definite and deliberate movement of electric charges (called *drift*) such that the free electrons (majority carriers) drift through the crystal from left to right, while holes (minority carriers) drift from right to left. The movement of charges is explained with reference to Fig. 1-6.

When an electron, supplied by the battery, enters the crystal via the negative contact, it collides with the hole in atom A and recombines. To maintain equilibrium within the crystal, a new electron-hole pair is generated. The free electron of atom A drifts to the right under the influence of the electric field and collides with the hole in atom B, causing recombination and generation of another electron-hole pair. The free electron of B drifts to the right and collides with the hole in atom C, where it recombines. The free electron of C then drifts to the right and leaves the crystal through the contact, to the positive battery terminal.

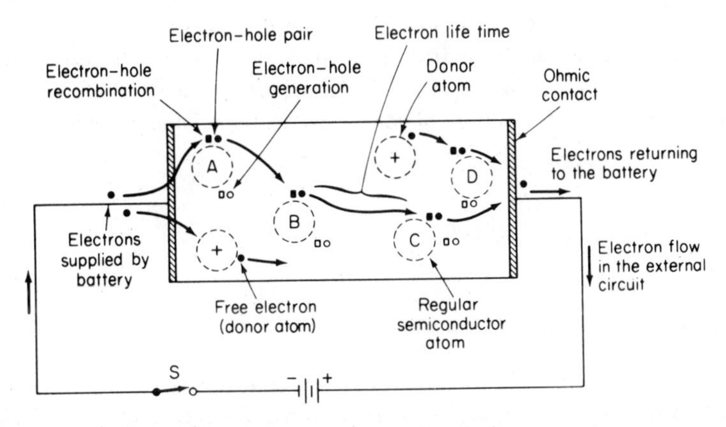

Fig. 1-6 The conduction process in an n-type semiconductor.

As shown in Fig. 1-6, donor electrons move through the crystal, taking part in the conduction process through recombination and generation, the same as the thermally generated electrons. Their number increases the conductivity of the crystal considerably, and thereby increases the electron flow.

The holes contribute fully to the conduction process, acting as charge carriers. The hole activity can best be followed by starting from the right-hand side of the crystal. When the free electron of atom *C*, attracted by the positive battery terminal, leaves the crystal through the contact, the hole remaining in the structure of *C* is filled by the free electron arriving from *B* (recombination). This leaves a hole in the structure of *B*, which in turn is filled by the free electron arriving from *A*. The resulting hole in atom *A* is filled by an electron arriving from the battery. The *apparent* hole movement is therefore from the right to the left, opposite in direction to the electron flow.

Electrons move around the *external* circuit, from the negative battery terminal through the crystal to the positive battery terminal. The hole movement is from positive to negative, but only *inside* the crystal, not in the external circuit.

1-2.4 Conduction in the p-type Semiconductor

The conduction process in the *p*-type crystal is explained with reference to Fig. 1-7. The majority carriers in the *p* crystal are holes, contributed by the acceptor impurities and by thermal generation of electron-hole pairs. The initial movement of electrons and holes, prior to closing the switch, is random (diffusion). When the switch is closed, the resulting electric field in the crystal causes the positive and negative charges to drift through the crystal structure. As in the case of the *n*-type crystal, the electrons flow from the negative battery terminal, through the crystal by the process of recombination and generation, back to the positive battery terminal. The mechanism of transporting electrons through the crystal is again through the availability of holes.

The hole movement is from positive to negative and is restricted to the crystal. The acceptor atoms, contributing free holes to the structure, increase the conductivity by making additional charge carriers available, as indicated in Fig. 1-7.

1-3 pn JUNCTION

1-3.1 Formation of the pn Junction

A *pn* junction is "*formed*" when *p*-type semiconductor material is brought into contact with *n*-type semiconductor material. We use the word "formed" instead of "joined," since the *pn* junction is not made by joining two separate crystals

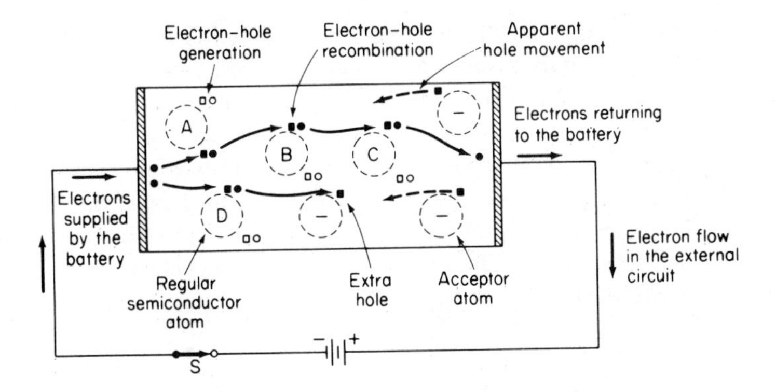

Fig. 1–7 The conduction process in a *p*-type semiconductor.

together, but by abruptly changing the doping characteristics of a single semiconductor crystal from *p*-type to *n*-type during the formation of the crystal, thus forming a *continuous* crystal structure whose characteristics change from *p*-type, through the junction, to *n*-type.

The actual construction of a *pn* junction is a rather complex chemical process which may involve one of several techniques, for example, alloying, diffusion, or chemical deposition. A single crystal is produced whose doping characteristics change abruptly from *p*-type to *n*-type, as indicated in Fig. 1–8(a). The separation between the *p* region and the *n* region is the *pn junction*, which forms the basis of all semiconductor devices, including junction diodes, bipolar junction transistors, and even integrated circuits.

Considering the two regions in Fig. 1–8(a) separately, before the formation of the *pn* junction, we observe that each acceptor atom in the *p*-type material contributes a hole, and each donor atom in the *n*-type material contributes a free electron. Holes predominate in the *p* material, and electrons predominate in the *n* material; so the two materials are at different charge levels.

When the junction between the two dissimilar materials is first formed, the energy levels at both sides of the junction equalize by an exchange of electric charges. Free electrons in the *n* region, attracted by the positive charges in the *p* region, cross the junction to fill holes in the *p* region. This process is illustrated in Fig. 1–8(b).

The donor atoms in the *n* region become positive ions as the free electrons cross the junction into the *p* region. Similarly, the holes in the *p* region accept the electrons from the *n* region, and the acceptor atoms become negative ions. On either side of the junction, there is a region where the free electrons from the *n* side fill the holes in the *p* side, producing an area *empty of free charge*, as shown in Fig. 1–8(c).

The exchange of free charge carriers ceases when the electric field, created by the exposed ions on either side of the junction, builds up and prevents free electrons from moving across the junction. The limited area free of charge carriers on either side of the junction is called the *depletion region*. The negative ions in the *p* region and positive ions in the *n* region create a potential difference across the junction, called *barrier potential*, *junction potential*, or *contact potential*. This potential is sufficiently large to oppose further movement of charges across the junction. The depletion region is then formed and the *pn* junction is at equilibrium. The situation is shown in Fig. 1–8(d), where the

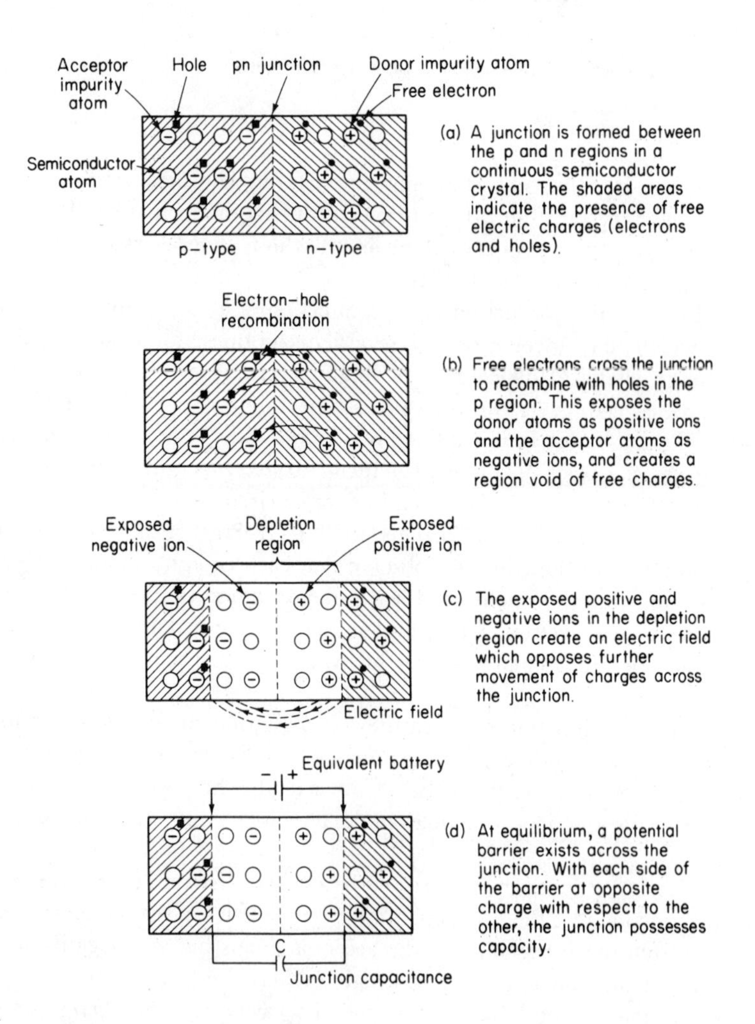

Acceptor impurity atom
Hole
pn junction
Donor impurity atom
Free electron
Semiconductor atom
p–type
n–type
(a) A junction is formed between the p and n regions in a continuous semiconductor crystal. The shaded areas indicate the presence of free electric charges (electrons and holes).
Electron–hole recombination
(b) Free electrons cross the junction to recombine with holes in the p region. This exposes the donor atoms as positive ions and the acceptor atoms as negative ions, and creates a region void of free charges.
Exposed negative ion
Depletion region
Exposed positive ion
(c) The exposed positive and negative ions in the depletion region create an electric field which opposes further movement of charges across the junction.
Electric field
Equivalent battery
(d) At equilibrium, a potential barrier exists across the junction. With each side of the barrier at opposite charge with respect to the other, the junction possesses capacity.
C
Junction capacitance

Fig. 1–8 Formation of the *pn* junction.

barrier potential is represented by a small battery across the depletion region, in the polarity shown.

It is found that at room temperature ($T = 25°C$) the barrier potential is approximately 0.3 V for a germanium *pn* junction and 0.7 V for a silicon *pn* junction. The contact potential of a *pn* junction can play an important role in the performance of a circuit, especially in low-voltage applications.

The minority carriers present in both halves of the *pn* junction diffuse through the crystal in a random motion. When they approach the depletion region, they are swept across the junction under the influence of the contact potential. At equilibrium of the junction, however, the minority electron flow from *p* to *n* is balanced by an equal but opposite minority hole flow from *n* to *p*, and there is no net flow of charge carriers across the junction.

1–3.2 Conductivity of the pn Junction

Consider the *pn* junction in Fig. 1–9(a). The positive battery terminal is connected to the *p* region and the negative terminal to the *n* region. This connection is called *forward bias* of the *pn* junction. When the forward bias voltage exceeds the barrier potential, an electric field is created in the crystal and a drift of charge carriers occurs.

Electrons, injected into the *n* region by the negative battery terminal, move toward the junction under the influence of the applied bias voltage. The greater part of these injected electrons *cross* the depletion region and recombine with holes in the *p* region. For every electron-hole recombination thus made, an electron leaves the *p* region through the contact, and another hole is created in the *p* region to maintain the overall balance of charges. Hence, we find a continuous flow of electrons, originating from the negative battery terminal, drifting through the *n* region across the junction, and via the recombination process in the *p* region back to the positive battery terminal. The apparent hole movement in the *p* region is limited to that region, from left to right as indicated in Fig. 1–9(a).

A fundamental property of the *pn* junction therefore is that it *conducts under forward-bias conditions* (positive battery to *p* region, negative battery to *n* region). The magnitude of this *forward current* is exponentially related to the magnitude of the applied forward-bias voltage.

Some of the injected electrons identify themselves with the exposed positive donor ions on the *n* side of the depletion region, and some of the holes in the *p* region identify themselves with the exposed negative acceptor ions on the *p* side of the depletion region. As a result, the depletion region becomes narrower and the potential across the barrier decreases. This is also indicated in Fig. 1–9(a).

When the polarity of the bias voltage is reversed, so that the positive battery terminal is connected to the *n* region and the negative to the *p* region, we speak of *reverse bias*. This condition is indicated in Fig. 1–9(b). The electric field established by the reverse bias voltage now aids or reinforces the field in the depletion region. As a result, the free electrons in the *n* region move away from the junction toward the positive device contact, and the holes in the *p* region move away from the junction toward the negative device contact. Hence, the depletion region widens, the barrier potential increases, and there is no movement of charges across the junction. The *pn* junction therefore "blocks" the flow of electrons under reverse-bias conditions.

This holds only under ideal conditions. In fact, with reverse bias applied, there is a small current through the *pn* junction, known as *reverse current* or *leakage current*. This leakage current is due to the minority carriers present in the *pn* junction. The reverse bias sweeps them across the junction, but since relatively few minority carriers are available, the leakage current is small (in the

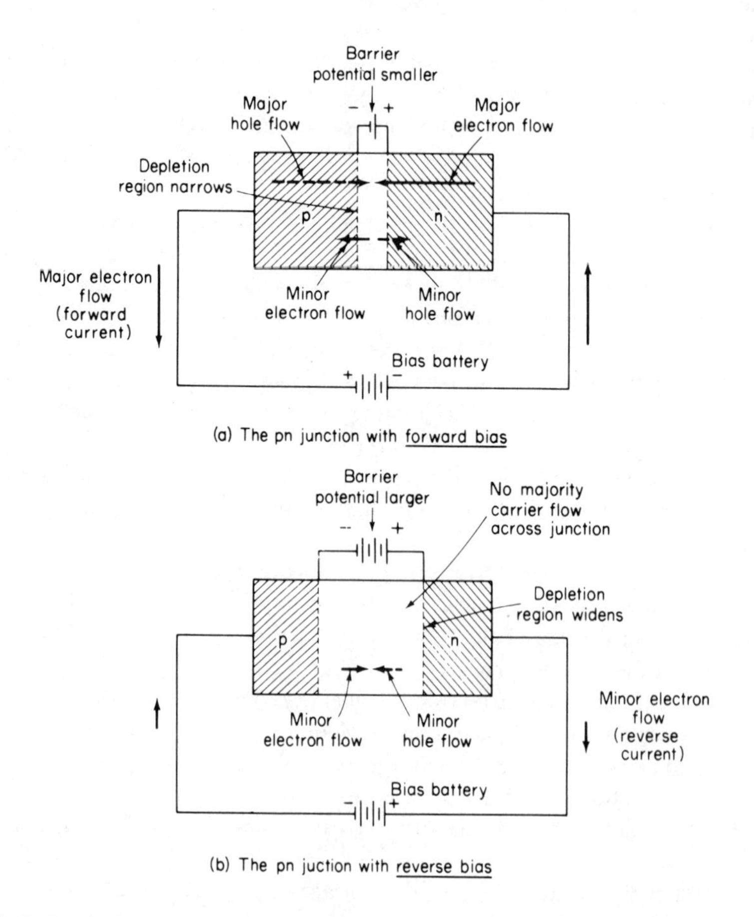

Fig. 1–9 Conductivity of the *pn* junction under forward and reverse bias conditions.

microampere range). We recall, however, that the number of minority carriers increases with temperature; therefore, the leakage current of the *pn* junction increases with an increase in temperature.

1-3.3 Current Direction

It is important not to be confused by the directions in which the various currents flow. Basic dc circuit theory says that electric current has both magnitude and direction. The *magnitude* is defined as electric charge (electrons) flowing past a point in the circuit over a given time interval (coulombs per second), and it can be measured by placing an ammeter in series with the circuit. The *direction* or *polarity* of the current is more difficult to define. *Conventional* current direction is from the positive battery terminal, through the external circuit, to the negative battery terminal. On the other hand, the *electron-flow* direction is from the negative battery terminal, through the external circuit, to the positive battery terminal, through the external circuit, to the positive battery terminal. Figure 1-10 shows that these two current directions are opposite to each other.

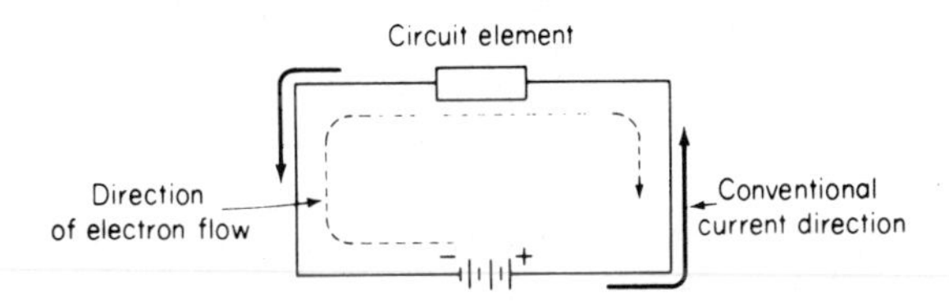

Fig. 1-10 Current directions: electron flow and conventional current.

Solid-state devices, whose action is frequently explained in terms of electron flow and hole flow, compound the problem. New terminology is introduced, such as majority and minority current, forward and reverse current, leakage current, and it becomes essential that the directions of current flow are clearly identified. Either the conventional current direction (from + to −), or the electron-flow direction (from − to +) can be chosen, but once a choice has been made it must be strictly applied, or confusion results.

In practical semiconductor circuits, the conventional current direction is generally used. To illustrate this, consider the forward-biased *pn* junction of Fig. 1-11(a). This *pn* junction conducts, passing a *forward current*, when the positive voltage is applied to the *p* material and the negative voltage to the *n* material. The direction of the forward current is from positive to negative in the circuit and from *p* to *n* in the crystal itself.

When the *pn* junction is reverse-biased, as in Fig. 1-11(b), the junction blocks current flow almost completely, except for the small minority carrier

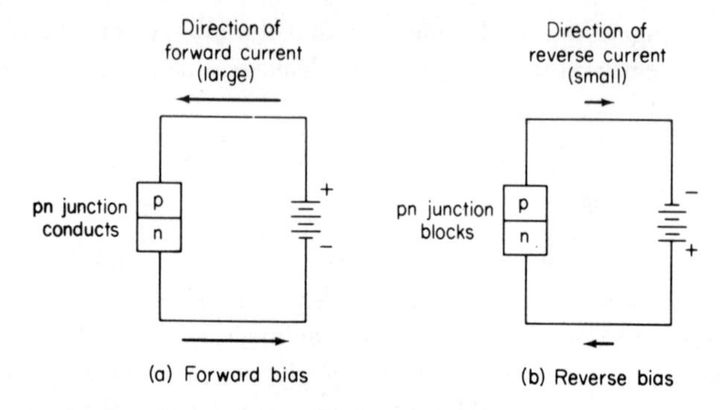

Fig. 1–11 Current direction in a forward-biased and in a reverse-biased *pn* junction.

movement across the junction (Section 1–3.2). The direction of this *reverse current* is again from positive to negative in the circuit, but now from *n* to *p* in the crystal.

The conventional current direction is used in all circuits and problems discussed in this textbook.

QUESTIONS

1–1 What is the relationship between the number of electrons and protons in an isolated silicon atom?

1–2 Why is an intrinsic semiconductor a poor conductor at low temperatures?

1–3 What is meant by a covalent bond?

1–4 What is the effect of doping on the conductivity of a semiconductor?

1–5 Name four impurities used in the manufacture of semiconductor crystals and state whether they produce *n*-type or *p*-type materials.

1–6 What are the majority carriers in *p*-type semiconductor material?

1–7 What are the minority carriers in *n*-type semiconductor material?

1–8 Why does the resistance of a semiconductor decrease as the temperature increases?

1–9 Explain how the contact potential is formed in a *pn* junction.

1–10 What is the approximate contact potential for a silicon and a germanium *pn* junction?

1–11 What is meant by leakage current and how does temperature affect leakage current?

1–12 What is the significance of the minority carriers in a reverse-biased *pn* junction?

1–13 What is the effect of temperature on the current in a forward-biased *pn* junction?

Chapter 2

The Junction Diode

2–1 JUNCTION DIODES

2–1.1 Basic Properties

A junction diode is basically a germanium or silicon *pn* junction, encapsulated and provided with connecting leads.

The *schematic symbol* for the junction diode is shown in Fig. 2–1. The *p*-type semiconductor region is called the anode (*a*) and is represented by the arrowhead of the symbol. The *n*-type semiconductor region is called the cathode (*k*) and is represented by the bar across the arrowhead. The arrowhead points in the direction in which conventional current is easily passed, from anode to cathode, or from plus (+) to minus (−).

The *polarity* of the junction diode is either marked on the body (case) of the device, or indicated by the shape or construction of the case. Low-power

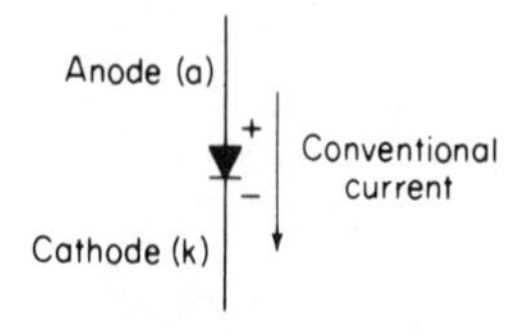

Fig. 2–1 The schematic symbol for the junction diode.

diodes are usually packaged in plastic, and the cathode is identified by a band across the body of the device, close to the cathode lead. In some cases, plus and minus signs are used to indicate the polarity. High-power diodes are generally mounted on a metal base and packaged in a metal container. The case of the device is then the cathode, and the insulated lead emerging from the top of the case is the anode. It is recommended practice, however, to consult the diode manual or the data sheet to verify the polarity before connecting the diode into a circuit.

The junction diode is *forward-biased* when the anode is connected to positive potential and the cathode to negative potential, as in Fig. 2–2(a). In this condition, the diode conducts easily and passes a forward current in the direction shown. It is *reverse-biased* when the anode is connected to negative potential and the cathode to positive potential, as in Fig. 2–2(b). In this case the diode blocks and prevents current in the circuit, as explained in Section 1–3.

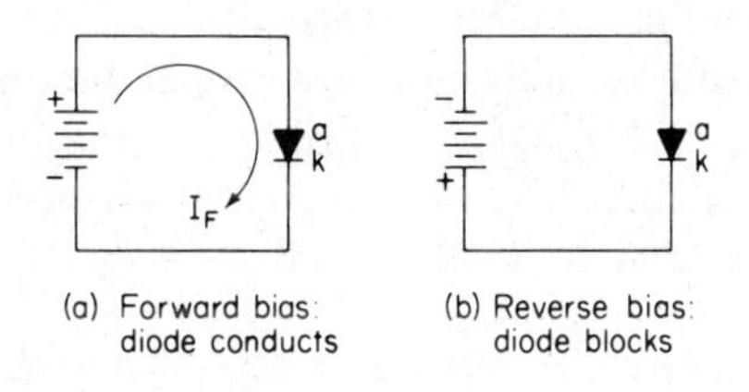

Fig. 2–2 Biasing the *pn* junction diode.

2–1.2 Diode Construction

The search for improved device performance and lower production cost has led to the development of different techniques to make *pn* junctions. Junction diodes can be classified as point-contact, grown-junction, fused- or alloy-junction, diffused-junction, and epitaxial.

Since the *pn* junction is basic to the construction of both diodes and transistors, some of the processes and techniques described here also apply to the construction of other solid-state devices.

The *point-contact* process to manufacture germanium diodes starts with a base wafer or *substrate* of *n*-type monocrystalline germanium (usually antimony-doped to give it the *n* characteristics). An indium-plated *whisker* of tungsten, platinum or molybdenum is brought into contact with the surface of the *n*-type material, as shown in Fig. 2–3. The whisker is then electrically pulsed so that the indium melts and fuses to the germanium substrate. Sudden quenching of the crystal produces thermal shock which converts the heated contact region from *n*-type to *p*-type germanium. A small hemispherical *p* region is formed in the *n* substrate around the tip of the whisker, and a *pn junction* is produced right

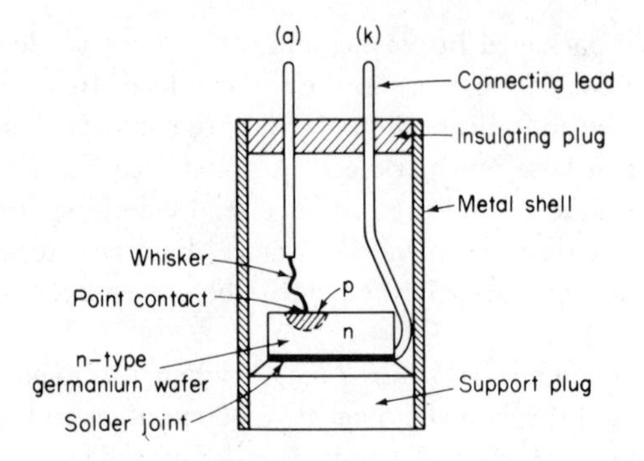

Fig. 2–3 Diagrammatic view of a germanium point-contact junction diode.

at the surface of the crystal. The resulting device is then placed in a metal can or shell to protect it against mechanical damage, and lead-in wires are attached.

This method, used mainly to make germanium diodes, produces low-power junction diodes with low forward resistance but limited reverse-voltage characteristics (Section 2–2). The low forward resistance gives the diode high rectification efficiency, and it is often used as a detector in high-frequency applications.

The *grown-junction* technique is a natural extension of the basic melting and crystal pulling process used to obtain intrinsic monocrystalline semiconductor material. It is mainly used to produce germanium diodes and bipolar transistors. The basic process is shown in Fig. 2–4.

To manufacture a junction diode, a *seed crystal* is inserted into a molten mass of weakly doped *p*-type germanium, and then gradually pulled out of the melt. As the seed crystal emerges from the melt, the molten material *crystallizes* on it, extending its structure and producing a large single germanium crystal. At a given time, a selected amount of *n*-type *impurity* is added to the melt as the crystal is pulled, and an *n* region is grown onto the *p* structure. When the *n* region has grown to the desired thickness, the crystal is removed from the melt

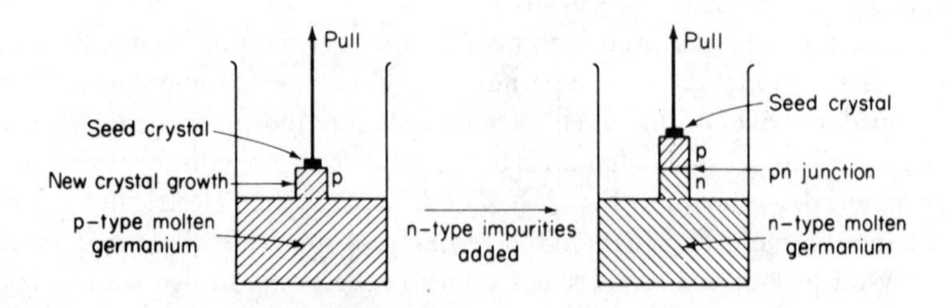

Fig. 2–4 Formation of a rate-grown *pn* junction.

and sliced across the junction into small sections. Each section is provided with contacts and lead-in wires and is sealed in a metal shell or plastic package.

The rate at which the crystal is pulled from the melt, and the rate of adding the n-type impurity, control the impurity concentration across the pn junction and determine the electrical characteristics of the grown junction. In general, grown-junction diodes have inherently high junction resistances and are often subject to wide tolerance spreads in their parameters.

The *alloy-junction* technique can be used to manufacture germanium diodes and transistors. To make a diode, the process starts with a single n-type germanium crystal, sliced up into many small *chips*. A small *pellet* of p-type impurity, such as indium, is placed on the surface of the chip, and the combination is heated in a reducing atmosphere to 500–600°C. The melting point of germanium is 968°C and that of indium is 155°C. Hence, the impurity pellet melts and *alloys* into the germanium chip. During subsequent cooling, the alloy crystallizes as a small p-type region on one side of the chip to form the pn junction, as shown in Fig. 2–5. Contacts and connecting leads are attached to the two elements to form the junction diode, and the device is sealed in an airtight envelope which is inflated with a dry gas such as nitrogen. A typical envelope for a low-power germanium junction diode is shown in Fig. 2–5. Silicon is not convenient with this process.

The *diffused-junction* process represents a major advance in the construction of semiconductor devices, because it gives the manufacturer a great deal of control over the desired performance characteristics of the device. The technique is suitable for both germanium and silicon semiconductor materials.

The formation of a pn junction starts with a wafer of homogeneously doped semiconductor material of either type, which is introduced to a *gaseous* doping process at elevated temperatures. Starting with a wafer of n-type silicon, for example, the doping agent is p-type silicon *vapor* which penetrates or diffuses into the surface of the solid semiconductor material to form a thin p-type layer.

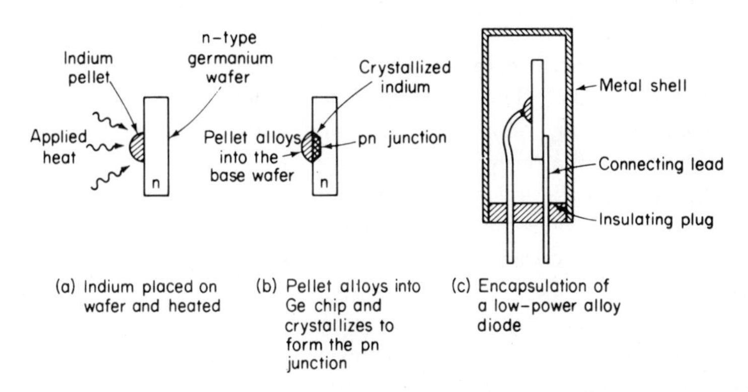

Fig. 2–5 Formation of an alloy junction diode.

The extent of penetration, and hence the thickness of the diffused layer, can be finely controlled by the duration of the vapor exposure, while the impurity concentration of the diffused layer can be controlled by the vapor concentration in the heating chamber. Figure 2–6 shows the construction of a diffused *pn* junction.

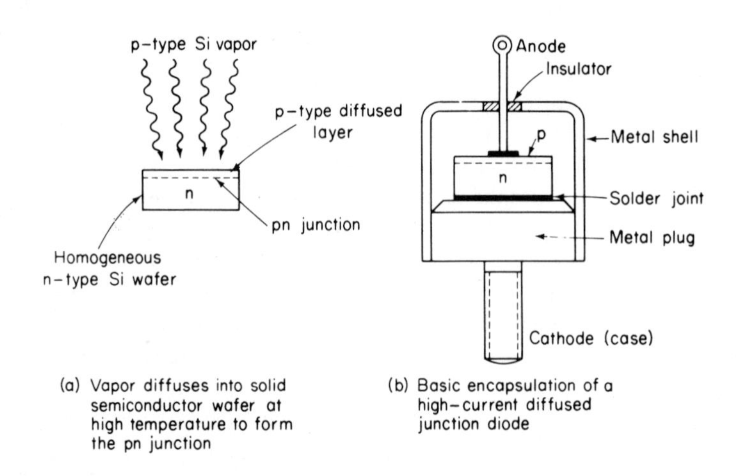

Fig. 2–6 Formation and encapsulation of a diffused *pn* junction diode.

After its formation, the *pn* junction is mounted and encapsulated. Large amounts of heat can be dissipated through the highly conductive solder joint between the semiconductor material and the device package, producing junction diodes with excellent power dissipation, suitable for use as high-current rectifiers.

Many high-performance silicon devices are manufactured by a process called *epitaxial growth.* (The word "epitaxial" is derived from the Greek words "epi," meaning "upon," and "taxis," meaning "form" or "structure." Epitaxial growth then means the "growth of a new structure upon an existing structure.")

The process starts with a single silicon crystal called the base wafer or *substrate,* which provides mechanical strength to the device. The substrate is a heavily doped n-type crystal with very low resistivity, shown as n^+-type silicon in Fig. 2–7.

By heating the substrate to a temperature of approximately 1200°C in a controlled atmosphere of silicon tetrachloride and hydrogen, a layer of n-type silicon can be grown *epitaxially* on the substrate. At this high temperature , the chlorine decomposes at the surface of the substrate and silicon atoms attach

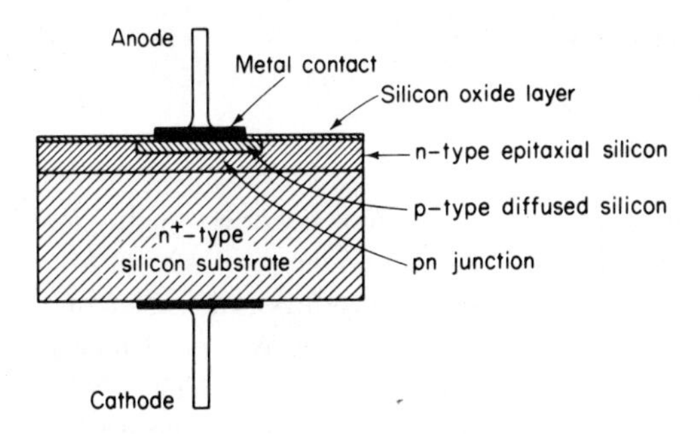

Fig. 2-7 Construction of an epitaxial *pn* junction diode.

themselves to the substrate as a *direct continuation* of the crystal lattice structure. The addition of phosphine to the atmosphere in which the base wafer is heated gives the epitaxial layer its *n*-type characteristics. The phosphine decomposes, and phosphorus atoms take positions in the lattice structure as impurity (donor) atoms. The concentration of phosphine in the atmosphere determines the number of phosphorus atoms in the epitaxial structure and hence controls the impurity concentration.

The substrate with its epitaxial layer is then exposed to the air, usually at an elevated temperature. Oxygen atoms in the air react with the silicon atoms on the surface of the grown crystal to form a layer of silicon dioxide (a form of glass). This oxide layer protects the crystal against mechanical damage and prevents chemical reactions at the surface.

By photographic etching methods a small portion of the oxide layer is then removed in a selected area. The wafer is heated in an atmosphere of boron, and boron atoms diffuse into the *n*-type epitaxial layer where it is exposed. If the boron concentration is sufficiently large, the boron atoms (as acceptor impurities) overcome the phosphorus atoms (donor impurities) in the epitaxial layer and *p*-type silicon is formed in the small exposed area. The depth of the diffused *p* layer is determined by the temperature at which the diffusion of boron atoms takes place and by the length of exposure to the boron atmosphere. The silicon dioxide layer is then regrown over the *p*-type material for protection.

The *pn* junction, formed entirely in the epitaxial layer, is a single silicon crystal with a small trace of impurity changing from donor (*n*-type) to acceptor (*p*-type) at the junction.

Finally, a metallic connection is made to the bottom of the base wafer to provide electrical contact through the low-resistance substrate to the *n*-type epitaxial layer. A second metallic contact is made with the diffused *p*-type silicon through a small hole etched into the regrown oxide. The device is then encapsulated.

2-2 DIODE CHARACTERISTICS

2-2.1 Forward Characteristic

Under forward-bias conditions, the diode conducts readily and passes a forward current I_F. The magnitude of I_F is exponentially related to the forward voltage V_F across the diode. The relation between forward current and forward voltage is the *volt–ampere characteristic* of the diode. A typical forward characteristic is shown in Fig. 2–8(a); the circuit to obtain this characteristic is given in Fig. 2–8(b). Note that V_F is measured directly across the diode and that the diode voltage is not equal to the applied battery voltage V.

When the diode voltage $V_F = 0$, the diode current $I_F = 0$, so the curve starts at the origin O. When the forward voltage is gradually increased, the diode current initially increases very slowly. When V_F is large enough to overcome the barrier potential of the *pn* junction, a significant increase in current occurs, and the characteristic curve rises steeply. The forward voltage at which this sharp increase in current occurs is called the *knee voltage* V_K. For germanium diodes V_K is approximately 0.3 V, and for silicon diodes V_K is approximately 0.7 V.

If V_F is increased much beyond V_K, the current may become very large and can easily overheat (and consequently destroy) the *pn* junction. In Fig. 2–8(b) a protective resistor R is placed in series with the diode to limit the current to the manufacturer's specification of maximum rated value.

EXAMPLE 2-1

The circuit of Fig. 2–8(b) is used to plot the forward characteristic of a 1N4150 general-purpose silicon diode. The diode manual lists the following maximum current ratings:

$$I_F = 300 \text{ mA (average or dc forward current)}$$
$$I_{FRM} = 600 \text{ mA (repetitive peak forward current)}$$
$$I_{FSM} = 4 \text{ A (nonrepetitive peak forward current)}$$

The dc supply voltage is adjustable from 0 to 12 V. Select the correct current-limiting resistor.

SOLUTION

Since the diode is connected in a dc circuit, and has no repetitive currents, the forward current should not exceed 300 mA. The approximate forward voltage drop across the diode during full conduction is 0.7 V. Hence, the maximum voltage drop across the current-limiting resistor R equals

$$V_R = 12 \text{ V} - 0.7 \text{ V} = 11.3 \text{ V}$$

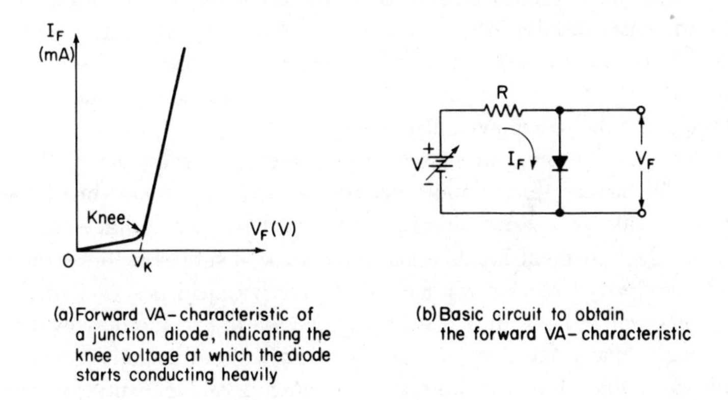

(a) Forward VA-characteristic of a junction diode, indicating the knee voltage at which the diode starts conducting heavily

(b) Basic circuit to obtain the forward VA-characteristic

Fig. 2–8 Forward VA characteristic of the junction diode.

The resistance of R then is

$$R = \frac{11.3\ \text{V}}{300\ \text{mA}} = 38\ \Omega\ \text{(approx.)}$$

The maximum power dissipation in R is

$$P = I^2R = 0.3^2 \times 38 = 3.4\ \text{W}$$

We select a commercially available resistor and specify $R = 40\ \Omega$ at 5 W.

Incidentally, the diode voltage drop at a forward current $I_F = 300$ mA is slightly larger than the knee voltage V_K. Assuming that $V_F = 1$ V, the power dissipation of the diode at $I_F = 300$ mA is $P = 300\ \text{mA} \times 1\ \text{V} = 300$ mW, a typical value for a low-power diode.

2–2.2 Reverse Characteristic

The voltage–current characteristic of the reverse-biased diode is shown in Fig. 2–9. Initially there is a very small reverse current I_R, caused by the minority carriers crossing the *pn* junction (Section 1–3.2). Since the number of minority carriers in a crystal depends solely on temperature, I_R should remain constant as the reverse voltage V_R is increased. Because of surface effects, however, I_R increases slightly with increasing reverse voltage. In practice, I_R is negligibly small and for most practical purposes the reverse-biased diode can be regarded as an open circuit.

When the reverse voltage reaches the *breakdown* value V_{BR}, the reverse current increases rapidly. At the breakdown point, the voltage across the diode is very nearly constant and is practically independent of the reverse current. This constant-voltage property leads to interesting applications, such as voltage-regulation circuits in dc power supplies.

The two physical processes responsible for breakdown of the junction diode under reverse-bias conditions are known as *Zener* breakdown and *avalanche* breakdown. *Zener breakdown* is based on the fact that electrons can be torn from their covalent bonds when the crystal is subjected to a strong electric field (large applied reverse voltage). If the electric field is sufficiently strong to break one covalent bond, it is also strong enough to break other covalent bonds. As a result, many covalent bonds are broken simultaneously and electrons become available in great numbers, causing a sudden increase in reverse current.

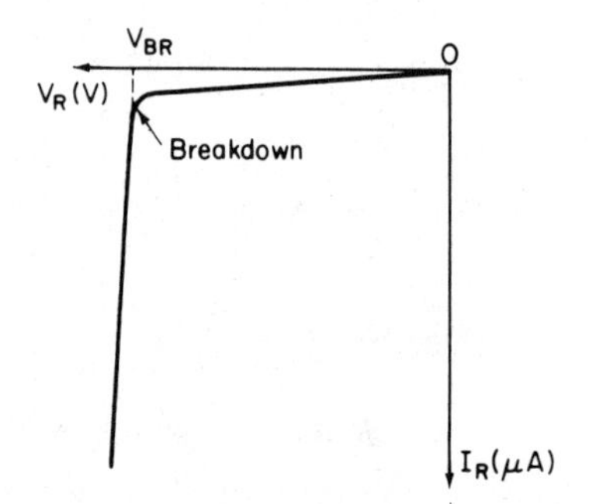

Fig. 2–9 Reverse VA characteristic of the junction diode.

The second breakdown mechanism is known as *avalanche breakdown*. At high reverse voltages the minority carriers gain sufficient energy to knock an electron out of a covalent bond, creating an electron-hole pair. These extra carriers are swept across the junction and obtain sufficient energy to cause other ionizing collisions, liberating additional electrons. This process builds up in the manner of an avalanche. Once the avalanche process starts, it cannot be stopped and the result is a very sudden increase in available charge carriers and hence a rapid increase in reverse current.

When the junction diode is used in the breakdown mode, we speak of a *breakdown diode* or, more commonly, a *Zener diode*. It should be noted that the diode is not destroyed if the reverse current is not allowed to exceed the maximum power rating of the device.

2-2.3 Complete Voltage-Current Characteristic

The complete voltage–current characteristics of a silicon and a germanium diode are shown in Fig. 2–10. The forward characteristics of both diodes are plotted in the first quadrant, with forward voltage (V_F) in volts and forward current (I_F) in milliamperes. The reverse characteristics are plotted in the third quadrant, with reverse voltage ($-V_R$) in volts and reverse current ($-I_R$) in microamperes. Different scales for forward and reverse current are used, because the reverse current is too small to be plotted to the same scale as the forward current.

Comparing the characteristics of both diodes, we observe several interesting differences. The knee voltage of the silicon diode is approximately 0.7 V and of the germanium diode approximately 0.3 V. The forward characteristic of the silicon diode is much steeper than that of the germanium diode, indicating that I_F of the silicon diode increases at a much faster rate than I_F of the germanium diode.

In the reverse direction, the reverse current I_R of the silicon diode is very much smaller than I_R of the germanium diode. Reverse current depends on the availability of minority carriers and hence is strongly dependent on temperature. For germanium diodes I_R approximately doubles for every 10°C rise in temperature; for silicon diodes the increase of I_R with temperature is much less pronounced. Since the reverse current for silicon diodes is so very much smaller than for germanium diodes, even at elevated temperatures, silicon diodes are generally preferred in applications where large temperature variations occur.

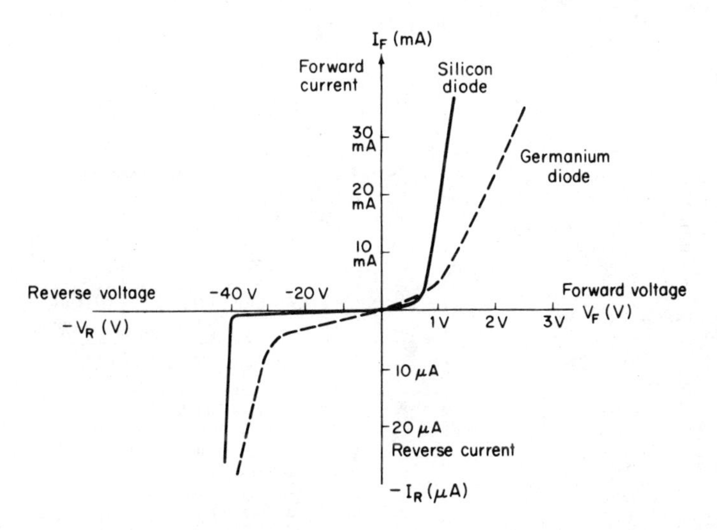

Fig. 2–10 Characteristic curves for silicon and germanium junction diodes (note the scale changes for voltage and current).

2-2.4 Graphical Solution of a Diode Circuit

The junction diode of Fig. 2–11(a) is connected in series with load resistor R_L and a battery. The battery supplies forward bias to the diode, so it conducts and passes a certain forward current I_F. The diode itself is specified by the characteristic curve of Fig. 2–11(b), which graphically relates V_F and I_F. The current in the circuit (I_F) can be determined by the *graphical method,* which makes use of the V_F–I_F characteristic of Fig. 2–11(b).

Kirchhoff's voltage law applied to the diode circuit yields the voltage equation

$$V = V_F + V_R = V_F + I_F R_L \tag{2-1}$$

where V_F = voltage drop across the diode
V_R = voltage drop across the load resistor

Equation (2–1) contains two unknowns, V_F and I_F, and can only be solved when the relation between V_F and I_F is known. The graph of Fig. 2–11(b) provides this information, so that *simultaneous solution* of the algebraic equation [Eq. (2–1)] and the graphical representation [Fig. 2–11(b)] yields the desired values of V_F and I_F.

Equation (2–1) is first-order, which can be represented by a straight line on the graph, called the *dc loadline,* and Eq. (2–1) is called the *loadline equation.*

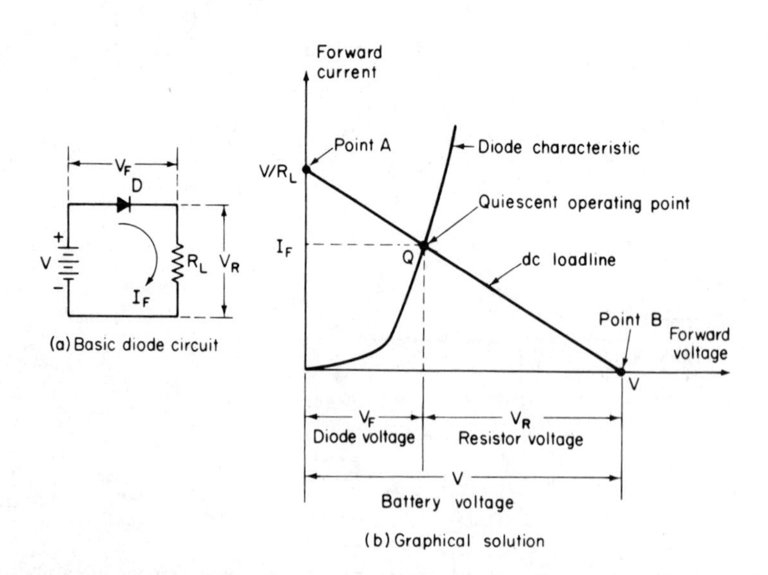

Fig. 2–11 Graphical solution of a simple diode circuit.

Two points on this line can be determined by imposing conditions on the circuit. If $V_F = 0$ V, Eq. (2–1) reduces to

$$V = I_F R_L \qquad \text{or} \qquad I_F = \frac{V}{R_L} \tag{2–2}$$

This yields point A of the dc loadline, where $V_F = 0$ and $I_F = V/R_L$.

Secondly, we let $I_F = 0$ so Eq. (2–1) reduces to

$$V_F = V \tag{2–3}$$

This yields point B of the dc loadline, where $V_F = V$ and $I_F = 0$. The straight line obtained by connecting points A and B represents all solutions to Eq. (2–1), and it is the dc loadline.

The diode circuit must satisfy Kirchhoff's voltage law, represented by the dc loadline. The diode also must operate somewhere on its voltage–current characteristic, because that curve specifically relates V_F and I_F for the diode. Hence, the only point satisfying both operating conditions must be the *intersection of the dc loadline and the V_F–I_F curve.* This point is called the dc or *quiescent operating point* of the diode, and is indicated as point Q on the graph of Fig. 2–11(b), which specifies the quiescent operating voltage and current of the diode. With both V_F and I_F known, the diode circuit of Fig. 2–11(a) is solved.

EXAMPLE 2-2

The junction diode of Fig. 2–12(a) is connected in series with a 240-Ω resistor across a 6-V battery. The forward characteristic of this diode is given in Fig. 2–12(b). Determine, by graphical means, the current in the circuit (I_F) and the voltage drops across the diode (V_F) and the resistor (V_R).

SOLUTION

The solution of the circuit is derived from the intersection of the dc loadline and the diode curve. The loadline equation is given by Kirchhoff's voltage law applied to the circuit, which yields

$$V = V_F + V_R = V_F + I_F R$$

One point of this loadline lies on the voltage axis, where $I_F = 0$, so

$$V = V = 6 \text{ V}$$

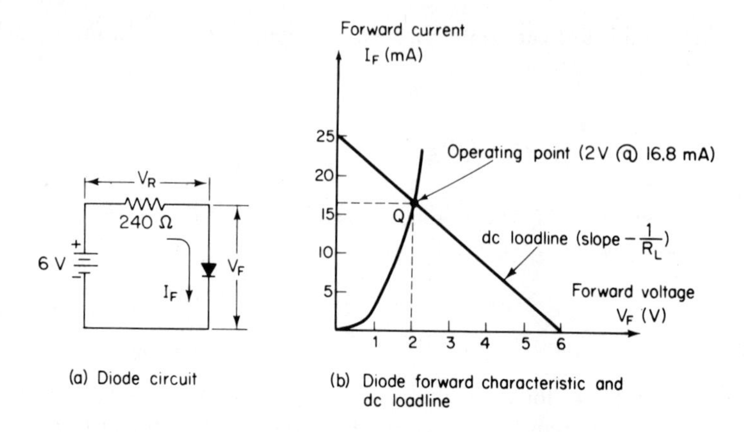

Fig. 2–12 Graphical solution of a simple diode circuit (Example 2-2).

The other point of the loadline lies on the current axis, where $V_F = 0$, so

$$I_F = \frac{V}{R} = \frac{6\text{ V}}{240\ \Omega} = 25\text{ mA}$$

Both points are plotted on the graph of Fig. 2–12(b) and connected by a straight line. This is the dc loadline. The intersection of the loadline and the diode curve yields quiescent operating point Q, where the diode current $I_F = 16.8$ mA and the diode voltage $V_F = 2$ V, so the voltage drop across the resistor $V_R = \text{V} - \text{V}_\text{F} =$ 4 V.

2–3 LARGE—SIGNAL APPROXIMATIONS

2–3.1 Equivalent Circuits

Information on diode characteristics and ratings, supplied by the manufacturer in graphical or tabular form, should always be regarded as *typical* rather than exact. The manufacturer usually guarantees that a given device will conform to certain minimum performance standards and states that certain maximum operating conditions should not be exceeded. But often an exact calculation of expected device behavior, based on information obtained from a manual or graph, does not accurately reflect the actual performance of the device in the circuit, and generally minor adjustments to the circuit are required to obtain the desired result.

To avoid complicated and lengthy calculations involving the published parameters, diode circuits can usually be analyzed quickly and fairly accurately by using *approximations* or *equivalent circuits*. There are several kinds of equivalent circuits. We speak of *large-signal* operation when the diode is driven by a signal source whose voltage is much larger than the knee voltage of the diode.

This gives insight into diode behavior and a practical method of solving basic diode circuits.

2-3.2 Diode as a Switch (First Approximation)

The volt-ampere characteristic of a junction diode, Fig. 2-10, indicates heavy conduction in the forward direction (provided that $V_F > V_K$), and practically no conduction in the reverse direction (provided that $V_R < V_{BR}$). This suggests that the diode, within these limitations, may be considered as a *switch* which is closed in the forward direction (heavy conduction) and open in the reverse direction (no conduction). In this first approximation, the diode is regarded as an *ideal diode,* opening and closing like a switch, depending on the polarity of the applied voltage. The actual diode does not behave like an ideal diode, because the small reverse current, the diode breakdown at a certain reverse voltage, and the fact that the actual diode does not conduct heavily until the knee voltage is exceeded, are neglected. In many applications the first approximation of a junction diode is adequate, especially in large-signal circuits, where great accuracy is not too important, e.g., the *positive clipper.*

EXAMPLE 2-3

A sinusoidal input signal with a peak voltage of 30 V is applied to the circuit of Fig. 2-13(a). Sketch the output waveform appearing across the diode, using the ideal diode approximation.

SOLUTION

During each *positive* half-cycle the diode is forward-biased and behaves like a closed switch, as in Fig. 2-13(b). Hence, the output terminals are short-circuited and the output voltage is zero volts. The current in the circuit is limited by the 10-kΩ resistor and has a peak value of 30 V/10 kΩ = 3 mA. During each *negative* half-cycle the diode is reverse-biased and behaves like an open switch, as in Fig. 2-13(c). The current in the circuit is zero and the applied voltage appears across the diode.

The total waveform appearing across the output terminals is shown in Fig. 2-13(d).

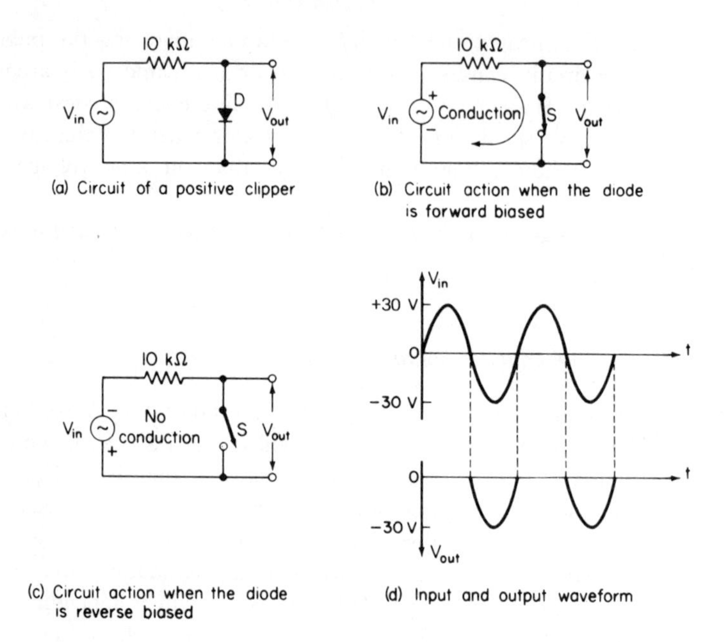

Fig. 2–13 **Circuit action for a positive clipper.**

Note that all the positive portions of the input waveform have been removed, which leads to the name "positive clipper." With reverse diode polarity, the circuit clips off the negative portions of the input waveform and we speak of a "negative clipper."

2–3.3 Knee Voltage (Second Approximation)

Assuming that the diode does not operate in the breakdown region ($V_R < V_{BR}$), we can refine our first approximation of the actual diode by realizing that conduction starts only when the forward voltage exceeds the knee voltage ($V_F > V_K$). Hence, the actual diode can be represented by an ideal diode in series with a small battery whose voltage equals the knee voltage of the actual diode (0.3 V for germanium and 0.7 V for silicon), as in Fig. 2–14. This means that during conduction, when the diode behaves like a closed switch, the voltage across the diode is 0.3 V or 0.7 V, depending on whether it is a silicon or germanium device.

In low-voltage circuits the small forward voltage drop across the diode can play a significant role. This is demonstrated in the simple *half-wave rectifier* of Fig. 2–15. The rectifying device in the circuit of Fig. 2–15(a) is a silicon diode,

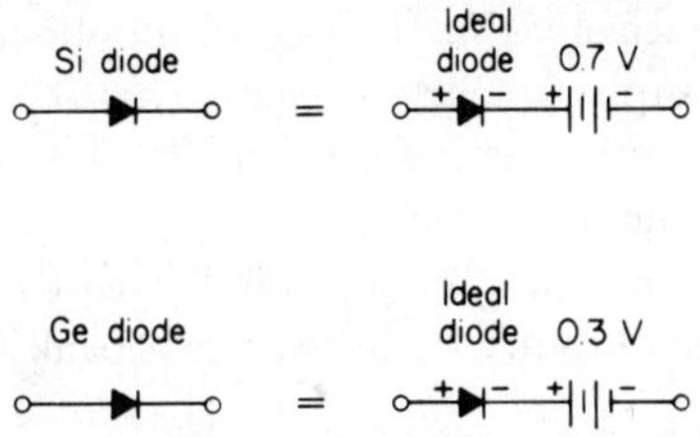

Si diode = Ideal diode 0.7 V

Ge diode = Ideal diode 0.3 V

Fig. 2–14 Second approximation of a Si diode and a Ge diode, taking the knee voltages into consideration.

whose forward voltage drop is assumed constant at 0.7 V. The input signal is sinusoidal with a peak voltage of 2 V.

The actual silicon diode can be replaced by an ideal diode in series with a 0.7-V battery, as in Fig. 2–15(b). During each *positive* half-cycle the diode is forward-biased but conducts only when the applied voltage exceeds the 0.7-V

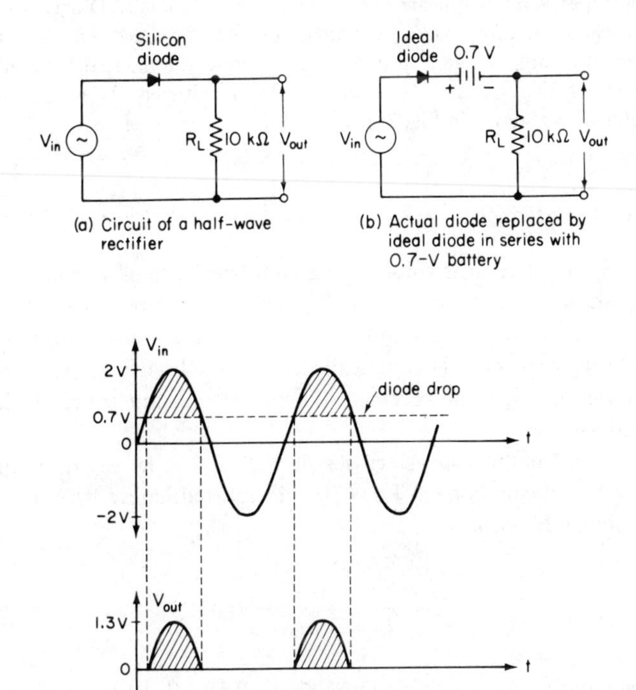

(c) Input and output waveforms

Fig. 2–15 Circuit action for a half-wave rectifier.

knee voltage. During each *negative* half-cycle the diode is reverse-biased and does not conduct. The output voltage on negative half-cycles is therefore zero.

The input and output waveforms are shown in Fig. 2–15(c). Diode conduction starts when the input voltage *exceeds* 0.7 V. The diode conducts for *less than half* of the positive input cycle, as indicated by the dashed lines in the waveform diagram. The output voltage reaches a peak value of 2 V − 0.7 V = 1.3 V.

EXAMPLE 2-4

The negative clipper of Fig. 2–16(a) uses a silicon diode whose forward voltage drop during conduction is assumed constant at 0.7 V. Sketch the output waveform.

SOLUTION

Replacing the silicon diode with an ideal diode in series with a 0.7-V battery yields the circuit of Fig. 2–16(b). During each *positive* half-cycle of the input voltage the diode is reverse-biased and acts like an open switch. Hence, the entire positive half of the input voltage appears across the output terminals. During each *negative* half-cycle of the input voltage the diode conducts, provided that the input voltage exceeds the knee voltage of 0.7 V. Since a conducting diode behaves like a closed switch, negative voltages exceeding 0.7 V are clipped. The input and output waveforms are shown in Fig. 2–16(c).

2–3.4 Reverse Resistance

Inspection of typical volt–ampere characteristics of silicon and germanium diodes shows that there is a small current under reverse-bias conditions. This implies that the diode exhibits a *reverse resistance.* Since the reverse characteristic in each case is practically a straight line from zero volts to the breakdown voltage V_{BR}, the reverse resistance is constant provided that V_{BR} is not exceeded.

Referring to the characteristics of Fig. 2–10, we see that the reverse current I_R of the silicon diode at $V_R = 30$ V is approximately 1 μA and hence the reverse resistance R_R equals

$$\mathrm{R_R} = \frac{V_R}{I_R} \cong \frac{30\text{ V}}{1\ \mu\text{A}} = 30\text{ M}\Omega$$

A similar calculation for the germanium diode of Fig. 2–10 shows that

$$R_R = \frac{V_R}{I_R} \cong \frac{30\text{ V}}{5\ \mu\text{A}} = 6\text{ M}\Omega$$

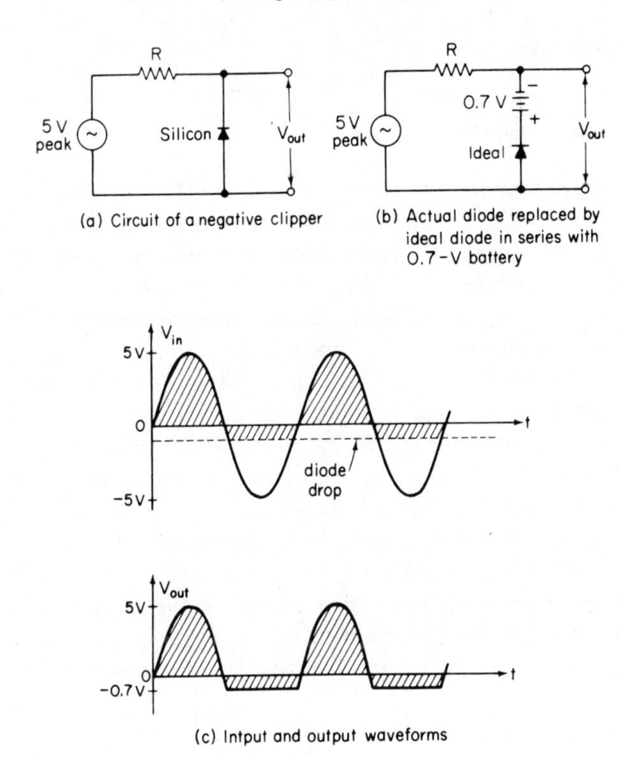

Fig. 2–16 Circuit action for a negative clipper.

Generally speaking, the reverse resistance of a silicon diode is very large and, as a result, the reverse current is very small, so that in many applications the silicon diode is preferred over the germanium diode.

The reverse resistance can easily be found from the information given in the diode manual. If, for example, the manual gives the reverse current of a certain silicon diode as $I_R = 100$ nA at a reverse voltage $V_R = 50$ V, the reverse resistance of that diode is $R_R = 50\text{ V}/100\text{ nA} = 500\text{ M}\Omega$.

The silicon diode more nearly approaches the ideal diode under reverse-bias conditions (an open switch) than the germanium diode. Although in many applications the diode can be regarded as an open circuit when it is reverse-biased, the reverse resistance must be taken into consideration when the diode operates in a high-resistance circuit.

2-4 ZENER DIODE

2-4.1 Basic Operation

A Zener diode is similar in construction to a regular low-power or rectifying diode, but it is especially manufactured to operate in the breakdown region. Practically all Zener diodes are made of silicon because of the superior thermal qualities (high junction temperature) and the low reverse current below breakdown (high reverse resistance). The breakdown or Zener voltage, V_Z, of a Zener diode is established during the manufacturing process and is sharply defined. Zener diodes are commercially available in a wide range of breakdown voltages, from approximately 1.4 V to several hundred volts.

Figure 2-17 shows the volt-ampere characteristic of a Zener diode. In the forward-bias direction, the Zener diode behaves like an ordinary diode and has characteristics identical to a normal silicon diode. In the reverse-bias direction, the Zener diode initially exhibits a relatively high resistance until the knee of the characteristic is reached. At the knee, the diode breaks down and the voltage across the device is essentially constant over a wide range of reverse current. This suggests that the Zener diode can be used as a *voltage reference* element (such as a voltage standard in a power supply or a voltage reference in a comparator

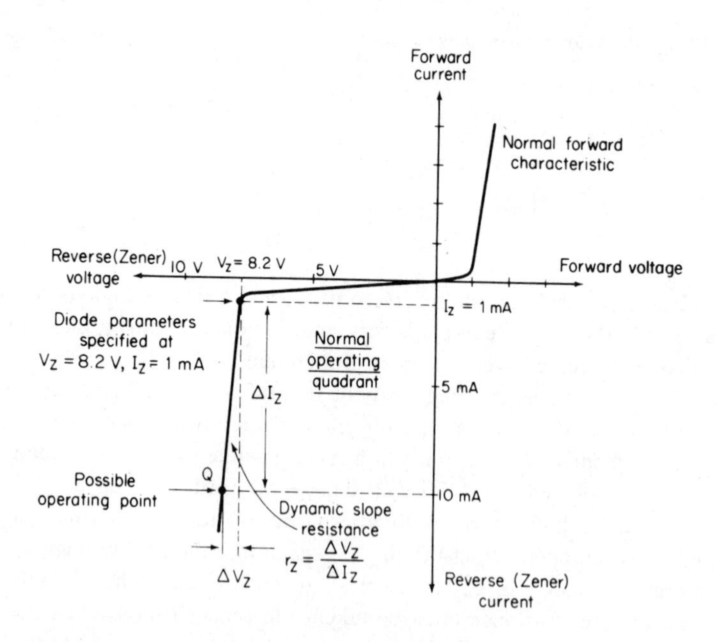

Fig. 2-17 Voltage-current characteristic of a Zener diode.

circuit), or as a *voltage-regulating* device (power supply regulation, meter-surge protection, relay arc suppression).

2–4.2 Device Parameters

The characteristics of the Zener diode, in both forward and reverse direction, are listed in the diode manual or data sheet. Since we are mainly interested in the reverse-bias characteristics, we list the following main parameters: V_Z = Zener voltage (operating voltage of the Zener diode), I_Z = Zener current (reverse current at the Zener voltage), S_Z = temperature coefficient of the Zener voltage, and r_Z = dynamic slope resistance.

The *Zener voltage,* V_Z, is the reverse breakdown voltage or normal operating voltage of the Zener diode. V_Z is always specified at a certain Zener current, I_Z. In Fig. 2–17, for example, the Zener voltage is specified as $V_Z = 8.2$ V at a Zener current $I_Z = 1$ mA. We note that at currents smaller than 1 mA the reverse characteristic enters the knee of the curve and the Zener voltage becomes difficult to define. At currents larger than 1 mA, the Zener voltage increases slightly with an increase in Zener current.

The *Zener current,* I_Z, is simply the reverse current at the Zener voltage. The reverse current must be limited by the external circuit so that the device operates within its maximum allowable power rating. The allowable power dissipation, P_{tot}, is listed in the diode manual or data sheet.

The fact that V_Z increases slightly with an increase in I_Z leads to the definition of another parameter, called *dynamic slope resistance,* r_Z. The dynamic slope resistance relates the variation of Zener voltage, ΔV_Z, to changes in Zener current, ΔI_Z.

By definition, the dynamic slope resistance

$$r_Z = \frac{\Delta V_Z}{\Delta I_Z} \qquad (2\text{-}4)$$

This parameter is particularly important when the Zener diode is used as a voltage regulator in a power supply (Section 8–1). It follows from Eq. (2–4) that a low dynamic slope resistance indicates an almost vertical characteristic; it is under these conditions that the Zener diode more nearly approaches the ideal of a constant voltage device.

In most practical cases, the Zener diode is required to pass a reverse current greater than the specified Zener current, so the voltage across the diode is slightly larger than the specified Zener voltage. The dynamic slope resistance allows us to calculate the actual operating voltage of the Zener diode in this case.

Suppose, for example, that the actual reverse current through the Zener diode, whose characteristics are given in Fig. 2–17, is 10 mA, and that its dynamic slope resistance $r_Z = 20\ \Omega$. The deviation from the Zener current then is $\Delta I_Z = 10\text{ mA} - 1\text{ mA} = 9\text{ mA}$, so the deviation from the Zener voltage (by definition of r_Z) is $\Delta V_Z = r_Z \times \Delta I_Z = 20\ \Omega \times 9\text{ mA} = 180\text{ mV}$. The actual operating voltage of the Zener diode has therefore increased from 8.2 V to 8.2 V + 0.180 V = 8.38 V.

The *temperature coefficient of the Zener voltage, S_Z,* indicates how the Zener voltage changes with variations in device temperature. S_Z is usually expressed in mV/°C and can be positive or negative, depending on the mechanism of breakdown. Generally speaking, at Zener voltages below 6 V, breakdown is due to the Zener effect and the temperature coefficient of the Zener voltage is *negative.* At Zener voltages above 6 V, breakdown is caused by the avalanche mechanism, and the temperature coefficient of the Zener voltage is *positive.*

EXAMPLE 2-5

A certain Zener diode has the following parameters at $T_{amb} = 25°\text{C}$: $I_Z = 20\text{ mA}$, $V_Z = 27\text{ V}$, $S_Z = 20\text{ mV/°C}$, and $r_Z = 18\ \Omega$. The diode is operated at an elevated temperature of 100°C. Calculate the actual Zener voltage at $I_Z = 20\text{ mA}$.

SOLUTION

The temperature difference between operating and ambient conditions

$$\Delta T = 100°\text{C} - 25°\text{C} = 75°\text{C}$$

Since $S_Z = 20\text{ mV/°C}$, the change in Zener voltage equals

$$\Delta V_Z = 20\text{ mV/°C} \times 75°\text{C} = 1.5\text{ V}$$

The actual Zener voltage at $T = 100°\text{C}$ is

$$V_{Z(100°\text{C})} = 27\text{ V} + 1.5\text{ V} = 28.5\text{ V}$$

2-4.3 Basic Zener Circuit

Figure 2-18(a) shows the commonly used symbol for a Zener diode; in Fig. 2-18(b) the Zener diode is connected in a dc circuit in its normal operating mode. If the battery voltage, V, exceeds the Zener voltage, V_Z, the diode breaks down and maintains a constant voltage across its terminals, equal to V_Z. If the battery voltage is less than the Zener voltage, the diode behaves like an ordinary reverse-biased diode and the voltage across its terminals equals the supply voltage.

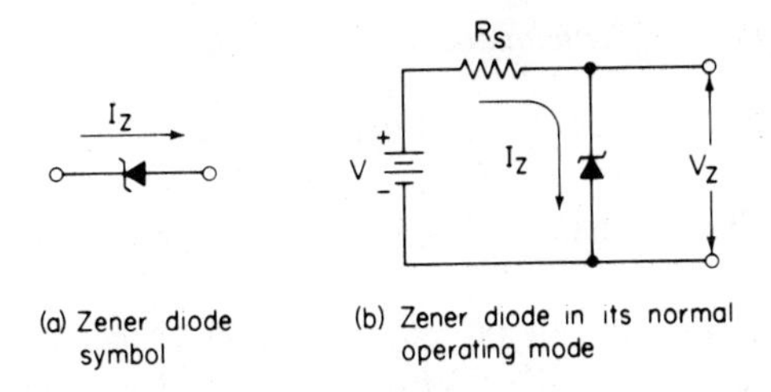

(a) Zener diode symbol

(b) Zener diode in its normal operating mode

Fig. 2-18 The Zener diode in a dc circuit.

Current-limiting resistor R_s is placed in series with the Zener diode to prevent excessive device current and to limit the device dissipation to its maximum rated value, P_{tot}. The range of power-handling capability of Zener diodes is very wide, from a typical value of 400 mW for low-power diodes to a maximum of approximately 100 W for large diodes.

EXAMPLE 2-6

The Zener diode of Fig. 2-18(b) has the following parameters at $T_{amb} = 25°C$: $V_Z = 8.2$ V, $I_Z = 20$ mA, and $P_{tot} = 1$ W. If the battery voltage $V = 12$ V, calculate the required minimum resistance of current-limiting resistor R_s.

SOLUTION

The maximum device current is limited by the total allowable power dissipation and equals

$$I_{Z(\max)} = \frac{P_{tot}}{V_Z} = \frac{1 \text{ W}}{8.2 \text{ V}} \cong 122 \text{ mA}$$

The voltage drop across R_s is constant and equals

$$V_R = V - V_Z = 12 \text{ V} - 8.2 \text{ V} = 3.8 \text{ V}$$

The minimum resistance of R_s then is

$$R_{s(\min)} = \frac{V_R}{I_Z(\max)} = \frac{3.8 \text{ V}}{122 \text{ mA}} = 31 \text{ }\Omega$$

2–4.4 Series and Parallel Connections

Zener diodes can be placed in series to increase the total Zener voltage to the sum of the individual Zener voltages as in Fig. 2–19(a). A single protective resistor in series with the diode string limits the current common to all Zener diodes.

Zener diodes can also be placed in parallel to increase the total power dissipation, as in Fig. 2–19(b). To ensure a balanced division of the total current, it is common practice to place a small resistor in series with each Zener diode. Alternatively, the diodes could be derated by at least 25 percent for each parallel path.

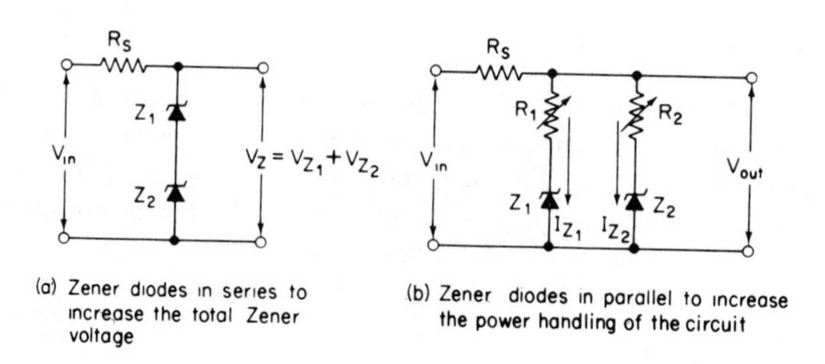

Fig. 2–19 Zener diodes in series and in parallel.

Applications of Zener diodes in voltage-regulation circuits are given in Section 8–1; the Zener diode as a voltage reference element is shown in the feedback regulator circuits of Section 8–3.

2–5 DIODE PARAMETERS

2–5.1 Definitions

The manufacturers of semiconductor devices publish information on device characteristics and ratings, either as a data sheet for each device, or as a manual. According to the recommendations of the International Electrotechnical Commission, the following definitions apply:

A *rating* of a semiconductor device is a value that establishes either a limiting condition (maximum *or* minimum), or a limiting capability. The rating is determined by specified values of environment and operation, and may be stated in any suitable terms.

Absolute maximum ratings are limiting values of operating and environmental conditons of the device, which should not be exceeded under the worst probable conditions. These values are specified by the manufacturer to provide acceptable serviceability of the device. The absolute maximum system of ratings is commonly used in published semiconductor data.

A *characteristic* is an inherent and measurable property of the semiconductor device. Such a property may be electrical, mechanical, thermal, electromagnetic, or nuclear, and can be expressed as a value for stated or recognized conditions. A characteristic may also be a set of related values in graphical form, such as the volt-ampere characteristic of a diode. Ratings and characteristics are expressed in terms of internationally accepted *symbols*.

2-5.2 Diode Specification Sheet

The published data sheet provides the essential device information needed to make sensible use of the device. It generally contains three categories of information:

1. A general statement of diode capabilities and recommended service.
2. Maximum ratings.
3. Electrical characteristics.

Consider the sample specification sheet of Fig. 2-20. The top of the sheet describes the most likely application and the important electrical characteristics of the diode. It is therefore useful in the initial comparison of diodes, and aids in the rapid selection of suitable devices for a specific purpose.

The maximum ratings of voltage, current, and temperature *must* not be exceeded. For ordinary rectifier diodes, the following ratings are usually specified:

Continuous reverse voltage, V_R (often called the *dc blocking voltage*): the continuous dc voltage which may be applied to the diode in the reverse direction. If V_R is exceeded, the diode may break down to become damaged or even utterly destroyed.

Total average forward current, $I_{F(AV)}$: the maximum, full-cycle, average value of forward current, allowable at a specified ambient temperature T_{amb} (for low-power diodes) or case temperature T_C (for medium and high-power rectifier diodes).

Nonrepetitive peak forward current, I_{FSM} (often called the *surge current*): the maximum permissible peak current applied as a nonrecurrent half-cycle, superimposed on maximum current and voltage. In many applications the surge current rating is more important in selecting the correct device than the maximum forward current rating.

1N3208 thru 1N3212

V_R — to 400 V
I_O = 15 A

CASE 42
(DO-5)

Medium-current silicon rectifiers. Cathode connected to case, but reverse polarity (anode-to-case connection) also available by adding suffix "R" to type number, e.g. 1N3208R. Supplied with mounting hardware.

MAXIMUM RATINGS

Rating	Symbol	1N3208 1N3208R	1N3209 1N3209R	1N3210 1N3210R	1N3211 1N3211R	1N3212 1N3212R	Unit
D C Blocking Voltage	V_R	50	100	200	300	400	Volts
RMS Reverse Voltage	V_r	35	70	140	210	280	Volts
Average Half-Wave Rectified Forward Current With Resistive Load	I_O*	15	15	15	15	15	Amp
Peak One Cycle Surge Current (60 cps & 25°C Case Temp)	$I_{FM(surge)}$	250	250	250	250	250	Amp
Operating Junction Temperature	T_J	-65 to + 175					°C
Storage Temperature	T_{stg}	-65 to + 175					°C

*T_C = 150°C

ELECTRICAL CHARACTERISTICS (All Types) at 25°C Case Temp.

Characteristic	Symbol	Value	Unit
Maximum Forward Voltage at 40 Amp D-C Forward Current	V_F	1.5	Volts
Maximum Reverse Current at Rated D-C Reverse Voltage	I_R	1.0	mAdc
Typical Thermal Resistance, Junction To Case	θ_{JC}	1.7	°C/W

Fig. 2–20 Diode specification sheet.

Some manufacturers specify maximum current and voltage ratings in addition to the ones shown in Fig. 2–20. In this case the reader is referred to the listing of symbols and descriptions contained in the Appendix.

The third paragraph of the specification sheet of Fig. 2-20 describes the electrical characteristics of the device. These characteristics are the important diode parameters, which are controlled to ensure interchangeability between devices. In the example of Fig. 2–20, the electrical characteristics are confined to two parameters: peak forward voltage drop, V_{FM}, and peak reverse current, I_{RM}.

2–5.3 Thermal Resistance

All diodes are designed to operate within a certain temperature range. When a diode is operated below the specified minimum temperature, problems may arise because of the excessive strain placed on the silicon wafer. Operating a diode above the specified maximum temperature may cause excessive device deterioration and catastrophic failure. The specified operating temperature range does not normally imply a derating of the reverse voltage, but forward current derating may be necessary.

For low-power diodes, the rating system generally specifies the absolute maximum of ambient temperature (T_{amb}) or the absolute maximum of junction temperature (T_j).

For high-current diodes, operating temperatures are not normally specified and it is assumed that the device is operated within the limits imposed by the *thermal resistance* ratings.

In general, the maximum permissible thermal resistance from junction to ambient, $R_{th(j-a)}$, is specified in the data sheet for the low-current devices, while for the high-current devices (where heatsinks are invariably used) the maximum thermal resistance from junction to case, $R_{th(j-c)}$, is specified. The reader is referred to Section 7–13, which presents detailed information on the evaluation of heatsinks and the use of thermal resistance parameters.

QUESTIONS

2–1 Name two outstanding advantages of silicon diodes over germanium diodes.

2-2 What is the cause of leakage current in a reverse-biased diode?

2–3 Does the leakage current increase or decrease with a rise in ambient temperature?

2–4 Is operation of a junction diode in the breakdown region always destructive? Explain.

2–5 Name the two mechanisms responsible for breakdown of the diode under reverse-bias conditions.

2–6 What is meant by large-signal operation?

2–7 Under what circumstances can the knee voltage of a diode generally be neglected?

2–8 What is the approximate knee voltage of a silicon diode? Of a germanium diode?

2–9 Explain what is meant by the clipping level of a positive clipper.

2–10 Using a simple circuit diagram, show how the clipping level of a positive clipper can be changed.

2–11 Sketch a simple diode circuit to obtain an approximate square-wave output signal from a sinusoidal input signal.

2–12 Explain why the Zener voltage is always specified at a certain Zener current.

2–13 Define temperature coefficient, S_Z, of a Zener diode.

2–14 Is the temperature coefficient of a 2.8-V Zener diode positive or negative?

2–15 What is the significance of the dynamic slope resistance, r_Z, of a Zener diode?

2–16 Does the Zener voltage increase or decrease when the Zener current increases?

2–17 What is the difference between a diode rating and a diode characteristic?

2–18 What is meant by current derating of a diode and under what circumstances must current derating be applied?

2–19 Why is it so important to keep the temperature of an operating diode below the rated value?

PROBLEMS

2–1 A general-purpose germanium diode has the following maximum ratings:

$$I_F = 100 \text{ mA} \qquad V_R = 30 \text{ V}$$
$$I_{FRM} = 500 \text{ mA} \qquad V_{RRM} = 50 \text{ V}$$

This diode is forward-biased by a 30-V battery. To limit the current to its maximum rated value, a protective resistor is placed in series with the diode. Calculate the resistance of the protective resistor.

2–2 A general-purpose silicon diode has the following characteristics:

I_F	at V_F
1 mA	700 mV
10 mA	900 mV
50 mA	1100 mV

V_R	at I_r
25 V	50 nA
50 V	200 nA

Calculate (a) The diode resistance at $I_F = 30$ mA.
(b) The reverse resistance at $V_R = 25$ V.

2–3 The diode of Prob. 2–2 is connected in the circuit of Fig. 2–21. Estimate the output voltage.

2–4 The diode used in the circuit of Fig. 2–22(a) has a typical forward volt-ampere characteristics as shown in Fig. 2–22(b).

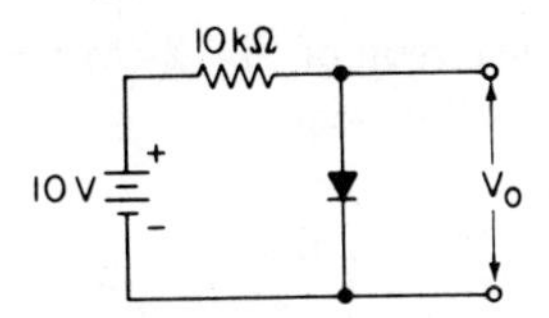

Fig. 2–21 Diode circuit for problem 2–3.

Determine (a) The diode current.
(b) The output voltage.

Hint: The difficulty in this problem is that the 20-Ω resistor in series with the diode is of the same magnitude as the diode forward resistance. The battery can therefore not be regarded as a current source. The problem is solved by construction a 20-Ω loadline on the volt-ampere characteristic in the manner discussed in Section 2–2.4.

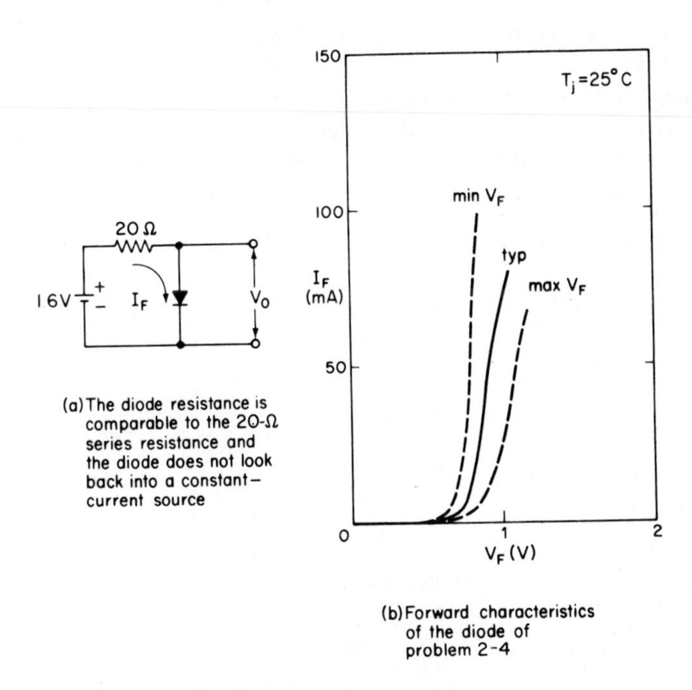

Fig. 2–22 Diode circuit and characteristics for problem 2–4.

2–5 Refer to the silicon diode circuit of Fig. 2–23 and *calculate*
(a) The dc voltage across the diode.
(b) The diode current.
(c) The power dissipated in the diode.

2-6 Refer to the basic regulator circuit of Fig. 2–24. The IN5732B voltage reference diode has the following characteristics at an ambient temperature of $T_{\text{amb}} = 25°\text{C}$:

$$V_Z = 6.8 \text{ V at } I_Z = 10 \text{ mA}$$
$$r_Z = 10\ \Omega$$
$$S_Z = 3 \text{ mV/°C}$$

The maximum ratings of the diode are

$$I_Z = 50 \text{ mA}$$
$$P_{\text{tot}} = 400 \text{ mW}$$
$$T_{\text{j(max)}} = 200°\text{C}$$

Supply voltage V is variable but is initially set at 12 V. The load resistor $R_L = 680\ \Omega$. It is desired to operate the diode at a quiescent current of 10 mA.
(a) Calculate the resistance of current-limiting resistor R_s.
(b) With R_s as determined in (a), calculate the maximum permissible input voltage without exceeding the maximum ratings of the diode.

2–7 Refer to Fig. 2–24. The load resistor $R_L = 1$ kΩ and the current-limiting resistor $R_s = 300\ \Omega$. Calculate the minimum voltage at which the regulation becomes inoperative.

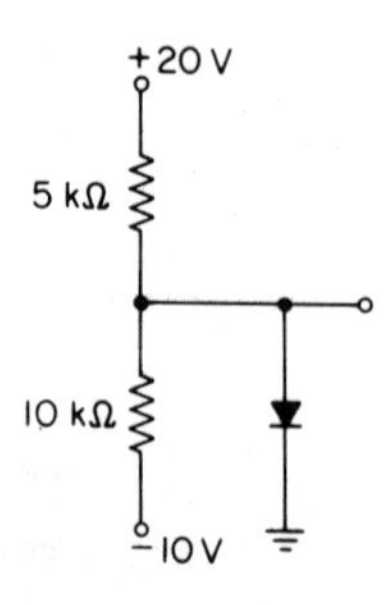

Fig. 2–23 Diode circuit for problem 2–5.

2–8 The Zener diode in Fig. 2–24 is operated at an elevated temperature of 100°C. The Zener current is constant at I_Z = 10 mA. Calculate the actual Zener voltage.

2–9 The Zener diode of Fig. 2–24 is operated at an ambient temperature of T_{amb} = 55°C, while the Zener current I_Z = 40 mA. Calculate the actual Zener voltage.

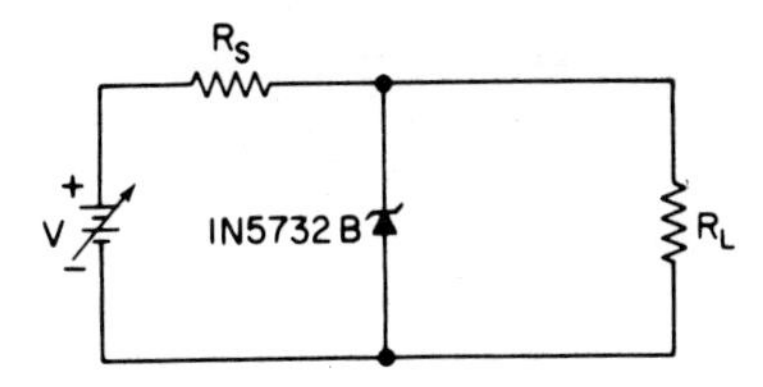

Fig. 2–24 A basic Zener diode regulator.

Chapter 3

The Bipolar Junction Transistor

3–1 BIPOLAR JUNCTION TRANSISTOR

3–1.1 Basic Properties

The bipolar junction transistor (BJT) is one of the most important solid-state devices because of its ability to provide current gain or *amplification*. The BJT derives its name from the fact that both majority and minority carriers take part in the physical processes occurring in the device, in contrast to the field effect transistor (FET), in which only the majority carriers are of importance.

There are two types of BJTs: the *pnp* and the *npn*. A *pnp* transistor consists of a thin slice of *n*-type semiconductor material sandwiched between two slices of *p*-type material, as shown in Fig. 3–1 (a). The larger region of *p*-type material is the *collector* (C), the thin center region of *n*-type material is the *base* (B), and the smaller *p*-type region is the *emitter* (E). The three regions are separated by two *pn* junctions: the *emitter–base* junction and the *collector–base* junction. The behavior of the BJT depends on the majority and minority currents across these junctions.

The schematic symbol for the *pnp* transistor is shown in Fig. 3–1 (b). The emitter is identified by an arrowhead, pointing in the direction of conventional current across the emitter–base junction, from the *p*-type emitter to the *n*-type base.

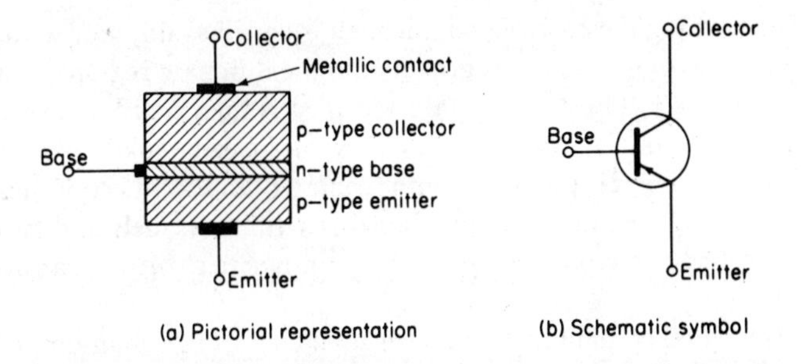

Fig. 3–1 *pnp* **bipolar junction transistor.**

In the *npn* transistor, a thin slice of *p*-type material is sandwiched between two slices of *n*-type material, Fig. 3–2 (a). The three elements, emitter (E), base (B), and collector (C), are separated by two *pn* junctions. The symbol for the *npn* transistor is shown in Fig. 3–2 (b). The emitter is identified by an arrowhead pointing in the direction of conventional current across the emitter-base junction, from the *p*-type base to the *n*-type emitter.

3–1.2 Grown-Junction BJT

The manufacturing techniques for making junction transistors are similar to those discussed for the junction diodes in Section 2–1.2, and are extended to produce a three-terminal device with two *pn* junctions. To manufacture a *pnp* grown-junction transistor, a seed crystal is inserted into a molten mass of weakly doped *p*-type germanium and then gradually pulled out of the melt. As the seed crystal emerges from the melt, the molten material crystallizes on it, extending its structure and producing a single *p*-type germanium crystal. At a certain time in the process, the melt is doped with an *n*-type impurity and a thin base region is

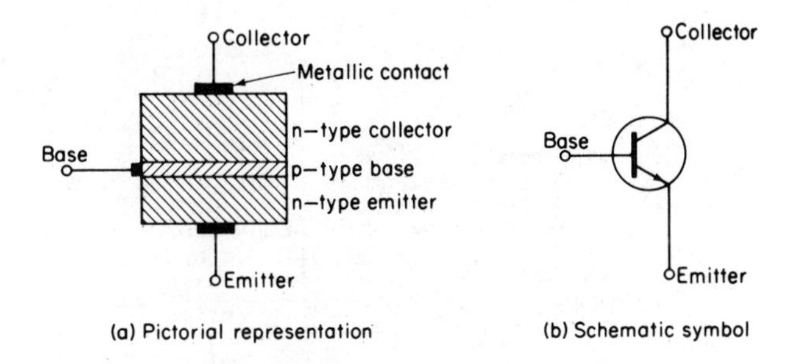

Fig. 3–2 *npn* **bipolar junction transistor.**

grown. After forming the base region, the melt is doped again, now with a *p*-type impurity, and a second *p* region is grown onto the thin base region. This process is shown in Fig. 3–3. The *pnp* crystal so formed is then sliced across the layers into individual sections, so many *pnp* sandwiches can be made from the same crystal. Each section is provided with contacts and connecting leads, and encapsulated. In an identical manner, starting with an *n* melt and successively adding controlled amounts of *p* and *n* impurities, *npn* transistors are constructed.

This method of manufacturing was used for early germanium transistors, but the devices had limited performance characteristics and wide tolerance spreads.

3–1.3 Alloy-Junction BJT

The alloy-junction technique is used mainly to produce germanium transistors. To manufacture a *pnp* transistor, for example, the process starts with a single *n*-type germanium crystal, cut up into many small chips. Pellets of *p*-type impurity, mainly indium, are placed on either side of the chip, and the combination is heated in a reducing atmosphere to 500–600°C. The impurity

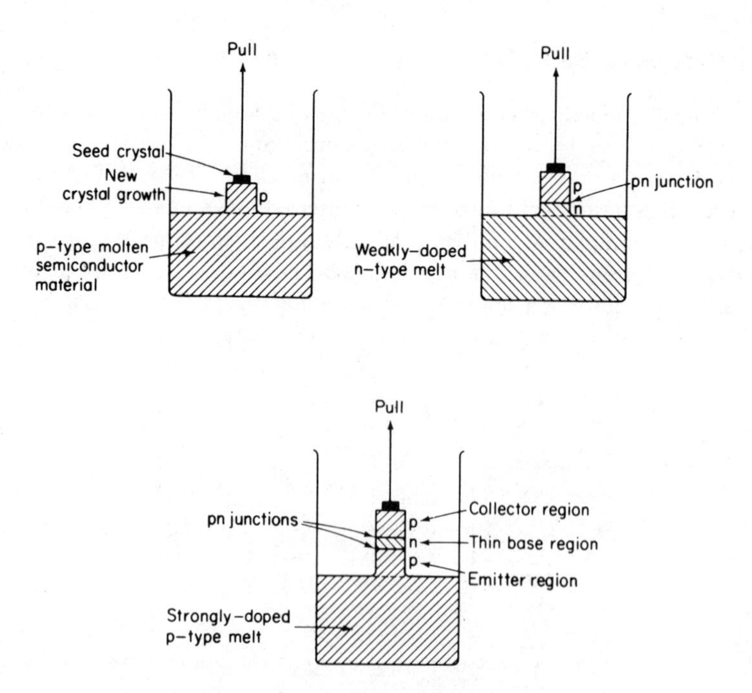

Fig. 3–3 Formation of a rate-grown junction transistor.

pellets melt and alloy into the germanium base chip. During subsequent cooling, the alloy crystallizes as small p-type regions on either side of the base chip to form the collector and emitter regions of the BJT, as shown in Fig. 3-4 (a).

Contacts and connecting leads are then attached to the three elements. The device is enclosed in an envelope inflated with nitrogen. A typical envelope for a low-power germanium transistor is shown in Fig. 3-4 (b).

For high-power germanium transistors, the case is shaped to provide better heat removal. After alloying, the collector pellet is cut flat and soldered to a solid copper base which can easily transfer heat to the chassis or the heatsink. The metal can is pressed tightly (or soldered) to the copper base and the envelope is filled with nitrogen. A typical high-power construction is shown in Fig. 3-4 (c).

Germanium alloy BJTs have excellent characteristics for low-frequency, low-voltage applications, but they do not perform well at the higher frequencies.

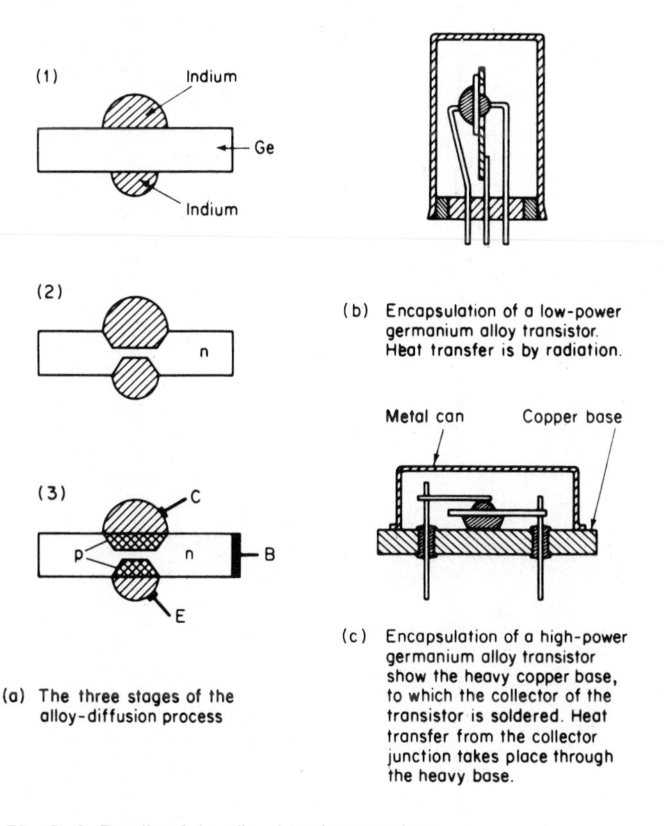

(a) The three stages of the alloy-diffusion process

(b) Encapsulation of a low-power germanium alloy transistor. Heat transfer is by radiation.

(c) Encapsulation of a high-power germanium alloy transistor show the heavy copper base, to which the collector of the transistor is soldered. Heat transfer from the collector junction takes place through the heavy base.

Fig. 3-4 Details of the alloy-junction transistor.

3-1.4 Diffused-Junction BJT

The diffusion process for BJTs uses the technique described in Section 2-1.2 for making diffused-junction diodes. In this process a p- of n-type impurity vapor penetrates into a solid n- or p-type semiconductor crystal (germanium or silicon) at elevated temperatures. A basic *double-diffusion* and etching process is shown in Fig. 3-5. Starting, for example, with a homogeneous silicon p-type substrate (the collector material), a gaseous doping agent of n-type silicon vapor diffuses into the surface of the solid crystal. This first diffusion is allowed to penetrate quite deeply to form a relatively thick n layer around the crystal. A second diffusion of p-type material is then applied, considerably reducing the thickness of the first diffused layer. Selected areas of the double-diffused substrate are then etched away so that the collector and base regions are exposed. The first diffusion produces the thin base region; the second diffusion creates the larger emitter region.

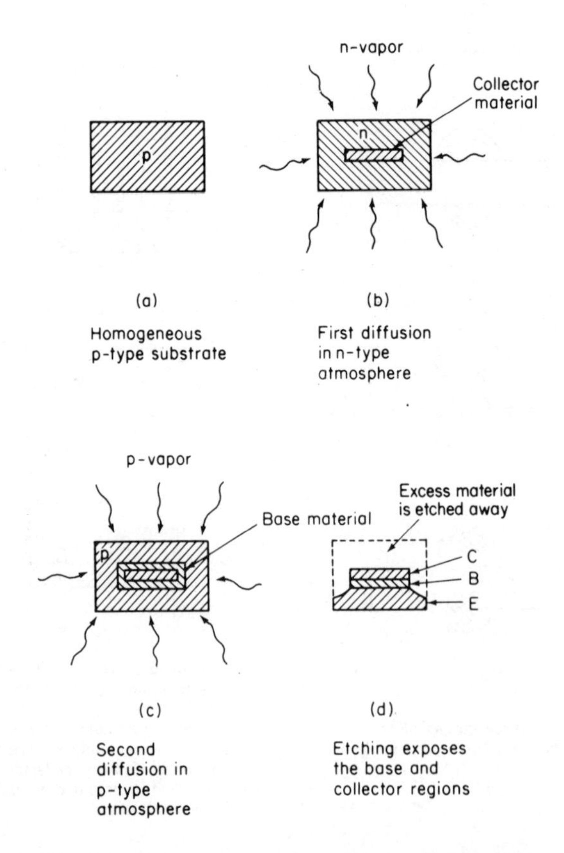

Fig. 3-5 Double diffusion and etching produces a diffused-junction BJT.

The possibility of masking selected areas of the substrate, and controlling the diffusion characteristics (impurity concentration of the vapor and duration of vapor exposure), gives the process great flexibility. Diffused layers of any thickness can be deposited on any desired area on the substrate, so almost any junction configuration can be produced.

3–1.5 Epitaxial BJT

The manufacturing process for epitaxial transistors follows the technique described in Section 2–1.2 for the fabrication of epitaxial *pn* junctions. Figure 3–6 shows the sequence of operations in making an epitaxial silicon planar transistor. The basis or substrate is a silicon *n*-type monocrystal. A silicon dioxide (SiO_2) layer of approximately 1 micrometer (1 μm) is deposited on the crystal by heating to 1000°C in steam. A rectangular window is then etched into the SiO_2 layer and the crystal is heated to 1200°C in a boron atmosphere. Boron atoms diffuse through the small rectangular opening into the crystal and form the *p*-type base region of the transistor. To protect the *pn* junction just formed against contamination and humidity, the oxide layer is regrown over the *p*-type base material in the same oven. A second window, of smaller dimensions, is then etched into this new oxide layer and a second diffusion in a phosphor atmosphere deposits the *n*-type emitter region within the *p*-type region. This completes the *npn* structure.

The oxide is then regrown and small windows are etched away for the terminal contacts. In the openings thus made, aluminum is deposited for the emitter and base connections, with the aluminum film extending over the SiO_2 layer to provide sufficient space to fasten the terminal wires. The bottom of the crystal is provided with a metallic coating for the collector terminal.

The original *p* layer (first diffusion) is approximately 4 μm thick. The second, *n*-type, diffusion penetrates approximately 2 μm into the *p* layer, which results in a base thickness of approximately 2 μm. Because of the slow diffusion process, the penetration depth and the degree of doping can be controlled quite accurately, and the technique yields transistors with similar characteristics.

The manufacturing process for a single transistor, with its many successive operations of etching and diffusion, is expensive. However, many transistors can be manufactured simultaneously on a small circular disk of silicon with a diameter of about 25–50 mm. Since the geometry of a single transistor is approximately 1 mm × 1 mm, such a slice may contain from 500 to 2000 transistors. Using this technique, the bottom of the slice is covered with a metal coating and the 1000 or so transistors on the slice are subjected to a preliminary measurement. A machine senses the transistor elements, measures the characteristics, and marks any faulty elements. A grid pattern is inscribed between the transistor elements on the slice, and it is broken up into separate transistor crystals. Terminal wires are attached to the three transistor elements under a microscope. At the correct spot a very thin gold wire is pressed against the contact plane, where it is heated and welded. Finally, the transistor is placed inside a metal envelope or pressed into plastic.

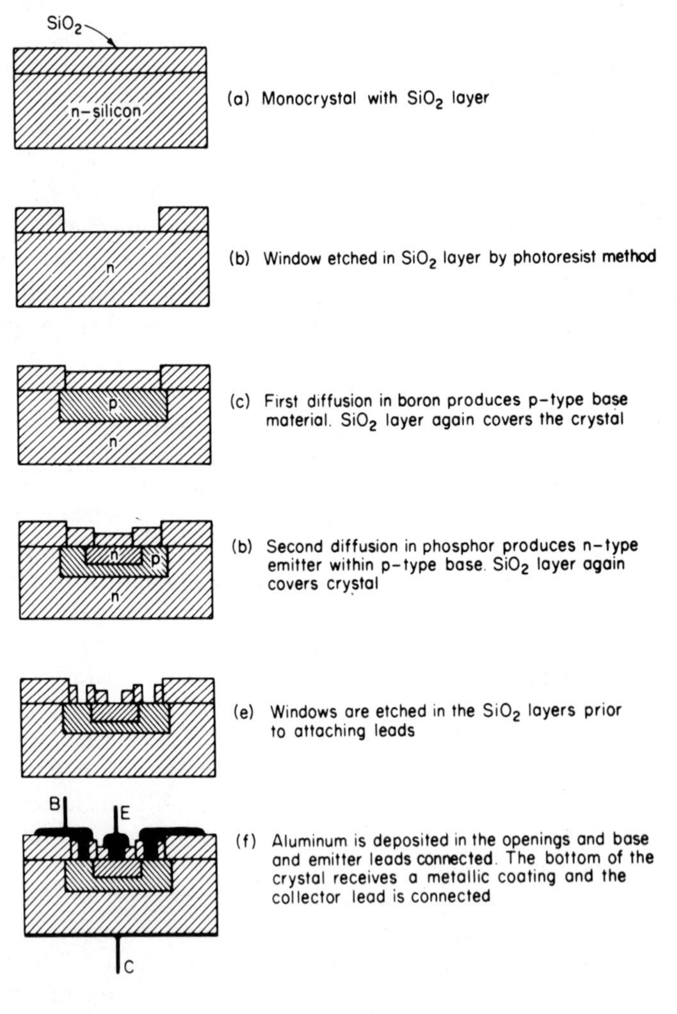

SiO2
n–silicon
(a) Monocrystal with SiO2 layer
n
(b) Window etched in SiO2 layer by photoresist method
p
n
(c) First diffusion in boron produces p–type base material. SiO2 layer again covers the crystal
n
p
n
(b) Second diffusion in phosphor produces n–type emitter within p–type base. SiO2 layer again covers crystal
(e) Windows are etched in the SiO2 layers prior to attaching leads
B
E
(f) Aluminum is deposited in the openings and base and emitter leads connected. The bottom of the crystal receives a metallic coating and the collector lead is connected
C

Fig. 3–6 Sequence of operations in manufacturing an epitaxial planar silicon transistor.

3-2 DC RELATIONS IN THE BJT

3-2.1 Basic Circuits

There are three ways of connecting the BJT in a circuit. The three *basic configurations* using an *npn* BJT are shown in Fig. 3-7. Figure 3-7 (a) shows the first configuration. In normal amplifier operation, an input signal (ac or dc) is applied to the emitter terminal, and the emitter-base circuit is therefore called the *input circuit.* The output signal is developed at the collector terminal, hence the collector-base circuit is the *output circuit.* Since the base of the BJT is the common element between the input and output circuits, the circuit is called the *common-base* (*CB*) *configuration*.

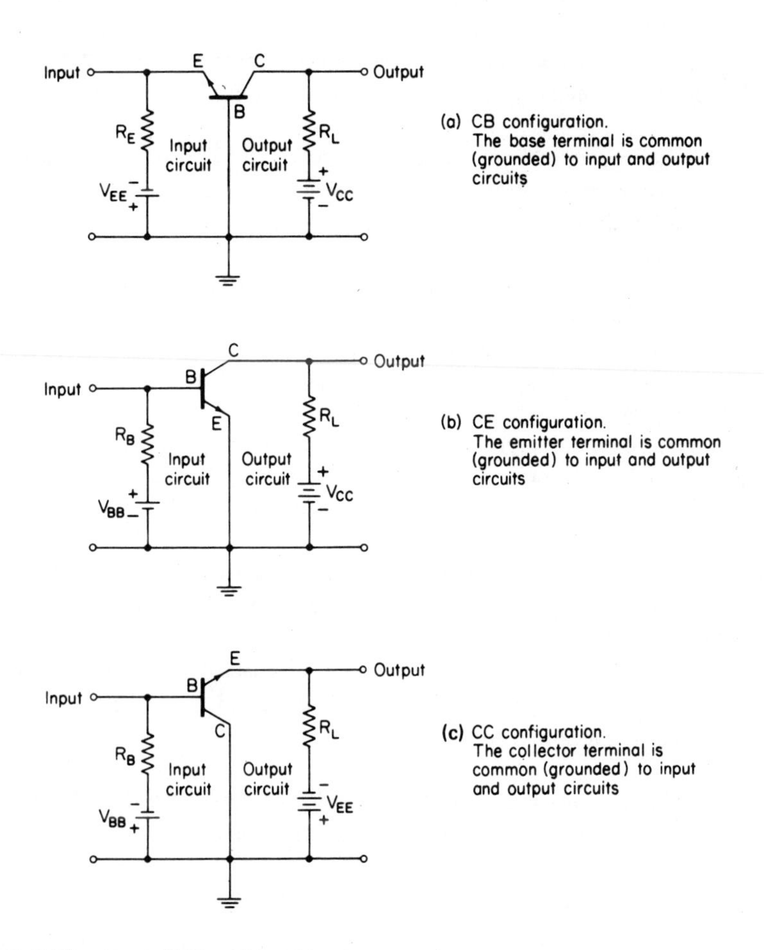

Fig. 3-7 Three basic BJT configurations.

Resistor R_E in the input or emitter circuit is called the *emitter resistor.* It limits the dc current in the input circuit. Resistor R_L in the output or collector circuit is the load resistor. It limits the dc current in the output circuit. Both R_E and R_L determine the dc operating conditions of the BJT in the *CB* configuration.

Two separate batteries *bias* the BJT: emitter supply V_{EE} and collector supply V_{CC}. In the normal mode of BJT operation, the *emitter-base junction is always forward-biased,* and the *collector-base junction is always reverse-biased.* The battery polarities in Fig. 3–7 (a) confirm this statement.

The second basic configuration, shown in Fig. 3–7 (b), is the *common-emitter (CE) configuration.* The input and output circuits are identified in the figure and the emitter of the BJT is the common element between the input and output circuits. Resistor R_B in the input or base circuit is called the *base resistor.* It limits the dc current in the input circuit. Resistor R_L in the output or collector circuit is the load resistor. It limits the current in the output circuit. Again, both resistors determine the dc operating conditions of the BJT in the *CE* configuration.

Two separate batteries are used to bias the BJT in its normal mode of operation: base supply battery V_{BB} supplies forward bias to the emitter-base junction, and collector supply battery V_{CC} reverse-biases the collector-base junction (provided that $V_{CC} > V_{BB}$).

The third basic configuration is the *common-collector (CC) configuration,* shown in Fig. 3–7(c). The input and output circuits are identified in this figure, and the collector is the common element between the two circuits. The two separate bias batteries, V_{BB} and V_{EE}, establish the correct bias conditions for the *CC* configuration.

To maintain the correct bias conditions when using BJT's of the opposite polarity (*pnp* instead of *npn*), the polarity of *both* bias batteries must be reversed. The rules for *bias polarity* are quite easy to remember:

1. For a *pnp* BJT, the *p*-type emitter must be connected to the positive battery terminal (forward bias) and the *p*-type collector must be connected to the negative battery terminal (reverse bias).
2. For an *npn* BJT, the *n*-type emitter must be connected to the negative battery terminal (forward bias) and the *n*-type collector must be connected to the positive battery terminal (reverse bias).

The three basic configurations of Fig. 3–7 possess distinctly different gain and impedance characteristics. These are explained below.

3–2.2 Transistor Action

The mechanism of bipolar junction transistor action is one of carrier *injection, diffusion,* and *collection.* Consider the *npn* BJT of Fig. 3–8(a), where the emitter–base junction is forward-biased by V_{EE} and the collector–base junction is reverse-biased by V_{CC}.

The forward emitter–base bias causes electrons (*majority carriers*) to drift from the negative terminal of V_{EE}, through the heavily doped emitter region, across the emitter–base junction, into the lightly doped base region. These electrons can be thought of as being emitted from the emitter and *injected* into the base region. Some of these injected electrons recombine with available holes in the *p*-type base region. For every electron-hole recombination in the base, another hole must be generated, so that a flow of electrons out of the base region toward the positive terminal of V_{EE} takes place. This electron flow is called the *base current,* I_B. Since the base width is rather narrow and the base material lightly doped, relatively few recombinations can take place, and the base current is relatively small.

Those electrons escaping recombination in the base diffuse through the base region as *minority carriers* (remember that electrons are minority carriers in *p* material), and are attracted across the reverse-biased base–collector junction to combine with holes in the collector region. For every electron-hole recombination occurring in the large and heavily doped collector region, an electron must leave the collector terminal to return to the positive battery terminal of V_{CC} as collector current I_C. In a sense, this recombination process can be thought of as a *collection* of majority carriers that have diffused across the junction into the collector region.

In this process majority carrier movement (electrons) has taken place across the forward-biased (low resistance) emitter–base junction and across the reverse-biased (high-resistance) base–collector junction by the process of injection, diffusion, and collection.

The transistor action of a *pnp* BJT is very similar to that described above, except the majority carriers in this process are holes. Holes are injected across the forward-biased emitter–base junction into the lightly doped base region. A small number of recombinations with electrons takes place in the base region, and a small base current I_B is established. The large majority of injected holes diffuse across the reverse-biased base–emitter junction, attracted by the relatively high collector potential, to be collected by the collector region through recombination. This action is shown in Fig. 3–8(b).

All the majority carriers entering the BJT at the emitter terminal take part in the conduction processes in the device. A small fraction of the injected carriers establish the base current, I_B, but by far the larger fraction crosses the base–collector junction to establish the collector current I_C. Hence, the emitter current must equal the sum of the base and collector currents, so

$$I_E = I_B + I_C \tag{3–1}$$

This equation is one of the fundamental transistor relations and should be memorized.

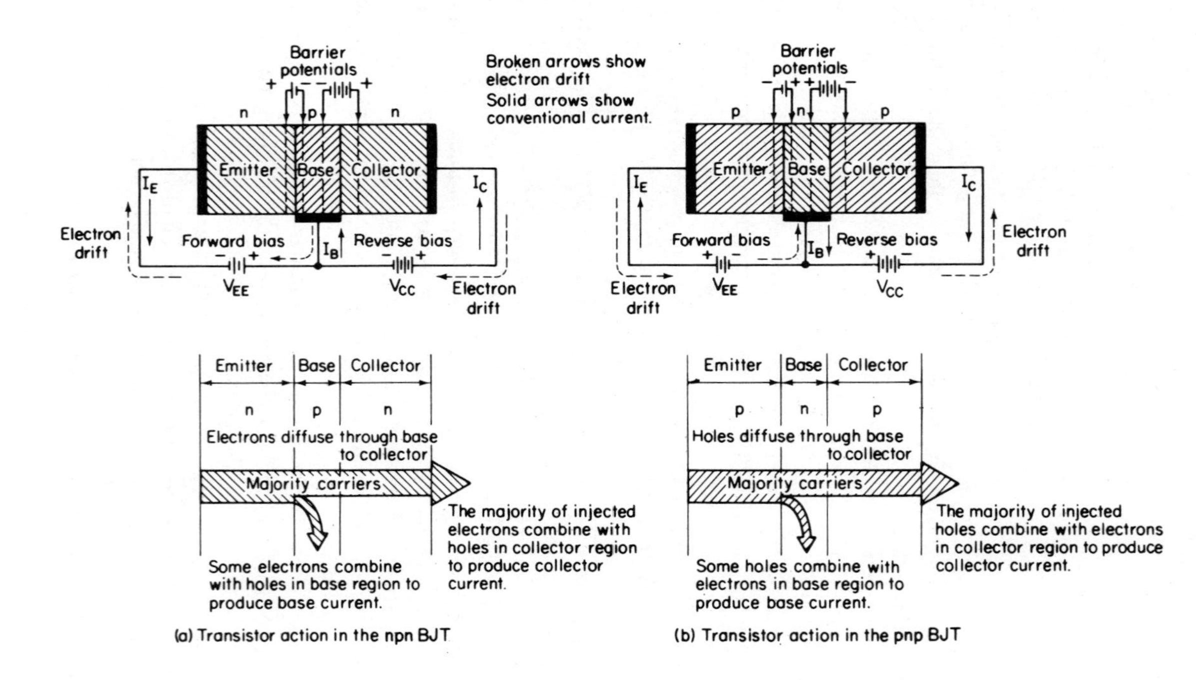

Fig. 3–8 Transistor action in the biased bipolar junction transistor, connected in the common-base configuration.

Figure 3–9 shows the three transistor currents. The *conventional* direction is used. Hence, for the *npn* transistor, I_B and I_C are directed into the base and collector terminals, respectively, and I_E is directed away from the emitter terminal. For the *pnp* transistor, I_B and I_C are directed away from the device and I_E is directed into the emitter terminal.

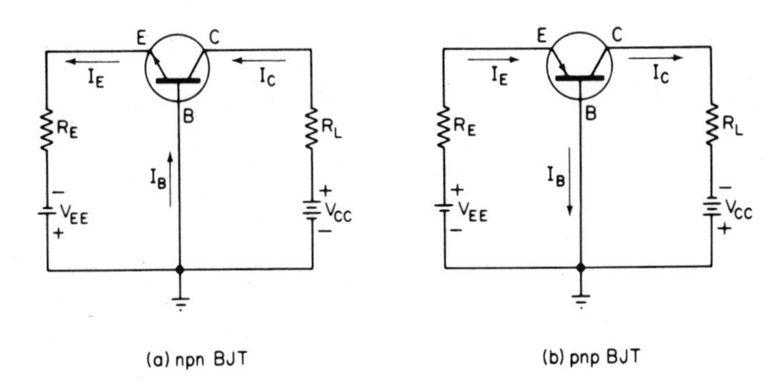

Fig. 3–9 Conventional currents in the BJT.

The relative magnitudes of I_B and I_C (the division of I_E into its component parts) are a function of several factors, having primarily to do with the physical construction of the BJT. A factor of great importance is the efficiency of the emitter region in injecting carriers into the base region. Great efficiency can be obtained by making the concentration of majority carriers in the emitter region very large compared to the concentration in the base region. A second factor is the recombination process in the base region, responsible for the magnitude of the base current. Few recombinations take place when the width of the base region is very narrow. The third factor concerns the level of carrier injection into the collector region and is a function of the relationship between the existing majority carriers in the collector and those that are being injected into the collector from the emitter region. All these factors can be controlled during the manufacturing process.

3–2.3 DC Current Gain

One of the important transistor parameters, always published in the manual or the data sheet, is the *current-gain* or *forward-current transfer ratio*. We speak of dc current gain when the transistor operates only under dc conditions, and of ac current gain when the transistor is subject to small-signal or ac conditions. There is very little difference between the values for dc and ac current gain at low frequencies, especially when the transistor operates in the linear part of its characteristics (Section 3–2.6). The two types of current gain are therefore often used interchangeably.

By definition, the dc current gain of a transistor is the ratio of the dc output current to the dc input current, with the dc output voltage held constant. To illustrate this, consider the three configurations of Fig. 3–10. In the *CB* configuration of Fig. 3–10(a), the dc input current is the emitter current I_E and the dc output current is the collector current I_C. The dc current gain is then defined as

$$CB \text{ current gain } \alpha_{\text{dc}} \text{ or } h_{FB} = \frac{I_C}{I_E}\bigg|\, V_{CB} = \text{constant} \tag{3–2}$$

Since fewer carriers reach the collector than leave the emitter, $I_C < I_E$ and $\alpha_{\text{dc}} < 1$. In practice, α_{dc} (or simply α) falls between 0.95 and 0.99.

In the *CE* configuration of Fig. 3–10(b), the dc input current is the base current I_B and the dc output current is again the collector current I_C. In this case, the dc current gain is defined as

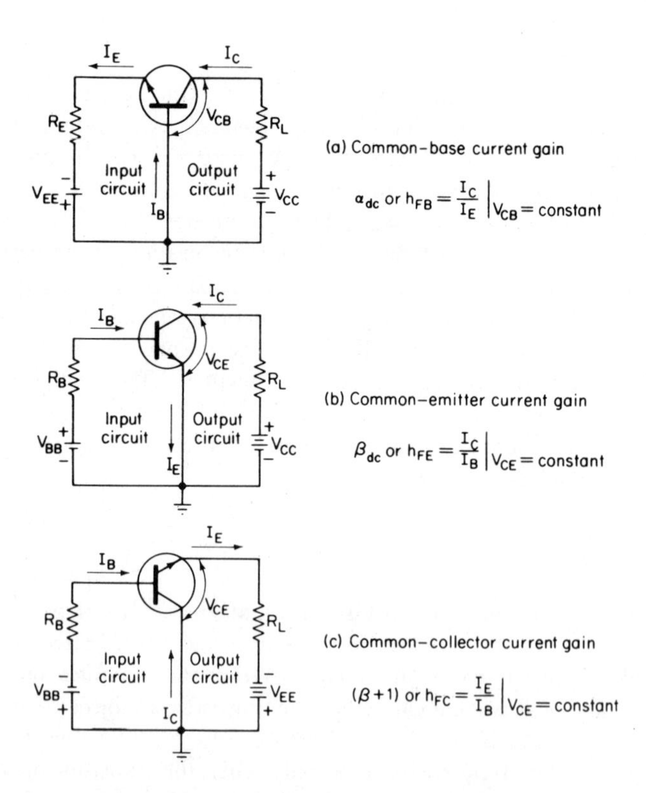

Fig. 3–10 The three BJT configurations and their dc current gains.

$$CE \text{ current gain } \beta_{dc} \text{ or } h_{FE} = \frac{I_C}{I_B}\bigg|_{V_{CE} = \text{constant}} \tag{3–3}$$

Since I_B is very much smaller than I_C, β_{dc} is much larger than α. In practice, β_{dc} (or simply β) ranges from 20 to several hundred.

In the *CC* configuration of Fig. 3–10(c) the dc input current is the base current I_B and the dc output current is the emitter current I_E. The dc current gain is then defined as

$$CC \text{ current gain } h_{FC} = \frac{I_E}{I_B}\bigg|_{V_{CE} = \text{constant}} \tag{3–4}$$

Since $I_E = I_B + I_C$ [Eq. (3–1)] and $I_C = \beta I_B$ [Eq. (3–3)], $I_E = I_B + \beta I_B = (\beta + 1)I_B$, so

$$CC \text{ current gain } h_{FC} = \beta + 1 \tag{3–5}$$

Since the three transistor currents are uniquely related by Eq. (3–1), current gains α and β can be expressed in terms of one another.

$$I_E = I_B + I_C \tag{3–1}$$

$$\alpha = \frac{I_C}{I_E} \quad \text{or} \quad I_E = \frac{I_C}{\alpha} \tag{3–6}$$

Substitution of Eq. (3–6) into Eq. (3–1) gives

$$\frac{I_C}{\alpha} = I_B + I_C \quad \text{or} \quad I_B = I_C\frac{1 - \alpha}{\alpha} \tag{3–7}$$

Rearranging the terms on both sides of Eq. (3–7) yields

$$\frac{I_C}{I_B} = \frac{\alpha}{1 - \alpha} \tag{3–8}$$

However, by Eq. (3–3), $I_C/I_B = \beta$, so

$$\beta = \frac{\alpha}{1 - \alpha} \tag{3–9}$$

Similarly, α can be derived in terms of β.

$$\alpha = \frac{\beta}{\beta + 1} \tag{3–10}$$

If a BJT with an $\alpha = 0.95$ is connected in the common-base configuration, it has a dc current gain of 0.95. This same BJT in the common-emitter configuration has a current gain of $\beta = \alpha/(1 - \alpha) = 0.95/(1 - 0.95) = 19$. The *CE* configuration is therefore more useful in that it produces more current gain than the *CB* circuit.

Current gain parameter β is determined by the base dimensions and the doping characteristics, and is established during the manufacturing process of the BJT. Since it is difficult to exercise the fine control required to produce transistors with absolutely identical base dimensions and characteristics, considerable spread exists in β values, even for devices of the same type. Typical values for β range from 20 to 500 or more. For good BJTs, α approaches unity ($I_C \cong I_E$) and often is assumed = 1.

EXAMPLE 3-1

In the *CE* configuration of Fig. 3–11 the base current is 1 mA. Assume that α = 0.98. Calculate the emitter and collector currents.

SOLUTION

According to Eq. (3–9),

$$\beta = \frac{\alpha}{1 - \alpha} \quad \text{so} \quad \beta = \frac{0.98}{1 - 0.98} = 49$$

Using Eq. (3–3), the collector current is

$$I_C = \beta I_B = 49 \times 1 \text{ mA} = 49 \text{ mA}$$

The emitter current equals the sum of the base and collector currents, so

$$I_E = I_B + I_C = 1 \text{ mA} + 49 \text{ mA} = 50 \text{ mA}$$

3–2.4 DC Voltages and Currents

Figure 3–12 shows an *npn* BJT in the *CE* configuration, with load resistor R_L in the collector circuit and base resistor R_B in the base circuit. The base–emitter voltage is shown as V_{BE}, and the collector–emitter voltage as V_{CE}.

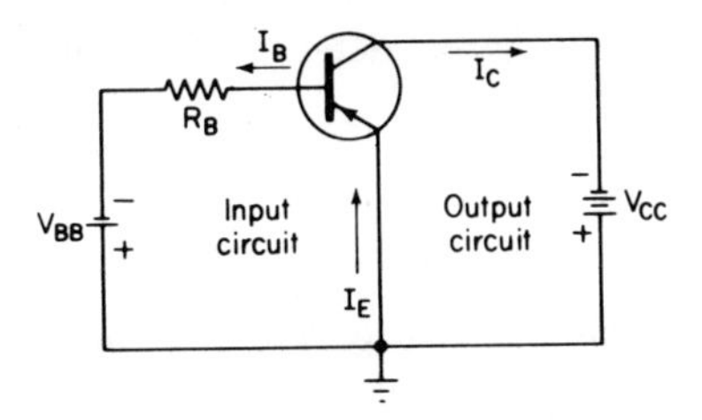

Fig. 3–11 A *pnp* transistor in the *CE* configuration.

Since the emitter is at ground potential, it is common practice to omit the second subscript and call the base voltage V_B and the collector voltage V_C, implying that they are measured with respect to the common or ground terminal.

The base–emitter junction is a forward-biased diode, with a forward voltage drop V_{BE} approximately equal to the knee voltage of a silicon or germanium diode. Hence, provided that the emitter–base diode conducts, V_{BE} is approximately 0.7 V for a silicon BJT and approximately 0.3 V for a germanium BJT.

To determine the dc voltages and currents in the *CE* circuit of Fig. 3–12, Kirchhoff's voltage law in the input circuit gives

$$V_{BB} = I_B R_B + V_{BE} \tag{3–11}$$

Solving Eq. (3–11) for I_B yields

$$I_B = \frac{V_{BB} - V_{BE}}{R_B} \tag{3–12}$$

Since V_{BE} is approximately constant (knee voltage), the magnitude of the base current is controlled by base battery V_{BB} and base resistor R_B.

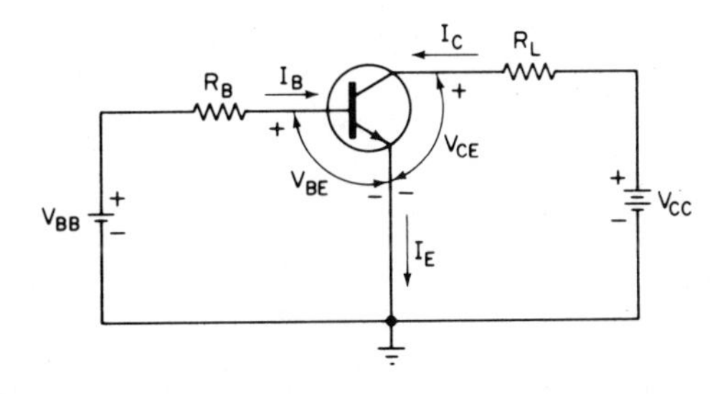

Fig. 3–12 dc currents and voltages in a *CE* configuration.

The collector current can be found if the dc current gain of the BJT is known, since

$$I_C = \beta I_B \tag{3–13}$$

The collector-to-emitter voltage V_{CE} can be calculated by applying Kirchhoff's voltage law in the collector circuit.

$$V_{CC} = I_C R_L + V_{CE} \tag{3–14}$$

The only unknown in Eq. (3–14) is V_{CE}. Rearranging and solving,

$$V_{CE} = V_{CC} - I_C R_L \tag{3–15}$$

Both terms on the right-hand side of Eq. (3–15) are known and V_{CE} can be calculated.

The above calculation is typical for a simple BJT circuit and can be applied to any configuration.

EXAMPLE 3-2

In Fig. 3–12, a silicon transistor has $\beta = 20$. $V_{BB} = 2$ V, $V_{CC} = 12$ V, $R_B = 10$ kΩ, and $R_L = 3$ kΩ. Calculate the dc currents and voltages.

SOLUTION

For a silicon transistor, $V_{BE} = 0.7$ V. Hence,

$$I_B = \frac{V_{BB} - V_{BE}}{R_B} = \frac{2\text{ V} - 0.7\text{ V}}{10\text{ k}\Omega} = 130\ \mu\text{A}$$

Since $\beta = 20$,

$$I_C = \beta I_B = 20 \times 130\ \mu\text{A} = 2.6\text{ mA}$$

and

$$I_E = I_B + I_C = 130\ \mu\text{A} + 2.6\text{ mA} = 2.73\text{ mA}$$

$$V_{CE} = V_{CC} - I_C R_L = 12\text{ V} - (2.6\text{ mA} \times 3\text{ k}\Omega) = 4.2\text{ V}$$

For the *CE* circuit of Fig. 3–12, base current I_B is set by base bias battery V_{BB} and base bias resistor R_B, and the magnitude of I_B is independent of the BJT itself. With I_B set to a certain value, the BJT, by virtue of its dc current gain β, determines the magnitude of collector current I_C. I_C, in turn, determines the dc voltage drop across load resistor R_L, and hence collector-to-emitter voltage V_{CE}. Any change in base current, caused by varying V_{BB} or R_B, changes the dc conditions in the circuit accordingly.

The CB circuit can be solved by writing the voltage equations for the emitter–base circuit and the collector–base circuit, and by using the basic current relationships of Eqs. (3–1), (3–2), and (3–3).

The *CC configuration* presents a slightly more complex problem since load resistor R_L is located in the emitter leg and hence forms part of both input and output circuits. In the *CC* configuration of Fig. 3–13, the various dc voltages and currents are identified. The BJT is a *pnp* device, hence the polarities of the supply batteries and the voltage drops across the transistor elements. In particular, emitter battery voltage V_{EE} is larger than base battery voltage V_{BB}, so that correct *forward* bias for the emitter–base junction is applied.

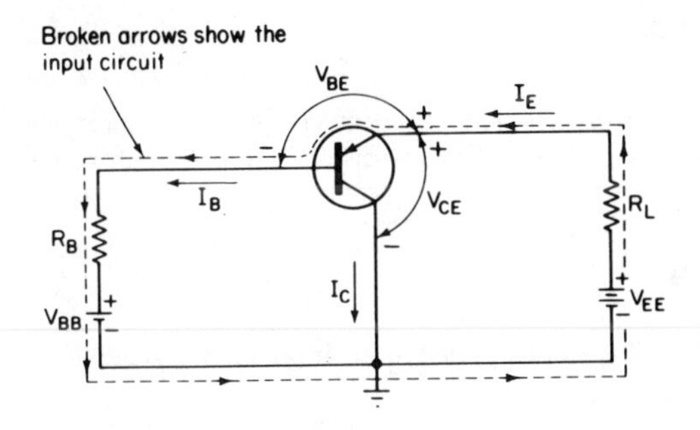

Fig. 3–13 dc analysis of the common-collector circuit.

The input circuit of the BJT includes the emitter–base junction and consists of battery V_{BB}, resistor R_B, emitter–base junction, load resistor R_L and emitter battery V_{EE}. Kirchhoff's voltage law in this circuit yields

$$V_{BB} - V_{EE} + I_E R_L + V_{BE} + I_B R_B = 0 \tag{3–16}$$

V_{BE} is the forward voltage drop across the emitter–base junction; it is about 0.7 V for a silicon device and 0.3 V for germanium. Battery voltages V_{BB} and V_{EE}, and resistors R_B and R_L are known circuit elements, so Eq. (3–16) contains two unknowns: I_E and I_B. But, $I_E = I_B + I_C$, Eq. (3–1), and $I_C = \beta I_B$, Eq. (3–3), hence

$$I_E = (\beta + 1)I_B \tag{3–17}$$

Substituting Eq. (3–17) into Eq. (3–16), and rearranging,

$$V_{EE} - V_{BB} - V_{BE} = I_B[R_B + (\beta + 1)R_L] \tag{3–18}$$

The only remaining unknown in Eq. (3–18) is I_B. Substitution of the known values yields I_B.

From I_B the collector current I_C and the emitter current I_E can be calculated, using Eqs. (3–1) and (3–3).

Kirchhoff's voltage equation for the collector–emitter circuit yields

$$V_{EE} = I_E R_L + V_{CE} \tag{3–19}$$

and the collector–emitter voltage drop V_{CE} can be found.

EXAMPLE 3–3

In the common-collector configuration of Fig. 3–13, $V_{BB} = 3.5$ V, $V_{EE} = 6$ V, $R_B = 10$ kΩ, and $R_L = 200$ Ω. The transistor is silicon, with a dc current gain $\beta = 39$. Calculate the transistor currents and voltages.

SOLUTION

Kirchhoff's voltage law applied to the input circuit yields

$$V_{BB} - V_{EE} + I_E R_L + V_{BE} + I_B R_B = 0$$

$I_E = (\beta + 1)I_B$, so the voltage equation changes to

$$V_{EE} - V_{BB} - V_{BE} = I_B[R_B + (\beta + 1)R_L]$$

Inserting the known quantities into this expression we obtain

$$6\text{ V} - 3.5\text{ V} - 0.7\text{ V} = I_B[10\text{ k}\Omega + (39 + 1)200\ \Omega]$$

Simplifying,

$$1.8\text{ V} = I_B(18\text{ k}\Omega)$$

or

$$I_B = \frac{1.8\text{ V}}{18\text{ k}\Omega} = 0.1\text{ mA}$$

$$I_C = \beta I_B = 39 \times 0.1 \text{ mA} = 3.9 \text{ mA}$$

$$I_E = I_B + I_C = 0.1 \text{ mA} + 3.9 \text{ mA} = 4 \text{ mA}$$

The voltage equation for the output circuit yields

$$V_{EE} = I_E R_L + V_{CE}$$

so

$$V_{CE} = V_{EE} - I_E R_L$$
$$= 6 \text{ V} - (4 \text{ mA} \times 200\ \Omega) = 5.2 \text{ V}$$

3–2.5 Loadline Concept

Examples 3–2 and 3–3 of Section 3–2.4 show that the dc conditions in the basic BJT configurations can be determined mathematically by solving Kirchhoff's voltage equations for the input and output circuits. In both examples, the current gain parameter β was used to calculate the collector and emitter currents. But the gain characteristics of the BJT are often presented in *graphical* rather than numerical form. One of the most important graphs is the *collector characteristic,* where collector current I_C is plotted as a function of collector-to-emitter voltage V_{CE} for various values of base current I_B. A typical *family* of collector characteristics is shown in Fig. 3–14. The graph of Fig. 3–14 shows that the collector current increases practically linearly with increasing base current, indicated by equal spacing between the curves marked I_B. This is in keeping with the definition of dc current gain in Eq. (3–3), where $I_C = \beta I_B$.

The family of collector characteristics can be used to determine the dc operating conditions of the basic BJT configurations by graphical means. The discussion here is limited to the *CE* configuration, but graphical solutions can also be used for the *CB* and *CC* configurations.

Consider the *npn* BJT in the *CE* configuration of Fig. 3–15. Kirchhoff's voltage law for the collector circuit yields the voltage equation

$$V_{CC} = I_C R_L + V_{CE} \tag{3–20}$$

which contains two unknowns: collector current I_C and collector-to-emitter voltage V_{CE}. I_C and V_{CE} are dependent on one another and the voltage equation can therefore only be solved when the relation between I_C and V_{CE} is known. The collector characteristics of Fig. 3–14, as a plot of I_C versus V_{CE}, provide this relationship. A simultaneous solution of the algebraic equations [Eq. (3–20)] and the graphical information (Fig. 3–14) therefore yields the desired values of I_C

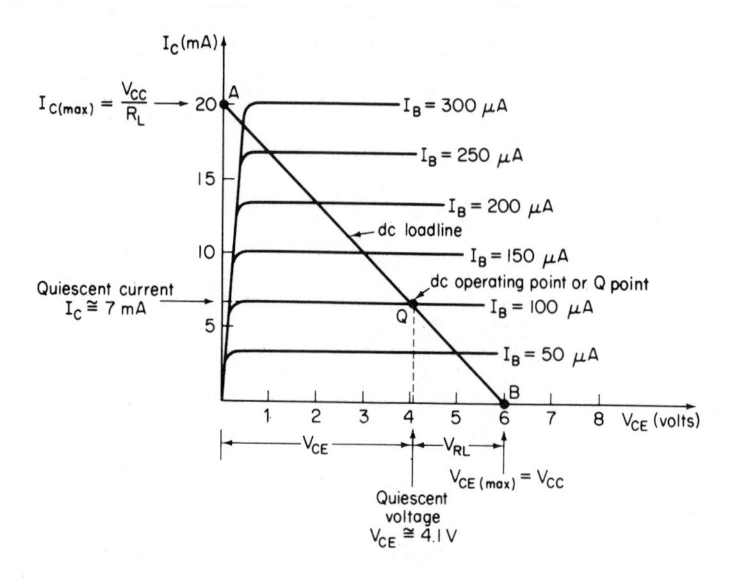

Fig. 3–14 Collector characteristics of a BJT and the construction of the dc loadline.

and V_{CE}, Eq. (3–20) being a first-order equation, which can be represented by a straight line on the graph of Fig. 3–14. It is called the *dc loadline* and Eq. (3–20) is the dc *loadline equation.* If two points on this line can be determined, the location of the loadline on the graph is known. To find these two points, conditions are imposed on the collector circuit. First, set the collector-to-emitter voltage $V_{CE} = 0$, so Eq. (3–20) reduces to

$$V_{CC} = I_C R_L \qquad \text{or} \qquad I_C = \frac{V_{CC}}{R_L} \tag{3–21}$$

This yields point A on the dc loadline, Fig. 3–14, where $V_{CE} = 0$ and $I_C = V_{CC}/R_L$. For the circuit values given, $I_{C(\max)} = 6\text{ V}/300\ \Omega = 20\text{ mA}$.

Secondly, set the collector current $I_C = 0$, so Eq. (3–20) reduces to

$$V_{CE} = V_{CC} \tag{3–22}$$

This yields a second point B on the dc loadline, where $I_C = 0$ and $V_{CE(\max)} = V_{CC} = 6$ V. The straight line obtained by connecting points A and B represents all solutions to Eq. (3–20), and is the dc loadline.

The collector circuit of Fig. 3–15 must, at all times, satisfy Kirchhoff's voltage law represented by the loadline equation, so that *the BJT must operate at some point on the loadline.* The exact location of this *operating point* depends on the value of the base current I_B.

Considering the input circuit of the BJT, we can write the voltage equation for the emitter–base circuit and obtain

$$V_{BB} = I_B R_B + V_{BE} \tag{3-23}$$

or

$$I_B = \frac{V_{BB} - V_{BE}}{R_B} \tag{3-24}$$

The right side of Eq. (3–24) contains known values, so that the magnitude of the base current is fixed. For the values given, and assuming that the BJT is a germanium device, the base current equals

$$I_B = \frac{1.5\text{ V} - 0.3\text{ V}}{12\text{ k}\Omega} = 100\ \mu\text{A}$$

If the transistor is to satisfy both the loadline equation [Eq. (3–20)] and the base current equation [Eq. (3–24)], the operating point of the device must be located at the *intersection* of the loadline and the collector characteristic marked $I_B = 100\ \mu\text{A}$. The point of intersection is indicated by point Q in Fig. 3–14. This Q point or *quiescent operating point* of the transistor identifies the dc or *quiescent current* I_C and the dc or *quiescent voltage* V_{CE}. Projecting the Q point onto the current and voltage axes, $I_C \simeq 7$ mA and $V_{CE} \simeq 4.1$ V. The voltage drop across the load resistor is therefore $V_{RL} = V_{CC} - V_{CE} \simeq 6\text{ V} - 4.1\text{ V} \simeq 1.9$ V, as indicated in Fig. 3–14.

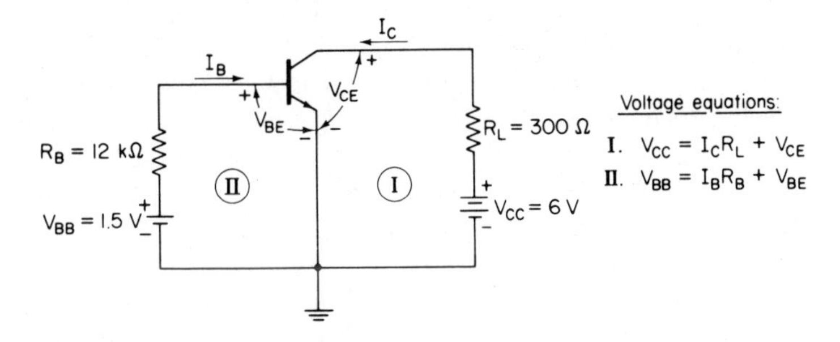

Fig. 3–15 A BJT in the common-emitter configuration. The loadline for this circuit is plotted on the collector characteristics of Figure 3–14.

EXAMPLE 3-4

The base resistor in Fig. 3–15 is changed from 12 kΩ to 6 kΩ, while the remaining circuit parameters are unchanged. The collector characteristics of the germanium BJT are given in Fig. 3–14. Determine the collector voltage V_{CE}, the collector current I_C, and the voltage drop V_{RL} across the load resistor.

SOLUTION

The conditions in the collector circuit have not changed and the dc loadline therefore retains its position on the graph of Fig. 3–14. The reduction in R_B from 12 kΩ to 6 kΩ increases the base current to

$$I_B = \frac{V_{BB} - V_{BE}}{R_B} = \frac{1.5\text{ V} - 0.3\text{ V}}{6\text{ k}\Omega} = 200\ \mu\text{A}$$

The quiescent operating point of the BJT therefore moves to the intersection of the dc loadline and the collector characteristic marked $I_B = 200\ \mu\text{A}$. The collector voltage and collector current corresponding to this new Q point are read off the graph as $I_C \simeq 13$ mA and $V_{CE} \simeq 2$ V. The voltage drop across the load resistor is equal to the difference between the collector supply voltage and the voltage drop across the BJT, so

$$V_{RL} \simeq 6\text{ V} - 2\text{ V} \simeq 4\text{ V}$$

3–2.6 Limits of Operation

Figure 3–16 shows a family of collector characteristics, where collector current I_C is plotted as a function of collector voltage V_{CE} for different values of base current I_B. This family of curves applies to a transistor in the common-emitter configuration of Fig. 3–15.

The characteristic curves of Fig. 3–16 are approximately equally spaced, indicating that I_C increases or decreases linearly with changes in I_B. However, the curves are not quite horizontal; they rise slightly to the right (higher values of V_{CE}). This means that, even with I_B held constant, I_C increases slightly when the collector voltage is increased. This increase in I_C is due to an increase in minority carriers crossing the reverse-biased base-collector junction at higher collector voltages. This effect is clearly visible at the $I_B = 0$ curve, which starts at the origin ($I_B = 0$, $I_C = 0$), and then gradually rises to give I_C some small finite value, even with zero base current.

The position of the operating point of the transistor is established by the intersection of the dc loadline [Eq. (3–20)] and the characteristic curve corresponding to the base current set by V_{BB} and R_B in Fig. 3–15 [Eq. (3–24)]. When I_B is increased (by decreasing R_B in Fig. 3–15), the Q point moves up, along the loadline, toward the I_C-axis, and I_C increases, while V_{CE} decreases. I_B can be increased to the point where the operating point enters the *saturation region,* indicated by the shaded area marked on the graph. In this region, the collector current is maximum and approaches V_{CC}/R_L, while the collector voltage is minimum and approaches zero.

The base current can be decreased by increasing R_B in Fig. 3–15. The Q point then moves down, along the loadline, toward the V_{CE}-axis, so that I_C decreases while V_{CE} increases. Eventually, I_B is reduced to zero and the transistor enters the *cutoff region*. In this region, the collector current is practically zero, and the collector voltage approaches the supply voltage.

For normal transistor operation, the Q point is placed in the *active region,* where the collector voltage is between the limits $V_{CE} = 0$ and $V_{CE} = V_{CC}$, and the collector current is between zero and the saturation value.

The three operating regions are shown in Fig. 3–16. In the active region the transistor operates as a linear amplifier, where small changes ΔI_B in input current cause relatively large changes ΔI_C in output current. In the saturation region the transistor behaves like a virtual short circuit, where V_{CE} is approximately zero volts and I_C is maximum, limited only by the resistance in the collector circuit. In the cutoff region the transistor behaves like an open circuit, where I_C is practically zero and V_{CE} equals the supply voltage. The saturation and cutoff regions are discussed in Ch. 13.

Operating the transistor practically in the *active region* means adjusting the base current so that the dc operating point is located approximately midway

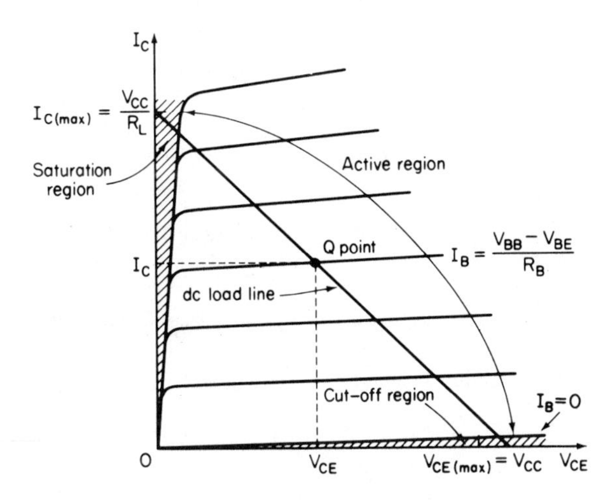

Fig. 3–16 Hypothetical collector characteristics of a BJT in the *CE* configuration. The intersection of the dc loadline and the selected base current curve determines the position of the *Q* point.

on the dc loadline, as indicated by point Q in Fig. 3–16. Here, the collector voltage is approximately one-half the supply voltage ($V_{CE} \simeq \frac{1}{2} V_{CC}$) and the collector current is approximately one-half the maximum (saturation) current [$I_C \simeq \frac{1}{2} I_{C(max)}$].

The BJT can be made to operate in the *cutoff region* simply by opening the base so that $I_B = 0$, or even applying reverse bias to the emitter–base junction.

On the other hand, the BJT can be made to operate in the *saturation region* by increasing the base current to the point where the collector current reaches its saturation value (Example 3–5).

EXAMPLE 3–5

The CE circuit of Fig. 3–15 uses a silicon BJT with $\beta = 50$. $V_{CC} = 10$ V, $V_{BB} = 4$ V, $R_L = 2$ kΩ, and R_B is variable. Calculate that resistance of R_B which causes the transistor to operate in the saturation region.

SOLUTION

The transistor is in saturation when the collector voltage $V_{CE} \simeq 0$ V. In that case, the collector current is

$$I_C \simeq \frac{V_{CC}}{R_L} = \frac{10 \text{ V}}{2 \text{ k}\Omega} = 5 \text{ mA}$$

$$I_B = \frac{I_C}{\beta} = \frac{5 \text{ mA}}{50} = 100 \ \mu\text{A}$$

I_B can be adjusted by variable resistor R_B. To calculate R_B, we write the voltage equation for the base circuit:

$$V_{BB} = V_{BE} + I_B R_B$$

Solving for R_B and substituting the known values yields

$$R_B = \frac{V_{BB} - V_{BE}}{I_B} = \frac{4 \text{ V} - 0.7 \text{ V}}{100 \ \mu\text{A}} = 33 \text{ k}\Omega$$

When $R_B = 33$ kΩ, the transistor enters into saturation. Hence, the *maximum* value of R_B which causes the transistor to saturate is 33 kΩ.

3-3 SINGLE-STAGE BJT AMPLIFIERS

3-3.1 Gain and Impedance

The primary function of an amplifier is to provide voltage gain, current gain, or power gain, depending on the application in which it is used. The gain of a single-stage BJT amplifier depends not only on the gain parameters α and β of the transistor itself, but also on resistances or impedances external to the BJT and internal resistances of the BJT.

Remember that the term "impedance" applies to a complex quantity that has both magnitude and phase angle. The total load in the collector circuit of a BJT in the *CE* configuration, for example, may consist of a resistor capacitively coupled to the collector circuit; in this case the total ac load is the combined impedance of these elements.

We now define the various gains of an amplifier. The *voltage gain* A_v of an amplifier is defined as the ratio of the ac voltage V_o developed across the load resistance R_L, to the ac input voltage V_i applied to the amplifier input terminals. Hence

$$A_v = \frac{V_o}{V_i} \tag{3-25}$$

It is quite possible that V_o and V_i are not in phase, in which case the voltage gain is a complex quantity.

The *current gain* A_i of an amplifier is defined as the ratio of the ac current I_o delivered to the load, to the ac current I_i supplied to the amplifier input. Hence

$$A_i = \frac{I_o}{I_i} \tag{3-26}$$

Here again, if I_o and I_i are not in phase, the current gain is a complex quantity.

The *power gain* A_p of an amplifier is defined as the ratio of the ac output power delivered to the load, to the ac input power supplied to the amplifier by the source. Hence

$$A_p = \frac{P_o}{P_i} \tag{3-27}$$

But $P_o = V_o I_o$, and $P_i = V_i I_i$, so Eq. (3-27) can also be written as

$$A_p = A_v A_i \tag{3-28}$$

indicating that the power gain is the product of the voltage gain and the current gain.

Another important parameter in amplifier gain calculations is the *input resistance* of the amplifier, defined as the ratio of ac input voltage V_i to ac input current I_i. Hence

$$R_i = \frac{V_i}{I_i} \tag{3-29}$$

If the input voltage and current are not in phase, their ratio is a complex number and we then speak of input impedance.

The gain of a single-stage BJT amplifier is closely related to the input resistance or impedance of the amplifier and the load resistance connected to its output terminals. Consider Fig. 3–17, which shows an ac amplifier with two input terminals and two output terminals. This amplifier can be a single BJT in

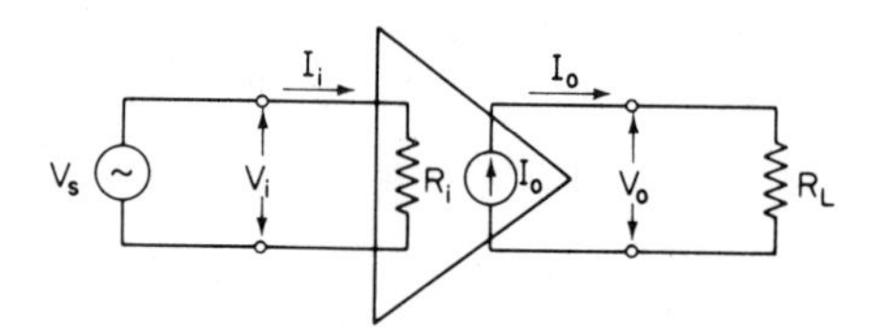

Fig. 3–17 General form of an ac amplifier.

the *CE* configuration, or it can be a number of BJTs connected in *cascade*. The input resistance of this amplifier is R_i. Signal source V_s develops input voltage V_i across the amplifier input terminals and supplies input current I_i. By Ohm's law,

$$V_i = I_i R_i \tag{3-30}$$

Representing the amplifier as a current generator I_o which develops an output voltage V_o across load resistor R_L, we can write

$$V_o = I_o R_L \tag{3-31}$$

Dividing Eq. (3–31) by Eq. (3–30) we obtain

$$\frac{V_o}{V_i} = \frac{I_o R_L}{I_i R_i} \tag{3-32}$$

The ratio V_o/V_i is the voltage gain A_v and the ratio I_o/I_i is the current gain A_i, so Eq. (3-32) reduces to

$$A_v = A_i \frac{R_L}{R_i} \tag{3-33}$$

Equation (3-33) is a *fundamental expression* for the voltage gain of any amplifier. It states that *the voltage gain of an amplifier is equal to the product of the current gain and the ratio of load resistance to input resistance.*

For a single-stage BJT amplifier, the voltage gain is easily established. In the first place, the current gain of a single BJT is a known quantity and depends only upon its configuration: α for *CB*, β for *CE*, and $(\beta + 1)$ for *CC*. The circuit load resistance is known. The input resistance of the BJT circuit can be calculated by using one of the equivalent circuits discussed below.

3-3.2 Introduction to Equivalent Circuits

The common-emitter collector characteristics of Fig. 3-18 show some nonlinear behavior. This nonlinearity is pronounced in the saturation and cutoff regions, identified in Fig. 3-18. To obtain approximately linear behavior, the BJT is biased so that its operating point falls in the active region. When the operation is limited to the active region, the linearity is acceptable and signal amplification with low distortion can be obtained.

When used as a linear amplifier, the BJT is biased roughly in the *center* of the active region, and the applied ac signal is *small* compared to the quiescent operating voltage and current. The BJT operates in the *small-signal mode.* BJTs used in this way are typically biased at a collector current ranging from 0.1 to 10 mA, and a collector voltage from 2 to 10 V. The ac signal applied to the BJT as a low-level amplifier typically ranges from 1 μV to 10 mV.

The small-signal BJT amplifier can be analyzed in two ways. The first method is *graphical,* and uses the characteristic curves of the device. The second method is mathematical, using an *equivalent circuit* which represents the behavior of the BJT under small-signal conditions.

Several small-signal equivalent circuits have been developed to meet various needs. Generally, two approaches are used. The first approach uses an equivalent circuit of parameters describing the transistor, the *T-equivalent circuit.* The second approach uses the *black-box concept,* in which the BJT is treated as a four-terminal network with certain measurable parameters, the *hybrid* or *h parameters.* The *T*-equivalent circuit and the hybrid equivalent circuit are primarily used for the purposes of device design or circuit design. But they are rather complex models, not well suited to rapid circuit analysis. With some approximations, they can be reduced to a simpler form, the *semihybrid* model.

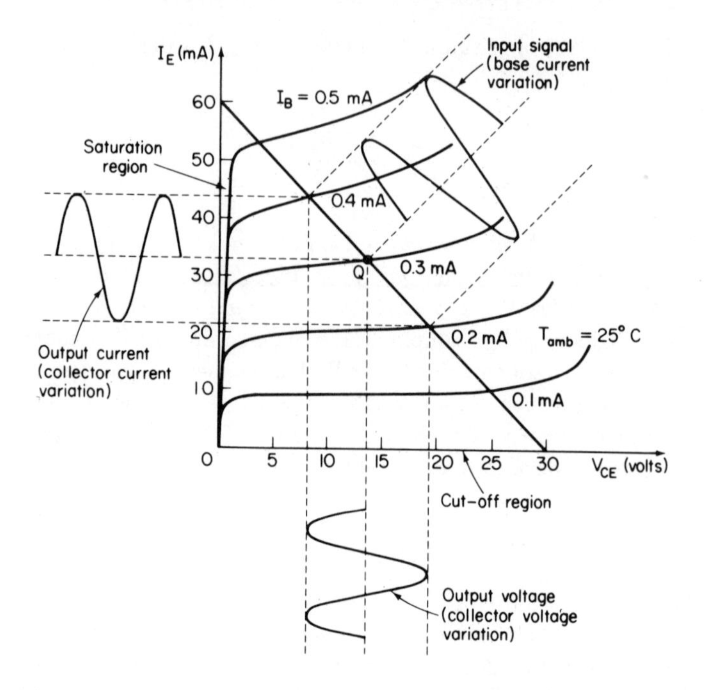

Fig. 3–18 Typical collector characteristics of a silicon planar BJT. The operating point is located in the center of the active region.

3–3.3 T-Equivalent Circuit

The *T*-equivalent circuit, composed of parameters identified with the transistor mechanisms, expresses the physical characteristics of the BJT under specified bias conditions and ambient temperature, and is shown in Fig. 3–19 in the *CB* configuration. The circuit consists of three resistive elements and two current generators, connected in the form of a *T*.

Emitter resistance $r_{e'}$ relates the current and voltage of the forward-biased emitter–base junction. This relation is essentially nonlinear and proportional to the dc emitter current. Shockley's relation identifies the emitter resistance of a forward-biased diode as $r_{e'} \simeq 26/I_E$, indicating that $r_{e'}$ changes with the dc bias current.

Base resistance $r_{b'}$ is associated with the flow of majority current from the base terminal to the internal active region of the BJT. In the *CB* connection of Fig. 3–19, $r_{b'}$ constitutes a feedback path between input and output circuits and causes a power loss.

Collector resistance $r_{c'}$ is the reverse resistance of the base-collector junction. It is generally very high, typically on the order of 1 MΩ or more.

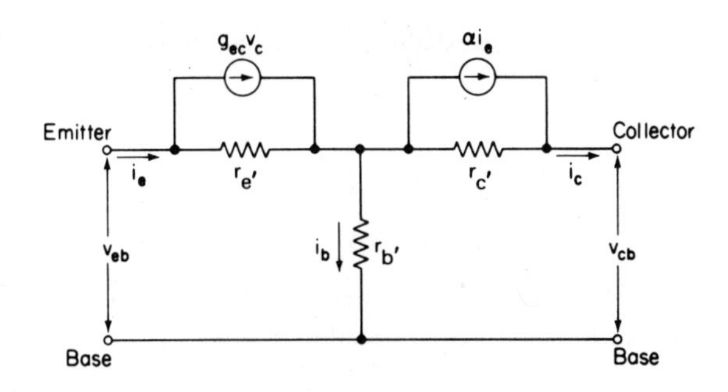

Fig. 3–19 Equivalent circuit of a bipolar junction transistor.

Emitter feedback conductance g_{ec} results from the fact that an increase in *collector* voltage causes an increase in *emitter* current, evidenced by the slow rise of the collector characteristics of Fig. 3–18. The equivalent circuit contains a small-signal current generator $g_{ec}v_c$ across the emitter junction to account for this phenomenon.

Current amplification factor α is defined as the ac current gain of the BJT in the *CB* configuration, so that $\alpha = i_c/i_e$. α is slightly smaller than 1 (typically from 0.95 to 0.995) and is sensitive to changes in bias and temperature.

Although the equivalent circuit accurately represents the transistor action, the two current generators make the circuit difficult to use in practical applications.

A simplified version of the *T-equivalent circuit,* shown in Fig. 3–20 for the *CB* configuration. Because there is only one current generator, αi_e, in this circuit, it is simpler to use and has found wide acceptance in BJT analysis.

The *T*-equivalent circuit shows three resistive elements representing the three BJT branches (indicated by r_e, r_b, and r_c), and a single current generator αi_e in the collector circuit. The terms r_e, r_b, and r_c are *not identical* to the resistances defined for Fig. 3–19. However, since both equivalent circuits

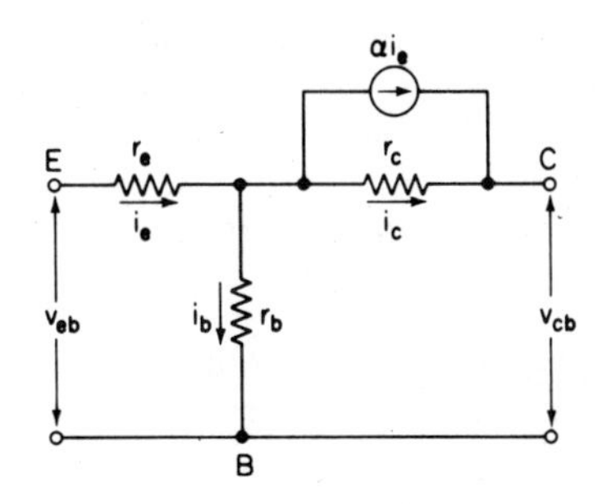

Fig. 3–20 *T*-equivalent circuit of a bipolar junction transistor in the *CB* configuration.

synthesize the BJT, the sets of resistances must be related, and can therefore be expressed in terms of each other:

$$r_{b(\text{tee})} = r_{b'} + \frac{\beta r_{e'}}{2} \quad \text{where } \beta = \frac{\alpha}{1-\alpha}. \qquad r_{e(\text{tee})} = \frac{r_{e'}}{2} \quad \text{where } r_{e'} \cong \frac{26}{I_E}$$

$$r_{c(\text{tee})} = r_{c'} \qquad \alpha_{(\text{tee})} = \alpha$$

Typical values for T-equivalent parameters, valid for both germanium and silicon junction transistors, are summarized in Table 3–1.

TABLE 3–1 TYPICAL VALUES FOR *T*-EQUIVALENT PARAMETERS.

Symbol	Typical value
$r_{b\,(\text{tee})}$	100 Ω to 500 Ω
$r_{e\,(\text{tee})}$	20 Ω to 50 Ω
$r_{c\,(\text{tee})}$	1 MΩ to 5 MΩ
$\alpha_{(\text{tee})}$	0.95 to 0.995

Figure 3–21 shows the schematic diagrams of the three amplifier configurations, together with their T-equivalent circuits. The simplified circuit diagrams do not contain the dc voltage sources necessary to establish the dc operating conditions, but only the ac signal source v_s with its internal resistance R_s. The internal BJT resistances r_e, r_b, and r_c are the T equivalents of Table 3–1.

The small-signal or ac performance of each circuit is evaluated in terms of the voltage, current, and power gains, and the input and output resistances. Consider the *common-base* amplifier of Fig. 3–21(a) and its T-equivalent circuit.

The ac *current gain* A_i is defined as the ratio of the ac output current i_c to the ac input current i_e, so

$$A_i = \frac{i_c}{i_e} \cong \alpha \tag{3–34}$$

α is less than 1, so the current gain of the CB amplifier stage is less than 1.

The *input resistance* r_i of the transistor, looking into its base–emitter terminals, is defined as the ratio of input voltage v_{eb} to input current i_e. It is common practice to use double subscript notation to identify the configuration in which the transistor is connected. For example, the input resistance of a transistor in the *CB* configuration is written as r_{ib}, with the second subscript b identifying the *CB* connection. Similarly, the input resistance of a transistor in the *CE* configuration is written as r_{ie}, with the second subscript e identifying the *CE* connection. For Fig. 3–21 (a)

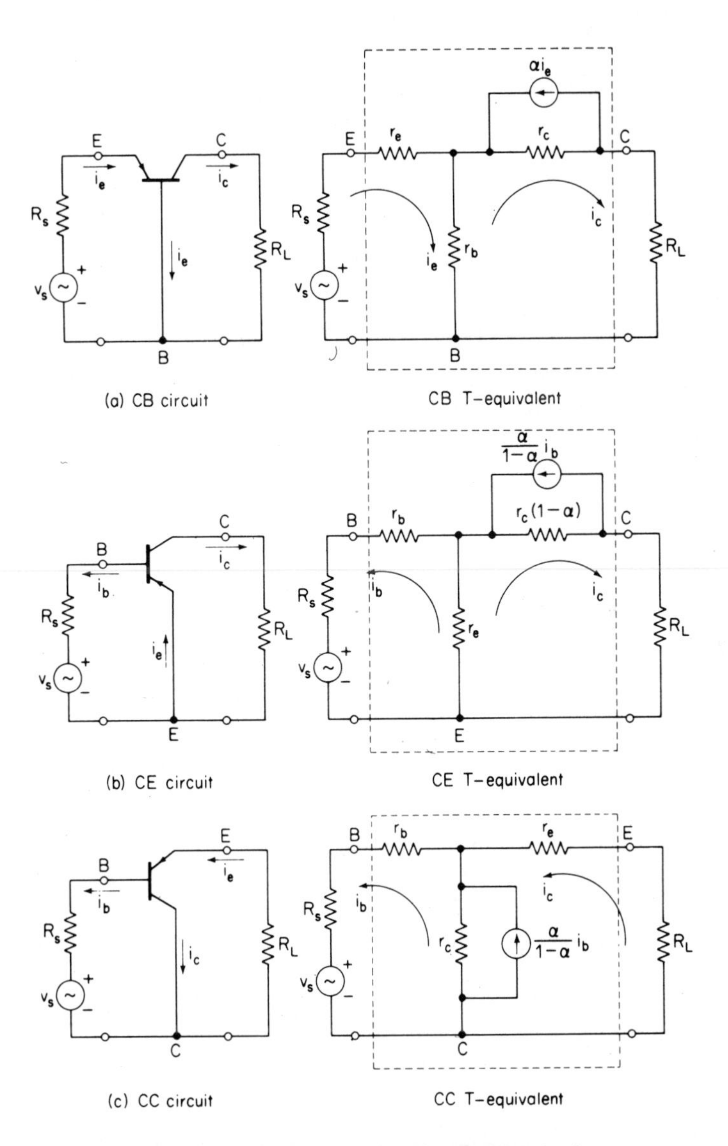

Fig. 3–21 Basic amplifier configurations and their *T*-equivalent circuits.

$$r_{ib} = \frac{v_{eb}}{i_e} \tag{3–35}$$

Writing now the voltage equation for v_{eb} we obtain

$$v_{eb} = i_e(r_e + r_b) - i_c r_b \tag{3–36}$$

Since $i_c = \alpha i_e$, Eq. (3–36) can also be written as

$$v_{eb} = i_e(r_e + r_b) - \alpha i_e r_b \tag{3–37}$$

$$= i_e[r_e + (1 - \alpha)r_b] \tag{3–38}$$

Dividing both sides of Eq. (3–38) by i_e yields

$$r_{ib} = \frac{v_{eb}}{i_e} = r_e + (1 - \alpha)r_b \tag{3–39}$$

Since α is only slightly smaller than 1, the second term on the right-hand side of Eq. (3–39) is relatively small, so r_{ib} approaches r_e for high-gain junction transistors.

The *voltage gain* of any amplifier,

$$A_v = A_i \frac{R_L}{R_i} \tag{3–33}$$

For the *CB* transistor, current gain $A_i = \alpha \simeq 1$ and input resistance $R_i = r_{ib}$, so Eq. (3–33) can be written as

$$A_v = \alpha \frac{R_L}{R_i} \simeq \frac{R_L}{r_{ib}} \tag{3–40}$$

Although α is slightly less than 1, considerable voltage gain can be realized by a favorable selection of the R_L/R_i ratio.

The power gain of the *CB* amplifier stage is defined as the product of voltage gain and current gain, so

$$A_p = A_v A_i = \frac{\alpha^2 R_L}{r_{ib}} \tag{3–41}$$

Since $\alpha \simeq 1$, $A_p \simeq A_v$.

The *output resistance* r_{ob} of the *CB* transistor, looking back into the collector–base terminals, can be evaluated by writing the mesh equations for the

T-equivalent circuit. This yields

$$r_{ob} = r_b + r_c - \frac{r_b(r_b + \alpha r_c)}{R_s + r_b + r_e} \tag{3–42}$$

Since r_c is large, $r_{ob} \approx r_c$. The *CB* stage therefore has a high output resistance.

The common-emitter and common-collector amplifiers can be analyzed similarly, using the *T* equivalents of Fig. 3–21. The mathematical expressions describing the performance of each circuit are summarized in Table 3–2.

If the typical *T*-equivalent parameter values of Table 3–1 are substituted into the equations of Table 3–2, we obtain a comparison of the characteristics of the three basic amplifier configurations.

The *CB* connection provides low input resistance and high output resistance with good voltage amplification. The current amplification is less than 1, but the power gain is good. The *CE* connection provides excellent voltage, current, and power gain. The input resistance is higher than that of the *CB* circuit; the output resistance is lower than that of the *CB* circuit. The common-emitter circuit is the most commonly used circuit. The *CC* connection has a voltage amplification of less than 1 but a high current amplification. The input resistance is very high, and the output resistance low. This circuit is often used as a buffer amplifier between stages or as an impedance-matching device.

EXAMPLE 3-6

The constants of a BJT in the *CE* connection of Fig. 3–21(b) are as follows: r_e = 30 Ω, r_b = 300 Ω, r_c = 1 MΩ, α = 0.98, R_L = 10 kΩ, and R_s = 500 Ω. Determine the ac performance of the circuit.

SOLUTION

(a) The current gain of the transistor is

$$A_i = \beta = \frac{\alpha}{1 - \alpha} = \frac{0.98}{1 - 0.98} = 49$$

(b) The input resistance of the transistor is

$$r_{ie} = r_b + r_e(1 + \beta) = 300 + 30(50) = 1800\ \Omega$$

(c) The voltage gain of the stage is

$$A_v = A_i \frac{R_L}{r_{ie}} = 49 \times \frac{10\ \text{k}\Omega}{1800\ \Omega} \simeq 272$$

TABLE 3–2 GAIN AND IMPEDANCE CHARACTERISTICS OF THE THREE BASIC AMPLIFIER CONFIGURATIONS, EXPRESSED IN TERMS OF THE *T*-EQUIVALENT PARAMETERS (TYPICAL VALUES).

Parameter	Symbol	Common-base	Common-emitter	Common-collector
Input resistance	r_i	$r_e + r_b(1-\alpha)$	$r_b + r_e(1+\beta)$	$r_c + r_b - \dfrac{r_c^2(1-\alpha)}{R_L + r_e + r_c(1+\alpha)}$
Current gain	A_i	α	β	$\dfrac{1}{1-\alpha}$ or $\beta + 1$
Voltage gain	A_v	$\alpha \dfrac{R_L}{r_i}$	$-\beta \dfrac{R_L}{r_i}$	$(\beta+1)\dfrac{R_L}{r_i}$
Power gain	A_p	$\alpha^2 \dfrac{R_L}{r_i}$	$\beta^2 \dfrac{R_L}{r_i}$	$(\beta+1)^2 \dfrac{R_L}{r_i}$
Output resistance	r_o	$r_b + r_c - \dfrac{r_b[r_b + \alpha r_c]}{R_s + r_b + r_c}$	$r_e + r_c(1-\alpha) - \dfrac{r_e(r_e - \alpha r_c)}{R_s + r_b + r_e}$	$r_e + r_c(1-\alpha) - \dfrac{r_c^2(1-\alpha)}{R_s + r_b + r_c}$

(d) The power gain of the circuit is

$$A_p = A_v A_i \simeq 272 \times 49 \simeq 13{,}320$$

(e) The output resistance of the stage is

$$r_{oe} = r_e + (1 - \alpha)r_c + \frac{r_e(\alpha r_c - r_e)}{r_b + r_e + R_s}$$

Substituting the appropriate values we obtain

$$r_{oe} \simeq 55\ \text{k}\Omega$$

The same junction transistor connected as a *CB* amplifier would have the following performance characteristics: $A_i = 0.98$, $r_{ib} = 36\ \Omega$, $A_v = 272$, $A_p = 267$, $r_{ob} = 670\ \text{k}\Omega$.

3-3.4 Hybrid Parameters

Transistor manuals and data sheets usually provide device characteristics in the form of *hybrid* or *h* parameters. These parameters are easy to measure at both low and high frequencies and they ideally fit the low input and high output impedances of a BJT.

The *h* parameters are derived from the general network equations of a four-terminal network or black box, having two input terminals and two output terminals. Although a BJT is a three-terminal device, the common terminal serves in both the input and output circuits, so four-terminal analysis can be applied to the three-terminal BJT as a special case.

Consider the black box shown in Fig. 3-22. Writing the network equations for the input and output circuits yields

$$v_1 = i_1 h_{11} + v_2 h_{12} \tag{3-43}$$

$$i_2 = i_1 h_{21} + v_2 h_{22} \tag{3-44}$$

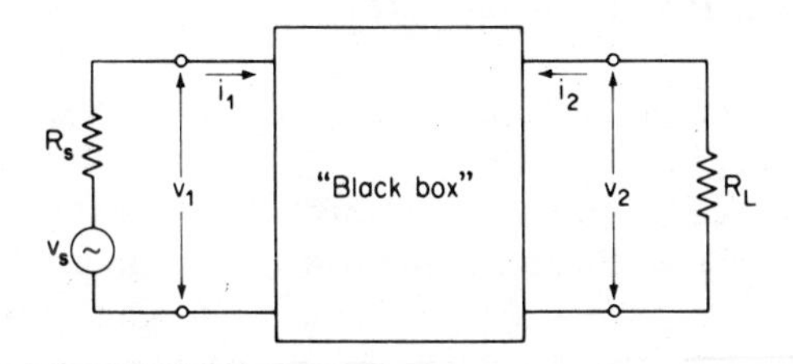

Fig. 3-22 Four-terminal network ("black box").

The hybrid or h parameters h_{11}, h_{12}, h_{21}, and h_{22} are a combination of impedance and admittance parameters. These h parameters are defined by imposing certain conditions on the network of Fig. 3–22. When the output terminals are short-circuited, $v_2 = 0$, so Eqs. (3–43) and (3–44) yield the values at left in Table 3–3.

TABLE 3-3. HYBRID PARAMETERS

$h_{11} = \frac{v_1}{i_1}$ = input impedance	$h_{12} = \frac{v_1}{v_2}$ = reverse voltage transfer ratio
$h_{21} = \frac{i_2}{i_1}$ = forward-current transfer ratio	$h_{22} = \frac{i_2}{v_2}$ = output admittance

When the input terminals are open-circuited, $i_1 = 0$, so Eqs. (3–43) and (3–44) yield the remainder of Table 3–3.

Under the conditions of short-circuited output and open-circuited input, the black box can be represented by an equivalent two-generator network, shown in Fig. 3–23.

When the black box contains a three-terminal transistor, with one terminal common to the input and output circuits, the h parameters describe the network characteristics of that transistor. In transistor analysis, it is customary to use descriptive letter subscripts instead of the numerical subscripts used above. According to IEEE standards, the first subscript indicates whether the particular parameter is an input (i), output (o), forward (f), or reverse (r) parameter. The second subscript then describes the transistor configuration (b for CB, e for CE, and c for CC). For the CE connection, the h parameters then become

h_{ie} = common-emitter input impedance

h_{fe} = common-emitter forward current transfer ratio

h_{re} = common-emitter reverse voltage transfer ratio

h_{oe} = common-emitter output admittance

For the CB configuration the h parameters are h_{ib}, h_{fb}, h_{rb}, and h_{ob}, and for the CC configuration h_{ic}, h_{fc}, h_{rc}, and h_{oc}.

Replacing the BJT in each of the three basic configurations with a hybrid model identical to that of Fig. 3–23 and using the appropriate letter subscripts provides the hybrid equivalent circuits of Fig. 3–24. The actual dynamic circuit parameters, such as A_i, A_v, A_p, r_i, and r_o, can be expressed in terms of the h parameters and the external resistors R_s and R_L. To illustrate the general

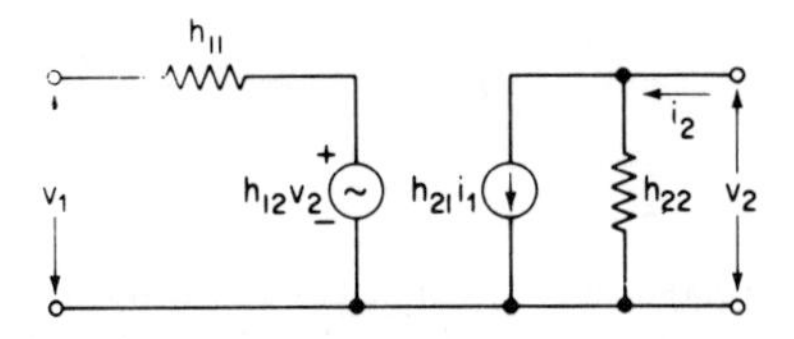

Fig. 3-23 Hybrid equivalent circuit for the four-terminal network.

method of obtaining these circuit parameters, we analyze the common-base circuit by writing the network equations for Fig. 3-24(a).

For the input circuit, the voltage equation reads

$$v_s = (h_{ib} + R_s)i_e + h_{rb}v_{cb} \tag{3-45}$$

For the output circuit, the voltage equation reads

$$O = h_{fb}i_e h_{ob} - i_c\left(R_L + \frac{1}{h_{ob}}\right) \tag{3-46}$$

Current gain A_i is defined as the ratio of the ac output current i_c to ac input current i_e. Equation (3-46) provides this ratio.

$$A_i = \frac{i_c}{i_e} = \frac{h_{fb}}{1 + h_{ob}\mathrm{R_L}} \tag{3-47}$$

When the output terminals are short-circuited, $R_L = O$ and the current gain is equal to the h parameter h_{fb}. The current gain decreases for increasing values of R_L and becomes zero when $R_L \Longrightarrow \infty$.

Input resistance r_i of the BJT is defined as the ratio of signal input voltage v_{eb} to signal input current i_e. The emitter-base voltage v_{eb} equals

$$v_{eb} = h_{ib}i_e + h_{rb}v_{cb} \tag{3-48}$$

Substituting $v_{cb} = -i_c R_L$ into Eq. (3-48) yields

$$v_{eb} = h_{ib}i_e - h_{rb}i_c R_L \tag{3-49}$$

Input resistance $r_{ib} = v_{eb}/i_e$, so that dividing both sides of Eq. (3-49) by i_e yields

$$r_{ib} = \frac{v_{eb}}{i_e} = h_{ib} - h_{rb}\frac{i_c}{i_e}R_L \tag{3-50}$$

or

$$r_{ib} = h_{ib} - h_{rb}A_iR_L \tag{3-51}$$

Here again, the input resistance, r_{ib} of the *CB* transistor equals h_{ib} when the output terminals are short-circuited ($R_L = 0$). Since h_{fb} has a negative value for the *CB* circuit, A_i is negative and the input resistance increase for increasing values of R_L.

Voltage gain A_v of the *CB* circuit is defined by Eq. (3–33) as

$$A_v = A_i\frac{R_L}{R_i} \tag{3-52}$$

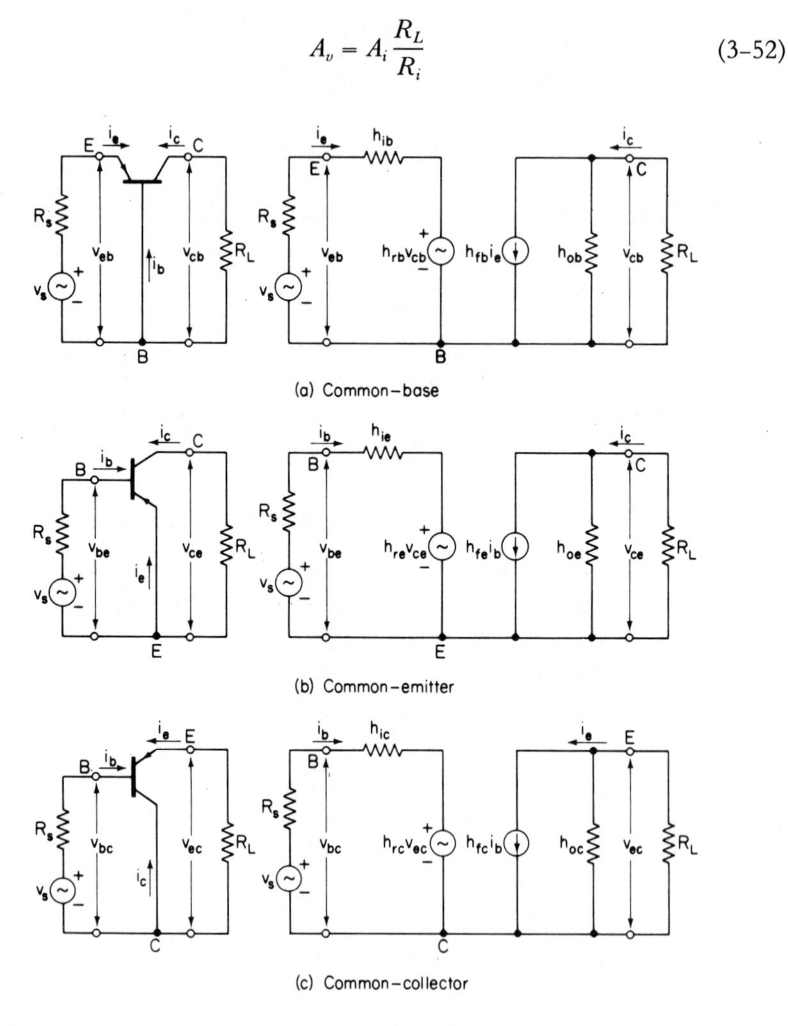

Fig. 3–24 Small-signal circuits and their hybrid equivalents for the three basic configurations.

where $A_i \cong h_{fb}$ and $R_i = r_{ib} \cong h_{ib}$. The power gain of the *CB* amplifier stage,

$$A_p = A_i A_v = \frac{A_i^2 R_L}{R_i} \tag{3-53}$$

The output resistance r_{ob} of the *CB* transistor, looking back into its collector–base terminals, is found by using the reciprocity theorem. Its application and solution yields

$$r_{ob} = \frac{1}{h_{ob} - \dfrac{h_{rb} h_{fb}}{R_s h_{ib}}} \tag{3-54}$$

The equations for input and output resistance, current and voltage amplification, and power gain, derived for the common-base circuit, are also applicable to the common-emitter and common-collector circuits, provided that the corresponding *h* parameters are used. Table 3–4 summarizes the operating equations for the three basic circuits. Numerical subscripts are used. Appropriate substitutions with letter subscripts must be made according to the type of circuit configuration being evaluated.

TABLE 3–4 SUMMARY OF THE OPERATING EQUATIONS FOR THE THREE TRANSISTOR CONFIGURATIONS. THE CORRECT LETTER SUBSCRIPTS FOR THE *h*-PARAMETERS SHOULD BE SUBSTITUTED FOR THE NUMERIC SUBSCRIPTS TO SUIT THE CIRCUIT CONFIGURATION.

Input resistance	$r_i = h_{11} - h_{12} A_i R_L$
Current gain	$A_i = \dfrac{h_{21}}{1 + h_{22} R_L}$
Voltage gain	$A_v = -A_i \dfrac{R_L}{r_i}$
Power gain	$A_p = A_i A_v$
Output resistance	$r_o = \dfrac{1}{h_{22} - \dfrac{h_{12} h_{21}}{R_s + h_{11}}}$

EXAMPLE 3-7

A silicon epitaxial BJT (RCA 40397), used as a high-current audio and video amplifier, has the following characteristics:

$$h_{fe} = 200; \quad h_{ie} = 600\ \Omega; \quad h_{oe} = 75\ \mu\text{mhos}; h_{re} = 125 \times 10^{-6}.$$

This BJT is connected in the *CE* configuration, with a load resistance $R_L = 1\text{k}\Omega$ and a source resistance $R_s = 200\ \Omega$.

Calculate (a) Current gain A_i.
(b) Input resistance r_{ie}.
(c) Voltage gain A_v.
(d) Power gain A_p.
(e) Output resistance r_{oe}.

SOLUTION

(a) $$A_i = \frac{h_{fe}}{1 + h_{oe}R_L} = \frac{200}{1 + (75 \times 10^{-6} \times 10^3)} \simeq 186$$

(b) $$r_{ie} = h_{ie} - h_{re}A_iR_L = 600 - (125 \times 10^{-6} \times 186 \times 10^3) \simeq 577\ \Omega$$

(c) $$A_v = -A_i\frac{R_L}{r_i} = -186 \times \frac{10^3}{577} \simeq -322 \text{ (phase reversal)}$$

(d) $$A_p = A_iA_v = 186 \times 322 \simeq 59{,}890$$

(e) $$r_{oe} = \frac{1}{h_{oe} - \dfrac{h_{re}h_{fe}}{R_s + h_{ie}}} = \frac{1}{75 \times 10^{-6} - \dfrac{125 \times 10^{-6} \times 200}{200 + 600}} \simeq 227\ \text{k}\Omega$$

For this high-gain transistor, the current gain A_i is almost equal to h_{fe}, and the input resistance $r_{ie} \simeq h_{ie}$. The *CE* circuit in this example has excellent current and voltage gains and very high power gains. The low source resistance R_s contributes to the high output resistance. It is also important to note that the input and output resistances of the *CE* configuration are least affected by the values of load resistance and source resistance, and assume values very close to the BJT hybrid parameters h_{ie} and h_{oe}, respectively.

3–3.5 Semihybrid Equivalent Circuits

The *T*-equivalent circuits of Section 3–3.3 have the disadvantage that the internal BJT resistances r_e, r_b, and r_c are difficult (and in some cases even impossible) to measure. In addition, the device manufacturers do not generally publish these parameters, so that the T-equivalent model is of limited practical value. On the other hand, the h parameters in the hybrid models of Section 3-3.4 are easy to measure and they are also published in the BJT data sheet, but the h-parameter circuit is still a rather complex model. For a really simple, but practical circuit, the T-equivalent and h-parameter models can be simplified considerably to yield the sufficiently accurate *semihybrid* model.

Consider the T equivalent for the transistor in the common-base configuration, Fig. 3–21(a). This equivalent circuit can be cast in the generalized form of an amplifier, with input resistance r_i, output resistance r_o, and output current generator i_o, as shown in Fig. 3–25. The input and output parameters of this amplifier can be derived from the T equivalent by making the following approximations.

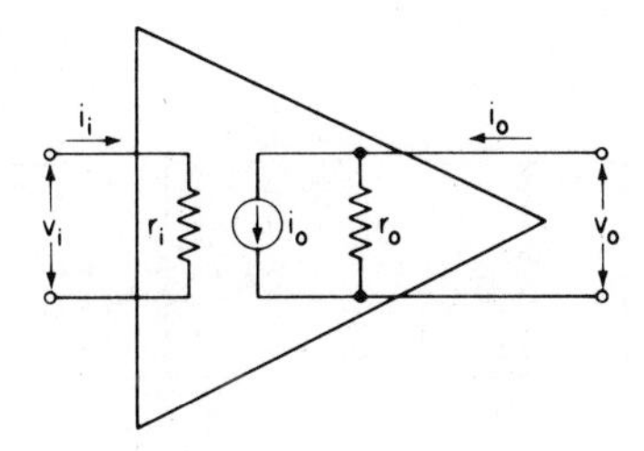

Fig. 3–25 General form of an amplifier.

According to the calculations of Section 3–3.3, the common-base T equivalent has an input resistance r_{ib} of

$$r_{ib} = r_e + (1 - \alpha)r_b \tag{3–39}$$

where r_e and r_b are the T-equivalent values defined in Section 3–3.3. Since r_e varies with the dc emitter current I_E, r_{ib} also depends on I_E. The input resistance of the CB transistor can be approximated by

$$r_{ib} \cong \frac{50}{I_E} \tag{3–55}$$

where I_E is the dc emitter current in milliamperes. It is convenient to assign a new symbol r_π to the quantity $50/I_E$, so that $r_{ib} = r_\pi \cong 50/I_E$.

Similarly, the output resistance of the T equivalent was given by Eq. (3–42) as

$$r_{ob} = r_b + r_c - \frac{r_b(r_b+\alpha r_c)}{R_s+r_b+r_c} \tag{3–42}$$

Since r_c in Eq. (3–42) is by far the largest resistance (typically from 1 to 5 MΩ), the output resistance of the amplifier can be approximated by

$$r_{ob} \cong r_c \tag{3–56}$$

The input current of the T equivalent is the ac emitter current i_e, and the output current is the ac collector current $i_c = \alpha i_e$, so the current generator of the amplifier can be represented by

$$i_o = \alpha i_e \tag{3–57}$$

These approximations then yield the extremely simple semihybrid model of Fig. 3–26 (a), consisting of input resistance r_{ib}, current generator αi_e, and

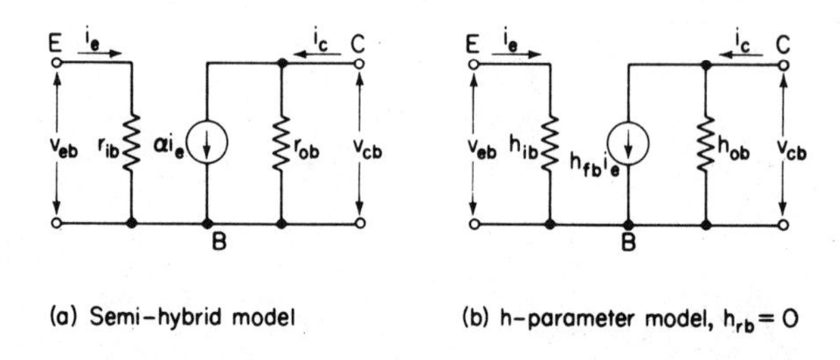

Fig. 3–26 Small-signal models for the BJT in the common-base connection.

output resistance r_{ob}. Since the h-parameter h_{rb} is almost always less than 5×10^{-4} it is justifiable to neglect its effect by setting $h_{rb} = 0$ in the h-parameter model. This then results in the simpler form of Fig. 3–26 (b), shown for comparison with the semihybrid circuit of Fig. 3–26 (a). By using the fact that $r_{ob} \gg r_{ib}$, it can be shown that the parameters in these two models are related as follows:

$$h_{ib} \cong r_\pi \cong \frac{50}{I_E} \cdot \qquad h_{fb} \cong \alpha \cdot \qquad h_{ob} \cong \frac{1}{r_{ob}} \cdot$$

Under typical operating conditions, the input resistance of the *CB* connection is low (a few ohms), the output resistance is high (approximately 1 MΩ), and the current gain is slightly less than 1. Considerable voltage gain can be realized by selecting the appropriate value of the load resistance.

A basic BJT amplifier in the common-base configuration is shown in Fig. 3–27. The dc conditions are established by the two bias batteries, emitter resistor R_E and collector resistor R_C, and the dc analysis proceeds in the usual manner. The ac conditions can be evaluated by replacing the BJT with its

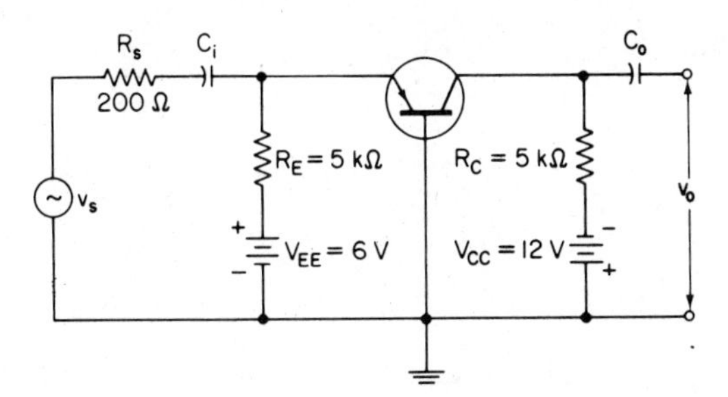

Fig. 3–27 Common-base amplifier.

semihybrid model, and then adding the remaining circuit components to obtain the complete ac equivalent of Fig. 3–28.

Often, $R_E \gg r_{ib}$, and the signal component of the emitter current is

$$i_e \cong \frac{v_s}{R_s + r_{ib}} \tag{3–58}$$

Hence, the output voltage is

$$v_o = \alpha i_e R_{ac} \cong \frac{R_{ac}}{R_s + r_{ib}} v_s \tag{3–59}$$

where R_{ac} is the parallel combination of BJT output resistance r_{ob} and collector resistor R_C. The voltage gain of the amplifier is

$$A_v = \frac{v_o}{v_s} \cong \frac{R_{ac}}{R_s + r_{ib}} \tag{3–60}$$

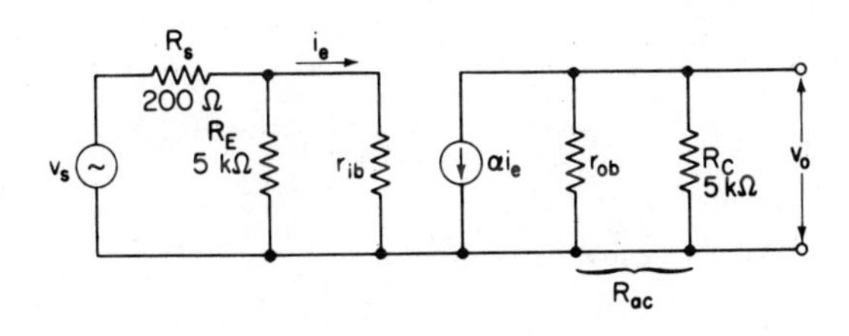

Fig. 3–28 Simplified small-signal model (Example 3.8).

Usually $r_{ob} \gg R_C$, and

$$A_v \simeq \frac{R_C}{R_s + r_{ib}} \tag{3-61}$$

EXAMPLE 3-8

The *pnp* silicon BJT used in the voltage amplifier of Fig. 3–27 has the following parameters: $h_{fb} = 0.98$, $h_{ob} = 0.6 \times 10^{-6}$ mho. The parameter r_{ib} in the small-signal model of Fig. 3–28 depends on the quiescent operating conditions and must be calculated. Determine the overall voltage gain of this amplifier.

SOLUTION

The dc operating conditions are determined as in Section 3–2.4.

$$I_E = \frac{V_{EE} - V_{BE}}{R_E} \simeq \frac{6\text{ V}}{5\text{ k}\Omega} \simeq 1.2\text{ mA}$$

$$I_C = \alpha I_E \simeq 1.2\text{ mA}$$

The dc collector voltage is

$$V_{CB} = V_{CC} - I_C R_C = 12\text{ V} - (1.2\text{ mA} \times 5\text{ k}\Omega) = 6\text{ V}$$

This places the BJT in the center of the active region. The small-signal parameter r_{ib} under these conditions equals

$$r_{ib} \simeq \frac{50}{I_E} \simeq \frac{50}{1.2\text{ mA}} \simeq 42\ \Omega$$

The small-signal parameter r_{ob} equals

$$r_{ob} \simeq \frac{1}{h_{ob}} = \frac{1}{0.6 \times 10^{-6}\text{ mho}} \simeq 1.67\text{ M}\Omega$$

Since $r_{ob} \gg R_C$, the ac load resistance equals

$$R_{ac} = r_{ob} // R_C \simeq 5\text{ k}\Omega$$

The 5-kΩ emitter resistor R_E is negligible in comparison with the 42-Ω BJT input resistance, so the voltage gain of the amplifier is given by Eq. (3–61) as

$$A_v \cong \frac{R_C}{R_s + r_{ib}} \cong \frac{5\ \text{k}\Omega}{242\ \Omega} \cong 20.6$$

The *common-emitter configuration* can be simplified in much the same manner as the common-base connection. This results in the semihybrid model of Fig. 3–29 (a). Table 3–2 gives the input resistance of the *T*-equivalent circuit as $r_{ie} = r_b + r_e(1+\beta)$. Comparing r_{ie} with the *CB* input resistance r_{ib}, we can show that $r_{ie} = (1+\beta)r_{ib}$, so the input resistance of the semihybrid model can be approximated by

$$r_{ie} \cong \beta r_\pi \cong \beta \frac{50}{I_E} \tag{3–62}$$

The input current of the *CE* connection is the ac base current i_b and the output current is the ac collector current $i_c = \beta i_b$, so the output current generator of the semihybrid model is

$$i_o = \beta i_b \tag{3–63}$$

Under small-signal conditions, the change in collector current is a linear function of the change in collector voltage, and this can be represented by output resistor r_{oe} shunting the terminals of the current generator, as in Fig. 3–29 (a). The *T*-equivalent value of r_{oe} is given in Table 3–2 and lies in the approximate range 20 to 200 kΩ.

For direct comparison, the *h*-parameter model of Fig. 3–24 (b) is placed alongside the semihybrid model of Fig. 3–29 (a). As in the previous case, it is justifiable to set the *h* parameter $h_{re} = 0$, and the simplified form of the *h*-parameter circuit in Fig. 3–29 (b) is identical to the semihybrid circuit. It can be shown that the parameters in these two models are related as follows:

$$h_{ie} \cong \beta r_\pi \cong \beta \frac{50}{I_E} \cdot \qquad h_{fe} \cong \beta \cdot \qquad h_{oe} \cong \frac{1}{r_{oe}} \cdot$$

Under typical operating conditions, the input resistance of the *CE* connection is in the intermediate range (from 500 Ω to 5 kΩ), the output resistance is also in the intermediate range (from 20 to 200 kΩ), and the current gain is on the order of 100. The voltage gain depends on the ratio of ac load resistance to ac input resistance, and can be considerable.

A basic BJT amplifier in the common-emitter configuration is shown in Fig. 3–30(a). The dc operating conditions are established by the bias batteries

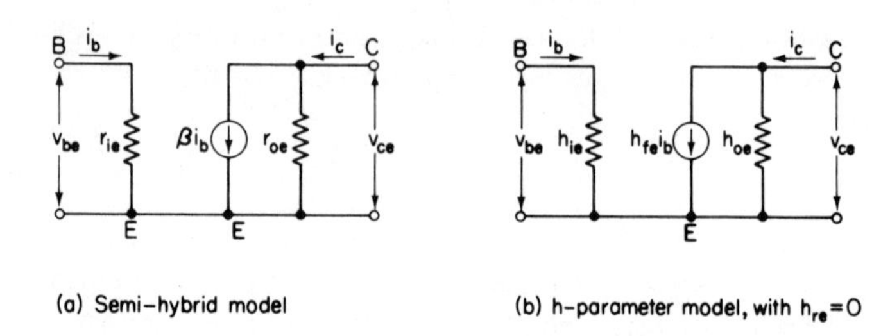

Fig. 3–29 Small-signal models for the BJT in the common-emitter connection.

V_{BB} and V_{CC} and by resistors R_B and R_C. The ac conditions can be evaluated by replacing the BJT with its semihybrid model, and then adding the remaining circuit components to obtain the complete ac equivalent circuit of Fig. 3–30(b).

In a practical circuit, $R_B \gg r_{ie}$ and the signal component of the base current is

$$i_b \cong \frac{v_s}{R_s + r_{ie}} \tag{3–64}$$

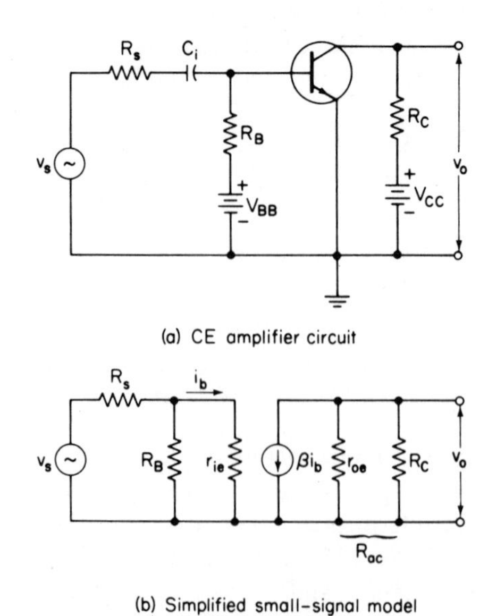

Fig. 3–30 Common-emitter amplifier and small-signal model.

The signal component of the output current is $i_c = \beta i_b$, so the output voltage equals

$$v_o = \beta i_b R_{ac} = \beta \frac{R_{ac}}{R_s + r_{ie}} v_s \tag{3-65}$$

where R_{ac} is the total ac load, consisting of r_{oe} in parallel with R_c. The over-all voltage gain of the amplifier is defined as v_o/v_s, so

$$A_v = \frac{v_o}{v_s} = \beta \frac{R_{ac}}{R_s + r_{ie}} \tag{3-66}$$

EXAMPLE 3-9

The *pnp* silicon transitor in the amplifier of Fig. 3–31 (a) receives its bias current directly from collector supply V_{CC} via base bias resistor R_B. The BJT parameters are $\beta = 80$ and $r_{oe} = 100$ kΩ. The small-signal parameter r_{ie} in the equivalent circuit of Fig. 3–31 (b) depends on quiescent operating conditions and must therefore be calculated. Determine the overall voltage gain of the amplifier.

SOLUTION

$$I_B = \frac{V_{CC} - V_{BE}}{R_B}$$

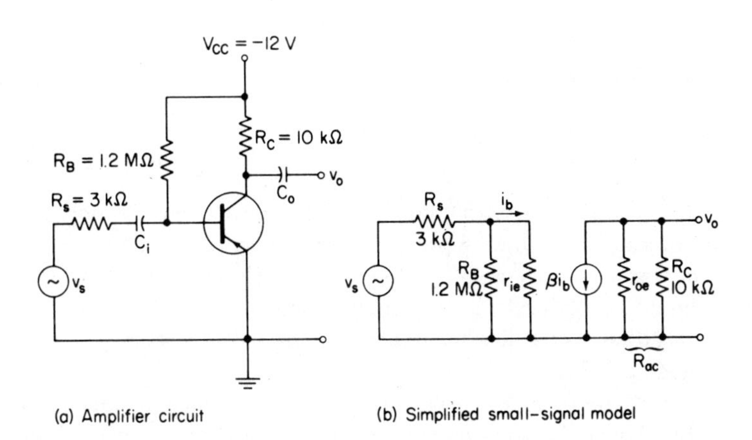

Fig. 3–31 Amplifier for Example 3.9.

Neglecting the base-emitter voltage drop,

$$I_B \simeq \frac{12\text{ V}}{1.2\text{ M}\Omega} = 10\ \mu\text{A}$$

$$I_C = \beta I_B = 80 \times 10\ \mu\text{A} = 0.8\text{ mA}$$

and the collector voltage is

$$V_{CE} = V_{CC} - I_C R_C = 12\text{ V} - (0.8\text{ mA} \times 10\text{ k}\Omega) = 4\text{ V}$$

The BJT therefore operates well within the active region of its characteristics.

The small-signal parameter r_{ie} under these conditions equals

$$r_{ie} \simeq \beta \frac{50}{I_E} = 80 \times \frac{50}{0.8} = 5\text{ k}\Omega$$

The ac load resistance consists of the parallel combination of r_{oe} and R_c, so

$$R_{\text{ac}} = \frac{100\ (10)}{110} \simeq 9.1\text{ k}\Omega$$

The 1.2-MΩ base resistor is very large compared to the 5-kΩ BJT input resistance, so the voltage gain of the amplifier is given by Eq. (3–66) as

$$A_v = \beta \frac{R_{\text{ac}}}{R_s + r_{ie}} = 80 \times \frac{9.1\text{ k}\Omega}{3\text{ k}\Omega + 5\text{ k}\Omega} = 91$$

3-4 GRAPHICAL METHODS

3-4.1 Types of Graphs

Manufacturers of transistors and other solid-state devices often supply information in graphical form. Graphs provide useful information about the *dynamic* behavior of transistors which is not directly or easily available from numerical data. It should be realized that the information presented in these graphs is *typical* rather than exact, just as numerical data on transistors always specify typical values.

Some of the more important graphs are presented in this section. In addition, the Appendix contains sets of typical graphs for silicon planar and germanium alloy transistors.

Figures 3–32 to 3–35 show sample graphs for a silicon diffused power transistor. Figure 3–32 shows the $I_C - V_{CE}$ characteristics (often called the *collector* or *output* characteristics). In this *family* of characteristic curves, collector current I_C is plotted as a function of collector-to-emitter voltage V_{CE}, for various values of base current I_B. The collector characteristics can be used to determine both dc and ac operating conditions of a transistor connected as a *CE* amplifier. The methods commonly used to find complete graphical solutions to basic amplifiers are explained in Sections 3–4.2 and 3–4.3.

Figure 3–33 shows the $I_C - V_{BE}$ characteristic of the same silicon power transistor. Collector current I_C increases practically linearly with increasing base-emitter voltage V_{BE}. The graph is useful in low-voltage circuits, where the magnitude of the base-emitter voltage can play an important role.

The $I_{CBO} - T_j$ characteristic of Fig. 3–34 displays the collector–base leakage current (I_{CBO} is the reverse current across the collector–base junction with the emitter open-circuited) as a function of the junction temperature. This graph shows that the reverse leakage current is negligible at room temperature but becomes significantly large at the higher temperatures at which power transistors often operate. At high junction temperatures the reverse leakage current may be sufficiently large to upset the normal quiescent conditions in the circuit.

The $P_{\text{tot}} - T_{\text{amb}}$ curve of Fig. 3–35 displays the maximum permissible collector dissipation as a function of the ambient temperature, with or without heatsink mounting of the transistor. This graph is useful when dealing with power transistors operating at elevated temperatures.

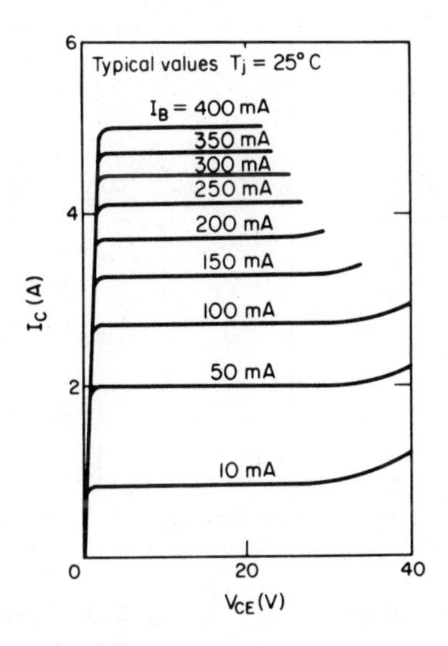

Fig. 3–32 Collector current I_C versus collector voltage V_{CE} for a silicon power transistor.

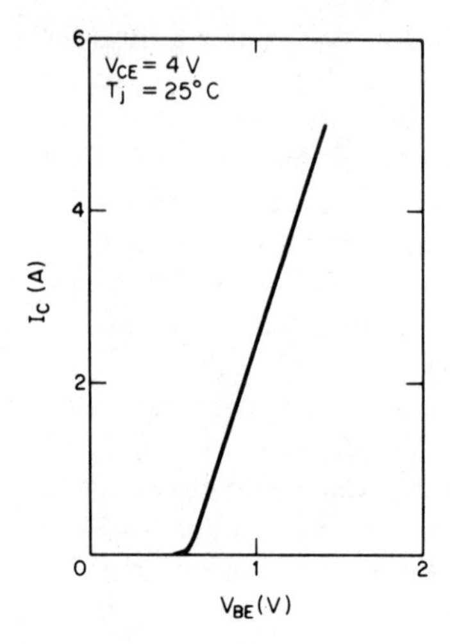

Fig. 3–33 Collector current I_C versus base-emitter voltage V_{BE} for a silicon power transistor.

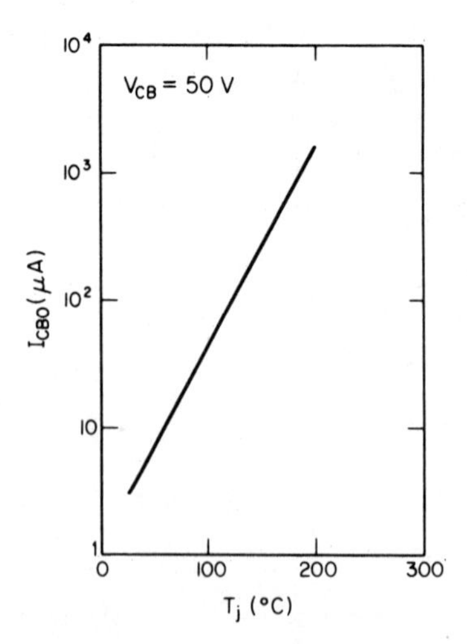

Fig. 3–34 Collector-base leakage current I_{CBO} versus junction temperature T_J for a silicon power transistor.

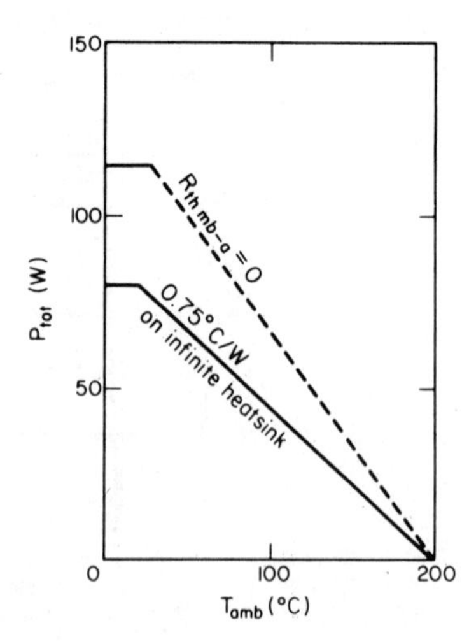

Fig. 3–35 Collector dissipation P_{tot} versus ambient temperature T_{amb} for a silicon power transistor.

Graphs of h parameters versus temperature are often provided to show the variations in gains or impedances as the temperature increases. Examples of these graphs are given in the Appendix.

3–4.2 Graphical Solution: A Standard Approach

One of the best-known graphs is the family of $I_C - V_{CE}$ or collector characteristics, which apply to a transistor in the common-emitter configuration. Assume that the curves of Fig. 3–36 represent the output characteristics of the BJT used in the amplifier stage of Fig. 3–37.

The dc operating conditions of this amplifier are established by collector supply battery V_{CC}, collector resistor R_C, and base resistor R_B. Note that V_{CC} not only serves as the collector supply voltage (positive battery terminal to the n-type collector), but also as the bias supply for the base–emitter junction (positive battery terminal to the p-type base). An ac signal is capacitively coupled to the input terminal of the amplifier and an ac output voltage is developed at the output terminal of the amplifier.

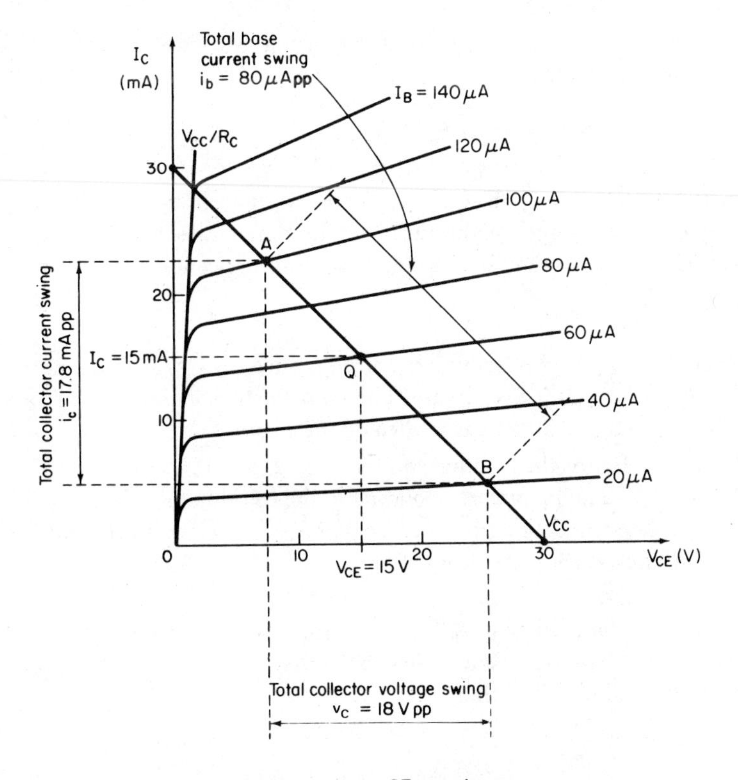

Fig. 3–36 Typical collector characteristics of a CE transistor.

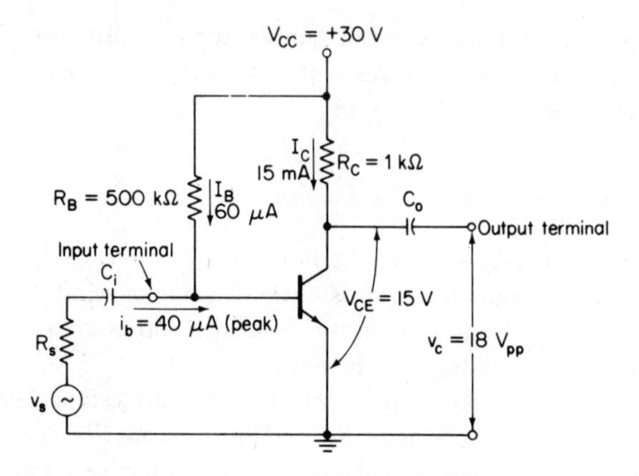

Fig. 3–37 A basic CE amplifier stage used in conjunction with the collector characteristics of Figure 3–36 to determine the quiescent and ac operating conditions.

Consider *dc conditions* in the circuit of Fig. 3–37. Kirchhoff's voltage law for the collector circuit yields

$$V_{CC} = V_{CE} + I_C R_C \tag{3–20}$$

which is also known as the *loadline equation*. It is first-order, hence can be represented by a straight line through the collector characteristics. Two points on this straight line are found by imposing two limiting conditions on the collector circuit:

1. When $V_{CE} = 0$ V, the loadline equation reduces to $V_{CC} = I_C R_C$ and $I_{C(\text{max})} = V_{CC}/R_C = 30\text{ V}/1\text{ k}\Omega = 30$ mA. The point corresponding to the coordinates $[V_{CE} = 0, I_{C(\text{max})} = 30\text{ mA}]$ is located on the graph.
2. When $I_C = 0$ mA, the loadline equation reduces to $V_{CE} = V_{CC}$ and $V_{CE(\text{max})} = V_{CC} = 30$ V. The point corresponding to the coordinates $[V_{CE(\text{max})} = 30\text{ V}, I_C = 0]$ is located on the graph. The two points so determined are connected by a straight line, which is the dc loadline.

The dc base current I_B into the transistor is established by collector battery V_{CC} and base resistor R_B. Neglecting the small forward voltage drop V_{BE} across the base-emitter junction,

$$I_B \cong \frac{V_{CC}}{R_B} = \frac{30\text{ V}}{500\text{ k}\Omega} = 60\ \mu\text{A}$$

The intersection of the dc loadline and the curve corresponding to $I_B = 60$ μA locates Q, the dc operating point, where the quiescent collector voltage $V_{CE} =$ 15 V and the quiescent collector current $I_C = 15$ mA, as indicated in Fig. 3–36. The Q point is in the center of the active region, allowing symmetrical signal excursions and approximately linear amplifier operation.

Next, consider the *ac conditions* in the circuit. Assume that signal source v_s delivers a sinusoidal base current with a peak value of $i_{bm} = 40$ μA. This ac base current is *superimposed* on the dc base current. On *positive* half-cycles, the maximum value of the instantaneous base current equals $i_{B(\max)} = I_B + i_{bm} = 60$ $\mu\text{A} + 40\ \mu\text{A} = 100\ \mu\text{A}$, and the operating point moves along the loadline to point A, where the collector current has a maximum value of 22.8 mA and the collector voltage has a minimum value of 7.4 V. On *negative* half-cycles, the minimum value of the instantaneous base current equals $i_{B(\min)} = I_B - i_{bm} = 60\ \mu\text{A} - 40$ $\mu\text{A} = 20\ \mu\text{A}$, and the operating point moves along the loadline to point B, where the collector current has a minimum value of 5 mA and the collector voltage has maximum value of 25.4 V.

This graphical analysis shows that an ac base current $i_b = 80$ μA peak to peak (pp) causes an ac collector current $i_c = 22.8\text{ mA} - 5\text{ mA} = 17.8$ mA pp, and a corresponding ac collector voltage $v_c = 25.4\text{ V} - 7.4\text{ V} = 18$ V pp. The current gain of the amplifier, defined as the ratio of output current (i_c) to input current (i_b), equals

$$A_i = \frac{i_c}{i_b} = \frac{17.8\text{ mA}}{80\ \mu\text{A}} = 222$$

The voltage gain of the amplifier can be determined only if the magnitude of the signal voltage and the internal resistance of the signal source are known.

3–4.3 AC Loadline

In many practical amplifiers the actual output circuit is composed of an RC network, consisting of collector resistor R_C, output coupling capacitor C_o, and load resistor R_L, as in the CE circuit of Fig. 3–38. In this case, the dc and ac current paths are *not* identical. The dc loadline, defined by V_{CC} and R_C, still fixes the Q point and determines the quiescent (no-signal) collector voltage and current. Under signal input conditions, however, the effective or ac load in the collector circuit consists of the $R_C - C_o - R_L$ network, and this requires the definition of an ac path of operation, called the *ac loadline.*

Assuming that Fig. 3–39 represents the collector characteristics for the BJT of Fig. 3–38, the dc loadline is determined by the dc loadline equation $V_{CC} = V_{CE} + I_C R_C$. The x intercept (V_{CE} axis) of this loadline is V_{CC}, the y intercept (I_C axis) is V_{CC}/R_C, and the slope of the loadline is $-1/R_C$. The intersection of the dc loadline and the curve corresponding to the dc base current

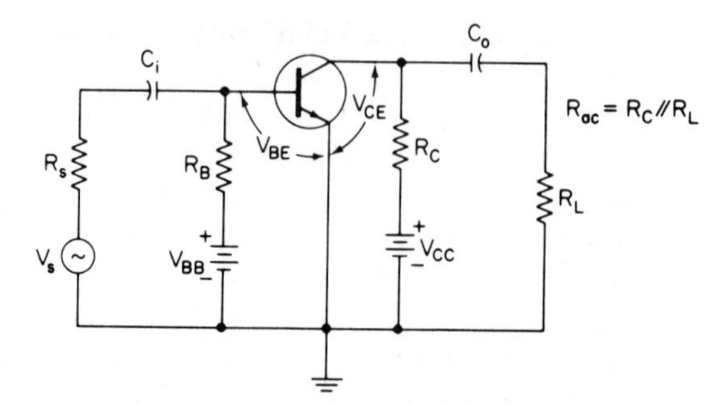

Fig. 3–38 Common-emitter amplifier with additional load resistor in the collector circuit.

$I_B = (V_{BB} - V_{BE})/R_B$ yields the location of the Q point. These facts are shown in Fig. 3–39.

For ac (signal) conditions, supply battery V_{CC} and output capacitor C_o behave like short circuits. This places R_C and R_L in parallel to produce an effective ac load resistance $R_{ac} = R_C \,//\, R_L$. Since the ac load resistance R_{ac} is smaller than the dc load resistance R_C, the slope of the loadline changes for signal conditions. A line, passing through the Q point with slope $-1/R_{ac}$ is the ac loadline. To determine the correct slope of the ac loadline, an auxiliary line can be drawn, as shown by the broken line in the corner of Fig. 3–39. This auxiliary line has an x intercept $V_{CE} = 10$ V and a y intercept $I = 10\ V/R_{ac}$ and therefore assumes the slope corresponding to $-1/R_{ac}$. The actual ac loadline is then drawn through the Q point, parallel to this auxiliary line. All points on the ac loadline

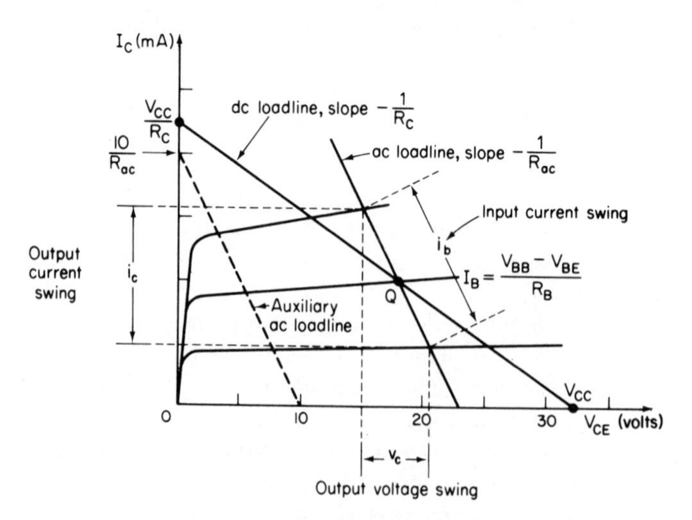

Fig. 3–39 Construction of the dc and ac loadlines. Signal excursion takes place along the ac loadline.

represent simultaneous solutions of the ac voltage equation of the collector circuit ($0 = V_{ce} + I_c R_{ac}$) and the collector characteristics. Signal source v_s in Fig. 3–38 causes the instantaneous base current into the transistor to vary. In response to these variations in base current, the Q point moves along the ac loadline, causing corresponding collector current and voltage variations. This is illustrated in Fig. 3–39, where input current swing i_b causes output swing i_c and collector voltage swing v_c.

EXAMPLE 3-10

The BC200 silicon planar epitaxial transistor is encapsulated in a microminiature plastic envelope and is designed for use in hearing aids, electronic watches, and other equipment in which very small size is of great importance. The collector characteristics of the BC200 are shown in Fig. 3–40. The transistor is connected as a class A common-emitter amplifier, as in Fig. 3–41. The circuit parameters are $V_{CC} = -6$ V, $R_B = 150$ kΩ, $R_C = 300$ Ω, and $R_L = 100$ Ω. The signal source delivers a current into the transistor whose peak value does not exceed 40 μA.

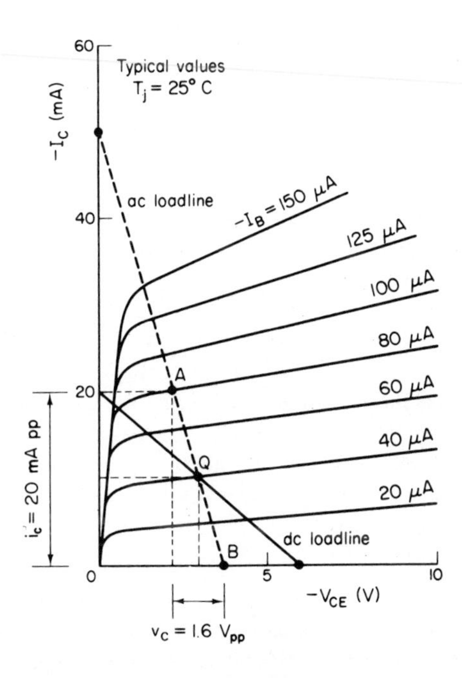

Fig. 3–40 Collector characteristics of the BC 200 silicon planar transistor.

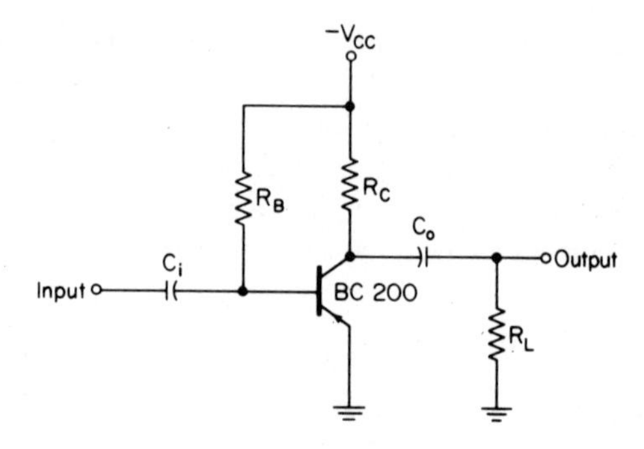

Fig. 3–41 Class-A common-emitter amplifier of Example 3–10.

Determine (a) The dc operating point of the transistor.
(b) The output voltage developed across the load.
(c) The output current delivered to the load.
(d) The collector dissipation of the transistor.

SOLUTION

(a) The dc operating point is determined by the intersection of the dc loadline and the base current curve. The dc loadline equation is

$$V_{CC} = V_{CE} + I_C R_C$$

To plot the loadline on the graph we locate the x and y intercepts. The x intercept is at $V_{CC} = -6$ V and the y intercept is at $I_C = V_{CC}/R_C = -6 \text{ V}/300\ \Omega = -20$ mA.

The two points located on the graph are connected by a straight line which represents the dc loadline. The quiescent base current is

$$I_B \cong \frac{V_{CC}}{R_B} = \frac{-6\text{ V}}{150\text{ k}\Omega} = -40\ \mu\text{A}$$

The intersection of the dc loadline and the curve for $I_B = -40\ \mu$A determines the Q point at $I_C = -10$ mA and $V_{CE} = -3$ V.

(b) For ac conditions, the output capacitor is short-circuited and load resistor R_L is placed in parallel with collector resistor R_C. This yields an effective load resistance

$$R_{ac} = R_C // R_L = 300\ \Omega // 100\ \Omega = 75\ \Omega$$

The *ac loadline* passes through the Q point with a slope corresponding to $-1/R_{ac} = -1/75\ \Omega$ and the ac loadline is plotted on the graph as indicated. The signal current has a peak value of 40 MA, which causes the Q point to move along the ac loadline to point A on the positive half-cycles and to point B on the negative half-cycles. The collector-to-emitter voltage corresponding to this signal excursion is read off the graph and equals

$$v_c = 1.6 \text{ V peak to peak}$$

which is also the voltage appearing across load resistor R_L.

(c) The base current excursion of 80 μA peak to peak (pp) causes a collector current excursion of 20 mA pp. This total signal current is divided between collector resistor R_C and load resistor R_L. The current in R_L is

$$i(R_L) = \frac{300}{400} \times 20 \text{ mA pp} = 15 \text{ mA pp}$$

(d) The collector dissipation is determined by the *quiescent* collector voltage and current, so

$$P_c = 10 \text{ mA} \times 3 \text{ V} = 30 \text{ mW}$$

The power dissipation of the BC200 transistor is rated as P_{tot} = maximum 50 mW, so the transistor operates well within its maximum power rating.

3–5 BIAS ARRANGEMENTS

3–5.1 Effect of Current Gain on Transistor Bias

The current gain of a transistor is not constant but increases as the junction temperature rises or as the emitter current increases. This is shown quite clearly in the $h_{FE}-T_j$ curves for the germanium and silicon transistors in the Appendix. The data sheets usually state a *typical* value for β at ambient temperature, and they often specify a rather wide *range* of gain values for the same transistor.

This varying nature of β can have disastrous effects on the dc operating conditions of the circuit, especially if the dc collector current is highly dependent on β. Consider, for example, the *fixed-bias* circuit of the common-emitter amplifier stage of Fig. 3–42(a). The value of the base current is *fixed* and is given by

$$I_B = \frac{V_{CC} - V_{BE}}{R_B} \cong \frac{V_{CC}}{R_B} \tag{3–67}$$

$$I_C = \beta I_B \tag{3–68}$$

$$V_{CE} = V_{CC} - I_C R_C \tag{3–69}$$

and is therefore also dependent on β.

If β increases by 10 percent, I_C also increases by 10 percent, resulting in a considerable shift of the dc operating point.

It is seen from Eqs. (3–67) to (3–69) that the base current is fixed and independent of current gain β. The collector current and the collector voltage, however, are directly proportional to β. In fact, if β is sufficiently large, the collector current may become so large that the transistor enters into saturation, when the collector voltage is practically zero and the collector current is maximum. In this case, Eq. (3–69) yields an expression for the saturation current

$$I_{C(\text{sat})} \cong \frac{V_{CC}}{R_C} \tag{3–70}$$

EXAMPLE 3-11

The fixed-bias circuit of Fig. 3–42(a) has the following circuit values: $V_{CC} = 10$ V, $R_B = 1$ MΩ, and $R_C = 10$ kΩ.

Calculate the minimum value of β at which saturation occurs.

SOLUTION

At saturation, the collector voltage is zero and the collector current maximum. According to Eq. (3–70),

$$I_{C(\text{sat})} \cong \frac{V_{CC}}{R_C} = \frac{10\text{ V}}{10\text{ k}\Omega} = 1\text{ mA}$$

$$I_B \cong \frac{V_{CC}}{R_B} = \frac{10\text{ V}}{1\text{ M}\Omega} = 10\ \mu\text{A} \tag{3–71}$$

Since $\beta = I_C/I_B$, saturation occurs when

$$\beta = \frac{1\text{ mA}}{10\ \mu\text{A}} = 100$$

The tabulation of Fig.3–42(b) shows that the BJT operates in the center of the active region when $\beta = 50$, but saturates when β is 100 or more.

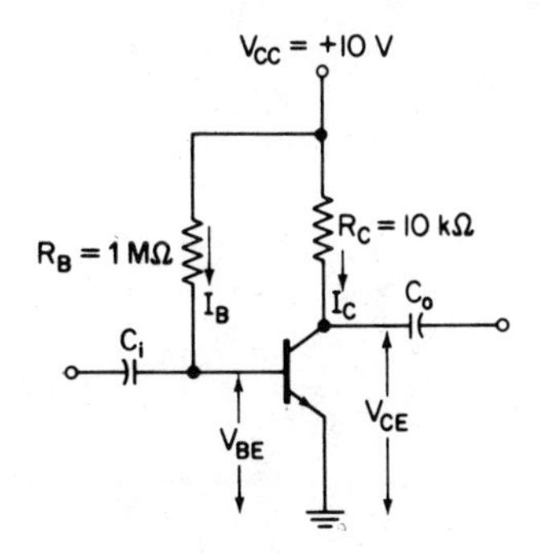

(a) A common-emitter stage

	$\beta = 50$	$\beta = 100$
$I_B \cong \dfrac{V_{CC}}{R_B}$	10 μA	10 μA
$I_C = \beta I_B$	0.5 μA	1 mA
$V_{CE} = V_{CC} - I_C R_C$	5 V	0 V
Type of operation	Active region	Saturation

(b) Circuit conditions for two values of β

Fig. 3–42 Fixed bias circuit, indicating the possible variations in dc operating conditions due to β variations.

This shows an inherent instability of the operating point due to β variations, and the biasing of the circuit is *β-dependent.* In some other circuits, called *β-independent,* the collector current is independent of the current gain, and the location of the Q point is not affected by the variations in β.

Some standard circuits help solve the problem. Figure 3–43 shows the basic arrangements. The β-dependent circuits are given in Fig. 3–43(a) to (d); the β-independent circuits (often called *universal* circuits) are given in Fig. 3–43 (e) and (f). These circuits are examined in some detail in the course of this section.

3–5.2 Stability Factor

The effect that variations in β have on the stability of the dc operating point (collector current) is often expressed in terms of the *stability factor S.* The stability factor indicates how sensitive the collector current is to changes in β and is defined as the ratio of the percentage change in collector current to the percentage change in β. Hence, by definition,

$$\text{stability factor } S = \frac{\Delta I_C / I_C}{\Delta\beta/\beta} \tag{3–72}$$

Ideally, $S = 0$. $S = 1$ is the worst case.

Suppose, for example, that the current gain of the transistor in Fig. 3–42 equals 50. For the circuit parameters given, $I_B = 10\ \mu\text{A}$ and $I_C = 50 \times 10\ \mu\text{A} = 0.5$ mA. If, for some reason, β increases from 50 to 60 ($\Delta\beta = 10$, or β increases by 20 percent), the collector current increases to $I_C = 60 \times 10\ \mu\text{A} = 0.6$ mA($\Delta I_C = 1$ mA, or I_C increases by 20 percent). Hence, a β variation of 20 percent causes

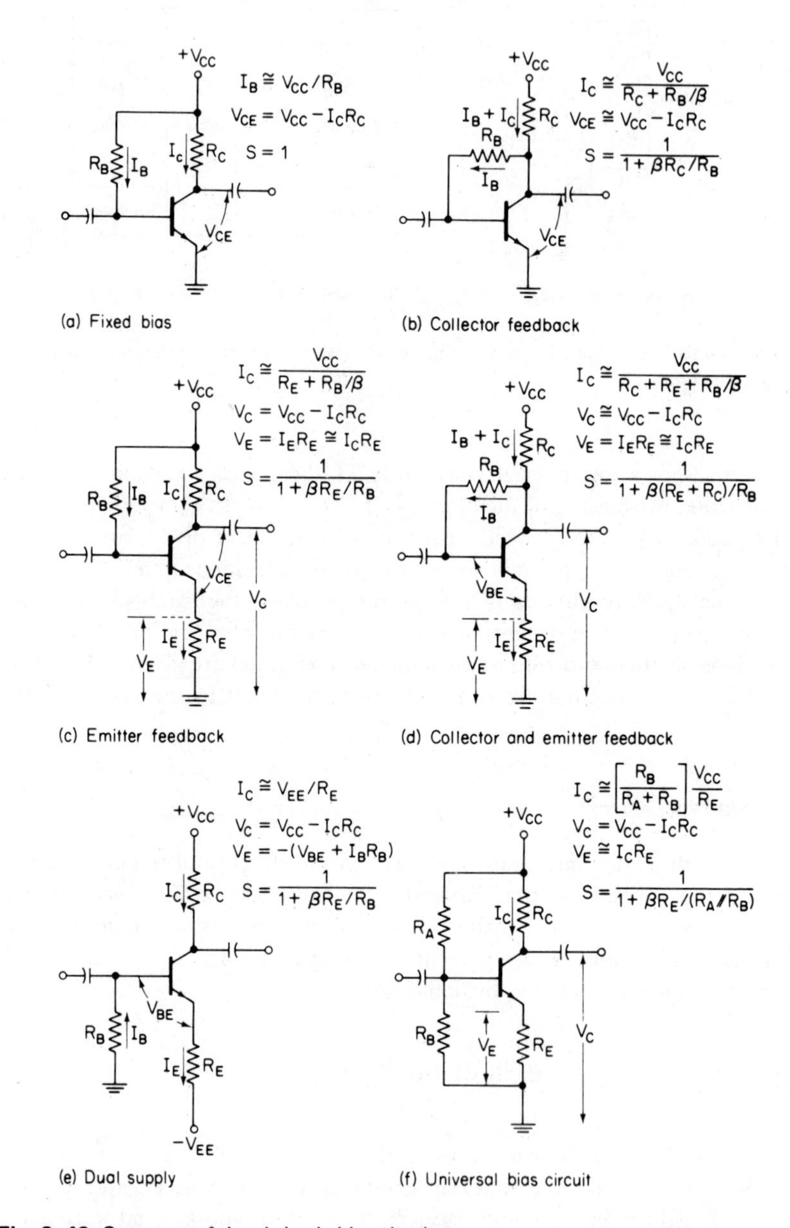

Fig. 3–43 Summary of the six basic bias circuits.

a collector current variation of 20 percent. The stability factor for the fixed-bias circuit is 1, because the percentage change in collector current equals the percentage change in β. Obviously, fixed bias is the worst possible way of biasing a transistor.

3–5.3 Collector Feedback

Figure 3–43(b) shows a collector feedback circuit, where base resistor R_B is connected to the collector instead of the supply voltage V_{CC}. The base current is derived from the collector voltage and therefore follows the collector variations.

To illustrate the concept of the *feedback action* in this circuit, assume that β increases. An increase in β causes an increase in I_C and a corresponding decrease in V_{CE}. The base current I_B therefore also decreases, since I_B depends on V_{CE}. Hence, the collector current I_C decreases and tends to oppose the original increase caused by the β variation.

Collector resistor R_C carries both the collector current and the base current. Hence,

$$V_{CE} = V_{CC} - (I_C + I_B)R_C \tag{3–73}$$

Since $I_B \ll I_C$ we can also write

$$V_{CE} \simeq V_{CC} - I_C R_C \tag{3–74}$$

The collector-to-emitter voltage V_{CE} equals the sum of the voltage drops from collector to base and from base to emitter, so

$$V_{CE} = V_{CB} + V_{BE} = I_B R_B + V_{BE} \tag{3–75}$$

Combining Eqs. (3–74) and (3–75) yields

$$V_{BE} + I_B R_B = V_{CC} - I_C R_C \tag{3–76}$$

Substituting $I_B = I_C/\beta$ into Eq. (3–76) and simplifying,

$$V_{BE} + \frac{I_C}{\beta} R_B = V_{CC} - I_C R_C \tag{3–77}$$

$$I_C\left(\frac{R_B}{\beta} + R_C\right) = V_{CC} - V_{BE} \tag{3–78}$$

$$I_C = \frac{V_{CC} - V_{BE}}{R_C + R_B/\beta} \tag{3–79}$$

Since, in general, $V_{BE} \ll V_{CC}$, Eq. (3–79) reduces to

$$I_C \cong \frac{V_{CC}}{R_C + R_B/\beta} \tag{3–80}$$

Equation (3–80) indicates that the collector current is not directly proportional to β and that its β dependence decreases for small values of R_B. Small values of R_B increase the base current and bring the transistor closer to the saturation point. Hence collector feedback improves the Q-point stability, especially when the transistor operates close to saturation.

EXAMPLE 3-12

The circuit of Fig. 3–43(b) has the following parameters: $V_{CC} = 12$ V, $R_C = 1$ kΩ, $R_B = 50$ kΩ, and $\beta = 50$.

(a) Determine the dc operating conditions in the circuit.

(b) If another transistor with a $\beta = 100$ is used in the circuit, calculate the new collector current and collector voltage.

(c) Express the stability of the circuit as the ratio of the percentage change in collector current to the percentage change in β.

SOLUTION

(a) By Eq. (3–80) the collector current is

$$I_C \cong \frac{V_{CC}}{R_C + R_B/\beta} \cong \frac{12\text{ V}}{1\text{ k}\Omega + 50\text{ k}\Omega/50} \cong 6\text{ mA}$$

The collector voltage is

$$V_{CE} = V_{CC} - I_C R_C = 12\text{ V} - (6\text{ mA} \times 1\text{ k}\Omega) = 6\text{ V}$$

The base current is

$$I_B \cong \frac{V_{CE}}{R_B} = \frac{6\text{ V}}{50\text{ k}\Omega} = 120\ \mu\text{A}$$

(b) For the new $\beta = 100$, the collector current increases and equals

$$I_C \simeq \frac{12 \text{ V}}{1 \text{ k}\Omega + 50 \text{ k}\Omega/100} \simeq 8 \text{ mA}$$

so the collector voltage is

$$V_{CE} = V_{CC} - I_C R_C = 12 \text{ V} - 8 \text{ mA} \times 1 \text{ k}\Omega = 4 \text{ V}$$

(c) The increase in collector current when β is increased from 50 to 100 equals $\Delta I_C = 2$ mA. The percentage change in collector current is

$$\frac{\Delta I_C}{I_C} \times 100\% = \frac{2 \text{ mA}}{6 \text{ mA}} \times 100\% \simeq 33\%$$

The percentage increase in transistor β,

$$\frac{\Delta\beta}{\beta} \times 100\% = \frac{50}{50} \times 100\% = 100\%$$

The stability factor of the circuit,

$$S = \frac{\Delta I_C/I_C}{\Delta\beta/\beta} = \frac{33\%}{100\%} = 0.33$$

Example 3–12 shows that collector feedback provides a notable improvement in circuit stability compared to the fixed-bias arrangement.

The stability factor of the collector feedback circuit can be expressed in terms of the circuit constants. Appropriate mathematical manipulation shows that

$$S = \frac{1}{1 + \beta R_C/R_B} \tag{3–81}$$

3–5.4 Emitter Feedback

Figure 3–43(c) shows a bias circuit with emitter feedback. The resistor R_E in the emitter circuit reduces the sensitivity of the collector current to changes in transistor β, and stabilizes the dc operating point by emitter feedback.

The effect of the emitter feedback can be analyzed qualitatively. Assume that current gain β increases. This causes an increase in both collector current and emitter current, with a corresponding increase in the voltage drop across emitter resistor R_E. This reduces the voltage across base resistor R_B ($V_{RB} = V_{CC} - V_{BE} - V_E$) and causes a reduction in base current. Since $I_C = \beta I_B$, the decrease in base current partly compensates for the increase in β and tends to stabilize the collector current. This stabilizing action is a direct result of the changes in the emitter voltage, i.e., emitter degeneration or emitter feedback.

$$V_C = V_{CC} - I_C R_C \tag{3–82}$$

$$V_E = I_E R_E \simeq I_C R_E \tag{3–83}$$

Collector-to-emitter voltage V_{CE} is the difference between V_C and V_E.

The collector current obviously depends on the base current and the current gain. To obtain an expression for I_C we can write

$$V_{CC} = I_B R_B + V_{BE} + I_E R_E \tag{3–84}$$

Since $I_B = I_C/\beta$ and $I_E \simeq I_C$, Eq. (3–84) can be solved for I_C.

$$V_{CC} \simeq \frac{I_C}{\beta} R_B + V_{BE} + I_C R_E \tag{3–85}$$

$$I_C \simeq \frac{V_{CC} - V_{BE}}{R_E + R_B/\beta} \tag{3–86}$$

In most cases $V_{BE} \ll V_{CC}$, so

$$I_C \simeq \frac{V_{CC}}{R_E + R_B/\beta} \tag{3–87}$$

Equation (3–87) indicates that the collector current still depends on β, although the stability of the operating point has improved.

EXAMPLE 3-13

The emitter feedback circuit of Fig. 3–43(c) has the following circuit values: $V_{CC} =$ 12 V, $R_C = 1$ kΩ, $R_E = 1$ kΩ, $R_B = 200$ kΩ, and $\beta = 100$.

Calculate (a) The actual value of the collector current.

(b) The percentage change in collector current if β increases by 20 percent.

$$I_C \simeq \frac{V_{CC}}{R_C + R_E + R_B/\beta} = \frac{20\text{ V}}{10\text{ k}\Omega + 10\text{ k}\Omega + 500\text{ k}\Omega/100} = 0.8\text{ mA}$$

$$V_C \simeq V_{CC} - I_C R_C = 20\text{ V} - (0.8\text{ mA} \times 10\text{ k}\Omega) = 12\text{ V}$$

$$V_E \simeq I_C R_E = 0.8\text{ mA} \times 10\text{ k}\Omega = 8\text{ V}$$

The collector-to-emitter voltage

$$V_{CE} = V_C - V_E = 12\text{ V} - 8\text{V} = 4\text{ V}$$

(b) If β increases by 20 percent, from 100 to 120,

$$I_C \simeq \frac{20\text{ V}}{10\text{ k}\Omega + 10\text{ k}\Omega + 500\text{ k}\Omega/120} = 0.83\text{ mA}$$

$$\Delta I_C = 0.03\text{ mA}$$

The percentage increase in collector current

$$\frac{\Delta I_C}{I_C} \times 100\% = \frac{0.03\text{ mA}}{0.8\text{ mA}} \times 100\% = 3.75\%$$

Stabilization of the dc operating point against variations in β has improved considerably: a 20 percent variation in β only produces a 3.75 percent variation in the collector current.

3-5.6 Emitter Bias (Dual Supply)

Figure 3-43 (e) shows an emitter bias circuit which requires two separate supplies: V_{CC} and V_{EE}. The emitter bias current is established by V_{EE} and emitter resitor R_E. By carefully choosing the values of the various circuit components, the collector current can be made practically independent of the transistor current gain.

An exact expression for the collector current can be derived from Kirchhoff's voltage equation for the emitter–base circuit, where

$$V_{EE} - I_E R_E - V_{BE} - I_B R_B = 0 \tag{3-94}$$

Taking $I_C \simeq I_E$ and $I_B = I_C/\beta$,

$$I_C \simeq \frac{V_{EE} - V_{BE}}{R_E + R_B/\beta} \tag{3-95}$$

Equation (3–95) still contains a β term, but the dependence of I_C on β can be minimized drastically by choosing $R_E \gg R_B/\beta$. Also, in most practical cases, the emitter supply voltage is much greater than the base-emitter voltage, so $V_{EE} \gg V_{BE}$ and Eq. (3–95) then simplifies to

$$I_C \simeq \frac{V_{EE}}{R_E} \qquad \left(R_E \gg \frac{R_B}{\beta} \quad \text{and} \quad V_{EE} \gg V_{BE}\right) \tag{3–96}$$

Hence, provided that the circuit components are properly selected, the collector current is virtually independent of β and the dc operating point changes little with variations in β.

The stability factor of the dual-supply circuit is

$$S = \frac{1}{1 + \beta R_E/R_B} \tag{3–97}$$

indicating that this circuit has the same Q-point stability as the emitter feedback circuit of Fig. 3–43 (c).

Neglecting the small base-emitter voltage drop V_{BE} and also the small voltage drop $I_B R_B$ across the base resistor, the emitter is found to be approximately at ground potential. The exact expression for the emitter voltage is

$$V_E = -(V_{BE} + I_B R_B) = -\left(V_{BE} + \frac{I_C R_B}{\beta}\right) \tag{3–98}$$

The collector-to-ground voltage is

$$V_C = V_{CC} - I_C R_C \tag{3–99}$$

When the transistor is in saturation, the collector-to-emitter voltage is approximately zero volts, and the saturation current is

$$I_{C(\text{sat})} \simeq \frac{V_{CC}}{R_C} \tag{3–100}$$

Under normal operating conditions the quiescent collector current should be approximately one-half of the saturation value.

3–5.7 Universal Bias Circuit

The universal bias circuit of Fig. 3–43(f) provides excellent Q-point stability and uses only a single power supply. Base voltage V_B is derived from collector supply voltage V_{CC} by voltage divider network $R_A - R_B$. If the resistances of R_A and R_B are chosen so that the voltage divider current is much larger than base current I_B into the transistor, there is practically no loading effect and the base voltage then equals

$$V_B = \frac{R_B}{R_A + R_B} V_{CC} \tag{3-101}$$

Neglecting V_{BE},

$$V_E \cong V_B \tag{3-102}$$

$$I_E = \frac{V_E}{R_E} \cong \frac{R_B}{R_A + R_B} \frac{V_{CC}}{R_E} \tag{3-103}$$

The universal circuit is therefore a modified form of the emitter bias circuit in the sense that the emitter current is established by the voltage divider.

$$I_B \ll I_C \qquad I_C \cong I_E = \frac{V_E}{R_E} \tag{3-104}$$

A precise calculation of the collector current actually indicates a small β component, but this may be neglected if $R_E \gg (R_A//R_B)/\beta$. This is reflected in the expression for the stability factor S, where

$$S = \frac{1}{1 + \beta R_E/(R_A//R_B)} \tag{3-105}$$

EXAMPLE 3-15

The universal bias circuit of Fig. 3-43(f) has the following circuit parameters: $V_{CC} = 12$ V, $R_C = 1$ kΩ, $R_E = 2$ kΩ, and $R_A = R_B = 10$kΩ.

Calculate
- (a) The collector current I_C.
- (b) The saturation value of the collector current $I_{C(sat)}$.
- (c) The collector voltage V_C, the emitter voltage V_E, and the collector-to-emitter voltage V_{CE}.
- (d) The stability factor S, if $\beta = 100$.

SOLUTION

(a) The base voltage is provided by the voltage divider and is

$$V_B = \frac{R_B}{R_A + R_B} V_{CC} = \frac{10\ \text{k}\Omega}{20\text{k}\Omega} \times 12\ \text{V} = 6\ \text{V}$$

Neglecting V_{BE}, $V_E = 6$ V. Hence,

$$I_C \simeq I_E = \frac{V_E}{R_E} = \frac{6\ \text{V}}{2\ \text{k}\Omega} = 3\ \text{mA}$$

(b) The transistor is in saturation when the collector-to-emitter voltage $V_{CE} \simeq 0$ V. The collector current is maximum and limited by R_C and R_E only, so

$$I_{C(\text{sat})} \simeq \frac{V_{CC}}{R_C + R_E} = \frac{12\ \text{V}}{1\ \text{k}\Omega + 2\ \text{k}\Omega} = 4\ \text{mA}$$

(c)

$$V_C = V_{CC} - I_C R_C = 12\ \text{V} - (3\ \text{mA} \times 1\ \text{k}\Omega) = 9\text{V}$$

$$V_E = I_E R_E \simeq 3\ \text{mA} \times 2\ \text{k}\Omega = 6\text{V}$$

The collector-to-emitter voltage

$$V_{CE} = V_C - V_E = 9\ \text{V} - 6\ \text{V} = 3\ \text{V}$$

(d) The stability factor

$$S = \frac{1}{1 + \beta R_E/(R_A /\!/ R_B)}$$

$$= \frac{1}{1 + (100 \times 2\ \text{k}\Omega)/(10\ \text{k}\Omega /\!/ 10\ \text{k}\Omega)}$$

$$\simeq 0.025$$

This result indicates that a 100 percent change in β causes only a 2.5 percent change in the collector current. The universal bias circuit provides excellent Q-point stability.

3-6 AC CIRCUIT EXAMPLES

3-6.1 CE Stage with Emitter Resistor

Generally speaking, transistor circuits with some form of emitter feedback bias arrangement provide improved dc circuit stability and guard against variations in dc operating conditions that arise from changes in temperature or device parameters. The addition of an emitter resistor, however, degrades the ac circuit performance in terms of the voltage and power gains, unless corrective measures are taken to cancel this detrimental effect.

Consider, for example, the BJT in the common-emitter connection and with the simple emitter feedback bias of Fig. 3-43(c). The dc conditions were derived in Section 3-5.4, Eqs. (3-82, 3-83, and 3-87).

The ac conditions can be analyzed by first replacing the BJT with its semihybrid model (see Fig. 3-29) and then adding the remaining circuit elements, remembering that the capacitors are considered to be ac short circuits. This then results in the ac equivalent circuit of Fig. 3-44. The ac input voltage to the circuit is

$$v_i = i_b r_{ie} + i_e R_E \tag{3-106}$$

$$v_i = i_b \beta r_\pi + (\beta + 1) i_b R_E \tag{3-107}$$

$$v_i \simeq \beta i_b (r_\pi + R_E) \tag{3-108}$$

The ac output voltage equals

$$v_o = i_c R_C = \beta i_b R_C \tag{3-109}$$

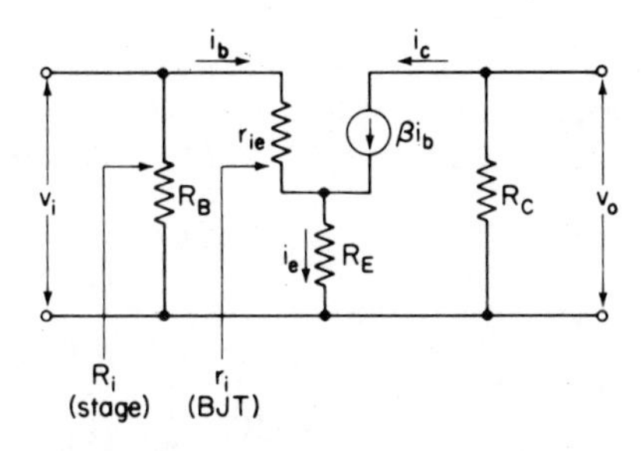

Fig. 3-44 Ac equivalent circuit for the CE stage of Figure 3-43(c).

The voltage gain of the stage is equal to the ratio of v_o and v_i, so

$$A_v = \frac{v_o}{v_i} \cong \frac{\beta i_b R_C}{\beta i_b(r_\pi + R_E)} \cong \frac{R_C}{r_\pi + R_E} \tag{3–110}$$

In general, R_E is much larger than r_π and we say that the emitter resistor *swamps out* the internal junction resistance of the BJT. In this case, then, the voltage gain of the *CE* stage with emitter resistor is given by the ratio of the collector load resistance and the emitter resistance, or

$$A_v \cong \frac{R_C}{R_E} \text{ (for } R_E \gg r_\pi) \tag{3–111}$$

It is well to memorize this important relationship, since Eq. (3–111) gives the voltage gain of the common-emitter stage.

The input resistance, looking into the base of the BJT, is derived from the ac input voltage and ac input current.

$$r_{i(\text{BJT})} = \frac{v_i}{i_i} = \frac{\beta i_b(r_\pi + R_E)}{i_b} = \beta(r_\pi + R_E) \tag{3–112}$$

Again, since $R_E \gg r_\pi$

$$r_{i(\text{BJT})} \cong \beta R_E \tag{3–113}$$

The input resistance of the amplifier stage consists of the parallel combination of bias resistor R_B and transistor input resistance βR_E, so

$$Ri_{(\text{stage})} \cong R_B // \beta R_E \tag{3–114}$$

EXAMPLE 3-16

Estimate the voltage gain A_v and input resistance $R_{i(\text{stage})}$ of the *CE* amplifier of Fig. 3–44 if $V_{CC} = 12$ V, $R_C = 1$ kΩ, $R_E = 500$ Ω, $R_B = 1$ MΩ, and $\beta = 100$.

SOLUTION

The approximate voltage gain is

$$A_v \cong \frac{R_C}{R_E} = \frac{1\ \text{k}\Omega}{500\ \Omega} = 2$$

The input resistance of the BJT is

$$r_{i(\mathrm{BJT})} \simeq \beta R_E = 100 \times 500\ \Omega = 50\ \mathrm{k}\Omega$$

The input resistance of the amplifier stage is

$$R_{i(\mathrm{stage})} = R_B // r_{i(\mathrm{BJT})} = 1\ \mathrm{M}\Omega // 50\ \mathrm{k}\Omega \simeq 50\ \mathrm{k}\Omega$$

3–6.2 Emitter Bypass Capacitor

Example 3–16 shows that the voltage gain of the common-emitter stage with emitter feedback can be very low, and that, in fact, gain is sacrificed in favor of dc circuit stability. However, this can be overcome by shunting the emitter resistor with *emitter bypass capacitor* C_E, as in Fig. 3–45.

As far as the dc conditions in the circuit are concerned, the emitter bypass capacitor behaves like an open circuit and the emitter resistor therefore assists in dc stabilization in exactly the same manner as before.

The behavior of the circuit under ac conditions, however, has changed considerably. If the reactance of the bypass capacitor is very small compared to the resistance of the emitter resistor, R_E is practically short-circuited (bypassed) and the emitter of the BJT is placed at ac ground potential. The voltage gain of

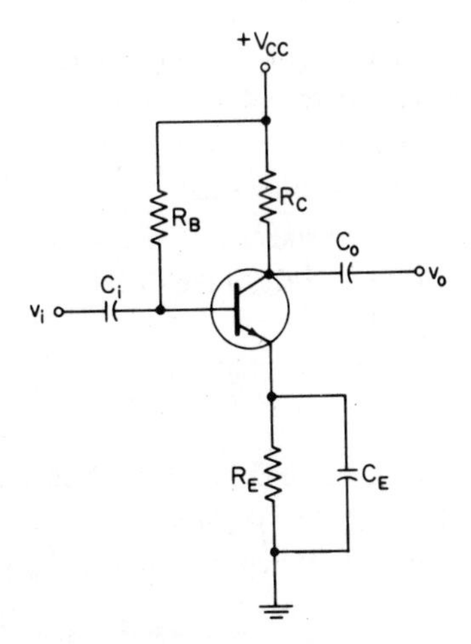

Fig. 3–45 Emitter bypass capacitor C_E places the emitter at ac ground potential and improves the voltage gain of the stage.

the CE stage is then derived from Eq. (3–110) by setting $R_E = 0$, and is

$$A_v \simeq \frac{R_C}{r_\pi} \tag{3–115}$$

This result, of course, is in keeping with Eq. (3–66), which expresses the voltage gain of the common-emitter amplifier in terms of the ac load resistance and the BJT input resistance.

Similarly, the input resistance of the BJT is derived from Eq. (3–112) by setting $R_E = 0$, which yields

$$r_{i(\mathrm{BJT})} = \beta r_\pi = \beta \frac{50}{I_E} \tag{3–116}$$

which is identical to the expression for the input resistance of the common-emitter BJT as stated in Eq. (3–62).

As a rule of thumb, the emitter resistor is considered effectively bypassed when the reactance of the bypass capacitor is approximately one-tenth the resistance of R_E at the lowest frequency of operation, or

$$X_{CE} < \frac{1}{10} R_E \tag{3–117}$$

EXAMPLE 3-17

Consider the common-emitter BJT in the universal bias circuit of Fig. 3–46 and draw the ac equivalent circuit. In addition, *calculate*

(a) The quiescent voltages and currents in the circuit.

(b) The voltage gain of the amplifier stage.

(c) The input resistance of the stage.

(d) The required capacitance of C_E, if the signal frequency ranges from 20 Hz to 20 kHz.

SOLUTION

The ac equivalent circuit is shown in Fig. 3–47. Note that R_E is bypassed and that the emitter of the BJT is at ground potential.

(a) The universal bias circuit develops a dc voltage V_B at the base of the transistor equal to

$$V_B = \frac{R_B}{R_A + R_B} V_{CC} = \frac{30\ \mathrm{k\Omega}}{120\ \mathrm{k\Omega}} \times 18\ \mathrm{V} = 4.5\ \mathrm{V}$$

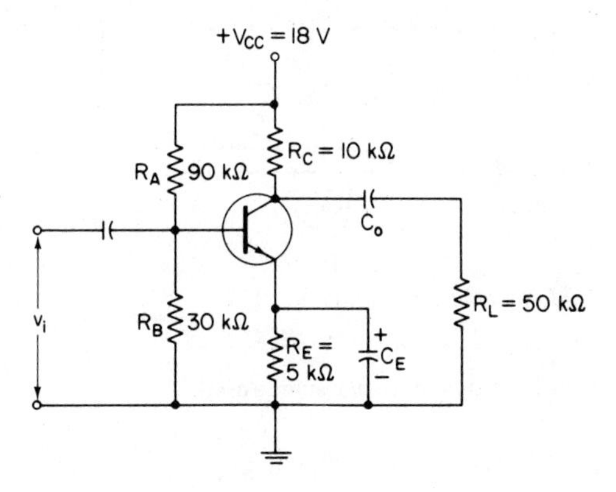

Fig. 3–46 CE stage with bypassed emitter resistor.

Neglecting the small voltage drop across the base-emitter junction, the dc emitter voltage equals

$$V_E \simeq V_B = 4.5 \text{ V}$$

$$I_E = \frac{V_E}{R_E} = \frac{4.5 \text{ V}}{5 \text{ k}\Omega} = 0.9 \text{ mA}$$

$$V_C = V_{CC} - I_C R_C \simeq 18 \text{ V} - (0.9 \text{ mA} \times 10 \text{ k}\Omega) = 9 \text{ V}$$

(b) The ac load resistance consists of R_C in parallel with R_L and equals

$$R_{ac} = R_C // R_L = 10 \text{ k}\Omega // 50 \text{ k}\Omega \simeq 8.33 \text{ k}\Omega$$

The transistor parameter r_π equals

$$r_\pi \simeq \frac{50}{I_E} = \frac{50}{0.9 \text{ mA}} \simeq 55.5 \ \Omega$$

It is assumed that emitter capacitor C_E effectively bypasses R_E, so the voltage gain of the amplifier stage is

$$A_v \simeq \frac{R_{ac}}{r_\pi} = \frac{8.33 \text{ k}\Omega}{55.5\Omega} \simeq 150$$

(c) The input resistance of the BJT equals

$$r_{i(\text{BJT})} = \beta r_\pi = 120 \times 55.5 \ \Omega \simeq 6.67 \text{ k}\Omega$$

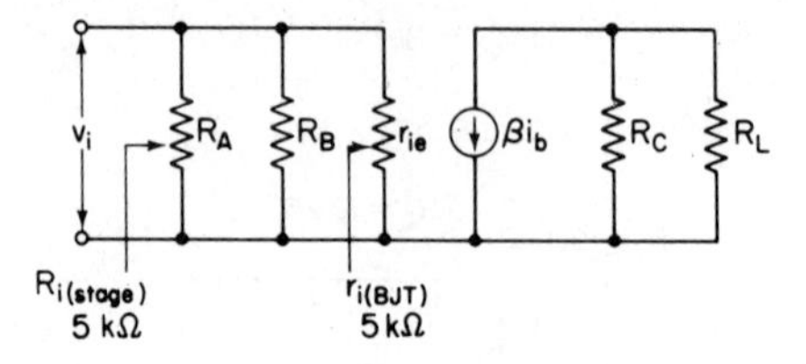

Fig. 3–47 CE stage and ac equivalent circuit.

The input resistance of the amplifier stage equals

$$R_{i(\text{stage})} = R_A//R_B//r_{i(\text{BJT})}$$

$$= 30\ \text{k}\Omega//90\ \text{k}\Omega//66.7\ \text{k}\Omega \simeq 5\ \text{k}\Omega$$

(d) To bypass the ac signal current to ground, the reactance of the emitter bypass capacitor should be less than one-tenth the resistance it shunts. Hence,

$$X_{CE} < \frac{1}{10} R_E \quad \text{or} \quad X_{CE} < 500\ \Omega$$

The reactance of the bypass capacitor is

$$X_{CE} = \frac{1}{2\pi f C_E}$$

so that $f = 20$

$$C > \frac{1}{2\pi f X_{CE}} = 16\ \mu\text{F}.$$

Example 3–17 illustrates that the voltage gain of the *CE* stage with bypassed emitter resistor can be very high ($A_v = 150$). If the bypass capacitor is removed from the circuit, the voltage gain is approximately

$$Av \simeq \frac{R_{\text{ac}}}{R_E} = \frac{8.33\ \text{k}\Omega}{5\ \text{k}\Omega} \simeq 1.67$$

3–6.3 CC Stage (Emitter Follower)

In the *CC* connection, the load resistance is placed in the emitter leg of the BJT, while the collector terminal is connected directly to the V_{CC} supply. Various bias arrangements can be used with the *CC* connection, but the most common one is perhaps the universal circuit shown in Fig. 3–48. The input signal is coupled to the base of the BJT and the output signal, developed across emitter resistor R_E, is capacitively coupled to the load resistor R_L. The output voltage is in phase with the input voltage and the voltage gain of the circuit is approximately 1.

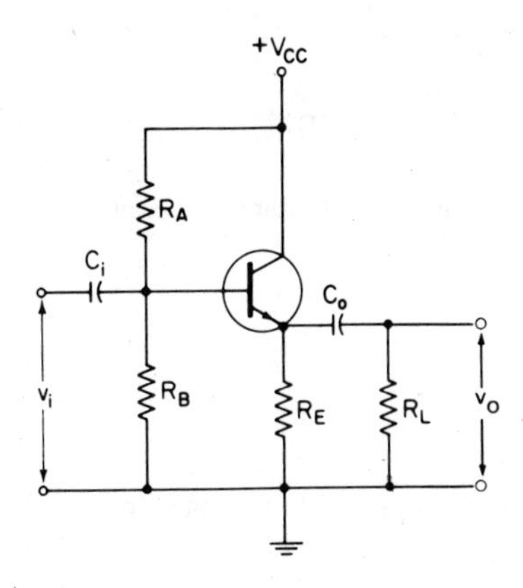

Fig. 3–48 CC stage (emitter-follower).

For the dc conditions, C_i and C_o behave like open circuits, so the source and the load are removed from the circuit. Inspection of the circuit shows that the dc voltage at the base of the BJT is

$$V_B = \frac{R_B}{R_A + R_B} V_{CC} \tag{3–118}$$

Neglecting V_{BE},

$$V_E \simeq V_B \tag{3–119}$$

Hence,

$$I_E = \frac{V_E}{R_E} \simeq \frac{R_B}{R_A + R_B} \frac{V_{CC}}{R_E} \tag{3–120}$$

The ac equivalent circuit can be developed by replacing the BJT with its semihybrid model and then adding the remaining circuit components, remembering that the capacitors behave like short circuits. This yields the ac equivalent circuit of Fig. 3–49.

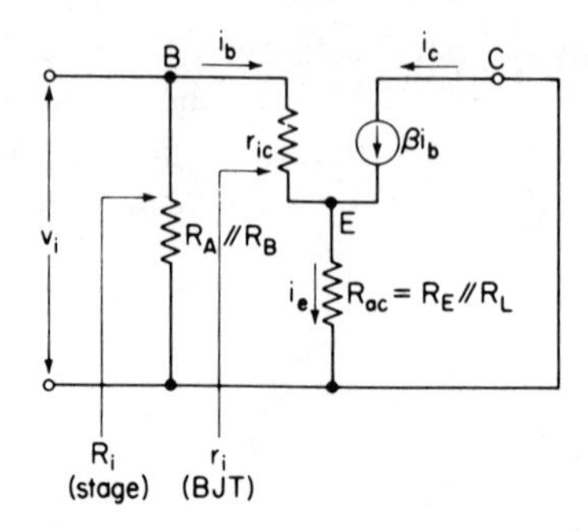

Fig. 3–49 Ac equivalent circuit of the emitter follower of Figure 3–48.

The ac input voltage applied to the base terminal

$$v_i = i_b r_{ic} + i_e R_{ac} \tag{3–121}$$

where $R_{ac} = R_E // R_L$ represents the total ac resistance at the output terminal of the BJT. Since $i_e = (\beta + 1)i_b$, Eq. (3–121) can also be written as

$$v_i = i_v \beta r_\pi + (\beta + 1) i_b R_{ac}$$

or

$$v_i \cong \beta i_b (r_\pi + R_{ac}) \tag{3–122}$$

The ac output voltage developed across the total load resistance

$$v_o = i_e R_{ac} \cong \beta i_b R_{ac} \tag{3–123}$$

The voltage gain of the *CC* stage, as the ratio of v_o and v_i,

$$A_v = \frac{v_o}{v_i} = \frac{\beta i_b R_{ac}}{\beta i_b (r_\pi + R_{ac})} = \frac{R_{ac}}{r_\pi + R_{ac}} \tag{3–124}$$

Since $r_\pi \ll R_{ac}$,

$$A_v = \frac{v_o}{v_i} \cong 1 \tag{3–125}$$

The input resistance of the BJT in the CC connection equals

$$r_{i(\mathrm{BJT})} = \frac{v_i}{i_i} = \frac{\beta i_b(r_\pi + R_{\mathrm{ac}})}{i_b}$$

$$r_{i(\mathrm{BJT})} = \beta(r_\pi + R_{\mathrm{ac}}) \tag{3–126}$$

Usually,

$$r_{i(\mathrm{BJT})} \cong \beta R_{\mathrm{ac}}, \text{ since } r_\pi << R_{\mathrm{ac}} \tag{3–127}$$

The input resistance of the CC stage of Fig. 3–48 is

$$R_{i(\mathrm{stage})} = R_A//R_B//r_{i(\mathrm{BJT})} \tag{3–128}$$

Although the voltage gain of the CC connection is slightly less than 1, the input resistance looking into the base of the BJT is β times the resistance in the emitter circuit (ac load). The circuit can therefore be used as an impedance transformer, because it can transform the low-impedance load in the emitter circuit to a high-impedance load looking into the base of the transistor.

EXAMPLE 3-18

Calculate the approximate input resistance of the emitter follower of Fig. 3–50. The β of the transistor is between 100 and 200.

SOLUTION

The ac load resistance in the emitter circuit is

$$R_{\mathrm{ac}} = R_E//R_L = 3\ \mathrm{k\Omega}//2\ \mathrm{k\Omega} = 1.2\ \mathrm{k\Omega}$$

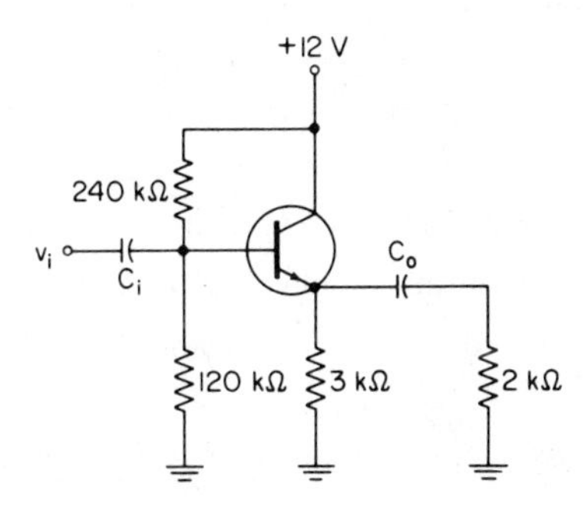

Fig. 3–50 CC circuit for Example 3–18.

The input resistance of the BJT, looking into the base terminal,

$$r_{i(\text{BJT})} \simeq \beta(r_\pi + R_{\text{ac}})$$

Resistance r_π depends on the dc emitter current and must be calculated. $V_B = 4$ V, and $I_E = 1.3$ mA. According to Eq. (3–55), r_π is approximately 40 Ω. Since $r_\pi \ll R_{\text{ac}}$,

$$r_{i(\text{BJT})} \simeq \beta R_{\text{ac}}$$

When $\beta = 100$, $r_{i(\text{BJT})} = 100 \times 1.2$ kΩ $= 120$ kΩ. Also, when $\beta = 200$, $r_{i(\text{BJT})} = 200 \times 1.2$ kΩ $= 240$ kΩ.

The input resistance of the *CC* stage consists of R_A, R_B, and $r_{i(\text{BJT})}$ in parallel. Hence, when $\beta = 100$,

$$R_{i(\text{stage})} = 120 \text{ k}\Omega // 120 \text{ k}\Omega // 240 \text{ k}\Omega = 48 \text{ k}\Omega$$

When $\beta = 200$,

$$R_{i(\text{stage})} = 240 \text{ k}\Omega // 120 \text{ k}\Omega // 240 \text{ k}\Omega - 60 \text{ k}\Omega$$

The significance of Example 3–18 is that a 1.2-kΩ ac load is transformed into a 120-kΩ or 240-kΩ resistance (depending on β), viewed from the input terminal.

QUESTIONS

3–1 Derive the relation between α and β.

3–2 Using the collector characteristics of the 2N929 silicon transistor given in the Appendix, explain the difference between dc current gain h_{FE} and ac current gain h_{fe}.

3–3 Why is the base region of a bipolar junction transistor lightly doped?

3–4 Which factors limit the maximum collector current a transistor can carry?

3–5 Does the collector current of a silicon transistor increase or decrease when the temperature increases? Why?

3–6 Why is it so important to operate the transistor within its rated temperature range?

3–7 Explain why fixed bias is a poor way of biasing a transistor.

3–8 What is the function of the emitter resistor in a common-emitter circuit? Why is the emitter resistor often bypassed with a capacitor?

3–9 What are the reasons a common-emitter configuration is so popular?

3–10 Which BJT configuration has a power gain approximately equal to the current gain?

3–11 Name two applications of the emitter follower.

3–12 What is meant by "loading" an amplifier stage?

3–13 Define the "stability factor" of a transistor stage.

3–14 Why is emitter bias practically independent of transistor current gain?

PROBLEMS

3–1 A certain BJT has an emitter current I_E = 10 mA and a collector current I_C = 9.9 mA
Calculate (a) The base current I_B.
(b) The *CB* current gain α or h_{FB}.
(c) The *CE* current gain β or h_{FE}.

3–2 A certain BJT has a dc current gain h_{FE} = 100. Calculate h_{FB}.

3–3 A silicon BJT is connected in the *CB* configuration of Fig. 3–51, where V_{CC} = 20 V, V_{EE} = 10 V, R_C = 10 kΩ, and R_E = 10 kΩ.
Calculate (a) The approximate emitter and collector currents I_E and I_C.
(b) The collector-to-ground voltage V_C.

3–4 Assume that in the circuit of Fig. 3–51, V_{EE} = 12 V, V_{CC} = 20 V, and R_C = 10 kΩ. Calculate the resistance of R_E which would produce a collector voltage V_C = 6 V.

3–5 Assume that in the circuit of Fig. 3–51, V_{CC} = 20 V, R_C = 8 kΩ, and R_E = 20 kΩ. Calculate the voltage of the emitter supply battery V_{EE} which would cause the transistor to enter into saturation.

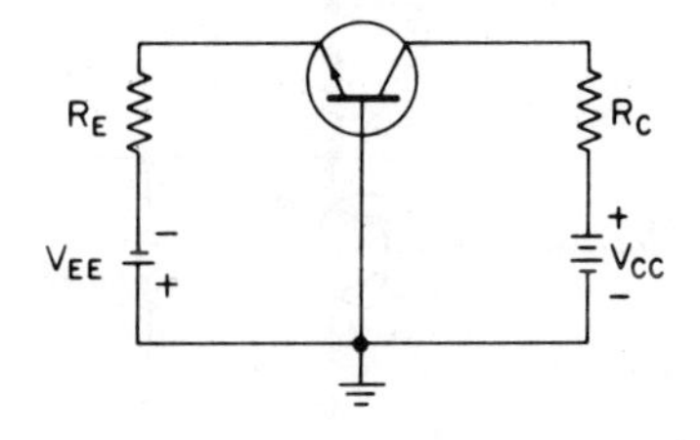

Fig. 3–51 Common-base configuration for Problems 3–3, 3–4 and 3–5.

3–6 In the common-emitter circuit of Fig. 3–52, a silicon BJT with current gain $\beta = 100$ is used. $V_{CC} = 20$ V, $R_B = 1$ MΩ, and $R_C = 10$ kΩ.
Calculate (a) The three transistor currents I_B, I_C, and I_E.
(b) The collector voltage V_C.

3–7 In the circuit of Fig. 3–52, a silicon BJT with current gain $\beta = 80$ is used. $V_{CC} = 5$ V, $R_B = 400$ kΩ, and $R_C = 5$ kΩ. Allowing 0.5 V for the emitter-to-base voltage, calculate the collector voltage.

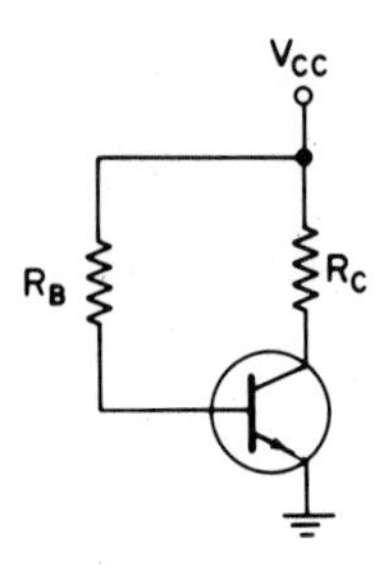

Fig. 3–52 Common-emitter configuration for Problems 3–6, 3–7 and 3–8.

3–8 Using the circuit parameters of Prob. 3–7, calculate the resistance of R_B which would cause the transistor to operate in saturation.

3–9 Figure 3–53 shows a common-emitter amplifier stage with fixed base biasing.
(a) Calculate the dc voltages and currents in the circuit.
(b) Draw the complete ac equivalent circuit, using the hybrid parameters.

3–10 Figure 3–54 shows a common-base amplifier stage.
(a) Calculate the dc voltages and currents in the circuit.

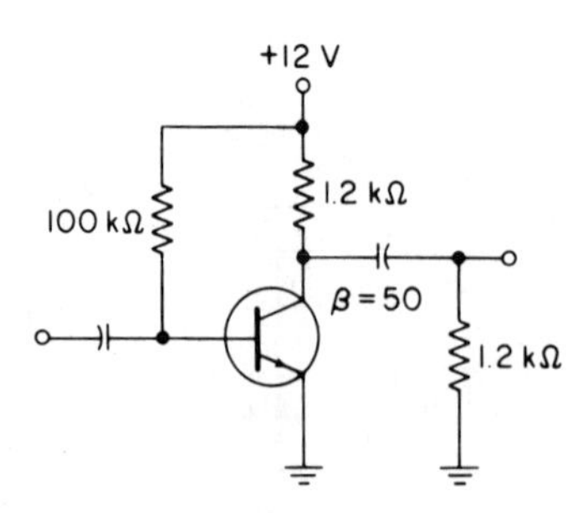

Fig. 3–53 Common-emitter circuit with fixed base bias for Problem 3–9.

(b) Draw the ac equivalent circuit for the complete stage, using the semihybrid equivalent for the BJT.

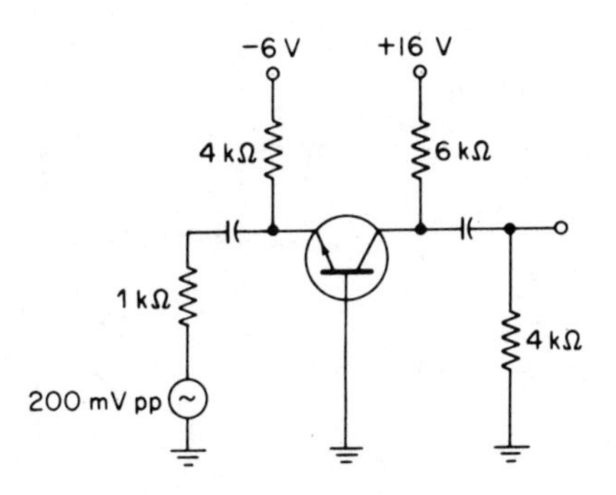

Fig. 3-54 Common-base circuit for Problem 3-10.

3-11 Figure 3-55 shows a common-collector amplifier stage.
(a) Calculate the dc voltages and currents in the circuit.
(b) Draw the ac equivalent circuit for the complete stage, using the semihybrid equivalent for the BJT.

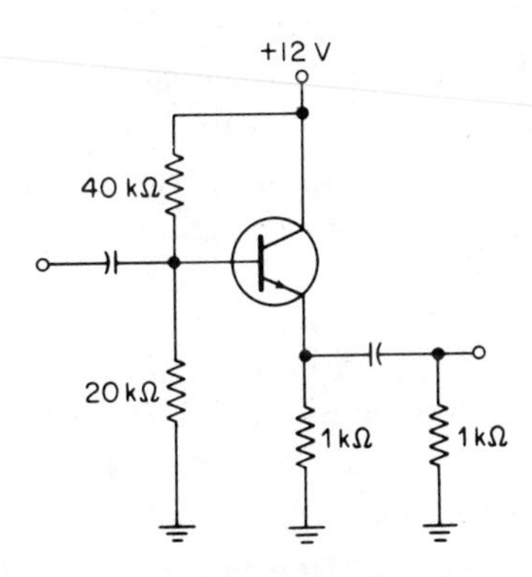

Fig. 3-55 Common-collector circuit for Problem 3-11.

3-12 Figure 3-56 shows a *CE* amplifier stage, using the 2N929 silicon planar BJT, whose characteristics are given in the Appendix.
(a) Draw the ac equivalent circuit.
(b) Determine the current gain A_i of the stage.
(c) Calculate the approximate input resistance of the stage.
(d) Calculate the voltage gain A_v of the stage.

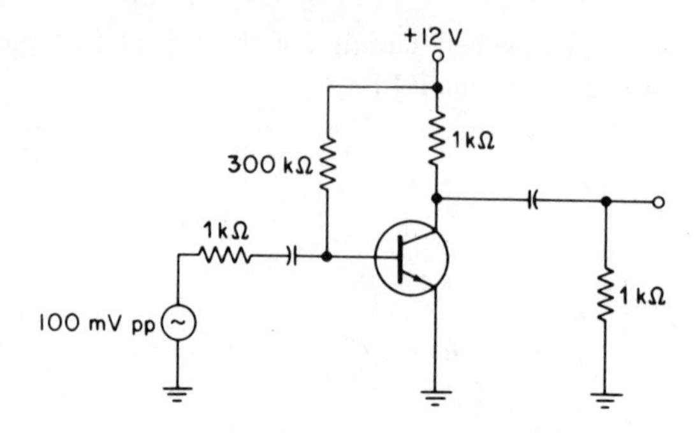

Fig. 3–56 Common-emitter circuit for Problem 3–12.

3–13 The signal source in Fig. 3–56 applies a sine wave of 100 mV pp to the circuit. Taking the internal resistance (1 kΩ) of the source into consideration, calculate the ac voltage appearing at the base terminal of the BJT.

3–14 The silicon planar transistor in the common-emitter configuration of Fig. 3–57(b) has collector characteristics as shown in Fig. 3–57(a). Draw the dc loadline and show the dc operating point for the following conditions:
(a) $V_{CC} = 20$ V, $R_C = 200$ Ω, $R_B = 50$ kΩ.
(b) $V_{CC} = 12$ V, $R_C = 100$ Ω, $R_B = 60$ kΩ.
(c) $V_{CC} = 16$ V, $R_C = 400$ Ω, $R_B = 10$ MΩ.

3–15 Refer to Fig. 3–57(a). Determine the ac current gain h_{fe} of the transistor at a collector-to-emitter voltage $V_{CE} = 10$ V and a base current $I_B = 400$ μA.

3–16 The 2N929 silicon planar transistor, whose characteristics are given in the Appendix, is used in the common-emitter circuit of Fig. 3–57(b). The collector supply voltage $V_{CC} = 25$ V and the collector resistor $R_C = 500$ Ω. The external load resistor $R_L = 500$ Ω. Base bias resistor R_B is chosen so that the dc base current $I_B = 60$ μA.
(a) Draw the dc loadline on the collector characteristics.
(b) Draw the ac loadline on the collector characteristics.
(c) Calculate the resistance of R_B for $I_B = 60$ μA.

3–17 Refer to the circuit and characteristics of Prob. 3–16. Assume that the ac signal source drives a base current $i_b = 80$ μA pp into the transistor.

Determine (a) The collector current swing.
(b) The collector voltage swing.
(c) The ac current gain of the circuit.
(d) The quiescent collector dissipation of the transistor.

3–18 Refer to Fig. 3–58 and *determine*
(a) The input resistance of the amplifier stage.

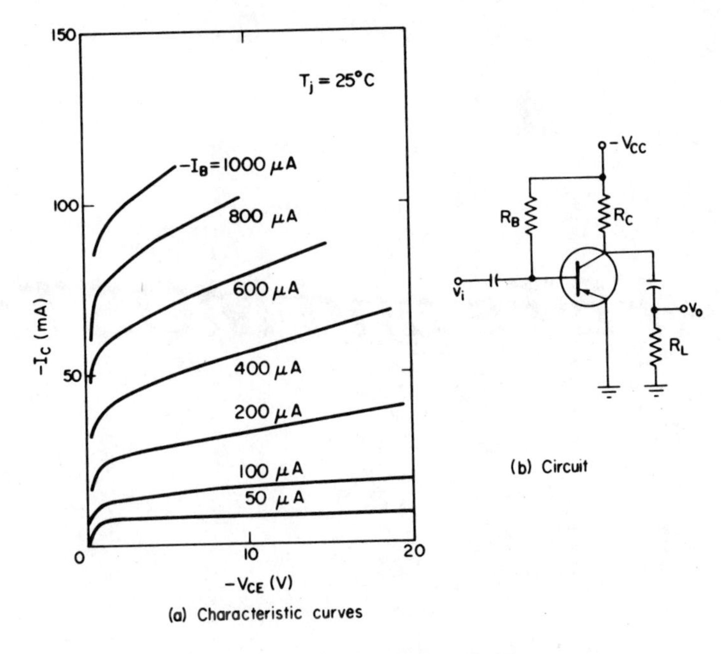

Fig. 3–57 Characteristic curves and circuit for Problem 3–14.

(b) The voltage gain of the circuit.
(c) The stability factor S.

3–19 If the ac input voltage to the amplifier of Fig. 3–58 is 50 mV rms, determine the ac voltage developed across the 10-kΩ load resistor.

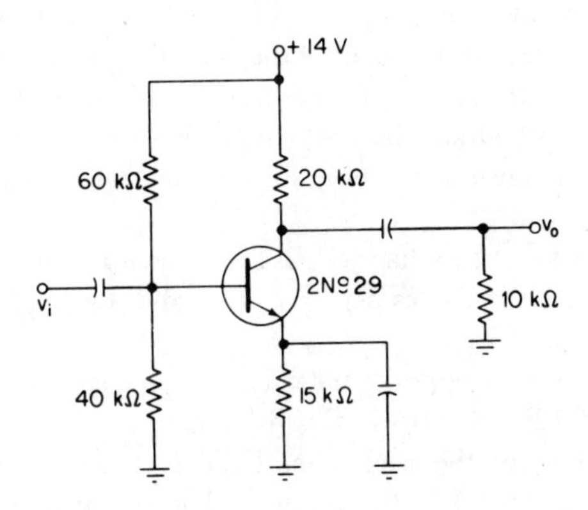

Fig. 3–58 Common-emitter circuit for Problem 3–18.

Chapter 4

FETs and MOSTs

4–1 CONCEPTS AND CHARACTERISTICS OF THE FIELD EFFECT TRANSISTOR

4–1.1 FET Operation

In contrast to the bipolar junction transistor of Chapter 3, the field effect transistor (FET) is a *unipolar* device in which only majority current carriers take part in the processes in the transistor.

Consider the field effect transistor of Fig. 4–1 (a), where a region of lightly doped n-type silicon is sandwiched between two regions of heavily doped p-type silicon. Terminals at both ends of the n-type region are *source* S and *drain* D. The two p-type regions or *gates* are internally connected to a single terminal G. The n-type region bounded by the two gates is called the *channel.* Since in this case the channel consists of n-type silicon, the device is called an n-channel FET.

The symbol for the n-channel FET is shown in Fig. 4–1(b), where the arrowhead on the gate indicates a p-type gate and, by implication, an n-type channel.

An FET may have opposite polarity, a p-type channel and n-type gates. This is a p-channel FET, symbol in Fig. 4–1(c).

The operation of the n-channel FET is based on the movement of majority carriers (electrons) in the channel. The conduction of the channel is controlled by two effects, acting simultaneously.

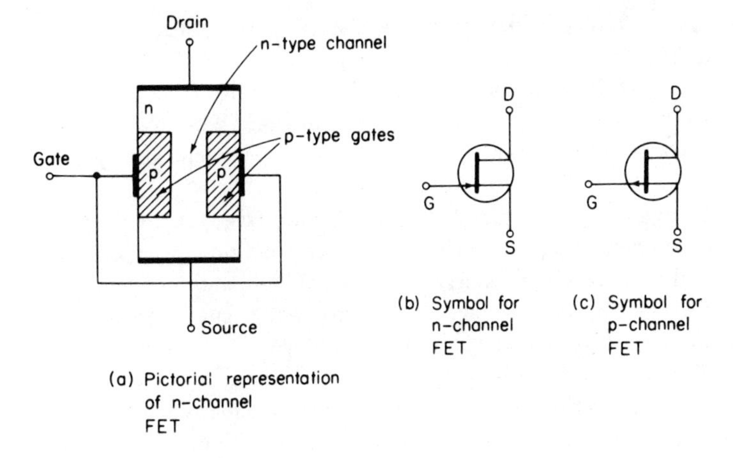

Fig. 4–1 Pictorial representation and FET symbols.

Under normal operating conditions, a negative bias voltage $-V_{GS}$ is applied between the gate terminal and the source terminal, as in Fig. 4–2. Since the left gate and the right gate are connected, this bias voltage produces *two* reverse-biased diodes: one from the left gate to the channel, and one from the right gate to the channel. In *pn* junctions (Section 1–3) a reverse voltage across a junction produces a *depletion region* at the junction, where no free charge carriers are available.

In Fig. 4–2(a), gate voltage V_{GS} is initially set at zero volts and there are no depletion regions at the junctions. Hence, the cross-sectional area of the

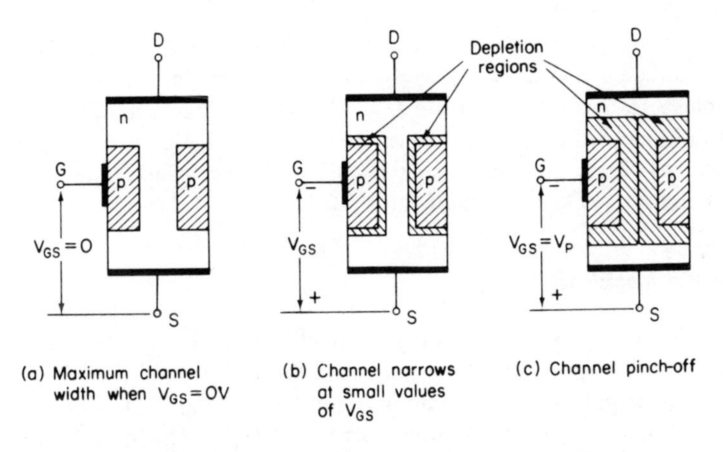

Fig. 4–2 The electric field applied to the gates controls the width of the channel.

channel is maximum in the sense that charge carriers are available across the entire channel area.

In Fig. 4–2(b), a small reverse voltage is applied from gate to source, and narrow depletion regions are formed at the junctions. The cross-sectional area of the channel is reduced and hence fewer charge carriers are available in the channel.

In Fig. 4–2(c), the gate bias is sufficiently large for the depletion regions to extend completely across the channel. In this condition the channel is cut off in the sense that there are no charge carriers available and the channel becomes a nonconductor. The gate voltage at which channel cutoff occurs is called the *pinchoff voltage* $V_{P(GS)}$ (gate-source cutoff voltage).

When a dc voltage is applied across the channel itself, the second effect is obtained. Consider Fig. 4–3(a), where the gate is connected to the source ($V_{GS} = 0$), and a positive voltage V_{DS} is applied from drain to source. Since the n-channel is a semiconductor, a certain channel or *drain current* I_D flows, as indicated. V_{DS} produces a *voltage gradient* along the channel, with the top of the channel positive with respect to the bottom. The gates are connected to the source and the two *pn* junctions are therefore reverse-biased by V_{DS}. Because of the voltage gradient, however, the depletion regions vary in width along the channel, as shown in Fig. 4–3(a).

When V_{DS} is increased, channel current I_D increases almost linearly. The voltage gradient along the channel becomes steeper and the depletion regions widen to the point where they eventually touch, as in Fig. 4–3(b). This condition is called *channel pinchoff*. It occurs at the top of the channel and only when V_{DS} is sufficiently large and equal to the pinchoff voltage V_P. Note that this voltage has the same magnitude as the gate-source cutoff voltage. The channel current has now reached its maximum value I_{DSS} (drain current with source short-circuited to gate). Note that I_{DSS} must flow to maintain the voltage gradient which produces pinchoff.

When V_{DS} is increased beyond V_P, the depletion regions expand at the top of the channel, as indicated in Fig. 4–3(c). The channel acts as a current limiter and holds I_D at a constant value of I_{DSS}. The relation between drain current and drain-to-source voltage V_{DS} is shown in the graph of Fig. 4–3(d). I_D increases practically linearly with increasing V_{DS} until channel pinchoff occurs and the drain current reaches its pinchoff value I_{DSS}.

4–1.2 Characteristic Curves

The two effects controlling channel width and pinchoff usually act together. Figure 4–4 shows an n-channel FET in a typical connection, where battery V_{GG} supplies negative gate bias and battery V_{DD} establishes the drain-to-source voltage V_{DS} and the drain current I_D. Gate voltage $-V_{GS}$ causes the depletion layers to expand along the entire length of the channel. Drain voltage

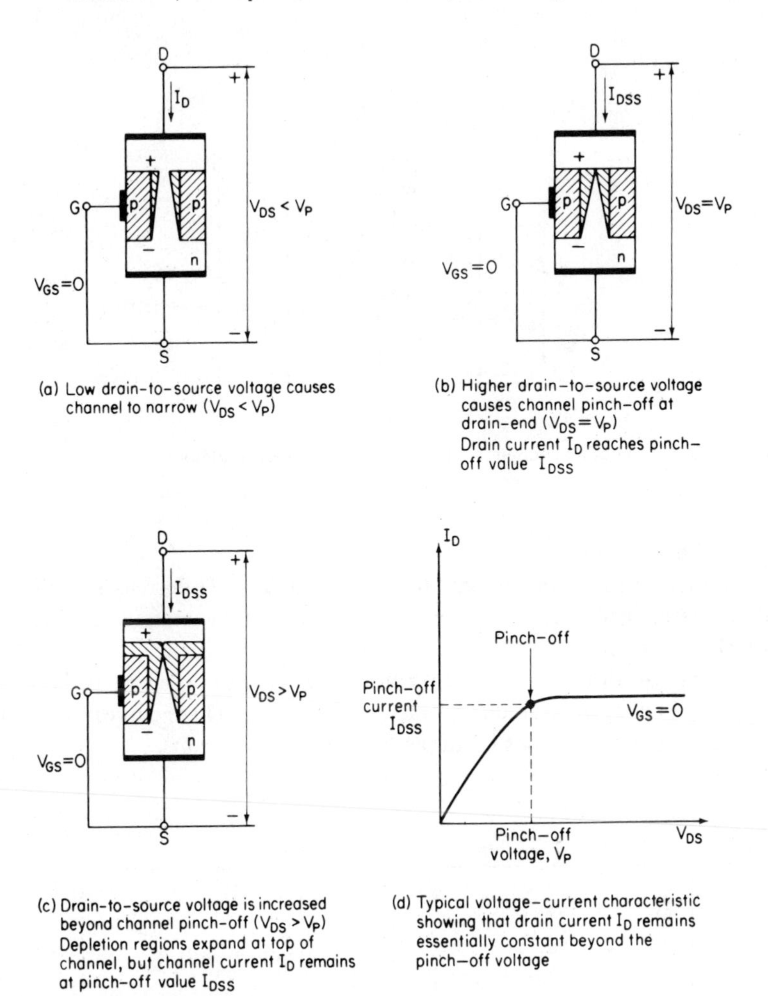
D
I_D
+
G
p
p
n
$V_{DS} < V_P$
$V_{GS}=0$
S
(a) Low drain-to-source voltage causes channel to narrow ($V_{DS} < V_P$)
D
I_{DSS}
G
p
p
n
$V_{DS}=V_P$
$V_{GS}=0$
S
(b) Higher drain-to-source voltage causes channel pinch-off at drain-end ($V_{DS}=V_P$) Drain current I_D reaches pinch-off value I_{DSS}
D
I_{DSS}
G
p
p
n
$V_{DS} > V_P$
$V_{GS}=0$
S
(c) Drain-to-source voltage is increased beyond channel pinch-off ($V_{DS} > V_P$) Depletion regions expand at top of channel, but channel current I_D remains at pinch-off value I_{DSS}
I_D
Pinch-off
Pinch-off current I_{DSS}
$V_{GS}=0$
Pinch-off voltage, V_P
V_{DS}
(d) Typical voltage-current characteristic showing that drain current I_D remains essentially constant beyond the pinch-off voltage

Fig. 4–3 Effect of drain-to-source voltage on the width of the channel in a field effect transistor.

V_{DS} sets up a voltage gradient along the channel which may cause channel pinchoff when V_{DS} is sufficiently large. Voltages V_{DS} and $-V_{GS}$ produce pinchoff. As $-V_{GS}$ is increased, channel pinchoff occurs at smaller values of V_{DS} and correspondingly smaller values of I_D. This is shown in the *drain characteristics* of Fig. 4–5, where I_D is plotted against V_{DS} for various values of gate bias $-V_{GS}$. When $-V_{GS}$ is sufficiently large, the channel is cut off completely and the drain current reduced to zero.

The *transfer characteristic* of Fig. 4–5 shows a plot of I_D versus negative gate voltage $-V_{GS}$ for a fixed drain-to-source voltage $V_{DS} > V_P$. When $V_{GS} = 0$,

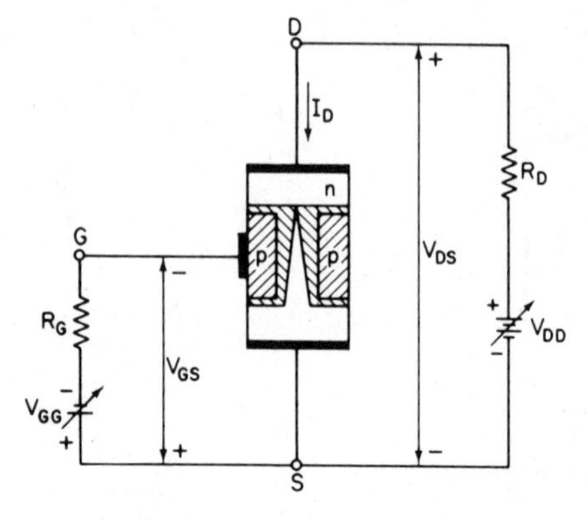

Fig. 4–4 Operation of an *n*-channel FET under normal bias conditions.

the drain current I_D is maximum and has a value I_{DSS}. I_D decreases nonlinearly with increasing gate bias $-V_{GS}$, which is evident from the unequal spacing of the drain characteristics.

Since the drain and transfer characteristics are so closely related, they are often presented together on one graph, as in Fig. 4–5. These curves can be used to obtain graphical solutions to simple FET amplifier circuits.

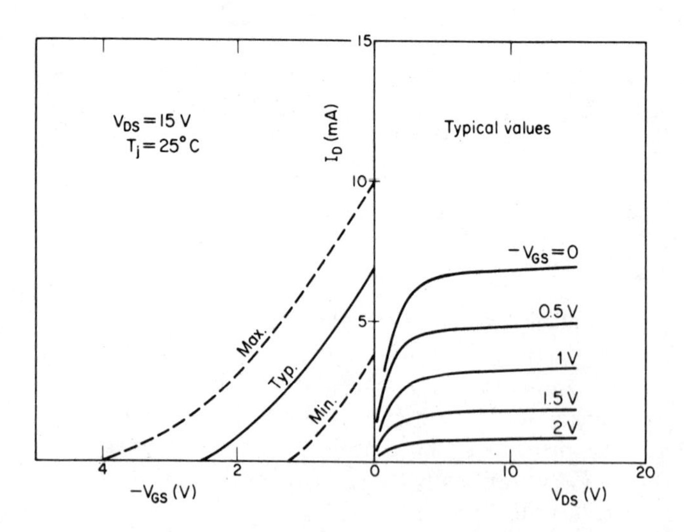

Fig. 4–5 Drain current as a function of drain-to-source voltage for various values of gate-to-source bias for an *n*-channel FET.

4–1.3 FET Parameters

Two parameters identifying the basic FET characteristics are the *zero-bias drain current* I_{DSS} and the *channel cutoff voltage* $V_{P(GS)}$. Both I_{DSS} and $V_{P(GS)}$ are inherent to a particular FET and are controlled during the manufacturing process.

I_{DSS} is defined as the drain current with the source short-circuited to the gate ($V_{GS} = 0$), and the drain voltage greater than the pinchoff voltage ($V_{DS} > V_P$). $V_{P(GS)}$ is defined as the gate-to-source cutoff voltage, where the channel is completely cut off and the drain current is zero. Both parameters are specified at FET operation in the constant-current region of the drain characteristics, where the drain current is virtually independent of the drain voltage.

The relation between I_{DSS} and V_P is shown in Fig. 4–6. The zero-bias pinchoff voltage V_P on the drain characteristic has the same magnitude as the channel cutoff voltage $V_{P(GS)}$ on the transfer characteristic.

The drain current at any value of gate voltage is a function of the two FET parameters I_{DSS} and V_P, and of the applied gate voltage.

$$I_D = I_{DSS}\left(1 - \frac{V_{GS}}{V_P}\right)^2 \tag{4–1}$$

where I_D = drain current at a given applied gate voltage
I_{DSS} = zero-bias drain current
V_{GS} = gate bias voltage
V_P = pinchoff voltage

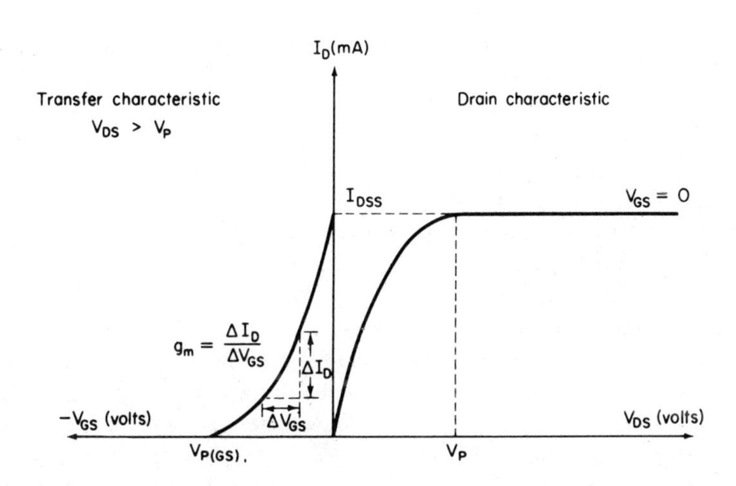

Fig. 4–6 Drain and transfer characteristics of a FET, showing the basic parameters I_{DSS}, V_P, and g_m.

An important FET parameter is the *transfer admittance* or *transconductance* g_m, often used in the calculation of FET voltage gain. The transconductance is a dynamic parameter, relating the change in drain current (ΔI_D) caused by a change in gate voltage (ΔV_{GS}). The $V_{GS} - I_D$ relationship is contained in the transfer characteristic and g_m is then defined as the slope of the transfer characteristic at the desired operating point, so

$$g_m = \frac{\Delta I_D}{\Delta V_{GS}} \bigg| V_{DS} > V_P \tag{4-2}$$

Since the transfer characteristic is not a straight line, the value of g_m varies along the characteristic. The FET data sheet lists transfer admittance as *y parameter* y_{fs} or as g_{mo}, at specified conditions of zero-bias drain current and operation in the constant-current region where $V_{DS} > V_P$. The transconductance at any other value of drain current is then given by

$$g_m = g_{mo} \sqrt{\frac{I_D}{I_{DSS}}} \tag{4-3}$$

Typical values of transconductance range from 0.5 to 30 mA/V.

4–2 FET AMPLIFIERS

4–2.1 Basic Circuit

Figure 4–7 shows an n-channel FET in a basic amplifier configuration. Bias supply V_{GG} provides a small negative gate voltage V_{GS}. Since the gate current is negligible (reverse-biased diode), there is pratically no voltage drop across R_G and hence $V_{GS} = V_{GG}$. Drain supply V_{DD} and drain resistor R_D provide the quiescent drain current I_D and establish the drain-to-source voltage V_{DS}. The circuit values are chosen so that the drain voltage V_{DS} exceeds the pinchoff value V_P, and the FET therefore operates in the constant-current region.

Signal source v_s is capacitively coupled to the gate and produces an instantaneous gate voltage $v_{GS} = v_s - V_{GS}$. On positive half-cycles of v_s the instantaneous gate voltage becomes less negative and the instantaneous drain current increases. On negative half-cycles of v_s the instantaneous gate voltage becomes more negative and the instantaneous drain current decreases. These drain current variations develop an ac output voltage v_o across drain resistor R_D.

Voltage amplification is obtained because the change in V_{DS} (output voltage) can be many times larger than the change in V_{GS} (input voltage). In a typical FET amplifier, a change in V_{GS} of 0.5 V can cause a change in I_D of 5

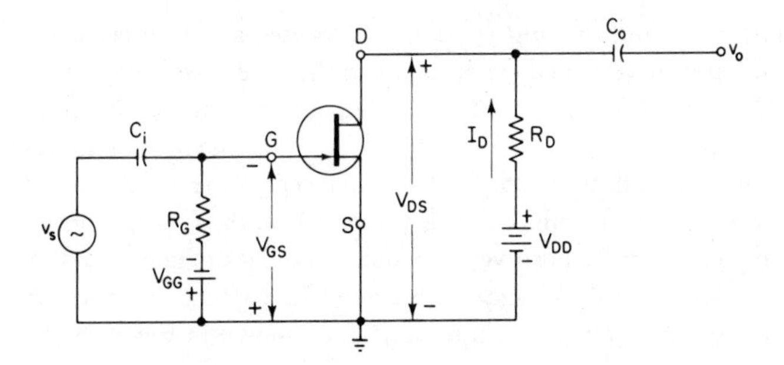

Fig. 4–7 A basic FET amplifier.

mA, which in turn can cause a change in V_{DS} of 10 V. The voltage amplification of the amplifier is defined as

$$A_v = \frac{v_o}{v_s} = \frac{\Delta V_{DS}}{\Delta V_{GS}} \tag{4–4}$$

and in this particular case equals 10 V/0.5 V = 20.

4–2.2 Graphical Solution

A graphical approach to the solution of a FET circuit uses the $V_{DS} - I_D$ or *drain characteristics* of the FET. The dc operating point is determined in a manner identical to that used in Section 3–2.5 for the junction transistor in the *CE* configuration.

The dc voltage equation for the drain circuit in Fig. 4–7 is

$$V_{DS} = V_{DD} = I_D R_D \tag{4–5}$$

which yields an expression for the quiescent drain current

$$I_D = \frac{V_{DD} - V_{DS}}{R_D} \tag{4–6}$$

Equation (4–5) is first-order, represented by a straight line on the drain characteristics called the *dc loadline*. It can be plotted on the drain characteristics of Fig. 4–8 by determining two points on the line. The first point (x intercept) is located on the V_{DS} axis, where $I_D = 0$ and $V_{DS} = V_{DD}$. The second point (y intercept) is located on the I_D axis, where $V_{DS} = 0$ and $I_D = V_{DD}/R_D$. The straight line obtained by connecting these two points is the dc loadline.

The *dc operating point* (Q point) is located at the intersection of the dc loadline and the curve which corresponds to the gate-to-source voltage provided by V_{GG}. If, in Fig. 4–7, $V_{GS} = -1$ V, the Q point is located at the intersection of the loadline and the $V_{GS} = -1$ V curve of Fig. 4–8. The quiescent drain current I_D and the quiescent drain voltage V_{DS} are marked on the graph.

If the input voltage has a peak value of 1 V, the Q point moves along the loadline to point A on the positive excursions, and back to point B on the negative excursions. This signal voltage causes a drain current variation ΔI_D and a corresponding drain voltage variation ΔV_{DS}, as shown in Fig. 4–8. The ratio of ΔV_{DS} and ΔV_{GS} is the voltage gain of the amplifier.

The drain voltage decreases when the signal voltage is positive-going, indicating a *phase reversal* of the output voltage.

EXAMPLE 4-1

The basic FET amplifier of Fig. 4–7 has the following circuit parameters:

$$V_{DD} = 20 \text{ V} \qquad R_D = 500\ \Omega$$
$$V_{GG} = -1.5 \text{ V} \qquad R_G = 1\ \text{M}\Omega$$
$$v_s = 1 \text{ V pp}$$

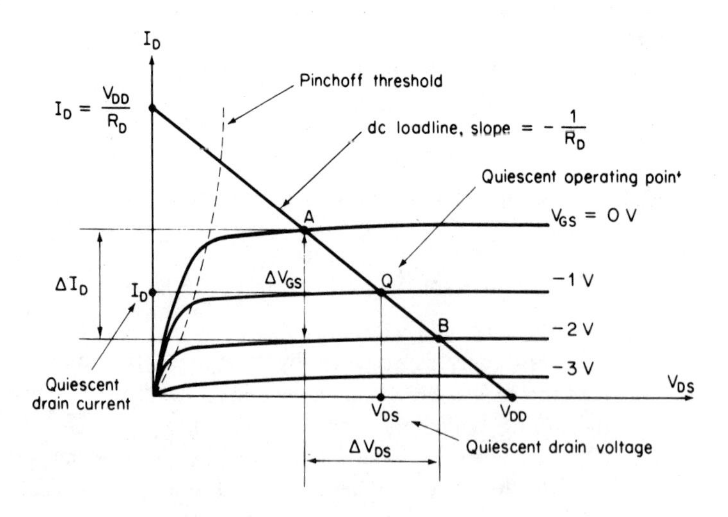

Fig. 4–8 Graphical analysis of the FET amplifier.

The drain characteristics of the FET are given in Fig. 4–9.
(a) Construct the dc loadline.
(b) Determine the quiescent drain current and drain voltage.
(c) Determine the ac output voltage and output current.
(d) Find the voltage gain of the amplifier.

SOLUTION

(a) One point of the dc loanline is located on the V_{DS} *axis,* when $I_D = 0$ and $V_{DS} = V_{DD} = 20$ V. A second point of the loadline is located on the I_D axis, when $V_{DS} = 0$ and $I_D = V_{DD}/R_D = 20\text{ V}/500\ \Omega = 40$ mA. The two points are connected to form the dc loadline, as shown on the graph of Fig. 4–9.

(b) Since the gate current is negligible, $V_{GS} = V_{GG} = -1.5$ V. The intersection of the $V_{GS} = -1.5$ V curve and the loadline locates the Q point, where $I_D = 20.8$ mA and $V_{DS} = 9.7$ V.

(c) On positive excursions of the signal voltage, point A is reached, where $I_D = 27.4$ mA and $V_{DS} = 6.3$ V. Negative signal excursions reach point B, where $I_D = 13.6$ mA and $V_{DS} = 13.1$ V. The total drain current variation, which constitutes the ac output current, is $\Delta I_D = 27.4\text{ mA} - 13.6\text{ mA} = 13.8$ mA pp. The total drain voltage variation, which is the ac output voltage, is $\Delta V_{DS} = 13.1\text{ V} - 6.3\text{ V} = 6.8$ V pp.

(d) The voltage gain of the amplifier is simply the output signal voltage divided by the input signal voltage, $A_v = v_o/v_s = 6.8\text{ V pp}/1\text{ V pp} = 6.8$

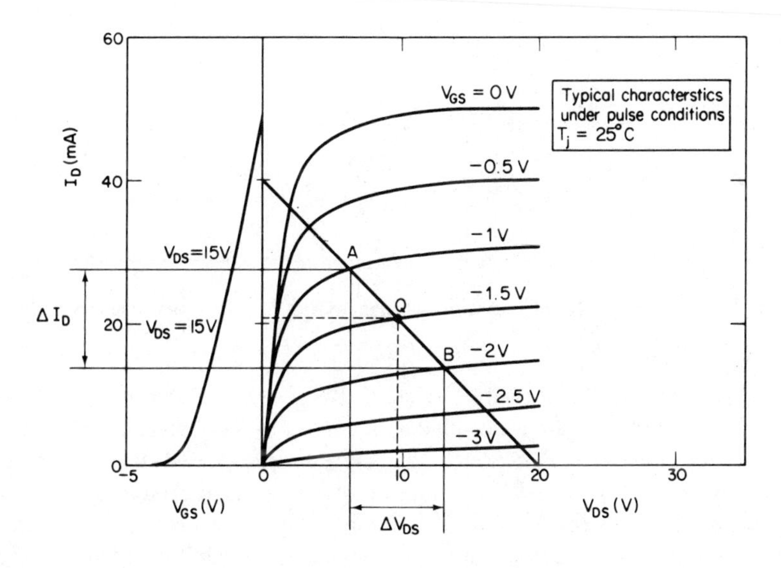

Fig. 4–9 Drain characteristics for the FET of Problem 4–1.

The graphical construction of Fig. 4–8 provides valuable information about possible *signal distortion* problems. The drain characteristics are not equally spaced, so symmetrical signal excursions around the *Q* point do not cause symmetrical drain current or drain voltage variations. To ensure reasonable linearity of the output waveform, it is good practice to limit the input signal to a relatively small amplitude. The operating point should not be located too close to the pinchoff threshold, indicated by the dashed line in Fig. 4–8. If the FET enters this threshold region, extreme distortion results because of the rather close spacing of the curves. The gate bias voltage should not be so high that the channel is completely pinched off on small negative signal excursions.

4–2.3 FET Amplifier with Self-bias

To avoid the need for a separate bias battery V_{GG}, the FET is often connected in a *self-bias* arrangement, as in Fig. 4–10. Resistor R_S, connected to the source terminal, is the bias resistor. The quiescent drain current I_D through R_S develops a dc voltage drop $V_S = I_D R_S$ across it, which then makes the source positive with respect to ground.

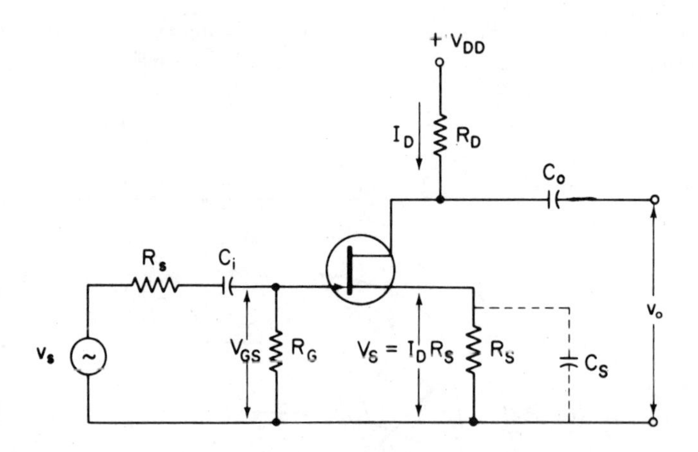

Fig. 4–10 *n*-channel FET amplifier with self-bias resistor R_S and bypass capacitor C_S.

Resistor R_G has a double function. Because the source is positive with respect to ground, the gate is negative with respect to the source, and the FET is therefore correctly biased. Since the gates are reverse-biased, they do not pass direct current and resistor R_G therefore places the gates at ground potential. R_G also provides a high-resistance load to the input signal source v_s. If R_G is large compared to the internal resistance R_s of the source (which is generally the case), the entire signal voltage appears at the gate terminal of the FET.

Consider dc conditions only. In the absence of a signal voltage, the gate-to-source voltage is

$$-V_{GS} = I_D R_S \tag{4–7}$$

Plotting (4–7) gives the *bias line*. For a given drain-to-source voltage, V_{GS} and I_D are related by the transfer characteristic of Fig. 4–11. The bias line with slope $-1/R_S$ intersects the transfer curve at operating point Q, which identifies the quiescent values of I_D and V_{GS}, as shown in Fig. 4–11.

The dc voltage equation for the drain circuit yields

$$V_{DD} = V_{DS} + I_D(R_S + R_D) \tag{4–8}$$

Equation (4–8) gives the *dc loadline*. It is a straight line with slope $-1/(R_D + R_S)$, which can be plotted on the drain characteristics of the FET. The x intercept of the loadline is $V_{DS} = V_{DD}$, and the y intercept is $I_D = V_{DD}/(R_D + R_S)$. The dc loadline, obtained by connecting these two points, is drawn on the graph of Fig. 4–11 and intersects the family of drain curves. The Q point on the transfer characteristic is then projected from the bias line onto the loadline, as shown in Fig. 4–11. This yields the quiescent drain-to-source voltage V_{DS} and the quiescent drain current I_D.

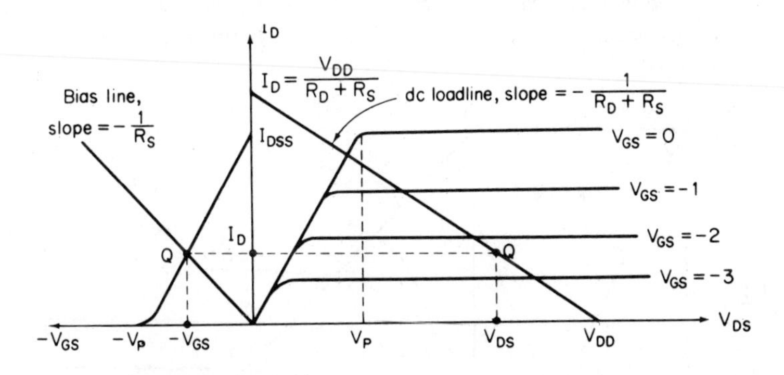

Fig. 4–11 Construction of the bias line and dc loadline to determine quiescent conditions in the self-bias circuit of Figure 4–10.

EXAMPLE 4-2

The self-bias circuit of Fig. 4–10 has the following dc circuit values: $V_{DD} = 20$ V, $R_D = 150\ \Omega$, and $R_S = 50\ \Omega$. The FET characteristics are supplied in Fig. 4–12.

Determine (a) The gate bias voltage, $-V_{GS}$.
(b) The quiescent drain current, I_D.
(c) The quiescent drain voltage, V_{DS}.

SOLUTION

(a) A bias line with slope $-1/R_S = -1/50\ \Omega$ is drawn on the transfer characteristics of Fig. 4–12, as shown. It intersects the transfer curve at $V_{GS} = -1.8$ V.
(b) The drain current is read off the graph as $I_D = 38$ mA.
(c) The dc loadline with shape $-1/(R_S + R_P) = -1/200\ \Omega$ is drawn on the drain characteristics of Fig. 4–12. The Q point on the transfer characteristic is projected onto this loadline, yielding the dc operating point corresponding to $I_D = 38$ mA and $V_{DS} = 12.4$ V. Checking our findings,

$$I_D R_D = 38 \text{ mA} \times 150\ \Omega = 5.7 \text{ V}$$

$$V_{DS} = 12.4 \text{ V}$$

$$I_D R_S = 38 \text{ mA} \times 50\ \Omega = 1.9 \text{ V}$$

The sum of these voltages equals the supply voltage $V_{DD} = 20$ V.

Positive signal excursions of input signal v_s in Fig. 4–10 increase the instantaneous total value of the drain current. This results in an effective increase in the negative gate voltage, and the FET then behaves as if the input voltage were smaller than v_s. In other words, bias resistor R_S causes *degeneration*, which reduces the voltage gain of the amplifier.

The degenerative effect of R_S can be overcome by placing a *bypass capacitor* C_S in parallel with R_S, as in Fig. 4–10. If the reactance of C_S at the

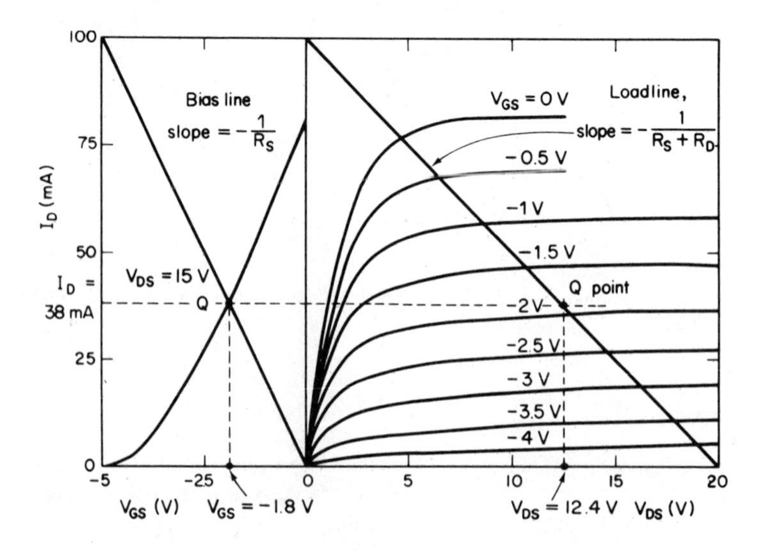

Fig. 4–12 Graphical solution to Example 4–2.

signal frequency is much smaller than the resistance of R_S (typically $X_C \leqslant R_S/10$), C_S provides a low-impedance path for the signal component of the drain current. In this case there is no signal current through R_S and the dc voltage drop across R_S is constant and equal to $I_D R_S$. It should be borne in mind that the ac load in this case consists of R_D only, and an ac loadline with slope $-1/R_D$ must be plotted on the drain characteristics to find the voltage gain of the amplifier.

4–3 FET CIRCUIT CALCULATIONS

4–3.1 Equivalent Circuits

The basic FET amplifier of Fig. 4–13 is subject to both dc and ac voltages and currents. The quiescent operating conditions are established by bias battery V_{GG}, drain battery V_{DD}, and drain resistor R_D. These provide the FET with a quiescent drain current I_D and a quiescent drain voltage V_{DS}.

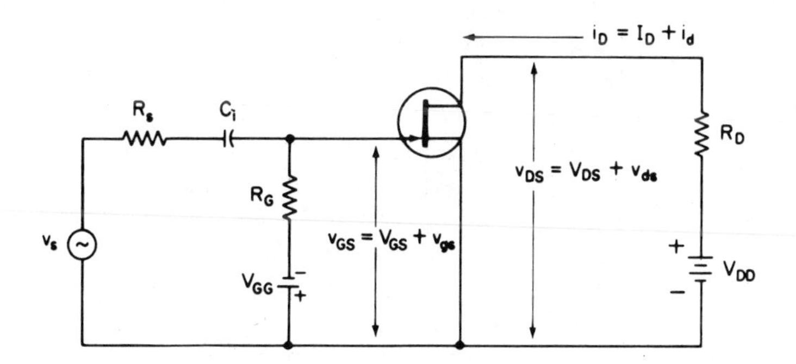

Fig. 4–13 A FET amplifier with instantaneous voltages and currents indicated.

Input signal v_s introduces voltage variations at the gate terminal and provides a total instantaneous gate voltage v_{GS} consisting of a dc component V_{GS} and an ac component v_{gs}, so

$$v_{GS} = V_{GS} + v_{gs} \tag{4–9}$$

Gate voltage variations result in drain current variations and corresponding drain voltage variations. The instantaneous drain current i_D consists of an ac component i_d superimposed on a dc component I_D, so

$$i_D = I_D + i_d \tag{4–10}$$

Similarly, the instantaneous drain voltage v_{DS} consists of a dc component V_{DS} and an ac component v_{ds}, so

$$v_{DS} = V_{DS} + v_{ds} \tag{4–11}$$

The transconductance g_m of the FET relates the variations in drain current (ac component i_d) to the variations in gate voltage (ac component v_{gs}). The transconductance is defined as

$$g_m = \frac{\Delta I_D}{\Delta V_{GS}} = \frac{i_d}{v_{gs}} \tag{4–12}$$

so the ac drain current can be written in terms of the transconductance as

$$i_d = g_m v_{gs} \tag{4–13}$$

The FET can therefore be represented by a simple *ac equivalent circuit* consisting of a current generator $g_m v_{gs}$, as in Fig. 4–14.

In this equivalent circuit the input resistance of the FET, looking into the gate terminal and the output resistance of the FET are not included because they are normally very large.

First, a dc equivalent circuit is drawn, and the quiescent currents and voltages are calculated. Then an ac equivalent circuit is drawn and the ac voltages and currents are determined. Finally, the dc and ac components are added algebraically to find the total currents and voltages in the circuit.

The dc equivalent circuit of the amplifier of Fig. 4–13 is simply determined. The input capacitor behaves like an open circuit for dc, or signal source v_s with internal resistance R_s is removed from the circuit.

Bias battery V_{GG} applies reverse bias to the gate terminal. Since the dc gate current is virtually zero, the gate voltage equals the supply voltage and $V_{GS} = V_{GG}$.

The voltage equation for the drain circuit yields

$$V_{DD} = V_{DS} + I_D R_D \tag{4–14}$$

Since the FET parameters are usually known, I_D can be calculated by using Eq. (4–1).

The ac equivalent circuit of the amplifier of Fig. 4–13 is shown in Fig. 4–15. All capacitors and all dc sources have been short-circuited. Signal source v_s, with internal resistance R_s, develops an ac voltage v_{gs} at the gate terminal

$$v_{gs} = \frac{R_G}{R_s + R_G} v_s \tag{4–15}$$

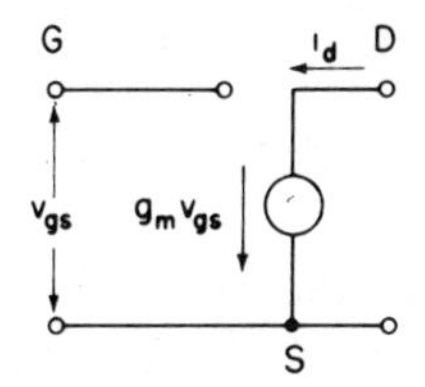

Fig. 4–14 Small-signal equivalent circuit for the FET.

The ac current generated by the FET equals

$$i_d = g_m v_{gs} = g_m v_s \frac{R_G}{R_s + R_G} \tag{4–16}$$

This current is delivered to drain resistor R_D and develops the ac output voltage

$$v_o = i_d R_D = g_m v_{gs} R_D = g_m v_s R_D \frac{R_G}{R_s + R_G} \tag{4–17}$$

The voltage gain of the amplifier, measured from the load to the source,

$$A_v = \frac{v_o}{v_s} = g_m R_D \frac{R_G}{R_s + R_G} \tag{4–18}$$

Usually, $R_G \gg R_s$, and

$$A_v \simeq g_m R_D \tag{4–19}$$

This expression states that the voltage gain of a FET amplifier stage equals the product of its transconductance and the load resistance.

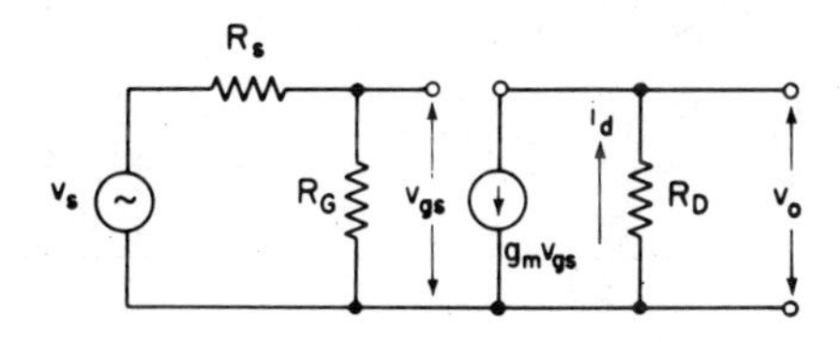

Fig. 4–15 Small-signal equivalent circuit for the amplifier of Fig. 4–13.

4-3.2 Circuit Examples

The configuration of Fig. 4-16 is a self-bias circuit with an additional load coupled to the drain terminal of the FET.

The parameters of the n-channel FET are $I_{DSS} = 5$ mA at $V_{DS} = 15$ V and $V_{GS} = 0$, $V_P = -2$ V at $V_{DS} = 15$ V, $y_{fs} = 5$ mA/V at $V_{DS} = 15$ V and $V_{GS} = 0$. The values of R_D and R_S are to be chosen to give a quiescent operating point at $I_D = 1.5$ mA and $V_{DS} = 10$ V. In addition, the output voltage and the voltage gain are to be calculated. The capacitors are considered to act like short circuits at the frequency under consideration. For dc, capacitors C_i and C_o, and source bypass capacitor C_S, act like open circuits, removing the source and the load from the circuit, but retaining R_S in the source leg of the FET. The quiescent drain current, given by Eq. (4-1), is

$$I_D = I_{DSS}\left(1 - \frac{V_{GS}}{V_P}\right)^2$$

Substituting the known values and solving for V_{GS},

$$1.5 = 5\left(1 + \frac{V_{GS}}{2}\right)^2$$

$$1 + \frac{V_{GS}}{2} = \sqrt{0.3} = 0.55$$

$$V_{GS} = -0.9 \text{ V}$$

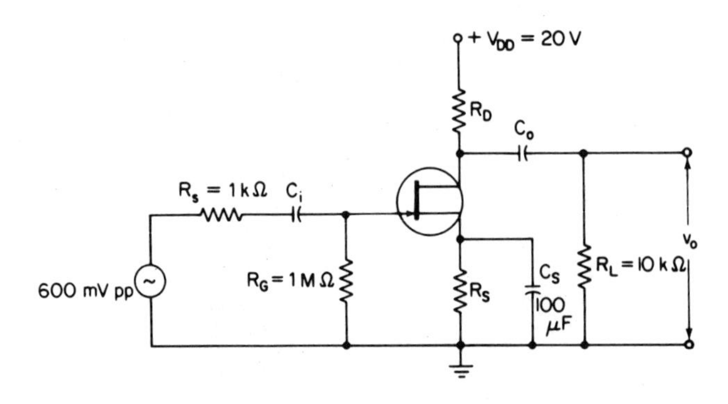

Fig. 4-16 Self-bias circuit with additional load resistor R_L in the drain circuit.

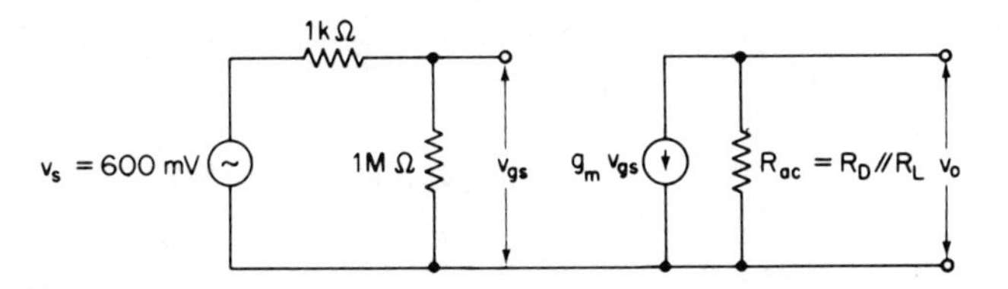

Fig. 4–17 Ac equivalent circuit for the self-bias circuit of Figure 4–16.

Since the gate is at ground potential, the bias voltage at the source terminal equals $V_S = -V_{GS} = 0.9$ V. This bias voltage must be developed across resistor R_S by drain current I_D, so

$$R_S = \frac{V_S}{I_D} = \frac{0.9\text{V}}{1.5\text{ mA}} = 0.6\text{ k}\Omega$$

The voltage equation for the drain circuit yields

$$V_{DD} = V_{DS} + I_D R_D + I_D R_S$$

and

$$R_D = \frac{V_{DD} - V_{DS} - I_D R_S}{I_D} = \frac{20 - 10 - 0.9}{1.5} = 6\text{ k}\Omega$$

The transfer admittance y_{fs} or transconductance g_{mo} is specified at zero-bias drain current. At the operating point of $I_D = 1.5$ mA the transconductance is given by Eq. (4–3) as

$$g_m = g_{mo}\sqrt{\frac{I_D}{I_{DSS}}} = 5\sqrt{\frac{1.5}{5}} = 2.75\text{ mA/V}$$

The ac equivalent circuit is shown in Fig. 4–17.

The signal voltage appearing at the gate terminal of the FET is

$$v_{gs} = \frac{1\text{ M}\Omega}{1\text{ M}\Omega + 1\text{ K}\Omega}600\text{ mV pp} \simeq 600\text{ mV pp}$$

The ac drain current is

$$i_d = g_m v_{gs} = 2.75\text{ mA/V} \times 600\text{ mV pp} = 1.65\text{ mA pp}$$

The total ac load resistance in the drain circuit consists of the parallel combination of R_D and R_L, so

$$R_{ac} = R_D // R_L = 6 \text{ k}\Omega // 10 \text{ k}\Omega = 3.75 \text{ k}\Omega$$

The output voltage then equals

$$v_o = i_d R_{ac} = 1.65 \text{ mA pp} \times 3.75 \text{ k}\Omega = 6.2 \text{ V pp}$$

The overall voltage gain of the amplifier is

$$A_v = \frac{v_o}{v_s} = \frac{6.2 \text{ V}}{600 \text{ mV}} \simeq 10$$

FETs of the same type and made by the same manufacturer are often subject to wide tolerance spreads in their basic parameters I_{DSS} and V_P. This may cause unacceptable changes in quiescent operating conditions when one FET is replaced by another one of the same type. Especially in mass-produced circuits, variations in FET parameters may cause large variations in the performance characteristics of the individual circuits.

In an effort to reduce variations in circuit characteristics, the Q point is often stabilized by the universal bias circuit of Fig. 4–18. This circuit is identical to that of Fig. 3–46. Resistors R_1 and R_2 form a voltage divider across drain supply V_{DD} and maintain the gate at a fixed positive potential V_{GG}. The gate-to-source voltage then equals

$$V_{GS} = V_{GG} - I_D R_S \tag{4–20}$$

and

$$I_D = \frac{V_{GG} - V_{GS}}{R_S} \tag{4–21}$$

Equation (4–20) shows that for V_{GS} to be negative (n-channel FET), $I_D R_S$ must be larger than V_{GG} and R_S usually has a fairly large resistance. This tends to maintain the drain current at a constant value and the dc operating point fixed regardless of FET parameters.

Consider the universal bias current of Fig. 4–18 and assume that V_{DD} = 30 V, R_1 = 1 MΩ, and R_2 = 500 kΩ. The parameters of the FET are I_{DSS} = 10 mA and $V_P = -5$ V. It is desired to provide quiescent conditions at I_D = 2.5 mA and V_{DS} = 8 V. We first calculate the required resistance of R_S and then investigate the change in dc operating conditions if the FET is replaced by one having parameters of I_{DD} = 20 mA and $V_P = -3$ V, assuming that the

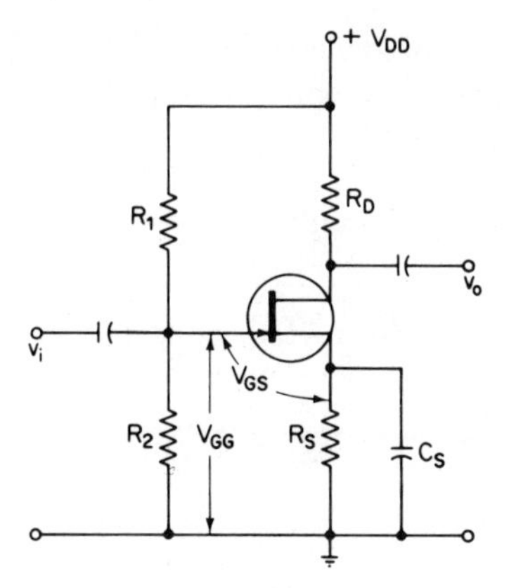

Fig. 4–18 *Q* point stabilization is achieved by the universal bias circuit consisting of resistors R_1 and R_2.

remaining circuit parameters are not changed. The drain current is given by Eq. (4–1) as

$$I_D = I_{DSS}\left(1 - \frac{V_{GS}}{V_P}\right)^2$$

$$2.5 = 10\left(1 + \frac{V_{GS}}{5}\right)^2$$

$$1 + \frac{V_{GS}}{5} = \sqrt{0.25} = 0.5$$

Solving for V_{GS},

$$V_{GS} = 5(0.5 - 1) = -2.5 \text{ V}$$

The gate voltage with respect to ground equals

$$V_{GG} = \frac{500 \text{ k}\Omega}{1.5 \text{ M}\Omega} 30 \text{ V} = 10 \text{ V}$$

Since $V_{GS} = V_{GG} - I_D R_S$ [Eq. (4–20)], we solve for R_S and find that

$$R_S = \frac{V_{GG} - V_{GS}}{I_D} = \frac{10 \text{ V} + 2.5 \text{ V}}{2.5 \text{ mA}} = 5 \text{ k}\Omega$$

When the FET is now replaced by one having $I_{DSS} = 20$ mA and $V_P = -3$ V, we again find the drain current from Eq. (4–1):

$$I_D = I_{DSS}\left(1 - \frac{V_{GS}}{V_P}\right)^2$$

$$I_D = 20\left(1 + \frac{V_{GS}}{3}\right)^2$$

The drain current is also given by Eq. (4–21) as

$$I_D = \frac{V_{GG}}{R_S} - \frac{V_{GS}}{R_S}$$

$$I_D = \frac{10}{5} - \frac{V_{GS}}{5}$$

Solving the two expressions for I_D simultaneously we obtain

$$20\left(1 + \frac{V_{GS}}{3}\right)^2 = 2 - \frac{V_{GS}}{5}$$

$$\left(1 + \frac{V_{GS}}{3}\right)^2 \cong \frac{1}{10}$$

$$V_{GS} = 3(0.33 - 1) = -2 \text{ V}$$

The drain current is found from Eq. (4–21) as

$$I_D = \frac{V_{GG}}{R_S} - \frac{V_{GS}}{R_S} = \frac{10}{5} - \frac{2}{5} = 2.4 \text{ mA}$$

Although the FET parameters have changed considerably, the drain current has remained practically constant, and the dc operating conditions are well stabilized against FET variations.

4–4 MOS TRANSISTORS

4–4.1 Basic Concepts of the MOS Transistor

The MOS transistor (MOS = metal/oxide/semiconductor) is a type of field effect transistor. Although the principles of the MOS transistor (MOST) have been known since about 1928, the first practical MOST was not made until about 1964, when the planar techniques of manufacturing bipolar junction transistors and junction field effect transistors became established.

In the planar technique a silicon substrate is selectively doped by diffusion of impurities through small windows etched into a silicon dioxide (SiO_2) layer deposited on the substrate. It was found that the SiO_2 layer could also act as an *insulator* between the semiconductor substrate and the gate electrode of the field effect transistor, and in this way the MOST was born.

MOS transistors are made on a relatively thick substrate of lightly doped monocrystalline material by means of the planar technique. Figure 4–19 shows a schematic cross section of a MOS transistor. Two wells of strongly doped n-type silicon are diffused into a p-type substrate through openings etched into the SiO_2 surface layer. A metal electrode called the *gate, G,* is deposited on the very thin oxide layer between the two wells. Since the gate is electrically insulated from the remainder of the structure by the SiO_2 layer, the MOST is also known as the *insulated gate* field effect transistor (IGFET). Electrodes are also provided to the two wells, which are known as the source S and the drain D.

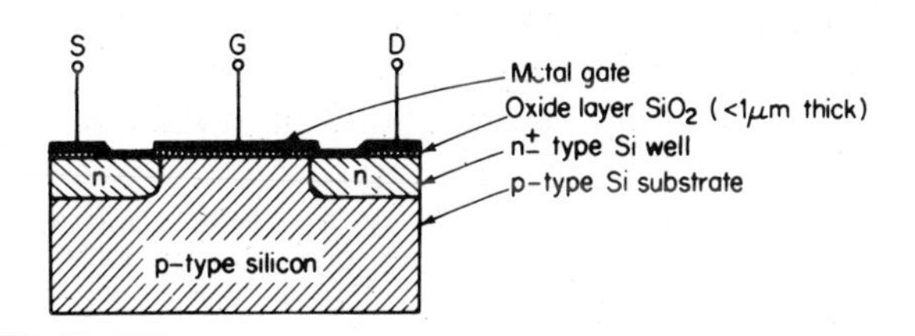

Fig. 4–19 Schematic cross-section of an n-channel MOS transistor. Electrons flow from source S to drain D through the substrate in an extremely thin layer, called the channel, situated directly under the oxide.

The source is connected to ground potential. The voltage between drain and source is called the drain voltage V_D and the voltage between gate and source is called the gate voltage V_G. In the MOST of Fig. 4–19 electrons flow from the source to the drain, through the substrate. This conduction takes place in an extremely thin layer situated right under the oxide called the *channel.* This channel is created by the voltage applied to the gate through a process called *inversion.*

If the substrate is of p-type silicon, the inversion process produces an n-type channel. The symbol for the n-channel MOST is shown in Fig. 4–20(a). If the substrate is of n-type material, conduction takes place by the flow of holes from source to drain and the channel is of p-type. The symbol for the p-channel MOST is shown in Fig. 4–20(b). Both symbols show that the substrate B is connected to the source S. In some cases this connection is made internally and the MOST is then a three-terminal device. In other instances, the substrate is brought out as a fourth terminal.

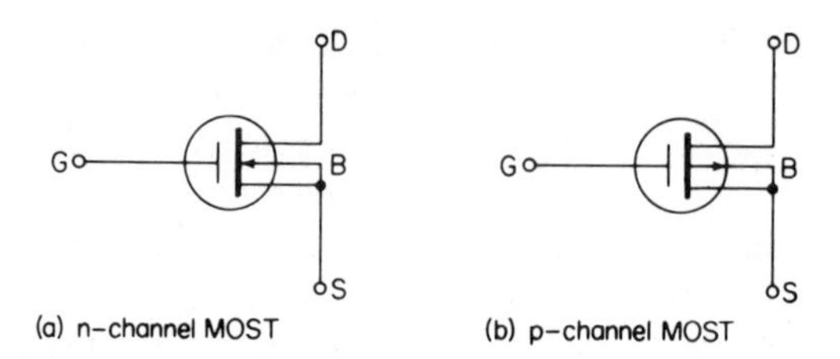

Fig. 4–20 Schematic symbols for MOS transistors. G = gate, S = source, D = drain, B = substrate. The diagram shows that the substrate is connected to the source; this connection is not always present.

4–4.2 Operation and DC Behavior of the MOST

Figure 4–21 shows a MOS transistor on a p-type substrate. Initially, the source is connected to ground (a normal connection) and the drain voltage is set to zero. The metal electrode G, the oxide layer, and the silicon substrate form a capacitor, where the substrate is the lower plate, the metal gate the upper plate, and the oxide layer the dielectric. When the gate voltage $V_G = 0$, the charge on both plates of this capacitor is zero and the substrate is electrically neutral. When $V_G > 0$, a certain positive charge appears on G and a negative charge of equal magnitude appears on the substrate in a layer next to the oxide. When V_G is small this negative charge is carried solely by the acceptor ions in the substrate and a thin depletion layer is formed. The thickness of the depletion layer increases as V_G increases.

When V_G is further increased, a point is reached where not only the acceptor ions, but also electrons supplied by the source and the drain carry the negative charge. This layer of electrons forms a conducting *channel* between S and D. The presence of these extra electrons changes the p-type substrate directly under the oxide layer to n-type and this process is called *inversion.* It should be noted that inversion only takes place when V_G is sufficiently large. The minimum gate voltage required to produce inversion is called the *threshold voltage* V_{th}. Since gate voltage V_G in effect determines the electron concentration in the channel, it also determines the conductivity of the channel.

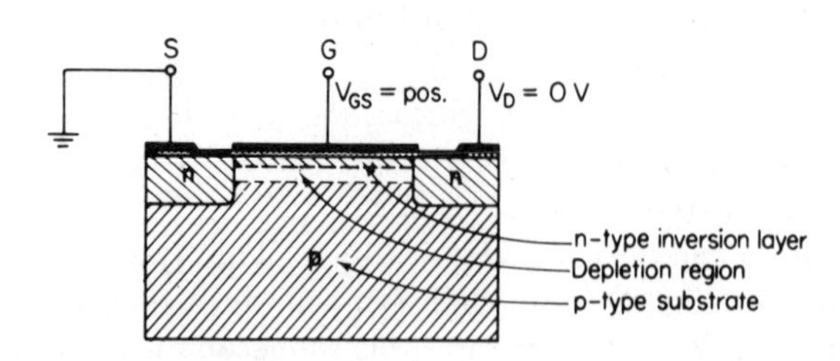

Fig. 4–21 Positive gate voltage V_{GS} changes the p-type substrate material under the oxide layer into an n-type channel by inversion. The minimum gate voltage to produce inversion is called the threshold voltage V_{th}.

In an operating MOST, shown in Fig. 4–22, gate voltage $V_G > V_{th}$ is applied to produce a channel, while in addition drain voltage $V_D > 0$ is applied from drain to source. The application of V_D accomplishes two things: First, since there is a conducting channel between drain and source, V_D maintains drain current I_D in the channel. Second, V_D sets up a voltage gradient V_x along the channel, linearly increasing from $V_x = 0$ at the source end ($x = 0$) to $V_x = V_D$ at the drain end ($x = l$). This voltage gradient causes the depletion region to widen from S to D, as shown in Fig. 4–22. The extent of inversion (channel width), however, gradually decreases from S to D, because the voltage producing inversion now equals $V_G - V_x$.

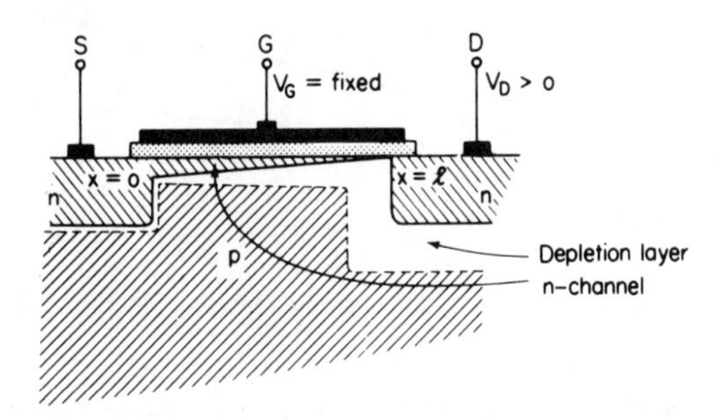

Fig. 4–22 Non-saturated MOS transistor. Immediately under the oxide is an extremely thin, inverted layer of the substrate through which the current flows and whose conductivity gradient *decreases* from *S* to *D*.

When gate voltage $V_G > V_{th}$ is held constant, and drain voltage V_D is gradually increased from zero, drain current I_D also increases. The increase in I_D however, becomes less pronounced as V_D increases. At a certain V_D, the effective gate voltage $V_G - V_x$ at the drain end of the channel has decreased to the value of the threshold voltage V_{th}. At this point inversion no longer exists and the channel is *pinched off*. The pinchoff voltage is usually called the *saturation voltage* $V_{D(sat)} = V_G - V_{th}$. If V_D is further increased, the pinchoff point shifts toward the source end of the channel, but the channel current remains essentially constant. In this case the MOST has entered the saturation or *constant-current* region.

In practice the gate voltage required to produce inversion is not always positive, because the MOST structure may already contain built-in charges. There are two types of MOS transistors: the *depletion* MOST and the *enhancement* MOST. Devices of the depletion type are normally open (conducting channel) and require a gate voltage to be closed. Those of the enhancement type are normally closed (nonconducting channel) and require a gate voltage to be opened. Each of these types may also be n-channel or p-channel, giving a total of four different possible MOS configurations. In general, however, the n-channel MOST is a depletion type and the p-channel MOST is an enhancement type.

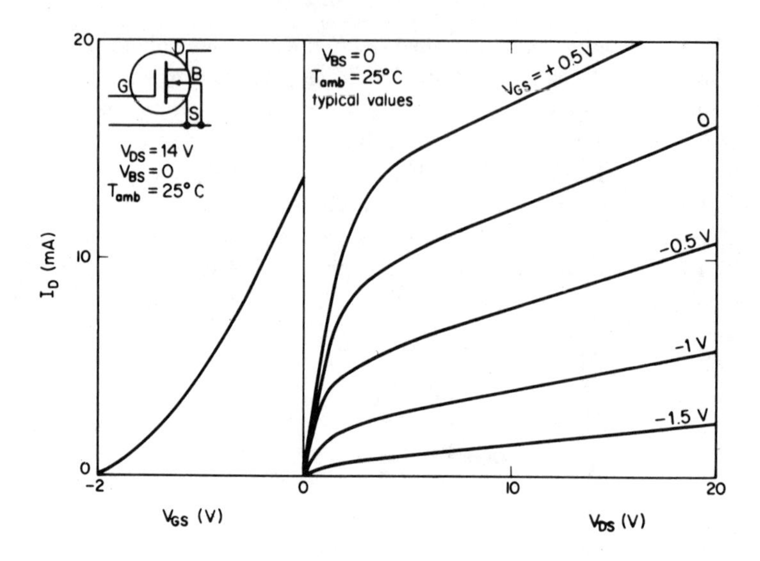

Fig. 4–23 Drain and transfer characteristics for *n*-channel depletion MOS transistor.

Typical drain and transfer characteristics for a depletion *n*-channel MOST are shown in Fig. 4–23. We observe that the drain characteristics of the *n*-channel MOST are very similar to those of the *n*-channel FET. The depletion MOS transistor is approximately interchangeable with the junction FET, and their circuit connections are identical. Since the gate of the MOST is *insulated* from the substrate, however, there is no gate current and the input resistance of the MOST is therefore extremely high, typically on the order of so many megohms.

The extremely small dimensions of the oxide layer under the gate terminal result in very low capacitances and the MOST has therefore a very low input capacitance. This characteristic makes the MOST useful in high-frequency applications.

4–4.3 Enhancement MOST Amplifier

A single-stage amplifier using a *p*-channel enhancement MOS transistor is shown in Fig. 4–24. The outstanding advantage of the enhancement MOST is that both drain voltage V_{DS} and gate voltage V_{GS} are of the same polarity and only one supply voltage is required. Gate bias voltage V_{GS} is derived from the main supply V_{DD} through series bias resistor R_G. To avoid signal currents, and hence undesirable feedback through the bias resistor, R_G is generally made very large, on the order of 10 MΩ or more.

Correct amplifier operation requires that the MOST operates in the saturation or constant-current region of its drain characteristics. This condition

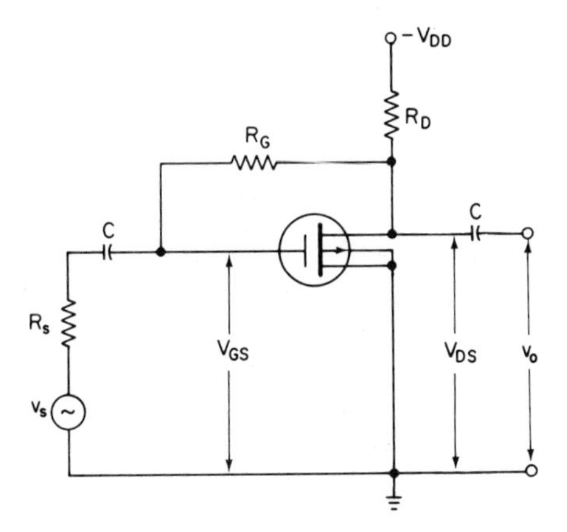

Fig. 4-24 Practical *p*-channel enhancement MOST amplifier.

is satisfied when the gate voltage V_G exceeds the MOST threshold voltage V_{th}, generally on the order of a few volts.

In the circuit of Fig. 4-24, the voltage equation for the drain circuit yields the expression

$$V_{DD} = V_{DS} + I_D R_D$$

which represents the dc loadline with slope $-1/R_D$. It is drawn on the hypothetical drain characteristics of Fig. 4-25. Since the gate voltage equals the drain voltage, $V_{GS} = V_{DS}$ is a second operating condition which must be satisfied. The line marked $V_{DS} = V_{GS}$ on the characteristics of Fig. 4-25 represents this condition. Note that the minimum drain voltage required to operate the MOST equals the threshold voltage and the bias line therefore starts at the point $V_{DS} = V_{th}$ on the V_{DS} axis. The intersection of the bias line and the dc loadline yields the dc operating point Q of the MOST.

Signal source v_s is capacitively coupled to the gate terminal of the MOST and a signal voltage is therefore superimposed on the dc gate voltage. The operating point of the MOST then moves along the loadline in accordance with the magnitude of the signal voltage. Provided that the signal excursion is within the region of the constant current curves of the characteristics, the drain current, and hence the output voltage, is an amplified replica of the input voltage.

The graphical solution of the circuit follows the methods outlined in previous sections.

An analytical solution of the MOST amplifier proceeds in the manner described for FET circuits. The ac equivalent circuit is identical to that of the FET and is given in Fig. 4-14.

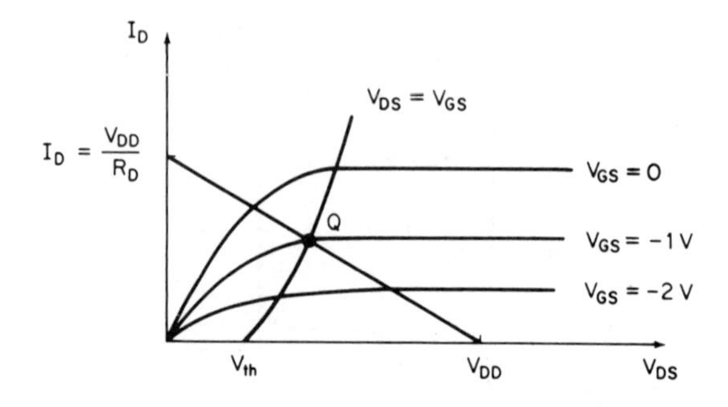

Fig. 4–25 Graphical analysis of the MOST amplifier of Figure 4–24.

QUESTIONS

4–1 Draw the circuit symbols for the following devices:
(a) n-channel FET.
(b) p-channel FET.
(c) n-channel enhancement MOST.
(d) p-channel depletion MOST.
Indicate the battery polarities for normal operation and identify the device terminals.

4–2 Why is the FET known as a unipolar transistor?

4–3 Are there any practical reasons why the drain and gate terminals of a FET cannot be interchanged? Explain.

4–4 What is meant by channel pinchoff? Which mechanisms are responsible for pinchoff? Explain the action of each of these mechanisms.

4–5 Does a p-channel FET require positive or negative gate bias to cause channel pinchoff? Explain.

4–6 Define the zero-bias drain current I_{DSS} in a FET. How is the drain current at any other bias condition related to I_{DSS}?

4–7 Draw the circuit diagram of a single-stage n-channel FET amplifier. Explain the effects of negative gate voltage and positive drain voltage on drain current.

4–8 Draw a commonly used self-bias circuit for a p-channel FET.

4–9 Draw a typical FET transfer characteristic and indicate how the transconductance g_m is related to this characteristic.

4–10 Define the FET parameter g_{mo} and write an expression for the transconductance g_m at any value of drain current.

4-11 Draw a typical set of drain characteristics for an n-channel FET, observing the correct polarities for drain and gate voltages. Indicate the region of constant current operation and explain why the FET must operate in this region when it is used as a linear amplifier.

4-12 What is the function of R_G in Fig. 4-10? What would happen to the voltage gain of the amplifier if C_S were removed from the circuit?

4-13 Explain the operation of the universal bias circuit of Fig. 4-18 and comment on the effects of this circuit with respect to stabilization of the Q point.

4-14 What is the fundamental difference between a depletion MOST and an enhancement MOST?

4-15 Explain the action of the gate voltage as it relates to channel conduction in a MOS transistor.

4-16 What is the significance of the threshold voltage in a MOS transistor?

4-17 What is the function of R_G in Fig. 4-24? Explain how correct gate potential is developed in this circuit.

4-18 Manufacturers of MOS transistors usually deliver their devices with a conductive strip covering the terminals. Why is this done?

4-19 Name some of the precautions one should observe when soldering a MOST into a circuit.

4-20 Explain why it is generally more attractive to use an enhancement MOST rather than a depletion MOST.

PROBLEMS

4-1 Identify and evaluate the FET parameters I_{DSS}, $V_{P(GS)}$, and g_{mo} of the n-channel FET whose characteristics are shown in Fig. 4-5.

4-2 The FET in the amplifier of Fig. 4-26 has a transconductance $g_m = 1$ mA/V. Neglect the internal impedance of the FET and calculate the voltage gain.

4-3 A certain FET has a transconductance $g_{mo} = 2$ mA/V, at a zero-bias drain current $I_{DSS} = 5$ mA, and a pinchoff voltage $V_P = -4$ V. Calculate the voltage gain of this FET in the circuit of Fig. 4-26 if the drain voltage $V_D = 12$ V.

4-4 The single-stage FET amplifier of Fig. 4-7 has the following circuit values: $V_{DD} = 12$ V, $V_{GG} = -0.5$ V, $R_D = 800\ \Omega$, and $R_G = 500\ \text{k}\Omega$. The drain characteristics of the FET are given in Fig. 4-5.

(a) Draw the dc loadline on the graph.

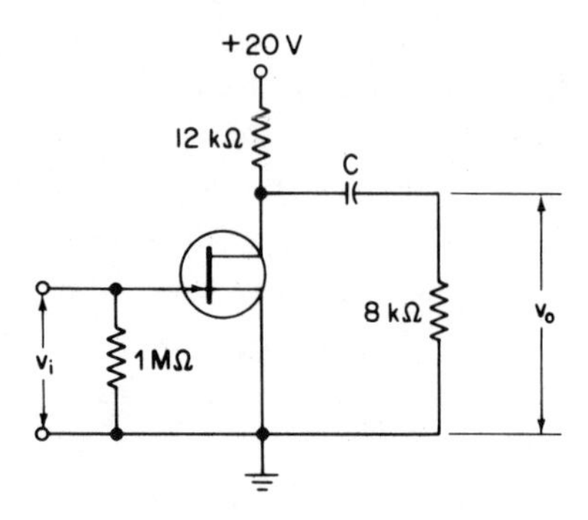

Fig. 4–26 FET amplifier stage for Problem 4–2.

(b) Determine the quiescent drain current I_D and the quiescent drain voltage V_D.

(c) Determine the change in drain current, ΔI_D, if the gate voltage is increased to -1.5 V.

4–5 The dc gate bias of the FET amplifier of Prob. 4–4 is adjusted to $V_{GG} = -1$ V. A signal source with negligible internal resistance delivers a sinusoidal voltage $v_i = 2$ V pp to this amplifier. Determine the voltage gain $A_v = v_o/v_i$ of this amplifier by graphical methods.

4–6 The self-bias amplifier of Fig. 4–10 has the following dc circuit values: $V_{DD} = 12$ V, $R_D = 1500\ \Omega$, $R_S = 500\ \Omega$, $R_G = 1$ MΩ, and $C_S = 40\ \mu$F. The drain characteristics of the FET are given in Fig. 4–27.

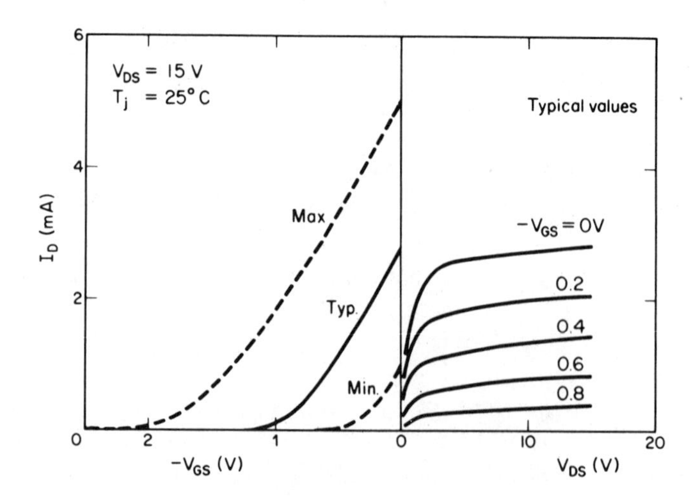

Fig. 4–27 Drain and transfer characteristics for Problem 4–16.

Determine (a) The gate bias voltage V_{GS}.
(b) The quiescent drain voltage V_{DS}.
(c) The quiescent drain current I_D.

4–7 A voltage source of 1.2 V pp with an internal resistance $R_i = 1\ \text{M}\Omega$ is connected to the input terminals of the amplifier of Prob. 4–6. Determine the overall voltage gain $A_v = v_o/v_i$.

4–8 An n-channel FET operates in the circuit of Fig. 4–28. The FET parameters are $I_{DSS} = 10$ mA, $V_P = -2$ V, and $g_{mo} = 10$ mA/V.

Determine (a) The gate-to-source voltage V_{GS}.
(b) The quiescent drain current I_D.
(c) The drain voltage V_D.

4–9 A voltage source with internal resistance $R_i = 1\ \text{M}\Omega$ delivers a peak sinusoidal voltage of 100 mV to the circuit of Fig. 4–28. Draw the ac equivalent circuit and determine the overall voltage gain $A_v = v_o/v_i$. Neglect the reactances of the coupling and bypass capacitors.

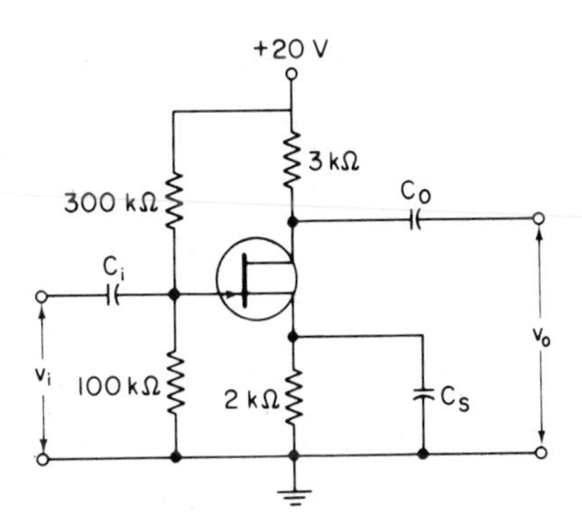

Fig. 4–28 Single-stage FET amplifier with self-bias circuit.

4–10 The FET in the universal bias circuit of Fig. 4–29 has the following parameters: $I_{DSS} = 10$ mA, $V_P = -5$ V, and $g_{mo} = 20$ mA/V.

Determine (a) The drain voltage V_D.
(b) The gate voltage V_G.
(c) The source voltage V_S.
(d) The drain current I_D.

4–11 Draw the ac equivalent circuit of the FET amplifier of Fig. 4–29. Determine the ac voltage developed across a 10-kΩ load resistor when a signal source of 50 mV and internal resistance of 25 kΩ is coupled to the amplifier input.

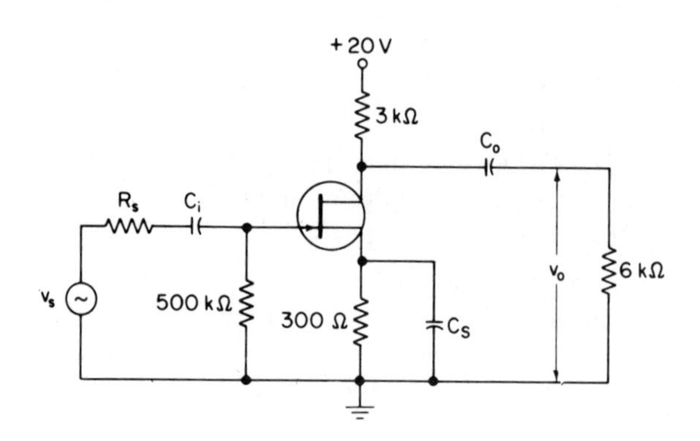

Fig. 4–29 Single-stage FET amplifier with universal bias circuit.

Chapter 5

Multistage Amplifiers

5-1 COUPLING METHODS

High-gain amplifiers are constructed by connecting two or more stages in *cascade* as indicated schematically in Fig. 5-1. The signal voltage at the output terminals of the first stage is coupled to the input of the next stage by an *interstage coupling network*. This network must be so designed that the dc operating conditions of the second stage are not disturbed by the first stage; at the same time the signal components must be transferred from the first stage to the second stage with minimum loss and distortion.

Several methods of interstage coupling are used. Perhaps the simplest and most widely used method of cascading stages is *resistance-capacitance* (*RC*) *coupling*, where the ac voltage developed across the load resistor of the first stage is applied to the input terminal of the next stage by means of a coupling capacitor. The coupling capacitor provides dc isolation between stages and thus maintains the bias conditions. The reactance of the coupling capacitor at the signal frequency must be low enough to transfer the signal component with practically zero loss and no phase distortion.

In *transformer coupling* the ac output voltage of the first stage is developed across the primary of an interstage coupling transformer and the signal is transferred to the next stage by the transformer secondary winding. The transformer provides dc isolation between stages. Transformer coupling is attractive because it can provide *impedance matching* between stages, but it has

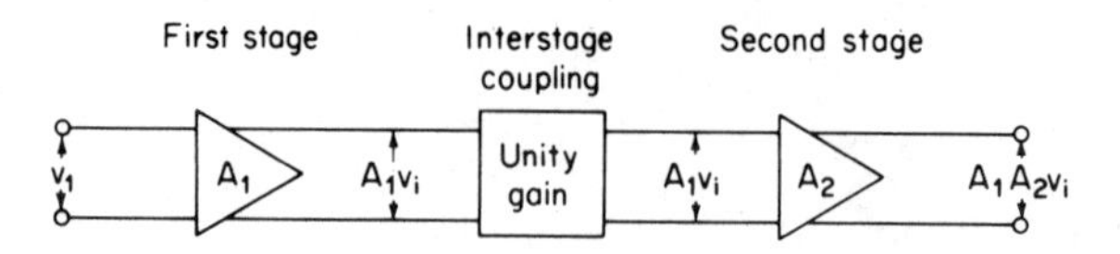

Fig. 5–1 Two amplifier stages connected in cascade to obtain a high-gain amplifier. Signal transfer takes place through an interstage coupling network.

disadvantages in terms of the frequency and phase response of the multistage amplifier. In addition, transformers are relatively large and heavy components and usually rather costly.

In the *direct-coupling* method the output of one stage is connected directly to the input of the next stage *without* using coupling elements. For example, in cascading two *CE* stages, the collector of the first stage is connected directly to the base of the next stage. This method is attractive because it requires fewer components and the frequency response of the amplifier is not affected by interstage coupling elements. It becomes more difficult, however, to establish the *Q*-point conditions required in each successive stage, because the dc output voltage of one stage determines the dc input voltage of the next stage.

High-grain direct-coupled amplifiers are readily available as integrated circuits (ICs), and they can be used in applications nearly as easily as a single-stage amplifier. In this case we are only concerned with the input and output impedances of the IC package, and its total gain characteristics. ICs with FET input (high input impedance and low-noise characteristics), driving conventional BJT amplifier stages in cascade, provide extremely useful and versatile circuits.

5–2 RC COUPLING

5–2.1 Two-Stage RC-Coupled Amplifier

Figure 5–2(a) shows a typical two-stage *RC*-coupled amplifier. Each stage consists of a BJT in the *CE* connection operating with a universal bias circuit. To increase the voltage gain of each stage, the emitter resistors of both transistors are shunted to ground by emitter bypass capacitors C_E. Coupling capacitor C_c connects the collector of *Q*1 to the base of *Q*2. As far as dc voltages are concerned, C_c behaves like an open circuit, so the dc collector voltage of *Q*1 does not disturb the bias voltage at the base of *Q*2. The quiescent operating conditions of each stage can therefore be evaluated independently, following the methods of Chapter 3.

Input signal v_i, capacitively coupled to the base of *Q*1, appears in amplified form as A_1v_i at the collector of *Q*1 and hence at the base of *Q*2. The

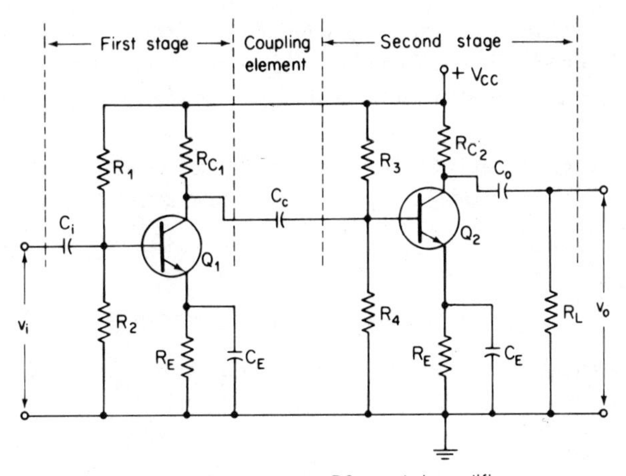

(a) Circuit diagram of a two-stage RC-coupled amplifier. Coupling capacitor C_C provides dc isolation between the stages.

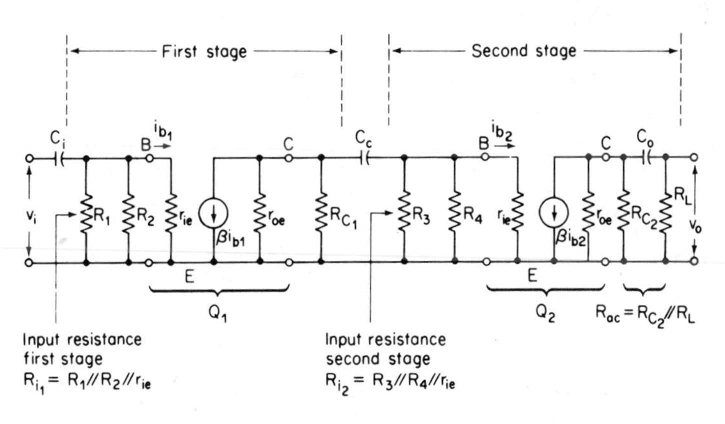

(b) ac equivalent circuit of the two-stage amplifier.

Fig. 5-2 Two-stage *RC*-coupled amplifier and its equivalent circuit.

second stage also provides voltage amplification, and the ac output voltage at the collector of $Q2$ equals

$$v_o = A_1 A_2 v_i \tag{5-1}$$

so the overall voltage gain of the circuit is

$$A_v = \frac{v_o}{v_i} = A_1 A_2 \tag{5-2}$$

Equation (5–2) states that the voltage gain of a multistage RC-coupled amplifier is equal to the product of the individual stage gains A_1 and A_2.

The ac conditions in the amplifier are evaluated with reference to the equivalent circuit of Fig. 5–2(b). Starting with the second stage, the input and output resistances are calculated. The ac resistance in the collector circuit of $Q2$ consists of collector resistor R_{C_2} in parallel with load resistor R_L and BJT output resistance r_{oe}, so

$$R_{ac} = R_{C_2} // R_L // r_{oe} \tag{5–3}$$

In practical circuits, r_{oe} is very much larger than either R_{C_2} or R_L, and can be neglected, so

$$R_{ac} \cong R_{C_2} // R_L \tag{5–4}$$

The input resistance R_{i_2} of the second stage consists of the parallel combination of bias resistors R_3 and R_4, and BJT input resistance r_{ie}, so

$$R_{i_2} = R_3 // R_4 // r_{ie} \tag{5–5}$$

The input resistance of a transistor in the CE connection is given by Eq. (3–62) as

$$r_{ie} = \beta r_\pi \cong \beta \frac{50}{I_E} \tag{5–6}$$

where I_E is the quiescent emitter current of $Q2$.

With both input and output resistance of the second stage known, the voltage gain can be calculated by using Eq. (3–36), which states that

$$A_v = A_i \frac{R_{out}}{R_{in}} \cong \beta \frac{R_{ac}}{R_{i_2}} \tag{5–7}$$

The voltage gain of the *first stage* is calculated in a similar manner. The ac load resistance in the collector circuit of $Q1$ consists of the parallel combination of BJT output resistance r_{oe}, collector resistor R_{C_1}, and the *input resistance R_{i_2} of the second stage.* It is quite important to realize that the second stage *loads* the output of the first stage in the sense that the input resistance of the second stage forms part of the total ac load for the first stage. Hence, we can write

$$R_{ac} = r_{oe} // R_{C_1} // R_{i_2} \tag{5–8}$$

In a practical circuit, r_{oe} is very much larger than R_{C_1} or R_{i_2}, and may therefore be neglected in the calculation. The load resistance for the first stage then is

$$R_{ac} \cong R_{C_1} /\!/ R_{i2} \tag{5-9}$$

The input resistance of the first stage consists of the parallel combination of bias resistors R_1 and R_2, and the input resistance r_{ie} *of* $Q1$, so

$$R_{i_1} = R_1 /\!/ R_2 /\!/ r_{ie} \tag{5-10}$$

where r_{ie} must be calculated from the quiescent emitter current of transistor $Q1$.

With the input and out resistances known, the voltage gain of the first stage can also be calculated. The overall gain of the amplifier is simply the product of the individual voltage gains of the two stages.

For the *RC*-coupled amplifier of Fig. 5–3, it is necessary to calculate the input resistances of $Q1$ and $Q2$, and we must therefore calculate their emitter currents. The two universal bias circuits are identical, so the calculation of the dc emitter current of $Q1$ applies to both transistors. The dc base voltage V_B is determined by the 30–60 kΩ voltage divider and $V_B = (30/90) \times 15\text{ V} = 5\text{ V}$. Neglecting the small voltage drop across the base–emitter junction, the emitter voltage V_E is also 5 V, making the emitter current I_E 5 V/5 kΩ = 1 mA. Both transistors operate under identical dc conditions so their input resistances are the same and equal to $r_{ie} \cong \beta 50/I_E = 5\text{ k}\Omega$.

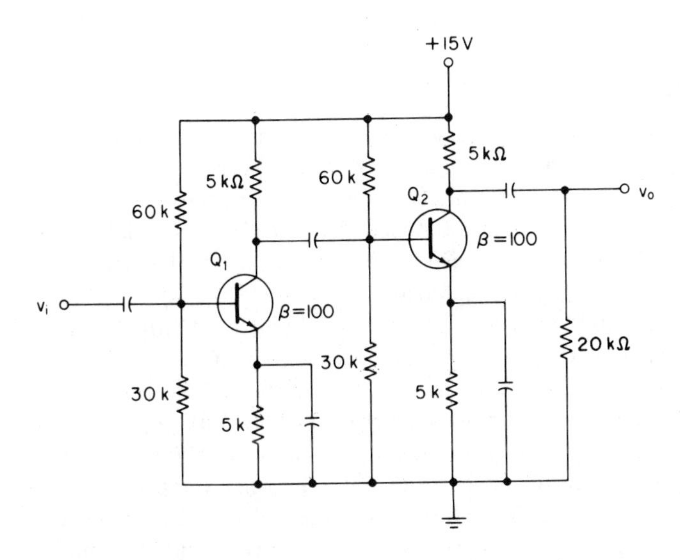

Fig. 5–3 *RC*-coupled amplifier.

The input resistance of the second stage is $R_{i_2} = 60\ \text{k}\Omega // 30\ \text{k}\Omega // 5\ \text{k}\Omega = 4\ \text{k}\Omega$. The total ac load resistance of the second stage is $R_{ac} \simeq 5\ \text{k}\Omega // 20\ \text{k}\Omega = 4\ \text{k}\Omega$, and the voltage gain of the second stage is

$$A_2 = \beta \frac{R_{ac}}{R_{i_2}} = 100 \times \frac{4\ \text{k}\Omega}{4\ \text{k}\Omega} = 100$$

The input circuit of the first stage is identical to that of the second stage, so $R_{i_1} = 4\ \text{k}\Omega$. The total load resistance of the first stage consists of the parallel combination of the 5-kΩ load resistor and the input resistance of the second stage, so $R_{ac} = 5\ \text{k}\Omega // 4\ \text{k}\Omega = 2.22\ \text{k}\Omega$. The voltage gain of the first stage is then

$$A_1 = \beta \frac{R_{ac}}{R_{i_1}} = 100 \times \frac{2.22\ \text{k}\Omega}{4\ \text{k}\Omega} = 55.5$$

The overall gain of the two-stage amplifier is equal to the product of the individual stage gains and

$$A_v = A_1 A_2 = 55.5 \times 100 = 5550$$

It should be noted that several approximations were made and that the result just obtained is approximate rather than exact. However, in view of the usual tolerances in component values and the spread in β values, these approximations are perfectly justifiable.

5-2.2 Low-Frequency Response

The amplitude and phase response of the *RC*-coupled amplifier may be determined by considering three frequency ranges: low, intermediate or midrange, and high. A typical amplitude or *gain response* curve is shown in Fig. 5-4. The amplifier gain as a function of frequency is essentially constant at the midrange frequencies but drops off rapidly at the low-frequency and the high-frequency ends of the spectrum.

There are two *corner frequencies*, where the amplifier gain is reduced by 3 dB relative to the midband gain. The 3-dB reduction in gain at the lower corner frequency (f_1) is mainly caused by signal loss in the series coupling capacitors and the emitter bypass capacitors. The 3-dB reduction in gain at the upper corner frequency (f_2) is mainly caused by internal transistor capacitances and stray wiring capacitances which shunt the signal components to ground. The *passband* of the amplifier refers to the range of frequencies between f_1 and f_2, where series and shunt capacitances have little effect and the gain is essentially constant. Amplifiers are generally designed to operate in the passband region.

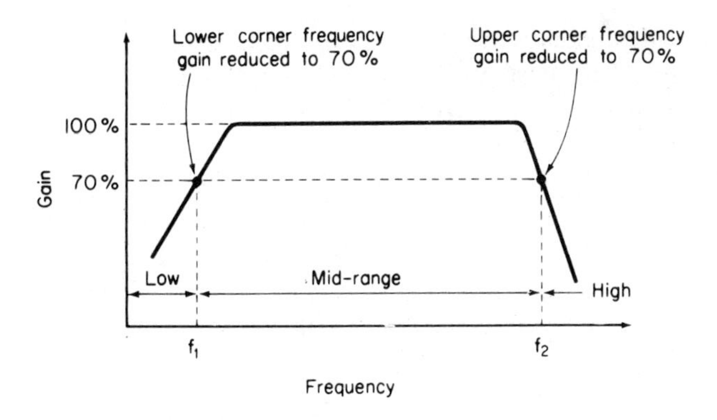

Fig. 5–4 Typical gain response curve of an *RC*-coupled amplifier. At the corner frequencies f_1 and f_2 the amplifier gain is reduced to approximately 70% of the mid-band gain.

The two-stage *RC*-coupled amplifier of Fig. 5–2 contains several series coupling elements. Input capacitor C_i connects the signal source to the base of $Q1$, interstage coupling capacitor C_c connects the collector of $Q1$ to the base of $Q2$, and output capacitor C_o connects the collector of $Q2$ to the load. In a well-designed amplifier operating in the passband region, the reactances of the coupling capacitors are negligibly small. However, when the frequency decreases, the capacitive reactances increase and the signal current is attenuated by the various coupling capacitors. In addition to signal *attenuation,* the coupling capacitors also introduce *phase shift,* which increases as the signal frequency decreases.

The response of the *RC*-coupled amplifier can best be analyzed by reference to Fig. 5–5, where a signal source is connected to a load via coupling capacitor C_c. The signal source is represented by current generator i_s with internal resistance R_s, while the load is represented by resistor R_L. At the midband frequencies, the reactance of C_c is negligibly small and the load is connected directly to the source. The load current then equals

$$i_L = i_s \frac{R_s}{R_s + R_L} \tag{5–11}$$

The midband current gain of the system is the ratio of output current to input current and equals

$$A_{i(\text{midband})} = \frac{i_L}{i_s} = \frac{R_s}{R_s + R_L} \tag{5–12}$$

Since there are no reactive elements in Eq. (5–12), the phase shift of the network is zero and the attenuation minimum (maximum gain).

When the signal frequency decreases, the reactance of C_c increases. This places a reactive element between source and load, which affects both attenuation and phase shift of the output current. Taking the reactance $X_C = j/\omega C$ of C_c into account, the load current now equals

$$i_L = i_s \frac{R_s}{R_s + R_L - j/\omega C_c} \tag{5-13}$$

where $\omega = 2\pi f$. The low-frequency current gain is

$$A_{i(\text{low frequency})} = \frac{i_L}{i_s} = \frac{R_s}{R_s + R_L - j/\omega C_c} \tag{5-14}$$

The low-frequency gain relative to the midband gain is

$$\frac{A_{i(\text{low frequency})}}{A_{i(\text{midband})}} = \frac{R_s + R_L}{R_s + R_L - j/\omega C_c} \tag{5-15}$$

When the reactance X_C of the coupling capacitor equals the total circuit resistance $(R_s + R_L)$, the low-frequency gain is reduced to 0.707 (−3 dB) of the midband gain, and the phase shift between output current and input current is +45 degrees (i_L leads i_s by 45 degrees). The frequency at which this occurs is called the *lower corner frequency* or *lower 3-dB frequency* f_1. Since f_1 is defined by the condition that $X_c = R_s + R_L$, the lower corner frequency can be expressed in terms of the circuit components as

$$f_1 = \frac{1}{2\pi C_c(R_s + R_L)} \tag{5-16}$$

Equation (5–16) indicates that large values of coupling capacitance reduce the lower corner frequency. In practical circuits, the dc supply voltages are low, so low-voltage capacitors can be used as the coupling elements. Capacitors of small physical dimension but large capacitance are commercially available, producing an *RC* coupling network of excellent low-frequency response.

EXAMPLE 5-1

The two transistors in the *RC*-coupled amplifier of Fig. 5–2 each have a current gain $\beta = 50$. The remaining circuit values are: $R_C = 5\ \text{k}\Omega$, $R_E = 2\ \text{k}\Omega$, $R_1 = R_3 = 500\ \text{k}\Omega$, $R_2 = R_4 = 100\ \text{k}\Omega$, and $V_{CC} = 12$ V. It is further assumed that the two emitter capacitors C_E effectively bypass emitter resistors R_E. It is desired to find the lower 3-dB frequency of the interstage coupling network if $C_c = 3.5\ \mu F$.

SOLUTION

This problem is solved by reference to Fig. 5–5, where R_s represents the output resistance R_o of the first stage, and R_L represents the input resistance R_i of the second stage. The output resistance of the first stage, looking back into the collector terminal of $Q1$, equals $R_o = R_c // r_{oe}$, where r_{oe} is the output resistance of $Q1$. Usually, r_{oe} is very large compared to the 5-kΩ collector resistance of $Q1$, so that it may be neglected. Hence, $R_o \simeq R_c = 5\ k\Omega$. The input resistance of the second stage equals $R_i = R_3 // R_4 // r_{ie}$, where r_{ie} is the input resistance of $Q2$. Since r_{ie} is determined by the dc emitter current of $Q2$, we must first estimate this current. The emitter voltage of $Q2$ equals $V_E \simeq [R_4/(R_3 + R_4)]V_{CC} = (100/600) \times 12\ V = 2\ V$. Hence, $I_E = V_e/R_E = 2\ V/2\ k\Omega = 1\ mA$ and $r_{ie} = \beta r_\pi \simeq 2.5\ k\Omega$. Therefore, the input resistance of the second stage equals $R_i = 500\ k // 100\ k // 2.5 \simeq 2.5\ k\Omega$.

According to Fig. 5–5, the lower 3-dB frequency of the $R_s - C_c - R_L$ network is determined by Eq. (5–16) and equals

$$f_1 = \frac{1}{2\pi C_c(R_s + R_L)}, \text{ where } R_s = R_o = 5\ k\Omega \text{ and } R_L = R_i = 2.5\ k\Omega.$$

Substituting the known values and solving for f_1, we find the lower 3-dB frequency as $f_1 \simeq 6$ Hz. Hence, at a signal frequency of approximately 6 Hz the attenuation of the interstage coupling network is 3 dB and the phaseshift between the input and output signals of the coupling network is 45 degrees.

The lower 3-dB frequencies of the input and output coupling capacitors can be calculated in a similar manner. We may, for example, find that input coupling capacitor C_i causes a lower corner frequency of 20 Hz, and that output

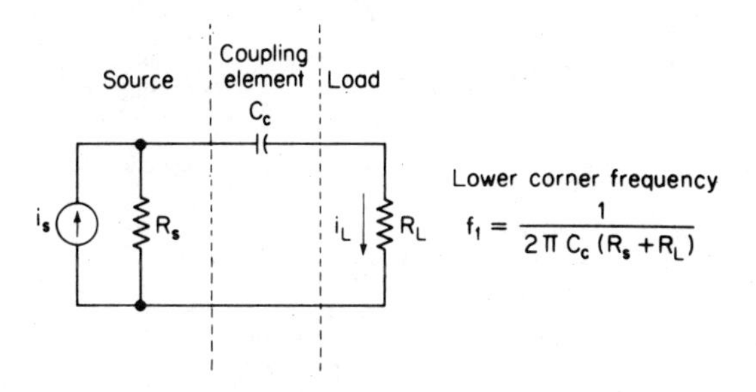

Fig. 5–5 Analysis of an *RC* coupling network.

capacitor C_o causes a lower corner frequency of 40 Hz. Each network attenuates the amplifier gain by 3 dB at the indicated corner frequency. The overall gain response curve of the amplifier at the low-frequency end then has the shape indicated in Fig. 5–6. At a frequency of 40 Hz (the corner frequency of C_o), the amplifier gain is attenuated by 3 dB and the gain response decreases at the rate of 6 dB per octave. At a frequency of 20 Hz (the corner frequency of C_i), the amplifier gain is reduced again by 3 dB and the gain response decreases an additional 6 dB per octave to attain a slope of 12 dB per octave. Finally, at 6 Hz (the corner frequency of C_c), the gain is attenuated once again by 3 dB, and the slope of the gain response curve becomes 18 dB per octave. The response curve is essentially flat for frequencies above the highest corner frequency (40 Hz in this case).

5–2.3 *High-Frequency Response*

The amplifier response at high frequencies is limited by two major factors. The first is caused by high-frequency limitations on the current gain of the transistor itself. The second is caused by capacitive effects inherent in the transistor and by stray wiring capacities in the circuit. Let us investigate these two factors separately.

Because of the inherent mechanism of charge transfer in the base region, the current gain of the transistor decreases as the frequency of the signal current increases. For a transistor in the common-base configuration, the short-circuit current gain α_o (i_c/i_e) remains fairly constant at low frequencies. When the frequency is increased, a point is reached where the *CB* current gain is reduced to 70 percent (-3 dB) of its original value. The frequency at which this occurs is

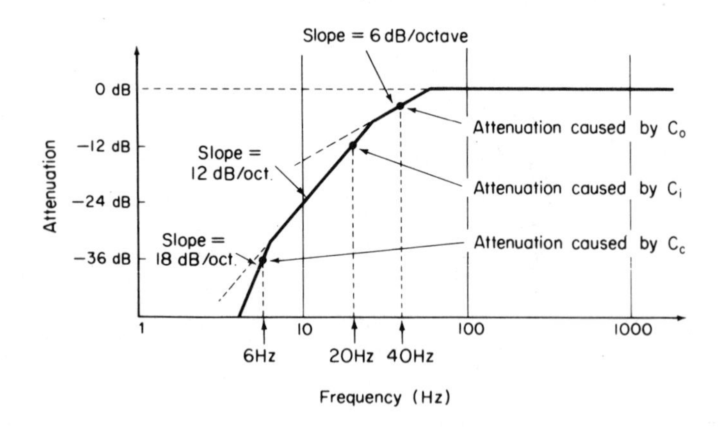

Fig. 5–6 Gain response of the *RC*-coupled amplifier of Figure 5–1.

called the *alpha cutoff frequency*, $f_{\alpha co}$. At this upper 3-dB point, the transistor output current i_c lags the input current i_e by approximately 45 degrees.

Similarly, for a transistor in the common-emitter configuration, the ac short-circuit current gain $\beta_o(i_c/i_b)$ remains approximately constant until the *beta cutoff frequency*, $f_{\beta co}$, is reached, where the current gain β has dropped 3 dB below its low-frequency value β_o. The relation between $f_{\alpha co}$ and $f_{\beta co}$ is given by

$$f_{\beta co} \cong \frac{f_{\alpha co}}{\beta} \tag{5-17}$$

Most manufacturers describe the high-frequency characteristics of their transistors in terms of a figure of merit called the *gain-bandwidth product* or *transition frequency*, f_T, specified at certain basic operating conditions. The transition frequency is approximately equal to the alpha cutoff frequency, so

$$f_T \cong f_{\alpha co} \tag{5-18}$$

These various cutoff frequencies are important in the high-frequency analysis of transistor circuits. Alpha cutoff frequency $f_{\alpha co}$ defines the high-frequency limitations of the transistor in the *CB* configuration; beta cutoff frequency $f_{\beta co}$ defines the high-frequency limitations of the transistor in the *CE* configuration. In many practical applications, the transistor itself is the limiting factor in terms of high-frequency circuit performance.

EXAMPLE 5-2

Transistor *A* has a transition frequency $f_T = 120$ MHz and a low-frequency current gain $\beta = 80$. Transistor *B* has an $f_T = 20$ MHz and $\beta = 20$. Calculate the beta cutoff frequency for each transistor.

SOLUTION

For transistor *A*: $f_{\beta co} = f_T/\beta = 120$ MHz/80 = 1.5 MHz. For transistor *B*: $f_{\beta co} = f_T/\beta = 20$ MHz/20 = 1 MHz. Transistor *A* has a better high-frequency performance in a *CE* circuit than transistor *B*.

The high-frequency response of an amplifier stage is also limited by the internal transistor capacitance and by stray wiring capacitances. Both of these effects are represented by the shunt capacitor C_L in Fig. 5-7, where $C_L = C_C + C_{stray}$. C_C is the transistor collector capacitance and C_{stray} represents the total stray wiring capacitance to ground. The high cutoff frequency, where the circuit gain

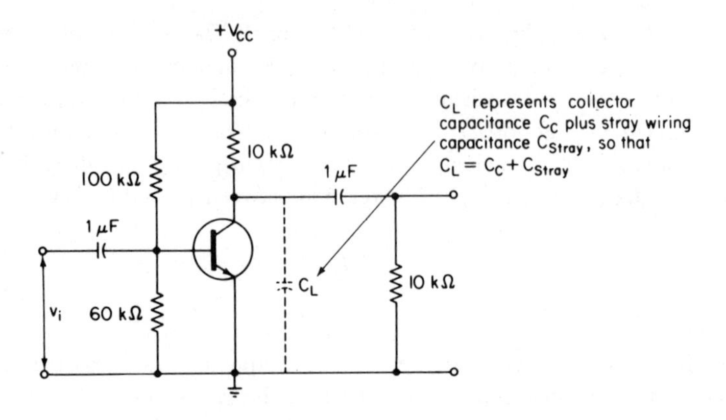

Fig. 5–7 Stray wiring capacitance and transistor collector capacitance adversely affect the high-frequency response of the CE amplifier stage.

is reduced by 3 dB, occurs when the capacitive reactance equals the total circuit resistance. This yields the upper 3-dB frequency f_2, where

$$f_2 = \frac{1}{2\pi C_L R_{ac}} \tag{5–19}$$

and R_{ac} represents the ac load resistance.

There are other high-frequency effects, but the two just mentioned are by far the most important ones. The alpha and beta cutoff frequencies are transistor parameters and they are generally quite different from the cutoff frequency determined by the shunt circuit capacitances. The lower of the two frequencies is the more critical one, since it determines the upper limit of high-frequency operation.

EXAMPLE 5-3

The transistor in Fig. 5–7 has the following parameters:

$$\text{collector capacitance, } C_C = 20 \text{ pF}$$
$$\text{transition frequency, } f_T = 100 \text{ MHz}$$
$$\text{current gain, } \beta = 40$$

The stray wiring capacitance is assumed to be 40 pF. Find the approximate value of the upper 3-dB frequency.

SOLUTION

The first limiting factor is given by the transition frequency which determines the beta cutoff frequency and we find that

$$f_{\beta co} = \frac{f_T}{\beta} = \frac{100 \text{ MHz}}{40} = 2.5 \text{ MHz}$$

The second limiting factor is given by the circuit shunt capacitances, where $C_L = C_C + C_{\text{stray}} = 60$ pF. The upper corner frequency then is

$$f_2 = \frac{1}{2\pi C_L R_{\text{ac}}} = \frac{1}{2\pi(60 \times 10^{-12})(10 \text{ k}\Omega//10 \text{ k}\Omega)} = 530 \text{ kHz}$$

The upper 3-dB frequency is determined by the lower of the two frequencies and equals 530 kHz. The transistor itself is therefore not the limiting factor.

5-3 TRANSFORMER COUPLING

5-3.1 Impedance Matching

Transformer coupling is frequently used to provide *impedance matching* between the various stages of an amplifier, thereby providing optimum gain and improved signal-handling capability. The transformer also takes care of dc isolation between the coupled stages; low winding resistances often result in good circuit stability.

Despite these advantages, transformer coupling can have serious disadvantages. In solid-state audio circuits, for example, where miniature transformers are often used, the frequency response of the circuit can be adversely affected, with low-frequency rolloff occurring at frequencies as high as 1 kHz. Such miniature transformers are also liable to introduce low-frequency waveform distortion.

Although correct impedance matching improves the overall voltage gain, an extra RC-coupled stage is often less costly than a transformer, and probably more attractive in terms of low-frequency response. However, in cases where impedances would be severely mismatched, transformer coupling provides a satisfactory solution.

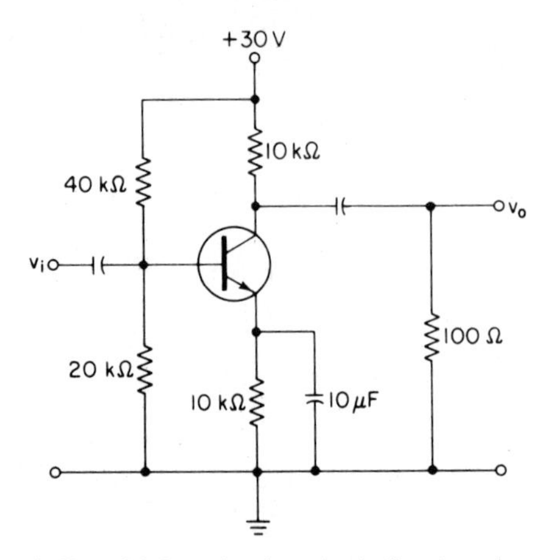

(a) The 100 Ω load resistor loads the stage down.

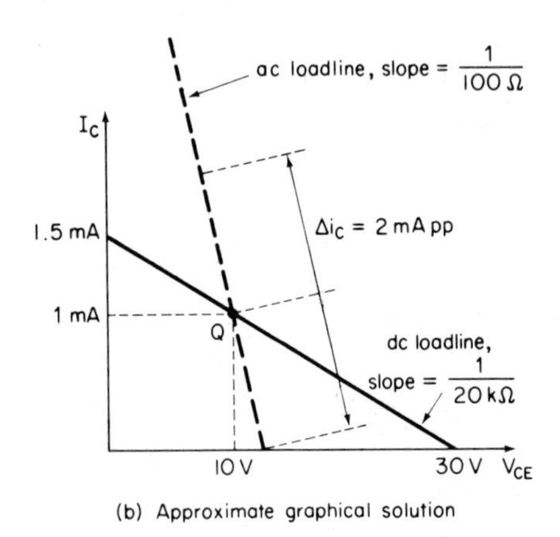

(b) Approximate graphical solution

Fig. 5–8 Loading effect is caused by connecting a low-resistance load to a high-resistance collector circuit.

To illustrate the effect of mismatching, consier the *CE* amplifier stage of Fig. 5–8(a), where a 100-Ω load is capacitively coupled to the 10-kΩ collector circuit of the transistor. For dc,

$$\text{base voltage, } V_B = \frac{20\text{ k}\Omega}{60\text{ k}\Omega} \times 30\text{ V} = 10\text{ V}$$

$$\text{emitter current, } I_E = \frac{V_E}{R_E} \cong \frac{V_B}{R_E} = \frac{10\text{ V}}{10\text{ k}\Omega} = 1\text{ mA}$$

$$\text{collector voltage, } V_C = V_{CC} - I_C R_C = 30\text{ V} - (1\text{ mA} \times 10\text{ k}\Omega) = 20\text{ V}$$

$$\text{collector-to-emitter voltage, } V_{CE} = V_C - V_E = 10\text{ V}$$

For ac,

$$\text{load resistance, } R_{ac} = R_C // R_L = 10\text{ k}\Omega\ //\ 100\Omega \simeq 100\ \Omega$$

$$\text{ac voltage gain, } A_v \simeq \beta \frac{r_{ac}}{r_{ie}} = \frac{R_{ac}}{50/I_E} = \frac{100\ \Omega}{50\ \Omega} = 2$$

The voltage gain of the amplifier is very low, because the output circuit of the BJT is *loaded* down by the 100-Ω load resistor. The maximum signal voltage which can be developed across the load resistor is limited by the maximum excursion of the collector voltage around the Q point. This is illustrated in Fig. 5–8(b), where a 100-Ω loadline is sketched through the Q point (1 mA, 10 V). The maximum peak-to-peak ac collector voltage equals

$$v_{o(\max)} = 2I_C R_{ac} = 2 \times 1\text{ mA} \times 100\ \Omega = 200\text{ mV pp}$$

Obviously, the 100-Ω load imposes severe restrictions on both voltage gain and signal-handling capability of the circuit. The problem here is that the load ($R_L = 100\ \Omega$) is not matched to the source (output resistance of the BJT stage is approximately 10 kΩ). The situation can be improved by transformer-coupling the 100-Ω load to the collector, as in Fig. 5–9.

The dc conditions in the circuit are almost identical to those of the RC-coupled circuit of Fig. 5–8, with $V_B = 10$ V and $I_E = 1$ mA. The dc collector load consists of the winding resistance of the transformer primary, which is negligibly small. Hence, the collector voltage is approximately equal to the supply voltage and $V_C \simeq V_{CC} = 30$ V. The collector-to-emitter voltage $V_{CE} = V_C - V_E = 20$ V. The ac conditions in the circuit have changed considerably. The ac load resistance R'_L is reflected from the secondary winding into the primary winding and equals

$$R'_L = a^2 R_L$$

where $a = N_1/N_2$ is the turns ratio of the transformer. For the transformer of Fig. 5–9, where $a = 10$, the reflected load resistance equals $R'_L = 10^2\ (100\ \Omega) = 10$ kΩ. Hence, as far as ac conditions are concerned, the BJT sees a 10-kΩ load in its collector circuit. The ac voltage gain, observed at the collector terminal, equals

$$A_v = \beta \frac{R_{ac}}{r_{ie}} = \frac{R_{ac}}{50/I_E} = \frac{10\text{k}\ \Omega}{50\ \Omega} = 200$$

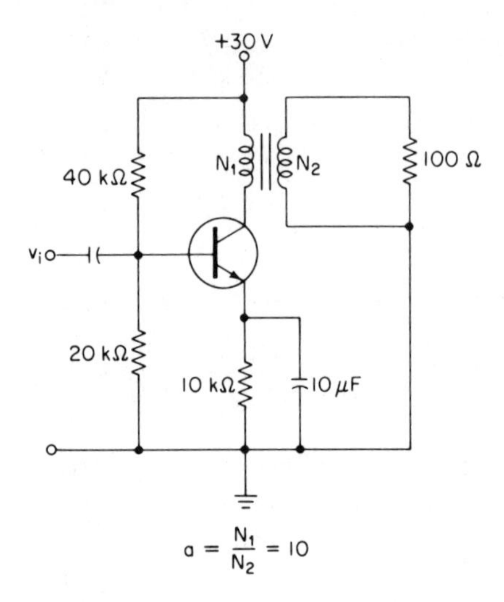

Fig. 5–9 The output transformer provides proper impedance matching between load and source.

The voltage across the secondary of the transformer (and hence across the 100-Ω load) is stepped down by a factor of 10 and the overall voltage gain of the circuit equals 20. The maximum signal voltage which can be developed at the collector terminal is again limited by the maximum permissible collector voltage swing and

$$v_{o(\max)} = 2I_C R'_L = 2 \times 1 \text{ mA} \times 10 \text{ k}\Omega = 20 \text{ V pp}$$

The transformer steps this voltage down by a factor of 10 to 2 V pp.

Comparing the performance of the two circuits, we observe that transformer coupling of the low-impedance load to the high-impedance source improves the voltage gain by a factor of 10 and the signal-handling capability by a factor of 100.

Transformers are often used to couple a power amplifier to a low-impedance load (impedance matching) or to couple a driver stage to a power amplifier (signal phasing). These applications are discussed in more detail in Chapter 6.

5–3.2 Frequency Response

A general representation of a transformer-coupled circuit is shown in Fig. 5–10(a), where voltage source v_s, with its internal resistance R_s, is coupled to load resistor R_L through a transformer whose turns ratio is $a = N_1/N_2$. The gain and phase response of a transformer-coupled amplifier can be derived from the equivalent circuit of the transformer. Figure 5–10(b) shows one of the common equivalent circuits, where the transformer is replaced by its mutual or magnetizing inductance L_m, its primary and secondary leakage inductances L_1 and L_2, and its primary and secondary winding resistances R_1 and R_2. To allow for voltage, current, and impedance transformation, the external load resistance R_L, connected to the secondary, is reflected back into the primary as $R'_L = a^2R_L$. The transformer secondary voltage v_2 is reflected back into the primary side as av_2.

In some transformers the coefficient of coupling, k, approaches unity, so that the mutual inductance is maximum ($L_m = kL$) and the leakage reactances are minimum [$L_1 = L_2 = L(1 - k)$]. The winding resistances are generally small compared to the circuit resistance and can often be neglected. This further simplifies the situation.

To determine the frequency response of the coupling network, we again consider three frequency ranges: low, intermediate or midrange, and high. At the midrange frequencies, the series leakage reactances are assumed to be sufficiently small compared to a^2R_L to be neglected. On the other hand, the reactance of the parallel magnetizing inductance is sufficiently large compared to a a^2R_L that it also can be neglected. This reduces the equivalent circuit of Fig. 5–10(b) to a simple resistive network for the midrange frequencies, as shown in Fig 5–10(c). The ac output voltage in terms of the source voltage is

$$av_2 = v_s \frac{a^2R_L}{R_s + a^2R_L}$$

The voltage gain of the circuit is given by the ratio of secondary output voltage to source voltage, so

$$A_{v(\text{mid})} = \frac{v_2}{v_s} = \frac{aR_L}{R_s + a^2R_L} \tag{5–20}$$

At the very low frequencies, the series leakage reactances are still negligible. The reactance of the magnetizing inductance is also smaller and parallels a^2R_L, as shown in Fig. 5–10(d). This has an effect on the output voltage and causes the low-frequency gain response to roll off. The frequency at which the gain is reduced to 70 percent of the midrange gain is called the *lower 3-dB frequency* or *lower corner frequency*. The lower 3-dB frequency is that frequency at which the magnetizing reactance equals the parallel resistance of R_s and a^2R_L (Thévenin resistance looking into the reactance terminals), and

$$X_L = R_s \,//\, a^2R_L \tag{5–21}$$

This yields an expression for the lower 3-dB frequency f_1, where

$$f_1 = \frac{R_s a^2 R_L}{2\pi L(R_s + a^2 R_L)} \tag{5–22}$$

The phase shift at the lower 3-dB point is again 45 degrees and the slope of the response curve at frequencies below f_1 is approximately −6 dB per octave.

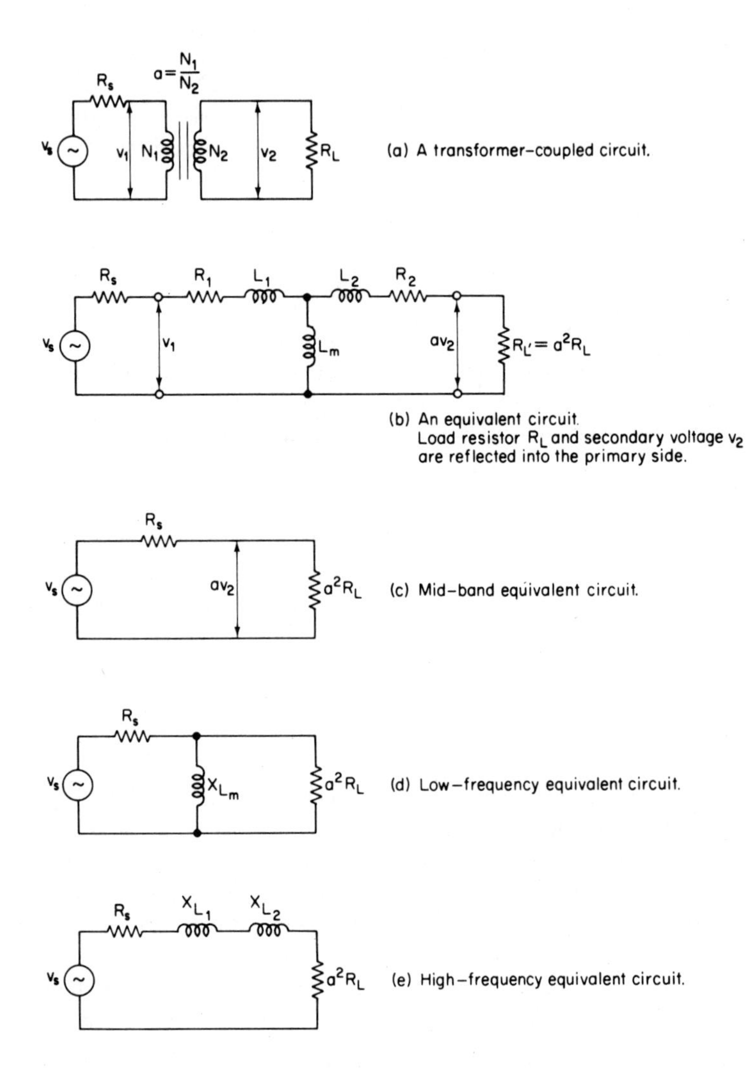

Fig. 5–10 Equivalent circuits for a transformer-coupled network.

At the high-frequency end of the spectrum, the reactance of the parallel magnetizing inductance is sufficiently large not to affect the reflected load resistance, and it can be neglected [see Fig. 5–10(e)]. The primary and secondary leakage reactances, however, also increase as the frequency increases and they eventually become sufficently large to cause a drop in output voltage, and a definite phase shift. At this point the gain response rolls off. The *upper 3-dB frequency* or *upper corner frequency* is reached when the total series reactance equals the total series resistance. The output voltage is then reduced to 70 percent of the midrange value (or −3 dB) and the phase shifts is 45 degrees.

Good high-frequency response can only be obtained when the transformer has very low leakage reactances. In practice, small capacitive effects in the transformer windings may cause the response curve to have resonant "bumps," often resulting in a slight rise of the response curve toward the higher frequencies. These capacitive effects may also cause the response curve to roll off more steeply than the theoretical slope of −6 dB per octave.

EXAMPLE 5-4

In Fig. 5–11 the 8-Ω load is transformer coupled to the collector circuit of a *CE* output stage and is to be reflected as a 10-kΩ collector load. The transformer is assumed to be ideal (zero winding resistance and zero leakage reactance) and has a magnetizing inductance of 16 H. The output resistance r_{oe} of the BJT is assumed to be 30 kΩ.

Determine (a) The turns ratio of the transformer.
(b) The lower 3-dB frequency.

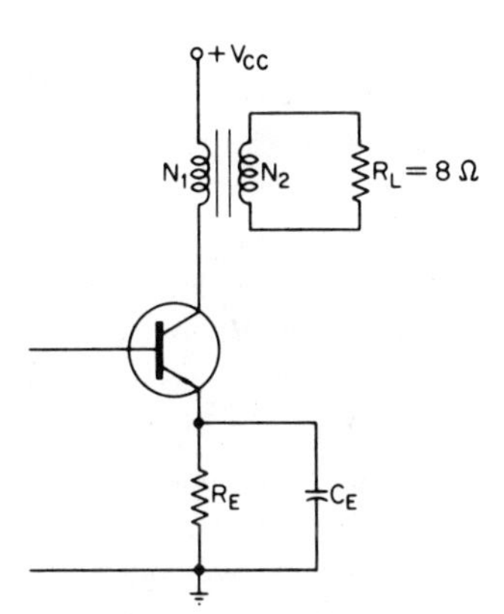

Fig. 5–11 Example 5–4. The output transformer couples the low-impedance load to the collector of the *CE* stage.

SOLUTION

(a) Since $R_L = 8\ \Omega$ and the desired reflected load $R'_L = 10\ \text{k}\Omega$, we obtain

$$R'_L = a^2 R_L \quad \text{or} \quad a^2 = \frac{R'_L}{R_L} = \frac{10\ \text{k}\Omega}{8\ \Omega} = 1250$$

and hence $a = \sqrt{1250} \simeq 35$.

(b) The lower 3-dB frequency occurs when the reactance of the magnetizing inductance equals the Thévenin resistance that the magnetizing inductance looks back into. This resistance equals

$$R = R'_L // r_{oe} = 30\ \text{k}\Omega // 10\ \text{k}\Omega = 7.5\ \text{k}\Omega$$

Hence,

$$X_L = 2\pi f L = 7.5\ \text{k}\Omega$$

Substituting $L = 16\text{H}$ and solving for f yields

$$f = 75\ \text{Hz}$$

The lower 3-dB frequency can be decreased by using a transformer with a very large magnetizing inductance, or by decreasing the reflected collector load.

5–4 DIRECT COUPLING

5–4.1 Direct-Coupled Amplifier

The low-frequency response characteristic of an RC-coupled or transformer-coupled amplifier drops off as a result of the reactances of the various coupling elements. These amplifiers are therefore not equipped to handle very low frequency, or even dc, signals.

The response of an amplifier can be extended to zero frequency (dc) by omitting the coupling elements altogether, and connecting the output of one stage directly to the input of the next stage. This leads to the *direct-coupled* or *dc amplifier,* whose basic circuit is shown in Fig. 5–12.

Note that this three-stage amplifier contains no reactive elements at all. The collector of $Q1$ is connected directly to the base of $Q2$, and the base voltage of $Q2$ is therefore identical to the collector voltage of $Q1$. Similarly, the collector of $Q2$ is connected to the base of $Q3$. Since there is no dc isolation between the stages, there must be dc as well as ac interaction. The dc interaction creates a difficulty in that it becomes progressively more difficult to provide the correct

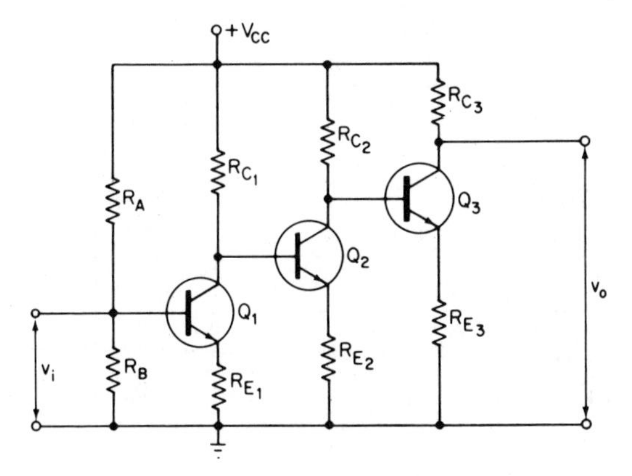

Fig. 5–12 Basic three-stage direct-coupled amplifier.

Q-point conditions for each succeeding stage. The ac interaction is favorable because it provides immediate signal transfer from one stage to the next without causing loss of gain or introducing phase shift. Any *change* in dc conditions, however, is interpreted as genuine ac signal, even if this change is caused by a fluctuation in the supply voltage or by an increase in temperature. For this reason, the dc amplifier must be firmly stabilized against dc variations which are not caused by the input signal. This is generally accomplished by *feedback.*

A practical two-stage dc amplifier is shown in Fig. 5–13. It is assumed that two identical silicon BJTs are used, each with a current gain $\beta = 50$. The base voltage of *Q*1 is determined by the 100 kΩ–10 kΩ voltage divider and equals $V_{B_1} = (10\text{ k}\Omega/110\text{ k}\Omega) \times 20\text{ V} \simeq 1.8\text{ V}$. Allowing for the 0.7-V drop across the base–emitter junction, the emitter voltage is $V_{E_1} = 1.8\text{ V} - 0.7\text{ V} = 1.1\text{ V}$. Hence, the emitter current is $I_{E_1} = 1.1\text{ V}/1\text{ k}\Omega = 1.1\text{ mA}$ and the collector voltage is $V_{C_1} \simeq 20\text{ V} - (1.1\text{ mA} \times 15\text{ k}\Omega = 3.5\text{ V}$. The base voltage of *Q*2 is equal to the collector voltage of *Q*1, so $V_{B_2} = V_{C_1} = 3.5\text{ V}$. Allowing again for the 0.7-V drop across the base–emitter junction, the emitter voltage of *Q*2 is $V_{E_2} = 3.5\text{ V} - 0.7\text{ V} = 2.8\text{ V}$. Hence, $I_{E_2} = 2.8\text{ V}/5\text{ k}\Omega = 0.56\text{ mA}$ and $V_{C_2} \simeq 20\text{ V} - (0.56\text{ mA} \times 20\text{ k}\Omega) = 8.8\text{ V}$. The overall voltage gain of the amplifier is determined by the individual stage gains. For the second stage, $A_{v2} \simeq 20\text{ k}\Omega/5\text{ k}\Omega = 4$. To determine whether the second stage loads the first stage, we find the input resistance of *Q*2 as $R_i \simeq \beta R_E = 50 \times 5\text{ k}\Omega = 250\text{ k}\Omega$. Since R_i is much larger than the collector load of *Q*1, the loading effect is negligible. The voltage gain of the first stage is $A_{v1} \simeq 15\text{ k}\Omega/1\text{ k}\Omega = 15$. The overall gain of the dc amplifier is $A_v = A_{v1}A_{v2} = 15 \times 4 = 60$.

It would make little sense to add a third stage to the amplifier of Fig. 5–13, because the base voltage of *Q*3 would be 8.8 V, which is almost one-half the supply voltage. This implies that the collector and emitter resistors of *Q*3

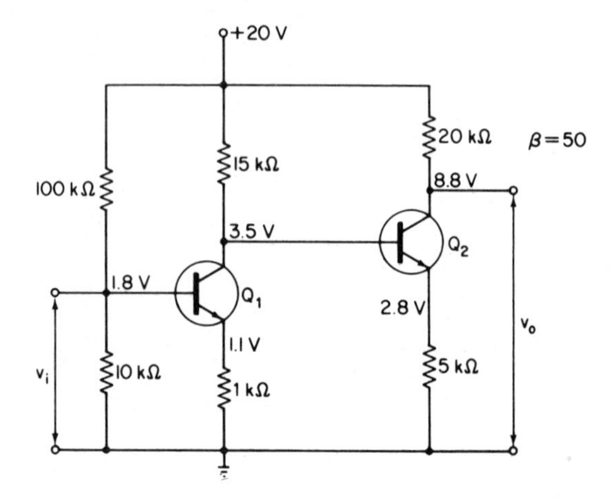

Fig. 5–13 Practical two-stage dc amplifier.

would be approximately equal and the voltage gain of the third stage would then be approximately unity.

5–4.2 Complementary Amplifier

An elegant method which overcomes the problem of increasing bias voltages for successive stages uses *complementary transistors,* as in Fig. 5–14. The first stage of this amplifier uses an *npn* transistor and the second stage a *pnp* transistor. We note that the emitter of $Q2$ is connected to the positive supply side; the collector is returned to the ground side. The base voltage of $Q1$ is $V_{b_1} = (1$ k$\Omega/16$ k$\Omega) \times 20$ V $\simeq 1.3$ V. Assuming silicon transistors, and allowing a 0.7-V drop across the BE junctions, the emitter voltage of $Q1$ is $V_{E_1} = 1.3$ V $- 0.7$ V $=$ 0.6 V. The emitter current is $I_{E_1} = 0.6$ V$/500\ \Omega = 1.2$ mA, and the collector voltage is $V_{C_1} = 20$ V $-$ (1.2 mA $\times$ 5 kΩ) $= 14$ V. The base voltage of $Q2$ is equal to the collector voltage of $Q1$, so $V_{B_2} = +\ 14$ V. The emitter voltage of $Q2$ is $V_{E_2} = 14$ V $+ 0.7$ V $= 14.7$ V. (Note that the base–emitter voltage drop has a *positive* sign.) The emitter current of $Q2$ is $I_{E_2} = (20 - 14.7)$ V$/500\ \Omega \simeq 1.06$ mA, and the collector voltage of $Q2$ is $V_{C_2} = 1.06$ mA $\times$ 5 k$\Omega = 5.3$ V. The voltage gain of each stage is approximately 10 and the overall gain of the circuit is about 100.

The outstanding advantage of complementary transistors is that an increased number of transistors can be cascaded without upsetting the bias voltages for each succeeding stage.

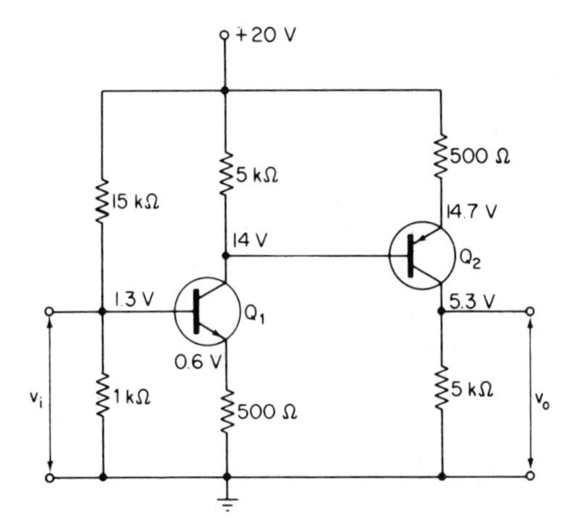

Fig. 5–14 Two-stage dc amplifier using complementary transistors.

5–4.3 Circuit Stability

The absence of coupling elements gives the dc amplifier its flat response down to dc. However, direct coupling produces some major problems in terms of circuit stability. Even minor changes in the dc conditions, especially when they occur in the first stages of the amplifier, can cause major output voltage variations, and these are indistinguishable from the true signal.

To begin with, the dc supply voltages must be well regulated so that line or load variations are not transmitted to the amplifier. Second, variations, or even slow changes in temperature (drift), upset the zero-signal condition of the amplifier. For example, the collector-to-base leakage current I_{CBO} increases with rising temperature. This increases the total collector current and hence changes the collector and base voltages. Also, the base–emitter voltage V_{BE} decreases at the rate of approximately 2 mV/°C. This affects the quiescent collector current and hence also changes the zero-signal output voltage. In addition, the transistor current gains vary with temperature and this again affects the total gain performance of the circuit.

Most of these effects are predictable and can be calculated from the given transistor data, but some other effects are very difficult to predict. Most circuit designers use feedback to stabilize the circuit against variations in quiescent conditions.

5–4.4 Darlington Compound

Figure 5–15 shows two *direct-coupled* transistors in an arrangement called the *Darlington compound.* The circuit consists of emitter follower $Q1$ driving a second emitter follower $Q2$, with external load R_L connected to the $Q2$-emitter terminal. Note that the circuit has three accessible terminals, marked E, B, and C, indicating that it behaves like a single device.

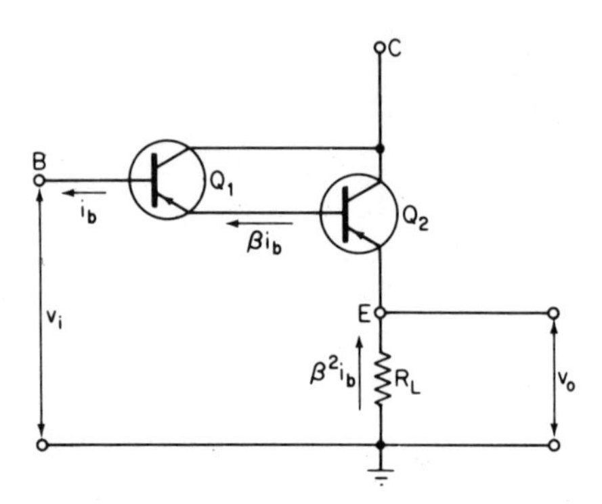

Fig. 5–15 Signal currents in the Darlington compound.

Assume for the moment that the dc bias conditions are satisfied, and that both transistors have identical current gains. An ac input current i_b into the base of $Q1$ produces a collector current $i_c = \beta i_b$, and an emitter current $i_e \simeq \beta i_b$. This emitter current drives the base of the second transistor and produces a $Q2$ emitter current $i_e \simeq \beta(\beta i_b) = \beta^2 i_b$. The Darlington pair therefore behaves like a single device with a *current gain* β_D of

$$\beta_D = \beta^2 \tag{5–23}$$

Since $Q1$ and $Q2$ are both emitter followers, each with a voltage gain of slightly less than 1, the overall *voltage gain* of the Darlington pair is also slightly less than 1.

The *input impedance* of a transistor in the common-collector configuration is approximately equal to the current gain times the ac load resistance. Since the current gain of the Darlington compound equals β^2, its input impedance equals

$$r_{i(D)} = \beta^2 R_L \tag{5–24}$$

The significance of the Darlington circuit lies in the fact that it has not only a very high current gain, but also a much higher input impedance than the single emitter follower. The Darlington compound can therefore be used as an *impedance matching* device, for example to connect a high-impedance source to a low-impedance load. Device manufacturers often produce the two transistors of the Darlington pair as a single package, although two separate transistors can easily be used to construct the circuit.

Figure 5–16 shows a Darlington compound as a practical high-gain amplifier. Supply battery V_{EE} provides emitter bias to $Q2$ through emitter resistor R_E. Considering the dc operating conditions only, we can write the following voltage equation for the input circuit:

$$V_{EE} = I_B R_B + V_{BE} + V_{BE} + I_E R_E \tag{5–25}$$

Since the current gain of the circuit equals β^2, we note that

$$I_B = \frac{I_E}{\beta^2} \tag{5–26}$$

The β^2 term is usually so large that I_B can be neglected. This reduces Eq. (5–25) to

$$I_E \simeq \frac{V_{EE} - 2V_{BE}}{R_E} \tag{5–27}$$

Usually, $V_{BE} \ll V_{EE}$ and we can also neglect the voltage drops across the base–emitter junctions, and the emitter current is

$$I_E \simeq \frac{V_{EE}}{R_E} \tag{5–28}$$

With the value of the dc emitter current established, the remaining dc voltages and currents are easily found.

EXAMPLE 5-5

The Darlington compound of Fig. 5–16 uses two identical transistors. The circuit parameters are

$$V_{EE} = 10 \text{ V} \qquad R_E = 1 \text{ k}\Omega$$

$$V_{CC} = -10 \text{ V} \qquad R_L = 1 \text{ k}\Omega$$

$$\beta = 50 \qquad R_B = 1 \text{ M}\Omega$$

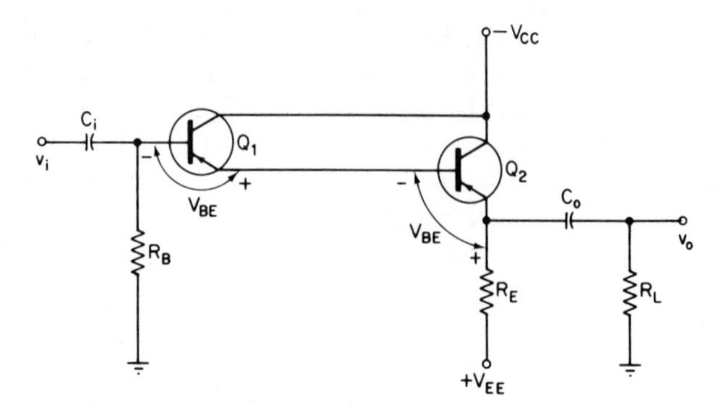

Fig. 5–16 Practical Darlington compound.

Calculate (a) The overall current gain of the circuit.
(b) The input impedance of transistor Q_1
(c) The input impedance of the entire stage.

SOLUTION

(a) The current gain per transistor is $\beta = 50$. Hence, the overall current gain of the transistor compound is

$$\beta_D = 50 \times 50 = 2500$$

(b) The ac load resistance consists of the parallel combination of R_E and R_L, so

$$R_{\text{ac}} = R_E // R_L = 1\ \text{k}\Omega // 1\ \text{k}\Omega = 500\ \Omega$$

The input impedance looking into the base of Q_1 equals

$$r_{i(D)} \simeq \beta^2 R_{\text{ac}} = 2500 \times 500\ \Omega = 1.25\ \text{M}\Omega$$

(c) The input impedance of the stage consists of the parallel combination of $r_{i(D)}$ and base resistor R_B, so

$$R_{i(\text{stage})} = r_{i(D)} // R_B = 1.25\ \text{M}\Omega // 1\ \text{M}\Omega \simeq 555\ \text{k}\Omega$$

5–5 FEEDBACK AMPLIFIERS

5–5.1 Principles of Feedback

The overall gain of an amplifier is defined as the ratio of output voltage to source voltage, or

$$\mathbf{A} = \frac{\mathbf{V}_o}{\mathbf{V}_s} \tag{5–29}$$

(Boldface type is used here to indicate quantities in complex notation, and we note that the input and output voltages have both magnitude and gain.)

Figure 5–17 shows an amplifier with feedback. A certain portion of the output voltage, determined by the $R_1 - R_2$ voltage divider, is returned or *fed back* to the input terminals of the amplifier. The modified input voltage changes the output voltage to a new value, $\mathbf{V}_o'$. The overall gain of the amplifier with feedback is defined, as before, as the ratio of output voltage to source voltage, and hence

$$\mathbf{A}' = \frac{\mathbf{V}_o'}{\mathbf{V}_s} \tag{5–30}$$

where $\mathbf{A}'$ represents the amplifier gain with feedback.

The voltage at the amplifier input terminals equals the sum of the original source voltage $\mathbf{V}_s$ and the feedback voltage $\beta\mathbf{V}_o'$, so

$$\mathbf{V}_i' = \mathbf{V}_s + \beta\mathbf{V}_o' \tag{5–31}$$

where β is defined as

$$\beta = \frac{\text{output voltage fed back}}{\text{output voltage}} = \frac{R_2}{R_1 + R_2} \tag{5–32}$$

The amplifier itself, with internal gain $\mathbf{A}$, produces an output voltage

$$\mathbf{V}_o' = \mathbf{A}\mathbf{V}_i' = \mathbf{A}(\mathbf{V}_s + \beta\mathbf{V}_o') \tag{5–33}$$

so

$$\mathbf{V}_o' = \mathbf{V}_s \frac{\mathbf{A}}{1 - \mathbf{A}\beta} \tag{5–34}$$

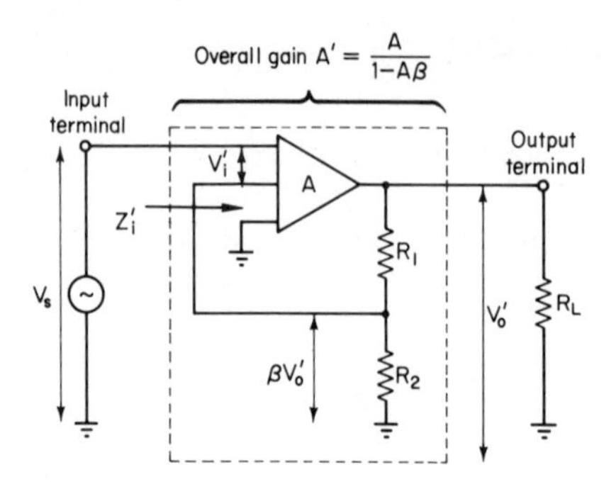

Fig. 5–17 Amplifier with feedback.

and the overall gain of the amplifier with feedback is

$$\mathbf{A}' = \frac{\mathbf{V}_o'}{\mathbf{V}_s} = \frac{\mathbf{A}}{1 - \mathbf{A}\boldsymbol{\beta}} \tag{5-35}$$

In other words, the overall gain of the amplifier with feedback is a function of the internal amplifier gain and the *feedback factor* $\mathbf{A}\boldsymbol{\beta}$.

Equation (5–35) leads to three possible feedback conditions. If

$$|1 - \mathbf{A}\boldsymbol{\beta}| < 1 \tag{5-36}$$

the gain with feedback is greater than the gain without feedback. There is *positive* or *regenerative* feedback, which increases the overall gain, produces more noise and higher waveform distortion, and reduces the circuit stability.

If

$$|1 - \mathbf{A}\boldsymbol{\beta}| > 1 \tag{5-37}$$

the gain with feedback is smaller than the gain without feedback. This is *negative* or *degenerative* feedback which reduces the overall gain, produces less noise and smaller waveform distortion, and it improves the circuit stability.

If

$$|1 - \mathbf{A}\boldsymbol{\beta}| = 0 \quad \text{or} \quad \mathbf{A}\boldsymbol{\beta} = 1 + j0 \tag{5-38}$$

the gain with feedback is infinite. In this case the output voltage is independent of the input voltage and the amplifier becomes an *oscillator* (see Chapter 9).

In a single amplifier stage, the phase shift between output and input voltage is either 0 or 180 degrees, depending on the circuit configuration. The feedback voltage in a multistage amplifier is therefore either in phase or out of phase with the input voltage. With *positive feedback,* the feedback voltage is *in phase* with the input voltage, and with *negative* feedback, the feedback voltage is *out of phase* with the input voltage.

5–5.2 Effects of Negative Feedback

Negative feedback provides a convenient method to predict the performance of an amplifier in terms of its *gain, stability,* and *impedance* levels.

If the feedback factor is very large, so that $|\mathbf{A}\boldsymbol{\beta}| \gg 1$, Eq. (5–35) reduces to

$$\mathbf{A}' \cong -\frac{1}{\boldsymbol{\beta}} \qquad (5\text{–}39)$$

showing that the *gain with feedback* is practically independent of the highly variable quantity **A**. Since $\boldsymbol{\beta}$ is generally determined by a resistive network, the gain $\mathbf{A}'$ is predictable and virtually independent of variations in components and supply voltages.

The *input impedance* of the amplifier without feedback is

$$\mathbf{Z}_i = \frac{\mathbf{V}_s}{\mathbf{I}} \qquad (5\text{–}40)$$

With feedback, as in Fig.5–17, the input voltage at the amplifier terminals is given as

$$\mathbf{V}'_i = \mathbf{V}_s + \boldsymbol{\beta}\mathbf{V}'_o \qquad (5\text{–}31)$$

Since $\mathbf{V}'_o = \mathbf{A}\mathbf{V}_i$=, Eq. (5–31) can be written as

$$\mathbf{V}'_i = \frac{\mathbf{V}_s}{1 - \mathbf{A}\boldsymbol{\beta}} \qquad (5\text{–}41)$$

The input current with feedback then is

$$\mathbf{I}' = \frac{\mathbf{V}'_i}{\mathbf{Z}_i} = \frac{\mathbf{V}_s}{(1 - \mathbf{A}\boldsymbol{\beta})\mathbf{Z}_i} \qquad (5\text{–}42)$$

so the input impedance of the amplifier with feedback equals

$$\mathbf{Z}_i' = \frac{\mathbf{V}_s}{\mathbf{I}'} = (1 - \mathbf{A}\boldsymbol{\beta})\mathbf{Z}_i \tag{5-43}$$

Hence, an amplifier with negative feedback has an input impedance which is increased by a factor $(1 - \mathbf{A}\boldsymbol{\beta})$ over that of an amplifier without feedback.

The gain of an amplifier without feedback is subject to *variations* caused by changes in temperature, transistor parameters, or other circuit constants. Negative feedback improves the *gain stability.* For a fractional change in amplifier gain, the percentage change in gain with feedback can be derived from Eq. (5–35) by differentiation. This yields

$$\Delta\mathbf{A}' = \frac{1}{1 - \mathbf{A}\boldsymbol{\beta}}\Delta\mathbf{A} \tag{5-44}$$

Equation (5–44) shows that a given percentage change in **A** causes a reduction in the percentage change in $\mathbf{A}'$ by a factor $1/(1 - \mathbf{A}\boldsymbol{\beta})$. Negative feedback, therefore, also improves the gain stability of the amplifier.

It can also be shown that the distortion components of the signal waveform are reduced by a factor $1/(1 - \mathbf{A}\boldsymbol{\beta})$ when negative feedback is applied. The reduction in distortion is in the same ratio as the reduction in gain, and hence feedback appears to be of no advantage. However, almost all amplitude distortion occurs in the final amplifier stage, where large signal swings take place. Local feedback is often applied to the final stage only, thereby reducing both its gain and its distortion. The loss in gain can be recovered in the earlier stages, which then do not contribute to the introduction of distortion.

Since feedback reduces the overall gain of the amplifier, the input signal $\mathbf{V}_s$ must be increased to a higher value $\mathbf{V}_s'$ in order to maintain the same output. If the output voltage with feedback must be the same as the output voltage without feedback, we can write

$$\mathbf{V}_o = \mathbf{V}_o' \quad \text{or} \quad \mathbf{A}\mathbf{V}_s = \mathbf{A}'\mathbf{V}_s' \tag{5-45}$$

where $\mathbf{V}_s'$ is the increased source voltage required to maintain the output.

Substituting Eq. (5–35) into Eq. (5–45) we obtain

$$\mathbf{V}_s' = (1 - \mathbf{A}\boldsymbol{\beta})\mathbf{V}_s \tag{5-46}$$

With negative feedback the source voltage must be increased by the feedback factor to maintain the same output voltage.

EXAMPLE 5-6

A certain amplifier has an input impedance $\mathbf{Z}_i = 10\ \text{k}\Omega$ and a voltage gain $\mathbf{A} = 500$. It requires a normal input signal of 10 m V rms. Its distortion in the midfrequency range is $D = 15$ percent. Temperature variations may cause the gain to vary by 20 percent. Find the effects of *negative* feedback if $\boldsymbol{\beta} = -0.1$.

SOLUTION

The feedback factor is

$$\mathbf{A}\boldsymbol{\beta} = 500 \times -0.1 = -50$$

The gain with feedback is

$$\mathbf{A}' = \frac{\mathbf{A}}{1 - \mathbf{A}\boldsymbol{\beta}} = \frac{500}{1 + 50} \simeq 10$$

The distortion is reduced to

$$D' = \frac{D}{1 - \mathbf{A}\boldsymbol{\beta}} = \frac{15\%}{1 + 50} = .3\%$$

The input impedance is increased to

$$\mathbf{Z}_i' = (1 - \mathbf{A}\boldsymbol{\beta})\mathbf{Z}_i = (1 + 50)10\ \text{k}\Omega = 510\ \text{k}\Omega$$

The variation in gain with feedback is reduced to

$$\Delta\mathbf{A}' = \frac{\Delta\mathbf{A}}{1 - \mathbf{A}\boldsymbol{\beta}} = \frac{20\%}{1 + 50} \simeq 0.4\%$$

The output voltage without feedback is

$$\mathbf{V}_o = \mathbf{A}\mathbf{V}_s = 500 \times 10\ \text{mV rms} = 5\ \text{V rms}$$

To maintain this output voltage with feedback applied we require a source voltage of

$$\mathbf{V}_s' = (1 - \mathbf{A}\boldsymbol{\beta})\mathbf{V}_s = (1 + 50)10\ \text{mV rms} = 510\ \text{mV rms}$$

5–5.3 RC-Coupled Amplifier with Feedback

Figure 5–18 shows a two-stage *RC*-coupled amplifier with negative feedback. The ac input signal is amplified and inverted by the first stage, capacitively coupled to the second stage, and inverted again. The output signal at the collector of $Q2$ is therefore in phase with the input signal at the base of $Q1$. Voltage feedback is provided by voltage divider R_F–R_{E_1}, which returns a certain fraction of the output signal back to the *emitter* of $Q1$. This causes *negative* feedback, because the feedback voltage *reduces* the instantaneous value of the base–emitter voltage of $Q1$.

Excellent dc circuit stability is provided by the universal bias circuits of $Q1$ and $Q2$. The function of the ac feedback loop then is to provide *gain stability* and to increase the input resistance of the amplifier.

The overall gain of the amplifier is the product of the individual stage gains, modified through feedback by a factor $1/(1 - \mathbf{A}\boldsymbol{\beta})$, where $\boldsymbol{\beta}$ is determined by the resistance ratio of the feedback network.

The overall gain of the amplifier with feedback is found by first determining the product of the stage gains and then modifying this gain to take account of the feedback provided by R_F and R_{E_1}. The collector current of $Q2$,

$$I_{C_2} = \frac{20\text{ V} - 8\text{ V}}{5\text{ k}\Omega} = 2.4\text{ mA}$$

Because $\beta = 100$, α is very nearly 1, and the emitter current $I_{E_2} \cong I_{C_2} = 2.4$ mA. The emitter resistance of $Q2$ is bypassed with a 40-μF capacitor, so the voltage

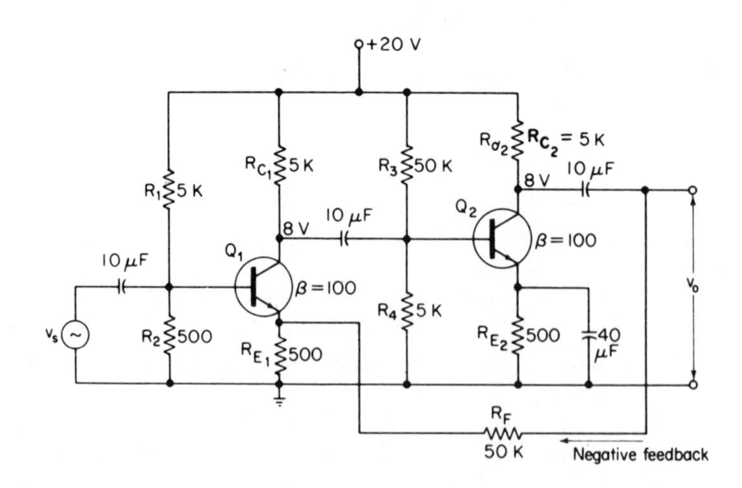

Fig. 5–18 Two-stage *RC*-coupled amplifier with ac feedback.

gain of the second stage is determined by the ratio of R_{C_2} and r_π. The transistor parameter r_π is determined by the dc emitter current.

$$r_\pi \cong \frac{50}{I_{E_2}} = \frac{50}{2.4\ \text{mA}} = 20.8\ \Omega$$

The gain of the second stage,

$$\mathbf{A}_2 \cong \frac{R_{C_2}}{r_\pi} = \frac{5\ \text{k}\Omega}{20.8\ \Omega} \simeq 240$$

The input resistance r_{ie} of $Q2$,

$$r_{ie} \simeq \beta r_\pi = 100 \times 20.8\ \Omega = 2.08\ \text{k}\Omega$$

The ac load in the collector circuit of $Q1$ consists of R_{C_1}, R_{C_2}, R_3, R_4, and r_{ie} in parallel, so

$$R_{\text{ac}} = 5\ \text{k}\Omega // 50\ \text{k}\Omega // 5\ \text{k}\Omega // 2.08\ \text{k}\Omega \simeq 1.1\ \text{k}\Omega$$

The gain of the first stage, without feedback,

$$\mathbf{A}_1 \cong \frac{R_{\text{ac}}}{R_{E_1}} = \frac{1.1\ \text{k}\Omega}{500\ \Omega} = 2.2$$

and the overall gain of the amplifier without feedback is

$$\mathbf{A} = \mathbf{A}_1 \mathbf{A}_2 = 2.2 \times 240 = 528$$

Next we determine the effect of the feedback loop. The fraction of output voltage returned to the input stage by the $R_F - R_{E_1}$, voltage divider equals

$$\boldsymbol{\beta} = \frac{R_{E_1}}{R_F + R_{E_1}} \simeq 0.01$$

The feedback factor $\mathbf{A}\boldsymbol{\beta}$ equals

$$\mathbf{A}\boldsymbol{\beta} = 528 \times -0.01 = -5.28$$

where the negative sign indicates that this is negative feedback.

The overall gain of the amplifier with feedback then is

$$\mathbf{A}' = \frac{\mathbf{A}}{1 - \mathbf{A}\boldsymbol{\beta}} = \frac{528}{1 + 5.28} = 84$$

The signal source sees an input resistance consisting of the parallel combination of bias resistors R_1 and R_2, and the input resistance r_{ie} of $Q1$. Without feedback, $r_{ie} \cong \beta R_{E_1} = 100 \times 500\ \Omega = 50\ \text{k}\Omega$. Negative feedback increases r_{ie} to a new value $r_{ie}' = r_{ie}(1 - \mathbf{A}\boldsymbol{\beta}) = 315\ \text{k}\Omega$. With feedback applied, therefore, the input resistance of the amplifier is

$$R_i' = 500\ \Omega // 5\ \text{k}\Omega // 315\ \text{k}\Omega \cong 500\ \Omega$$

Although feedback increases the input resistance of $Q1$, it does not affect the overall input resistance of the amplifier. This situation can be improved by increasing the resistances of R_1 and R_2 by a factor of 100, so that they are of the same magnitude as the input resistance of $Q1$.

The addition of the feedback loop has improved the gain stability of the circuit considerably, e.g., if one of the BJTs is replaced by one having a lower current gain, the gain of that particular stage decreases. Consequently, the gain of the amplifier without feedback decreases. Assume that the overall gain **A** decreases from 528 to 400 (24 percent). The feedback factor then also decreases and assumes the value $\mathbf{A}\boldsymbol{\beta} = 400 \times -0.01 = -4$. The amplifier gain with feedback then becomes

$$\mathbf{A}' = \frac{\mathbf{A}}{1 - \mathbf{A}\boldsymbol{\beta}} = \frac{400}{5} = 80$$

a decrease of less than 5 percent.

5-5.4 *Direct-Coupled Amplifier with Feedback*

A two-stage direct-coupled BJT amplifier with dc feedback is shown in Fig. 5-19. The dc feedback loop consists of voltage divider $R_F - R_1$, which connects the emitter of $Q2$ to the base of $Q1$. The object of this feedback arrangement is to stabilize the dc voltages in the circuit, despite possible variations in transistor current gains. This is accomplished by keeping the emitter voltage of $Q2$ constant. If V_{E_2} is held constant, I_{E_2} is also constant, and $Q2$ is stabilized against β variations. Since the transistors are dc-coupled, the collector voltage of $Q1$ equals V_{E_2} plus the (constant) base–emitter voltage V_{BE_1} of $Q1$. Hence, stabilization of V_{E_2} automatically stabilizes V_{C_1} and I_{C_1}.

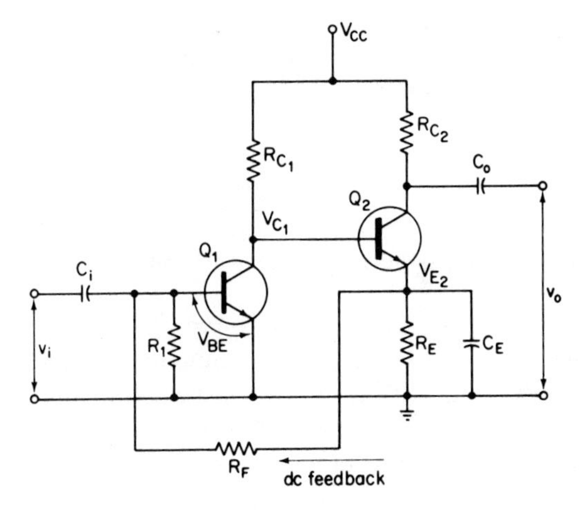

Fig. 5–19 Two-stage dc-coupled amplifier with dc feedback.

The dc analysis of this circuit is carried out by considering the Thévenin equivalent of the input circuit of $Q1$, shown in Fig. 5–20. The feedback network supplies a dc base voltage to $Q1$, equal to

$$V_{B_1} = V_{E_2}\frac{R_1}{R_1 + R_2} = V_{E_2}\frac{1}{1 + (R_F/R_1)} \tag{5–47}$$

This dc source looks into an equivalent resistance of

$$R_{\text{eq}} = R_1 /\!/ R_F = \frac{R_1 R_F}{R_1 + R_F} = \frac{R_F}{1 + (R_F/R_1)} \tag{5–48}$$

while it also sees the forward-biased base–emitter junction of $Q1$, with forward voltage V_{BE}.

The input circuit of $Q1$ can then be replaced by its Thévenin equivalent, as in Fig. 5–20. It follows from this circuit that

$$V_{E_2}\frac{1}{1 + (R_F/R_1)} = I_{B_1}\frac{R_F}{1 + (R_F/R_1)} + V_{BE} \tag{5–49}$$

or

$$V_{E_2} = \left(1 + \frac{R_F}{R_1}\right) V_{BE} + R_F I_{B_1} \tag{5–50}$$

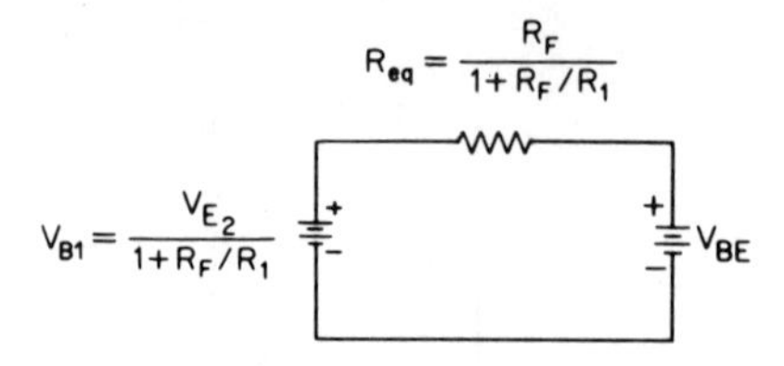

Fig. 5–20 Thévenin equivalent for the base circuit of $Q1$ under quiescent conditions.

Since the base–emitter voltage of a BJT is virtually constant, the first term on the right-hand side of Eq. (5–50) is practically constant. If it is made substantially larger than the second term, V_{E_2} is well stabilized. With V_{E_2} held constant, the remaining dc voltages and currents are also constant, and the circuit has stable quiescent characteristics.

5–6 DIFFERENTIAL AMPLIFIERS

5–6.1 Basic Differential Amplifier

Figure 5–21 shows two transistors in a symmetrical configuration called a *balanced* or *differential* amplifier. The circuit has two input terminals, marked v_1 and v_2, which gives the amplifier a degree of flexibility that is useful in many applications. Differential amplifiers, in fact, are extensively used as basic elements in integrated circuits.

In a perfectly balanced circuit (matched transistors and identical collector resistors), the total dc current through emitter bias resistor R_E divides equally between the two transistors. Under *quiescent* conditions, with both bases connected to ground ($v_1 = v_2 = 0$), the dc transistor currents are

$$I_{E_1} = I_{E_2} = \frac{1}{2} I_E \qquad (5\text{–}51)$$

The quiescent collector currents and voltages are then also equal, and the output voltage, measured from the collector of $Q1$ to the collector of $Q2$, is

$$v_o = V_{C_1} - V_{C_2} = 0 \qquad (5\text{–}52)$$

There are basically two modes of operation for the balanced amplifier. In the *common mode,* both input terminals are driven by *in-phase* signal voltages of the *same magnitude,* so

$$v_1 = v_2 = v_c \qquad \text{(common-mode signal)} \qquad (5\text{–}53)$$

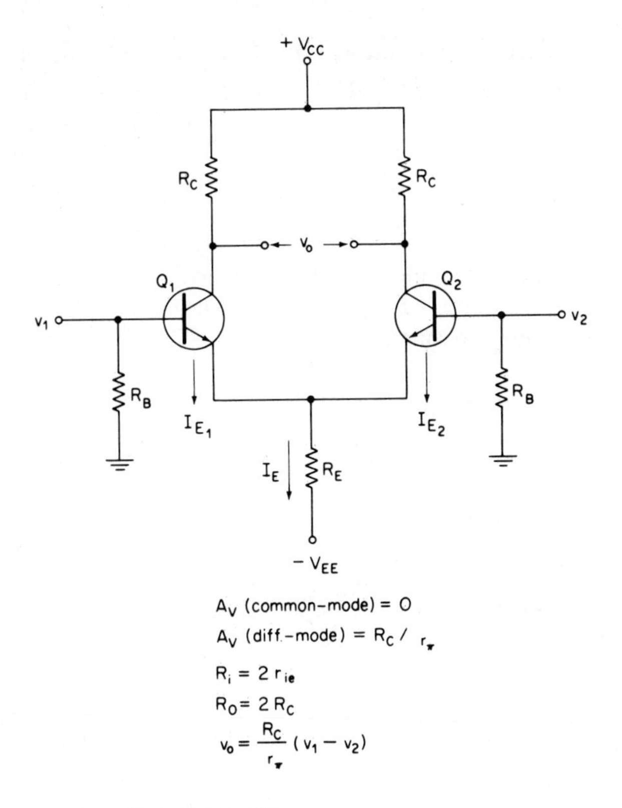

Fig. 5–21 Balanced or differential amplifier.

The two transistor currents vary equally in response to the common input conditions, and the balance of the circuit is not disturbed. The total instantaneous emitter current through R_E continues to be divided equally between the two transistors. The output voltage therefore remains at zero volts and the voltage gain of the amplifier

$$A_{v(\text{common-mode})} = 0 \tag{5–54}$$

In the *differential mode,* the two input signals are of equal magnitude but of *opposite phase,* so

$$v_1 = -v_2 = v_d \qquad \text{(differential-mode signal)} \tag{5–55}$$

In this case the current in one transistor increases while the current in the other transistor decreases by exactly the same amount. The unbalance in collector currents causes an output voltage between the collector terminals. The ac components of the total current through R_E are out of phase and cancel, so that the emitter resistor behaves like a short circuit for ac signals in the differential mode.

For a perfectly balanced circuit in the differential mode we can write

$$i_{C_1} = I_C + i_c \quad \text{and} \quad i_{C_2} = I_C - i_c \tag{5-56}$$

where i_C is the total instantaneous collector current, I_C is the dc or quiescent component, and i_c is the ac or signal component. The output voltage is simply the difference between the two collector voltages, so

$$\begin{aligned} v_o &= v_{C_1} - v_{C_2} \\ &= (V_{CC} - R_C i_{C_1}) - (V_{CC} - R_C i_{C_2}) \\ &= R_C(i_{C_2} - i_{C_1}) \\ &= 2R_C i_c \end{aligned} \tag{5-57}$$

If base resistor R_B is large, the input voltage at each base terminal is

$$v_1 = -v_2 = i_b r_{ie}$$

where $r_{ie} = \beta r_\pi$ is the input resistance of the BJT in the *CE* connection. The total applied input voltage is

$$v_i = v_1 - v_2 = 2i_b r_{ie} \tag{5-58}$$

The voltage gain of the amplifier in the differential mode is

$$A_{v(\text{diff mode})} = \frac{v_o}{v_i} = \frac{2R_c i_c}{2i_b r_{ie}} = \frac{R_C}{r_\pi} \tag{5-59}$$

Provided that $R_B \gg r_{ie}$, the input resistance of the amplifier is

$$R_i = \frac{v_i}{i_i} = \frac{2i_b r_{ie}}{i_b} = 2r_{ie} \tag{5-60}$$

and the output resistance is

$$R_o = \frac{v_o}{i_o} = \frac{2R_c i_c}{i_c} = 2R_C \tag{5-61}$$

Arbitrary input voltages v_1 and v_2 can be represented as the superposition of a common-mode pair of voltages and a differential-mode pair of voltages, so

$$v_1 = v_c + v_d \quad \text{and} \quad v_2 = v_c - v_d \tag{5-62}$$

If the two input voltages are known, the common and differential components can be calculated from

$$v_c = \tfrac{1}{2}(v_1 + v_2) \quad \text{and} \quad v_d = \tfrac{1}{2}(v_1 - v_2) \tag{5-63}$$

The common-mode components of the input signals do not contribute to the output voltage and the differential amplifier therefore *discriminates* against these signals. This is a very useful feature, since noise and hum introduced at the input terminals, together with genuine signal components, are eliminated in the amplifier output.

In practical circuits, the rejection of common-mode signals is not entirely perfect. Small variations in transistor characteristics cause a slight unbalance in the collector currents and a small common-mode output voltage is present. The rejection of common-mode signals is expressed in a figure of merit for the differential amplifier known as the *common-mode rejection ratio* (*CMRR*). The CMRR is defined as the ratio of the gain in the differential mode to the gain in the common mode, with both inputs short-circuited. When, for example, an amplifier has a common-mode rejection ratio of 1,000, a 1-mV differential input provides the same output voltage as a 1-V common input applied to both terminals. The CMRR is often expressed in decibels, and may be as high as 90 dB for differential amplifiers in IC form.

5-6.2 Types of Differential Amplifiers

Some variations to the dual-input differential amplifier of Fig. 5-21 are shown in Fig. 5-22. In the *single-input* differential amplifier of Fig. 5-22(a) only one input terminal is used for signal input. The second terminal is connected to ground, and hence $v_2 = 0$. The output voltage is determined by the voltage gain $A_v = R_C/r_\pi$ and the difference between the two input signals. Since $v_2 = 0$, the input voltage $v_i = v_1$ and the output voltage is

$$v_o = \frac{R_C}{r_\pi} v_i \tag{5-64}$$

The second variation of the basic differential amplifier, Fig. 5-22(b), has two input terminals and a *single-ended* output. The output voltage at the collector of $Q2$ is measured with respect to ground. The collector of $Q1$ is connected directly to V_{CC} and this affects the voltage gain of the circuit. It can be shown that the gain of this circuit is

$$A_v = \frac{R_C}{2r_\pi} \tag{5-65}$$

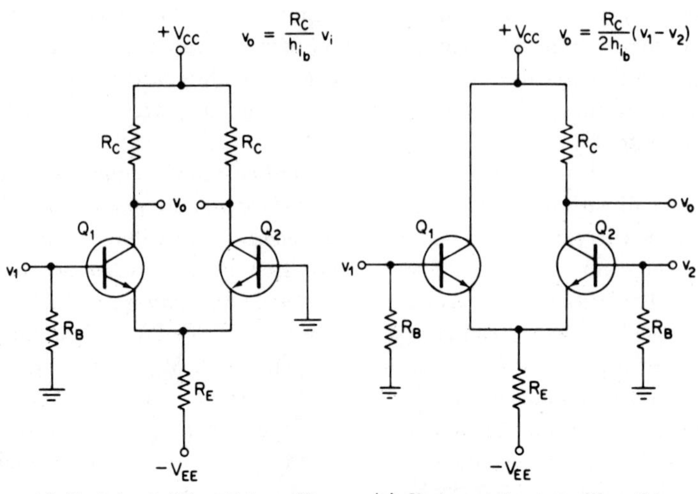

(a) Single-input differential amplifier.

(b) Single-ended output differential amplifier.

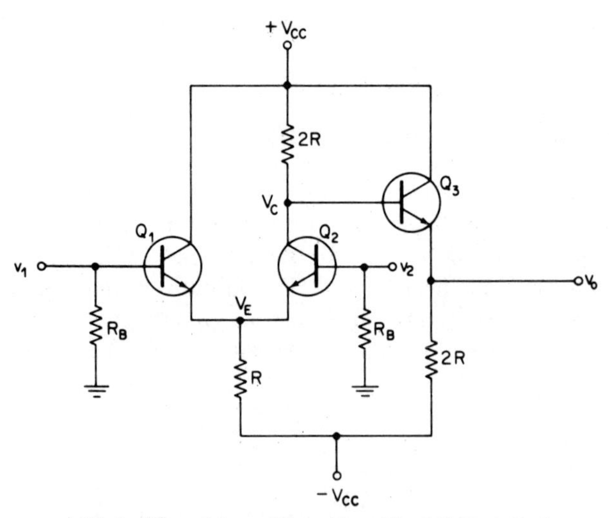

(c) A differential amplifier with emitter-follower output.

Fig. 5–22 Alternate forms of differential amplifiers.

and the output voltage therefore is

$$v_o = \frac{R_C}{2r_\pi}(v_1 - v_2) \tag{5–66}$$

Another differential amplifier often found in commercial ICs is shown in Fig. 5–22(c). Transistors $Q1$ and $Q2$ form the single-ended differential amplifier and $Q3$ is an emitter-follower output stage which provides a low output impedance. Assuming a perfectly balanced circuit, the quiescent currents in the difference amplifier are equal, so $I_{C_1} = I_{C_2}$. If all transistors are silicon with $V_{BE} = 0.7$ V, the emitter voltage of the difference amplifier is

$$V_E = V_{BE} = -V_{CC} + 2I_C R \tag{5–67}$$

The quiescent collector voltage of $Q2$ is

$$V_C = V_{CC} - 2I_C R \tag{5–68}$$

or, from Eq. (5–67),

$$V_C = V_{BE} = 0.7 \text{ V}$$

The quiescent collector-to-emitter voltage of $Q2$ is then

$$V_{CE_2} = V_C - V_E = 2V_{BE} = 1.4 \text{ V} \tag{5–69}$$

and the output voltage under quiescent conditions is

$$v_o = V_C - V_{BE} = 0 \text{ V}$$

Hence, zero input voltage provides zero output voltage. This condition is very useful in practical circuit applications.

EXAMPLE 5-7

For the single-input differential amplifier of Fig. 5–23, *calculate*

(a) The voltage gain.
(b) The input resistance.
(c) The output voltage.

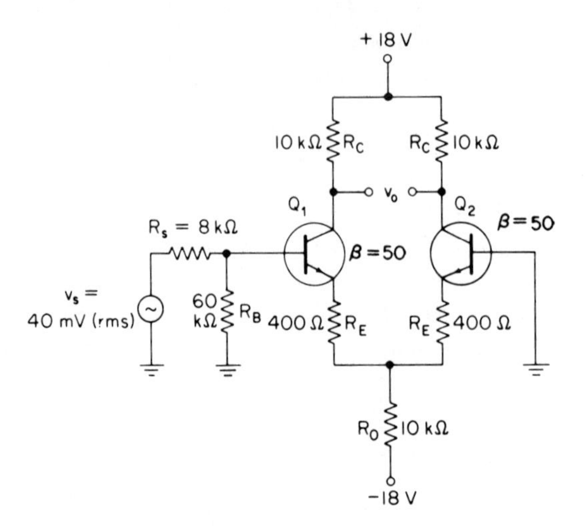

Fig. 5–23 Differential amplifier for Example 5-7.

SOLUTION

(a) Under quiescent conditions, with zero input, and neglecting the base-emitter voltage drops, the current in R_0 equals

$$I_0 \cong \frac{18\text{ V}}{10\text{ k}\Omega} = 1.8\text{ mA}$$

For a balanced circuit, I_0 divides equally into the two quiescent emitter currents, so

$$I_{E_1} = I_{E_2} = 0.9\text{ mA}$$

The dc collector voltages are

$$V_{C_1} = V_{C_2} = V_{CC} - I_C R_C = 18\text{ V} - (0.9\text{ mA} \times 10\text{ k}\Omega) = 9\text{ V}$$

$$r_\pi \cong \frac{50}{I_E} = \frac{50}{0.9} \cong 55\ \Omega$$

Since $R_E \gg r_\pi$, emitter resistors R_E swamp the internal BJT diode resistances and tend to stabilize the circuit.

$$A_v = \frac{R_C}{R_E} = \frac{10\text{ k}\Omega}{400\ \Omega} = 25$$

(b) The input resistance looking into the base of the transistor is $r_{i(\text{base})} \simeq 2\beta R_E = 40$ kΩ. This resistance is in parallel with the 60-kΩ base resistor, so the input resistance of the amplifier is

$$R_i = R_B // r_{i(\text{base})} = 60\ \text{k}\Omega // 40\ \text{k}\Omega = 24\ \text{k}\Omega$$

(c) The input voltage to the amplifier is

$$v_i = \frac{24\ \text{k}\Omega}{32\ \text{k}\Omega} \times 40\ \text{mV rms} = 30\ \text{mV rms}$$

The output voltage is

$$v_o = A_v v_i = 25 \times 30\ \text{mV rms} = 0.75\ \text{V rms}$$

5–7 MULTISTAGE MOST AMPLIFIERS

MOS transistors, with their high input impedance and low noise figure, are ideally suited for the construction of multistage voltage amplifiers, especially in integrated circuits. The basic circuit of a three-stage amplifier using identical MOSTs is shown in Fig. 5–24(a). This amplifier can be built from discrete components, and its overall voltage gain can be as high as several thousand.

Under quiescent conditions, the direct-coupled drains and gates are at the same potential. Although we may assume that the three MOSTs have very closely matched characteristics (a fair assumption, especially in ICs), there are slight differences between them. If a variation in ambient temperature causes the quiescent conditions of $Q1$ to vary so that its drain voltage deviates from the characteristic value common to all three MOSTs, the amplifier interprets this deviation as signal and produces an (unwanted) output variation.

Feedback resistor R_F, which provides gate bias for $Q1$, tends to stabilize the quiescent operating points for the three MOSTs by *negative feedback*. If, for example, the drain voltage of $Q1$ increases, the corresponding increase in output voltage is fed back to the gate of $Q1$. Its drain current then increases and its drain voltage decreases, thereby compensating for the initial rise in drain voltage.

Unfortunately, the feedback resistor reduces the input resistance of the amplifier to the extent that it may present an unrealistically low load to the source. If the overall voltage of the amplifier is designated as A_v, the output voltage v_o for an input voltage v_i is

$$v_o = -A_v v_i \tag{5–70}$$

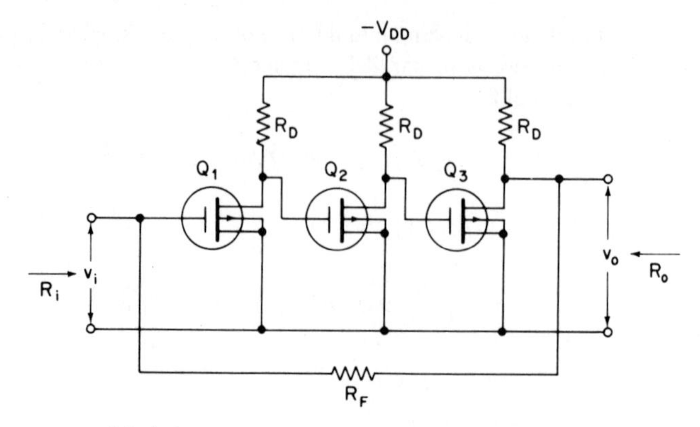

(a) A three-stage p-channel enhancement MOST amplifier with overall negative feedback

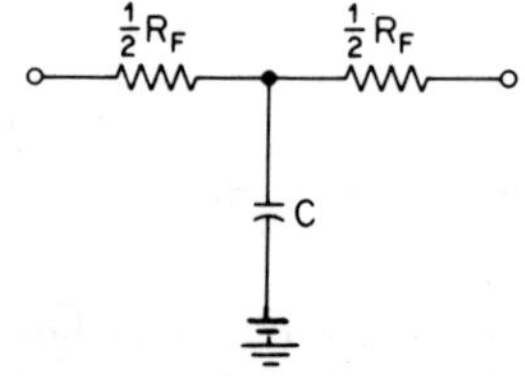

(b) Modified feedback network to remove ac signal feedback and increase the input resistance of the amplifier

Fig. 5–24 **Multistage enhancement MOST amplifier with dc stabilization.**

where the minus sign indicates the phase reversal between output and input. The voltage drop across the feedback resistor consists of the difference between v_i and v_o, so

$$v_{R_F} = v_i - v_o = v_i + A_v v_i = (1 + A_v)v_i \tag{5–71}$$

The signal current driven into the amplifier by the source equals

$$i_F = \frac{v_{R_F}}{R_F} = \frac{(1 + A_v)v_i}{R_F} \tag{5–72}$$

so the input resistance of the amplifier equals

$$R_i = \frac{v_i}{i_F} = \frac{R_F}{1 + A_v} \simeq \frac{R_F}{A_v} \tag{5–73}$$

Hence, if the feedback MOST amplifier of Fig. 5–24(a) has an overall voltage gain of $A_v = 1000$ and the feedback resistance is $R_F = 10$ MΩ, the amplifier input resistance is only $R_i = 10$ MΩ/1000 = 10 kΩ. This relatively low input resistance may well cause an undesirable loading of the source.

The reduction in input resistance can be eliminated by modifying the feedback path as in Fig. 5–24(b), where R_F is broken into two equal parts and the junction between the parts is connected to ac ground by a large capacitor. In this case the dc feedback which was required to establish quiescent stabilization is maintained, while the ac feedback which caused the low input resistance is removed. The amplifier input resistance with the modified feedback network of Fig. 5–24(b) is

$$R_i \cong \frac{1}{2} R_F \tag{5–74}$$

while the amplifier output resistance is

$$R_o \cong R_D \tag{5–75}$$

The overall gain of the three-stage amplifier is

$$A_v = (-g_m R_D)^3 \tag{5–76}$$

where $-g_m R_D$ is the individual stage gain.

EXAMPLE 5-8

The three-stage MOST amplifier of Fig. 5–25 uses identical MOS transistors each having a transconductance of $g_m = 3$ mA/V. The coupling capacitors can be treated as ac short circuits.

Determine (a) The gain of the amplifier $A_v = v_o/v_i$.
(b) The input resistance of the amplifier.
(c) The overall voltage gain $A_v = v_o/v_s$.
(d) The overall voltage gain, if R_F is split into two 5-MΩ resistors, with their junction ac bypassed, as in Fig. 5–24(b).

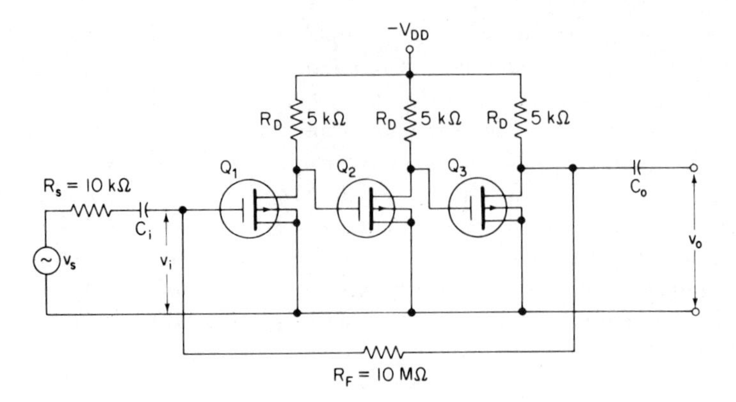

Fig. 5–25 Example 5-8. Three-stage *p*-channel enhancement MOST amplifier with negative feedback.

SOLUTION

(a) The gain of the three stage amplifier is

$$A_v = (-g_m R_D)^3 = \left(-\frac{3\text{ mA}}{\text{V}}5\text{ k}\Omega\right)^3 = -3375$$

(b) The amplifier input resistance is

$$R_i = \frac{R_F}{A_v} = \frac{10\text{ M}\Omega}{3375} \simeq 3\text{ k}\Omega$$

(c) The source, with its internal resistance of 10 kΩ, looks into an amplifier input resistance of 3 kΩ. This reduces the input voltage, and hence the overall gain, to

$$A_{v(\text{overall})} = \frac{R_i}{R_i + R_s}A_{v(\text{amplifier})}$$

$$= \frac{3\text{ k}\Omega}{13\text{ k}\Omega}3375 = 775$$

(d) When the feedback resistor is split into two 5-MΩ resistors, the amplifier input resistance is

$$R_i = \frac{1}{2}R_F = 5\text{ M}\Omega$$

In this case the amplifier does not load the source and the entire source voltage appears at the amplifier input terminals. The overall gain is therefore equal to the amplifier gain, so that

$$A_{v(\text{overall})} = 3375$$

In practical integrated circuits the drain resistors R_D usually take the form of integrated MOS devices, as shown in Fig. 5-26. Here MOS transistor $Q2$ acts as the drain resistor for $Q1$. The drain of $Q2$ is connected to its gate and the MOST is so biased that it always operates in the saturation or constant-current region. One of the advantages of using a MOST as the drain resistor is that its nonlinear resistance characteristic exactly cancels the nonlinear transfer characteristic of $Q1$ and this therefore produces distortionless amplification. Another advantage is that it is often easier and cheaper to manufacture an additional MOST on an IC chip than it is to manufacture a resistor.

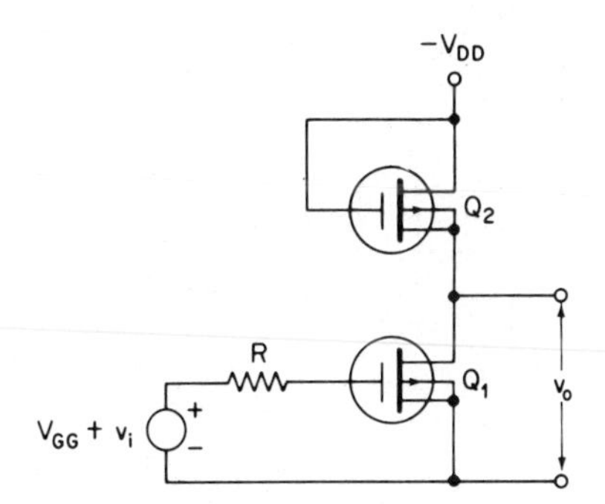

Fig. 5-26 MOST amplifier stage where $Q2$ acts as the load resistor for $Q1$.

5-8 FET-BJT COMPOUNDS

An amplifier to be used with a voltage source of high internal impedance, such as a crystal pickup, should have a high input impedance (typically 1 MΩ or more) and introduce a minimum of noise. Bipolar junction transistors do not meet these requirements, but field effect transistors are much better suited for this purpose. Although FETs have a very high input impedance their transconductance is rather low, generally on the order of a few mA/V, and they do therefore not provide sufficient drive for a conventional output stage. A typical low-level voltage amplifier, for example, should have a transconductance of about 50 mA/V. High input impedance and high gain can be achieved by cascading FETs and BJTs. A FET stage, with its high input impedance, followed by one or more high-gain BJT driver or output stages, provides an amplifier that combines the advantages of both devices.

Figure 5–27 shows a high input impedance FET-BJT compound which can easily be constructed from discrete components. The input stage of this amplifier consists of a general-purpose FET, having an input resistance in excess of 1 MΩ and a typical transconductance of 3 mA/V. The signal voltage developed at the FET drain terminal is dc-coupled to the base of the *Q*2 driver stage, which supplies adequate base current for the *Q*3 power output stage. The overall voltage gain of this amplifier is about 1,000, and its input impedance is approximately 1 MΩ. The quiescent operating conditions are stabilized by the 510-Ω feedback resistor, which also provides ac gain stabilization. The circuit is entirely dc-coupled, and its high dc gain limits this amplifier to operation over a restricted temperature range. Wide-range temperature stability can be achieved by increasing the feedback or by introducing temperature-compensating bias networks.

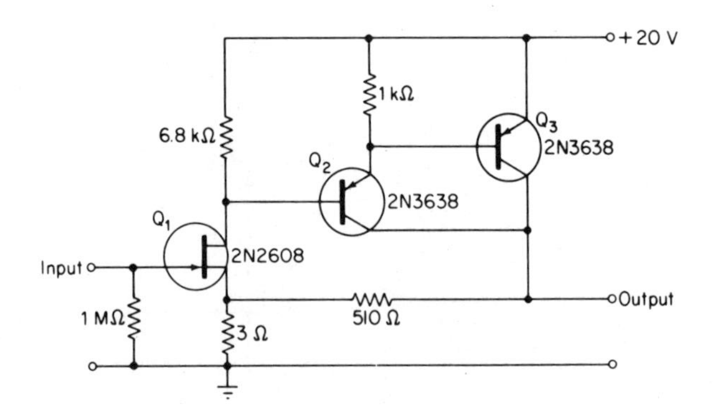

Fig. 5–27 High input impedance FET-BJT compound.

Increased input impedance and higher transconductance can be obtained effectively by using an insulated gate FET or a MOST as the input stage. Figure 5–28 shows a practical circuit for a record player amplifier, consisting of an IC preamplifier and an output driver stage. The input impedance of the IC amplifier is equal to that of the MOST, i.e., a parallel combination of a 10-pF capacitor and a 10^{12}-Ω resistor. This impedance is so high as to be of no practical significance at audio frequencies. The output impedance of the IC is virtually a pure resistance which depends on the operating current of the emitter-follower driver stage, and is typically about 2 kΩ. The emitter follower drives a conventional BJT output stage with a transformer-coupled load. The noise in the circuit is extremely low and is determined by the noise generated in the

MOST. Overall signal distortion is reduced by negative feedback from the speaker to the MOST input.

QUESTIONS

5-1 Name some of the factors limiting the gain response of a transformer-coupled amplifier.

5-2 Draw the circuit diagram of a two-stage BJT amplifier using input, interstage, and output transformers. Include the appropriate biasing arrangements.

5-3 What is the advantage of using complementary transistors in a multistage direct-coupled amplifier?

5-4 Why do emitter bypass capacitors limit the low-frequency response of an amplifier?

5-5 As a rule of thumb, how large should the reactance of an emitter bypass capacitor be?

5-6 Define the alpha and beta cutoff frequencies of a BJT.

5-7 What is the significance of the transition frequency of a bipolar transistor? How is the transition frequency related to the current gain?

5-8 What are the two major limitations on the high-frequency performance of a *CE* circuit? Which of these factors is the more critical one?

5-9 Name some of the advantages and disadvantages of direct coupling.

5-10 How is the passband of a dc amplifier defined?

5-11 What is meant by CMRR?

5-12 A certain laboratory oscilloscope has a vertical deflection amplifier with a differential input. What are the advantages of using a differential input in this application?

5-13 Explain how emitter feedback in an *RC*-coupled amplifier affects the voltage gain and the circuit stability.

5-14 The input and output capacitors in a common-base amplifier each produce a certain cutoff frequency. Which of these capacitors is the more critical in terms of the overall frequency response? Why?

5-15 Three identical *CE* stages are connected in cascade. Explain how the lower 3-dB frequency of the cascade is affected by the lower cutoff frequency of each individual stage.

5-16 Draw the circuit diagram of a two-stage amplifier with a FET input stage and an emitter-follower output stage. Comment on the impedance levels of input and output.

5–17 Draw the circuit diagram of a two-stage MOST amplifier with negative feedback from output stage to input stage.

5–18 If the feedback resistor in the MOST amplifier of Fig. 5–25 is reduced to one-half its original value, explain how this affects the overall voltage gain, the input impedance, and the distortion.

5–19 If the feedback resistor in the MOST amplifier of Fig. 5–25 is removed entirely, how would this affect the input impedance of the amplifier?

5–20 Explain how the feedback resistor in Fig. 5–28 affects the gain and the circuit stability.

PROBLEMS

5–1 Calculate the input resistance of the *RC*-coupled amplifier of Fig. 5–3.

5–2 A 10-mV signal source with an internal resistance of 10 kΩ is connected to the input terminals of the two-stage *RC*-coupled amplifier of Fig. 5–3. Determine the output voltage across the 20-kΩ load resistor.

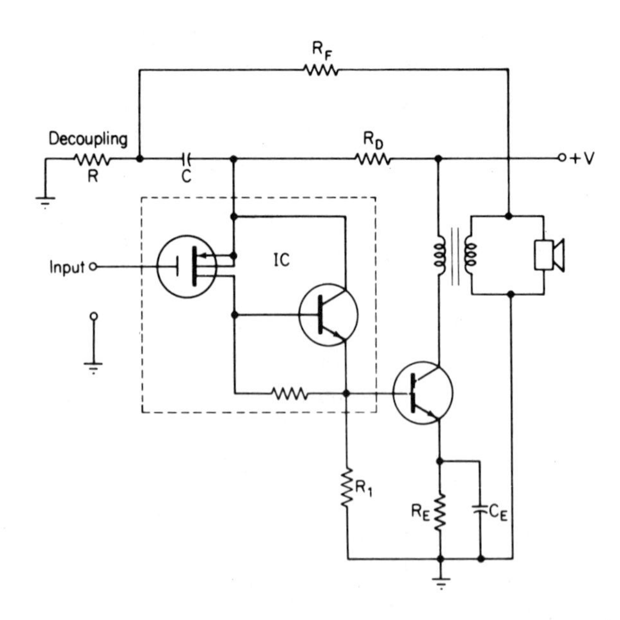

Fig. 5–28 Circuit of a record player amplifier consisting of an integrated pre-amplifier plus an output stage. The circuit inside the dashed line is an integrated monolithic amplifier.

5–3 Assume that all coupling capacitors in the amplifier of Fig. 5–3 are 0.5 μF and the source generator has an internal resistance of 5 kΩ. Determine the lower corner frequency of the amplifier.

5–4 A certain *RC*-coupled amplifier has upper and lower corner frequencies of 1 and 15 kHz, respectively. What is its bandwidth?

5–5 The BJT in the single-stage amplifier of Fig. 5–7 has the following parameters:

$$\text{collector capacitance, } C_C = 10 \text{ pF}$$
$$\text{transition frequency, } f_T = 120 \text{ MHz}$$
$$\text{current gain, } \beta = 240$$

The stray wiring capacitance is 20 pF. Determine the upper corner frequency of this amplifier.

5–6 It is desired to couple a 4-Ω loudspeaker to a *CE* output stage so that the reflected collector load is 5 kΩ. Calculate the transformer turns ratio.

5–7 Calculate the secondary load voltage in Fig. 5–9, when the turns ratio a = 8 and a signal source with an internal resistance of 1 kΩ applies a 10-mV signal to the amplifier input. The BJT has a $\beta = 100$.

5–8 Determine the approximate dc operating voltages and currents for the three-stage direct-coupled amplifier of Fig. 5–29 and calculate the overall voltage gain. Assume identical transistors with a current gain of 40.

5–9 A certain differential amplifier has a common-mode rejection ratio of 10,000 and a voltage gain of 100. Determine the output voltage for a common input of 1 V rms.

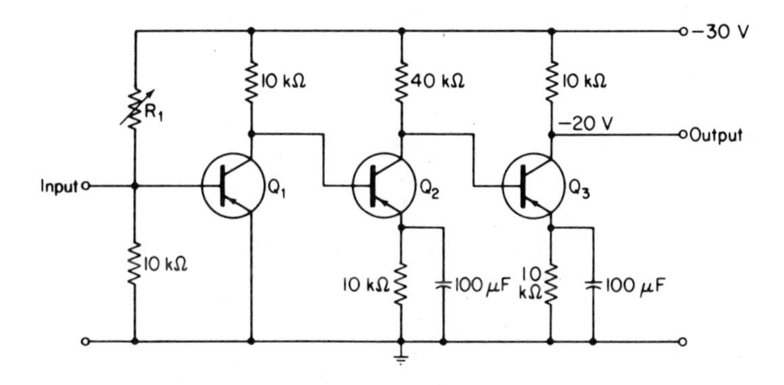

Fig. 5–29 Problem 5–8. Three-stage dc-coupled amplifier.

5–10 Determine the voltage gain of the two-stage dc amplifier of Fig. 5–14, if the output is taken from the emitter instead of from the collector.

5–11 What is the voltage gain of the amplifier of Fig. 5–14 if the emitter resistor of $Q1$ is bypassed with a very large capacitor?

5–12 Feedback resistor R_F in Fig. 5–18 is assumed to be 100 kΩ.
Calculate (a) The feedback factor.
(b) The amplifier gain.
(c) The amplifier input resistance.

5–13 The differential amplifier of Fig. 5–21 has a voltage gain $A_v = 60$. If $v_1 = +200$ mV and $v_2 = -100$ mV, determine the output voltage v_o.

5–14 A certain differential amplifier has a CMRR of 60 dB and a voltage gain of 40 dB. If the differential input consists of a 100-mV signal with a 10-mV in-phase noise component superimposed, determine the signal output and the noise output.

5–15 The differential amplifier of Fig. 5–30 uses identical transistors with a current gain $\beta = 20$ and a *CE* input resistance of 2 kΩ. Determine the differential gain of this amplifier.

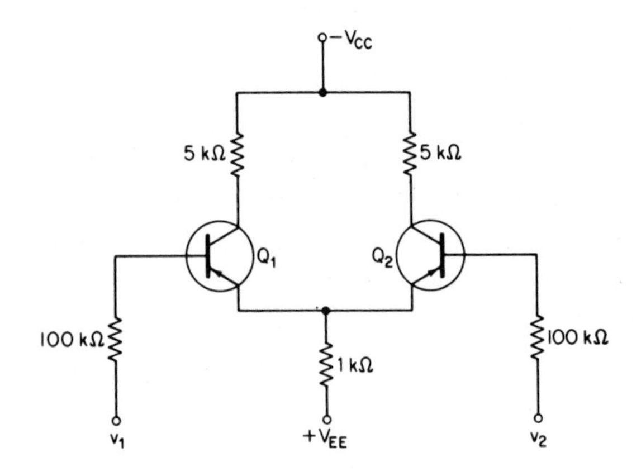

Fig. 5–30 Circuit for Problem 5–15.

5–16 Figure 5–31 shows a two-stage dc-coupled amplifier.
Determine (a) The collector and emitter voltages of $Q1$.
(b) The overall current and voltage gains.
(c) The input impedance of the amplifier.
(d) The required setting of bias resistor R_1.

5–17 Refer to Fig. 5–31 and determine the amplifier gain if the output terminal is shifted to the emitter of $Q2$.

5–18 Refer to Fig. 5–31 and determine the voltage gain if the emitter resistor of $Q2$ is bypassed with a large capacitor.

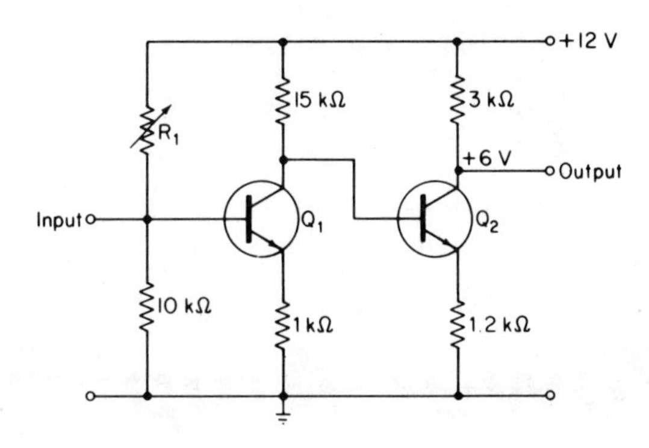

Fig. 5-31 Circuit for Problem 5-16.

5-19 A feedback amplifier produces an output of 10 V with an input of 0.5 V. Without feedback the amplifier requires 10 mV for the same output. Determine the value of the factor β and also the gain of the amplifier without feedback.

5-20 A certain amplifier has a voltage gain of 40 dB and produces an output of 5 V with 10 percent distortion. It is required to reduce the distortion to 2 percent by using negative feedback. Determine the factor β and the voltage gain with feedback.

Chapter 6

Power Amplifiers

6–1 INTRODUCTION

In many electronic systems substantial signal power must be delivered to an output device (load). The load may take several forms, such as a loudspeaker, an ac motor, a servomotor, or an electromechanical device.

The power amplifier should deliver signal power to the load with a minimum of distortion and with high efficiency. The efficiency depends on the circuit configuration and can, to a certain extent, be controlled by the circuit designer. The distortion, generated in the power amplifier, is a function of circuit components and circuit design.

Frequency distortion is defined as the variation in amplifier gain as a function of the frequency of the input signal. An amplifier can be designed with a linear gain response (within 1 dB) over a wide frequency band. This is helped by rejecting frequency-sensitive components (such as coupling capacitors) in favor of direct-coupled amplifier stages wherever possible.

Capacitors may also cause *phase distortion.* This results from unequal phase shifts in signals of different frequencies. To preserve the phase relationship of all frequency components in a complex waveform, the phase angle between output and input signal voltages should be linear with frequency. Phase distortion is of minor importance in audio circuits, where it is not perceptible to the ear. But it can be objectionable in systems requiring exact preservation of waveform characteristics, such as television or pulse circuits. The use of capacitors then should be restricted.

Nonlinear distortion may be of two kinds: harmonic or amplitude distortion, and intermodulation distortion. *Harmonic* distortion is caused by nonlinearity in the transistor characteristics or by nonlinear components. Overdriving a transistor, so that cutoff or saturation occurs, results in harmonic distortion. *Intermodulation* distortion is the result of interaction of two or more signal frequencies in an amplifier stage. The output signal of the stage may then contain frequency components consisting of the sum and difference of the original input signal frequencies.

6–2 CLASSIFICATION OF AMPLIFIERS

Depending on circuit bias conditions, power amplifiers are of three basic classes: A, B, and C.

In a class A amplifier, the bias conditions are such that the operating point (Q point) of the transistor is located well within the linear part of the collector characteristics. The applied signal never drives the transistor into cutoff or into saturation. The optimum bias is such that the Q point is approximately halfway between cutoff and saturation, as in Fig. 6–1.

Signal excursion is limited to the active region and avoids cutoff (point A) and saturation (point B). In Fig. 6–1, the quiescent collector voltage V_C is approximately one-half the supply voltage V_{CC}. The quiescent collector current I_C causes quiescent collector dissipation and this is wasteful power in the sense that it does not contribute to ac load power, so the class A amplifier is the least efficient of the three basic categories.

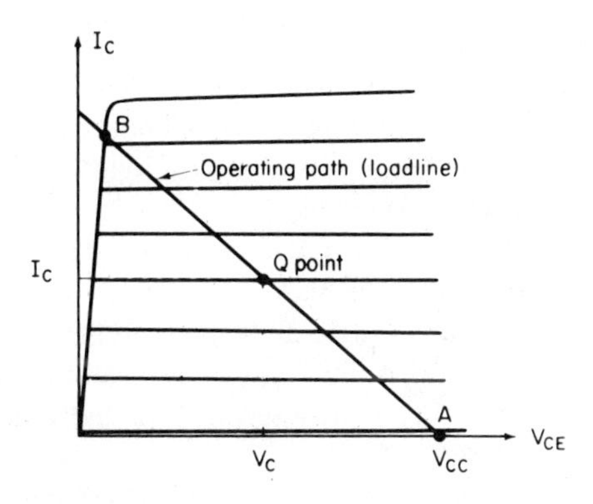

Fig. 6–1 In class A the operating point is placed half-way between saturation and cutoff.

In a class B amplifier, the transistor is biased at cutoff, with the *Q* point located on the V_C axis, Fig. 6–2. In the case of a sinusoidal input signal, ac collector current flows only during one-half of the cycle (180 degrees), and is zero for the other half of the input cycle. The quiescent collector current is zero and hence the quiescent power dissipation is zero. Because of zero power dissipation in the absence of an input signal, a class B amplifier is more efficient than a class A amplifier. Since only one-half of the input signal is subject to amplification, the class B amplifier is generally connected in a push–pull configuration, when two transistors alternately amplify one-half of the input signal (Section 6–4).

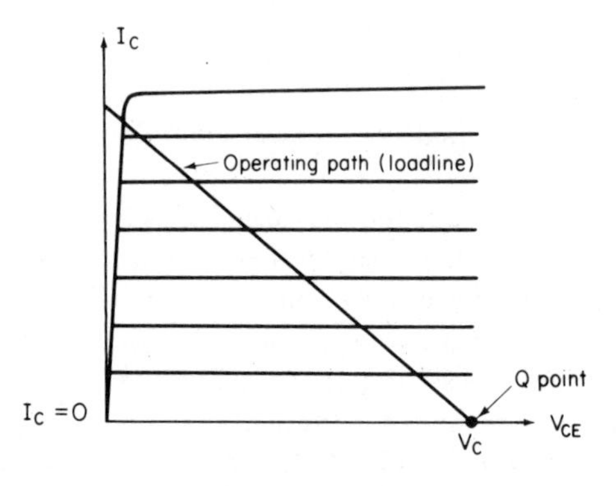

Fig. 6–2 In class B the operating point is placed on the V_{CE} axis, at the cutoff point.

A class AB amplifier operates between the two conditions described for class A and class B, with the *Q* point somewhere between cutoff and the center of the collector characteristics. Hence, some quiescent collector current flows and some dc collector dissipation is present. With an applied input signal, the transistor is driven into cutoff for less than half of the signal cycle and the collector current is therefore zero for less than half of the input signal cycle. Class AB amplifiers are also connected in push–pull configurations and tend to reduce crossover distortion (Section 6–5).

In a class C amplifier, the transistor is biased well beyond cutoff and the collector current is zero for more than one-half of the input signal cycle. Collector current tends to flow in short-duration pulses, only when the transistor is driven out of its cutoff condition into conduction. The class C amplifier has the highest efficiency, and is used primarily in tuned radio frequency power amplifiers.

Because of its relatively high current and power gain, the *CE* transistor configuration is generally used in power amplifiers.

6–3 CLASS A POWER AMPLIFIER

Figure 6–3 shows the circuit diagram of a transformer-coupled class A power amplifier. Resistors R_A and R_B supply bias to power transistor $Q1$ while emitter resistor R_E provides the necessary dc circuit stabilization (Section 3–5). R_E is usually bypassed with a fairly large emitter capacitor C_E to improve the stage gain. Load resistor R_L is transformer-coupled to the collector and is reflected into the collector circuit as

$$R'_L = \left(\frac{N_1}{N_2}\right)^2 R_L \tag{6–1}$$

There are several advantages in using a transformer to couple the load to the collector. First, with transformer coupling there is no direct current and hence no wasteful dc power dissipation in R_L. This is particularly important when R_L represents a loudspeaker or a servomotor, where direct current would degrade the performance of the output device. Second, if the resistance of the transformer primary is neglected, the collector voltage V_C equals the supply voltage V_{CC}, and hence a smaller supply voltage can be used. Third, the transformer provides impedance transformation and allows a low-impedance load, such as a loudspeaker, to be reflected as a higher impedance collector load necessary for correct ac operation of the power transistor. On the other hand, a transformer is relatively heavy and occupies valuable space.

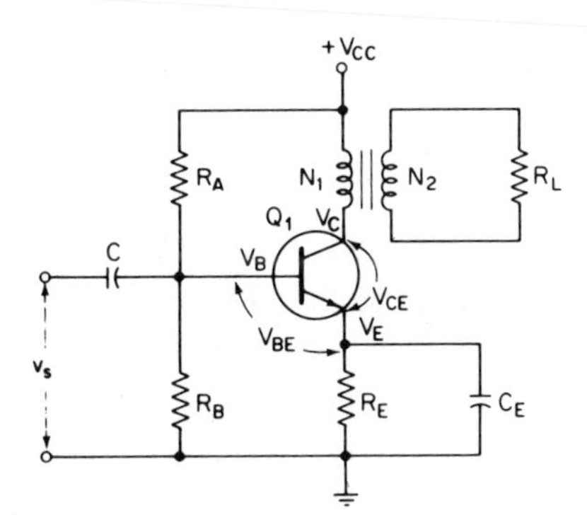

Fig. 6–3 Transformer-coupled class-A power amplifier.

Figure 6–4 represents a simplified circuit of the class A power amplifier of Fig. 6–3. In this equivalent circuit, the collector supply voltage is represented by $V_{CC'} = V_{CC} - V_E$. The base bias circuit is replaced by its Thévenin equivalent battery V_{BB} in series with resistor R_b, where

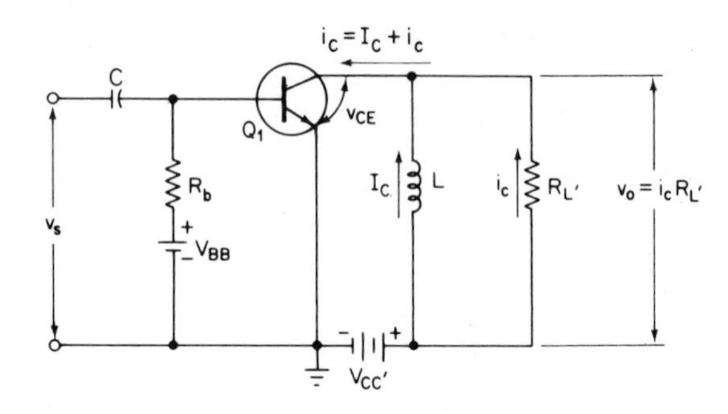

Fig. 6–4 Simplified equivalent circuit of the class-A power amplifier. The base bias circuit of Figure 6–3 is replaced by its Thévenin equivalent, while the load resistor is reflected into the collector circuit as R_L'.

$$V_{BB} = \frac{R_B}{R_A + R_B} V_{CC} \tag{6–2}$$

and

$$R_b = \frac{R_A R_B}{R_A + R_B} \tag{6–3}$$

The total collector current i_C consists of two components: a dc component I_C (quiescent collector current) and an ac component i_c (load current), so

$$i_C = I_C + i_c \tag{6–4}$$

The primary of the output transformer, represented by inductor L, acts as a short circuit to the dc component so that there is no dc voltage drop across $R_{L'}$. The ac voltage drop across $R_{L'}$ is the output voltage

$$v_o = i_c R_L' \tag{6–5}$$

In the circuit of Fig. 6–4, the instantaneous collector-to-emitter voltage equals

$$v_{CE} = V_{CC'} + v_o = V_{CC'} + i_c R_L' \tag{6–6}$$

Equation (6–6) is the expression for the *ac loadline,* plotted on the hypothetical collector characteristics of Fig. 6–5. The quiescent collector-to-emitter voltage equals $V_{CC'}$ and the quiescent collector current equals I_C. Since in a class A amplifier the operating point makes equal excursions on each side of the Q point, the ac collector voltage can reach a peak value $V_m = V_{CC'}$ and the total collector voltage swing can therefore extend from 0 to 2 $V_{CC'}$. To avoid nonlinearity and distortion, however, the signal excursion is generally kept well below this possible peak value.

When the input voltage v_s is sinusoidal, the average or dc value of the ac collector signal current is zero. Only the quiescent collector current contributes to the power drawn from the battery, and this power equals

$$P_{CC} = V_{CC'}I_C \tag{6–7}$$

Hence, in a class A power amplifier, the power drawn from the dc supply is constant and depends only on collector voltage $V_{CC'}$ and quiescent collector current I_C.

The ac power delivered to the load equals

$$P_L = R_L'(I_{\mathrm{rms}})^2 \tag{6–8}$$

where I_{rms} is the rms value of the ac load current in the transformer primary. The total power drawn from the collector supply equals the power delivered to the

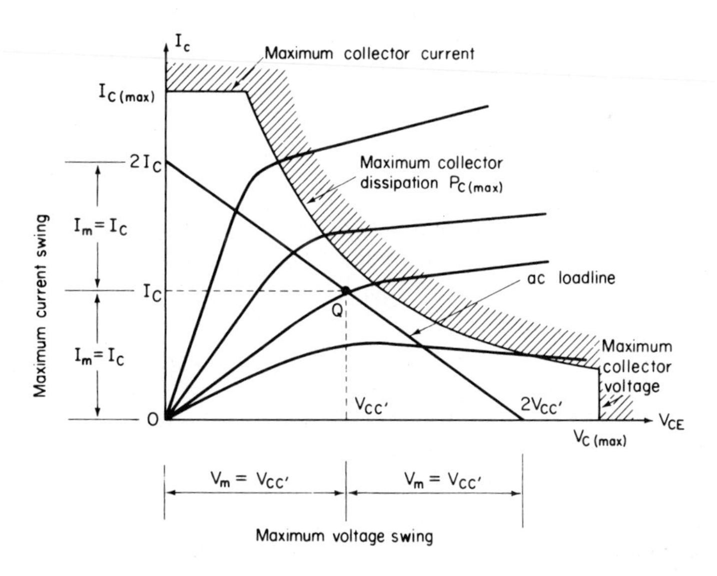

Fig. 6–5 Graphical analysis of the class-A power amplifier. The shaded area indicates the limitations imposed by maximum transistor ratings of collector voltage, collector current, and power dissipation.

load plus the power dissipated in the transistor as collector dissipation, so we can write

$$P_{CC} = P_L + P_C \tag{6–9}$$

where P_C represents the collector dissipation. Under quiescent conditions, when there is no input signal, $P_L = 0$ and hence $P_{CC} = P_C = V_{CC'}I_C$. This indicates that the collector dissipation is maximum when there is no input signal.

The maximum power that can be delivered to the load equals the product of the maximum ac collector voltage and the corresponding maximum ac collector current, so

$$P_{L(\max)} = V_{\text{rms}(\max)} I_{\text{rms}(\max)} = 0.5 V_{CC'} I_C \tag{6–10}$$

The maximum theoretical efficiency (sometimes called "conversion efficiency") of the class A power amplifier then equals

$$\eta = \frac{P_{L(\max)}}{P_{CC}} = \frac{0.5 V_{CC'} I_C}{V_{CC'} I_C} = 0.5 \quad \text{or} \quad 50\% \tag{6–11}$$

Equation (6–11) indicates that maximum ac load power is obtained only when the amplifier is designed for maximum quiescent collector dissipation. In practice, 50 percent efficiency is not reached, since the transistor would then operate in the nonlinear part of its characteristics and unacceptable distortion would be introduced.

The amount of power the transistor itself can dissipate is limited by the maximum ratings of collector voltage, current, and power. These maximum permissible ratings are indicated by the closed region bounded by $I_{C(\max)}$, $V_{C(\max)}$, and $P_{C(\max)}$ in Fig. 6–5.

When the power to be delivered to the load P_L is specified (with a given input signal waveform), the maximum collector dissipation $[P_{C(\max)}]$ can be determined. For a sinusoidal input signal,

$$P_L = P_{C(\max)} = 0.5 V_{CC'} I_C \tag{6–12}$$

Based on these parameters, a suitable power transistor with permissible collector dissipation at least as great as $P_{C(\max)}$, and suitable current and voltage ratings, can be selected.

EXAMPLE 6-1

A silicon power transistor, operating in the class A configuration of Fig. 6–3, has the following maximum ratings: $I_C = 4$ A, $V_{CE} = 45$ V, $P_c = 15$ W, $V_{BE} = 1$ V, and

$\beta > 30$. A 4-Ω loudspeaker is transformer-coupled to the collector circuit. The transistor is to operate within its ratings and is to deliver maximum power to the 4-Ω load. Specify reasonable operating conditions.

SOLUTION

The maximum ratings determine the maximum allowable values for quiescent collector voltage and current. These are $I_{C(\max)} = 4\text{ A}/2 = 2\text{ A}$ and $V_{CC'} = 45\text{ V}/2 = 22.5\text{ V}$. At these Q-point values, the collector dissipation is $P_{C(\max)} = I_{C(\max)} \times V_{CC'} = 2\text{ A} \times 22.5\text{ V} = 45\text{ W}$, which far exceeds the rated dissipation of 15 W. Hence, the actual quiescent current and/or voltage must be reduced. Suppose that we select a power supply voltage of $V_{CC} = 15$ V and assume that the emitter voltage V_E is 10 percent of the supply voltage (an average value). In that case, $V_E = 1.5$ V and $V_{CC'} = V_{CC} - V_E = 13.5$ V. The Q-point collector voltage is therefore 13.5 V and the maximum Q-point collector current which may be allowed without exceeding the rated dissipation is $I_{C(\max)} = 15\text{ W}/13.5\text{ V} = 1.1$ A. To assume operation below this limit, we select quiescent operation with $I_C = 1$ A. To give the ac loadline the correct slope, the load resistance reflected into the primary of the output transformer is $R'_L = V_{CC'}/I_C = 13.5\text{ V}/1\text{ A} = 13.5\ \Omega$. This establishes the required turns ratio of the transformer. The quiescent power dissipation equals $P_C = V_{CC'}I_C = 13.5\text{ V} \times 1\text{ A} = 13.5$ W. With a sinusoidal input signal applied, the maximum signal power the amplifier can deliver to the load equals $P_L = 0.5\ V_{CC'}I_C = 6.75$ W. The total power drawn from the battery under these conditions is $P_{CC} = P_C + P_L = 13.5\text{ W} + 6.75\text{ W} = 20.25$ W.

6–4 CLASS B PUSH–PULL POWER AMPLIFIER

In the class B mode of operation, the transistor is biased at cutoff, as was indicated in Fig. 6–2. If the power amplifier of Fig. 6–3 were biased at cutoff, the transistor would respond only to the positive portions of the input signal, when it would be forced into conduction. During the negative portions of the input signal the transistor would remain cut off and no collector current would flow. Clearly, if a class B amplifier is to reproduce the entire signal waveform, a different circuit configuraion must be considered. One such configuration, called a *transformer-coupled push–pull amplifier,* is shown in Fig. 6–6.

Input signal v_s is coupled by a center-tapped input transformer to a pair of identical *npn* transistors. The input or driver transformer provides two input signals in *antiphase,* which is evident from the polarity dots on the transformer windings.

The load, represented by load resistor R_L, is coupled to the collector circuits of the power transistors by a center-tapped output transformer. The use of transformers in this circuit has disadvantages such as distortion, frequency response and bandwidth; on the other hand, it provides for impedance transformation and matches the load to the collector circuit.

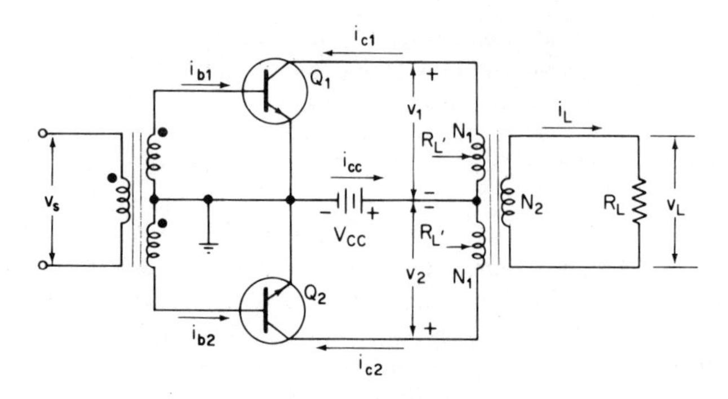

Fig. 6–6 Transformer-coupled class-B push-pull amplifier.

In the absence of an input signal (quiescent conditions), the base and emitter of each transistor are at ground potential and hence both transistors are cut off (class B operation). When input signal v_s goes positive, transistor $Q1$ becomes forward-biased and conducts, while transistor $Q2$ becomes reverse-biased and is driven deeper into cutoff. When $Q1$ conducts, it produces a collector current i_{c1} through the upper half of the output transformer with a direction as indicated in Fig. 6–6. When the input signal v_s goes negative, $Q2$ conducts and $Q1$ is driven into cutoff. Collector current i_{c2} flows through the lower half of the output transformer with a direction as indicated in Fig. 6–6. Hence, $Q1$ amplifies the *positive* portions of the input signal and $Q2$ amplifies the *negative* portions. The two transistors operate as switches delivering output current to the transformer-coupled load on alternate half-cycles. Figure 6–7 illustrates the action of the circuit with a sinusoidal input signal v_s.

Under quiescent conditions, since both transistors are cut off, the power dissipation is zero. With input signal v_s applied, the current drawn from the power supply V_{CC} is represented by Fig. 6–7(d), where

$$i_{cc} = i_{c1} + i_{c2} \tag{6–13}$$

If I_m represents the peak value of collector currents i_{c1} and i_{c2}, the average value of the total current waveform of Fig. 6–7(d) is

$$I_{(av)} = \frac{2}{\pi} I_m \tag{6–14}$$

and the power delivered by V_{CC} is $P_{CC} = V_{CC}I_{(av)}$

$$P_{CC} = \frac{2}{\pi} V_{CC} I_m \tag{6–15}$$

or

$$P_{CC} = \frac{2}{\pi} \frac{V_{CC}^2}{R_L'} \tag{6–16}$$

Since I_m represents the peak value of the collector current or the primary load current, the power delivered to the load equals

$$P_L = R_L' I_{\text{rms}}^2 = 0.5 R_L' I_m^2 \tag{6–17}$$

The collector-to-emitter voltages of $Q1$ and $Q2$ can be expressed as

$$v_{CE_1} = V_{CC} - v_1 = V_{CC} - i_{c1} R_L' \tag{6–18}$$

and

$$v_{CE_2} = V_{CC} - v_2 = V_{CC} - i_{c2} R_L' \tag{6–19}$$

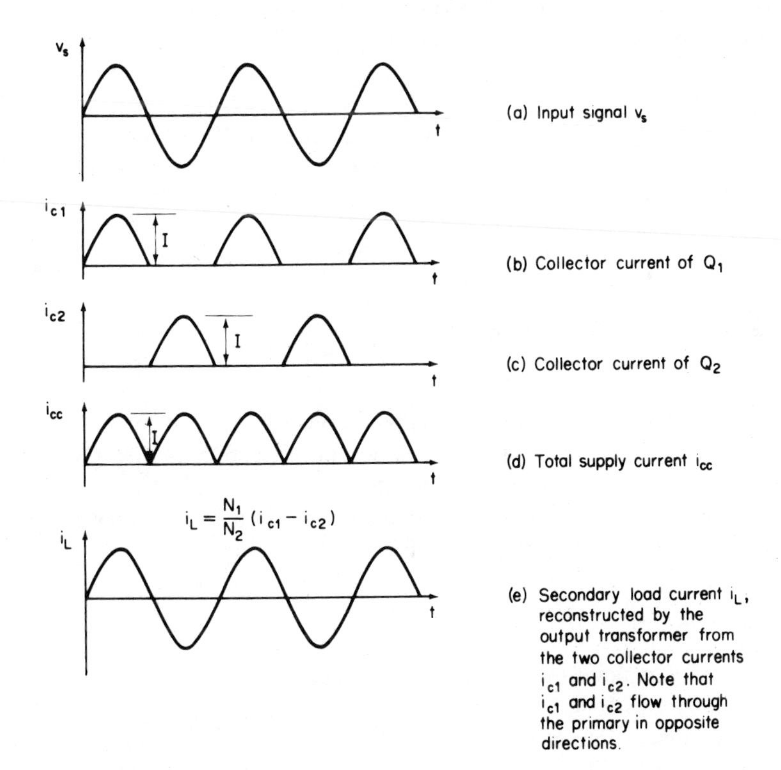

v_s
t
(a) Input signal v_s
i_{c1}
I
t
(b) Collector current of Q_1
i_{c2}
I
t
(c) Collector current of Q_2
i_{cc}
I
t
(d) Total supply current i_{cc}
$i_L = \frac{N_1}{N_2}(i_{c1} - i_{c2})$
i_L
t
(e) Secondary load current i_L, reconstructed by the output transformer from the two collector currents i_{c1} and i_{c2}. Note that i_{c1} and i_{c2} flow through the primary in opposite directions.

Fig. 6–7 Current waveforms for the push-pull amplifier of Figure 6–6.

It follows from Eqs. (6–18) and (6–19) that the peak value of the transformer primary voltages v_1 and v_2 can reach a maximum of V_{CC}. Hence, the maximum power which can be delivered to the load equals

$$P_L = \frac{0.5V_{CC}^2}{R_L'} = 0.5V_{CC}I_m \tag{6–20}$$

The maximum theoretical efficiency of the class B power amplifier therefore equals

$$\eta = \frac{P_L}{P_{CC}} = \frac{0.5V_{CC}^2/R_L'}{(2/\pi)(V_{CC}^2/R_L')} = 0.785 \quad \text{or} \quad 78.5\% \tag{6–21}$$

The maximum output power which can be delivered by the class B amplifier is limited by the maximum rated collector current and voltage of the transistors or, in some cases, by the rated collector dissipation of the transistors.

The maximum collector-to-emitter voltage $V_{CE(\max)}$ and the maximum collector current $I_{C(\max)}$ are listed as maximum ratings in the transistor data sheet. Referring to Fig. 6–6, when $Q1$ conducts, its collector current i_{c1} through the upper half of the output transformer induces a voltage $-v_2$ in the lower half of the output transformer. Since $v_1 = -v_2$ may reach a peak value of V_{CC}, the transistors must be able to withstand a maximum collector-to-emitter voltage $V_{CE(\max)}$ equal to twice the supply voltage V_{CC}. Entering this condition into Eq. (6–20) we obtain

$$P_L = 0.5V_{CC}I_m = \tfrac{1}{4}V_{CE(\max)}I_{C(\max)} \tag{6–22}$$

Equation (6–22) represents the maximum output power the class B amplifier can deliver without exceeding the maximum transistor current and voltage ratings.

In the second case the output power may be limited by the maximum collector dissipation of the transistors. In the class B amplifier the collector dissipation of each transistor is zero under quiescent conditions ($i_{c1} = i_{c2} = 0$). Hence, the maximum collector dissipation must occur when signal is applied. The instantaneous collector dissipation equals

$$p_c = v_{ce}i_c \tag{6–23}$$

As the operating point moves along the ac loadline (Fig. 6–8), from the Q point at $V_{CE} = V_{CC}$ and $I_C = 0$ to the other extreme on the I_C axis, where $V_{CE} = 0$ and $I_C = I$, the instantaneous values of v_{ce}, i_c, and p_c all change from point to point along this path. The maximum value of p_c occurs when the ac loadline is tangent to the hyperbola of maximum permissible collector dissipation. The point of

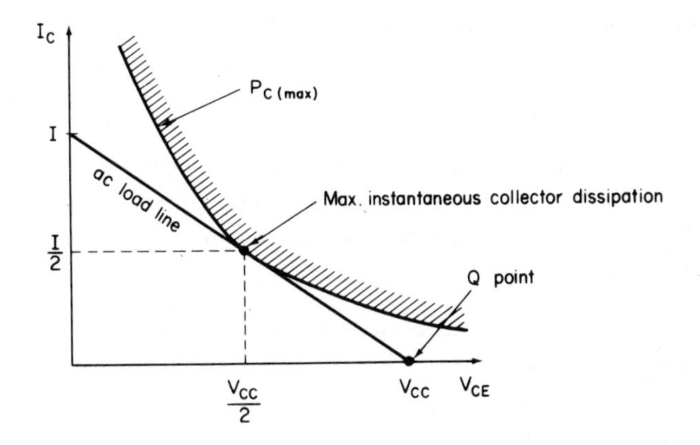

Fig. 6–8 Maximum instantaneous collector dissipation occurs when the ac loading is tangent to the hyperbola of maximum collector dissipation. The point of tangency occurs at $V_{CE} = \frac{1}{2} V_{CC}$.

tangency is located at $V_{CE} = V_{CC}/2$ and $I_C = I/2$. This reasoning holds for $Q1$ on positive signal excursions and for $Q2$ on negative signal excursions. Thus the maximum instantaneous collector dissipation for each transistor equals

$$P_{C(\max)} = \frac{V_{CC}}{2}\frac{I}{2} = \frac{V_{CC}^2}{4R_L'} \tag{6–24}$$

Hence, from Eq. (6–20), the output power the amplifier can deliver without exceeding the maximum instantaneous collector dissipation equals

$$P_L = \frac{0.5\,V_{CC}^2}{R_L'} = 2P_{C(\max)} \tag{6–25}$$

If the instantaneous maximum collector dissipation is indeed equal to the maximum permissible collector dissipation (as is the case when the ac loadline is tangent to the hyperbola of maximum collector dissipation), the output power under maximum signal conditions can be expressed as

$$P_L = 2P_{C(\max)} \tag{6–26}$$

Comparing this result with the maximum output power available from a class A amplifier, in class B operation the output power for a sinusoidal input signal is twice as great for class A operation.

The objective of a class B push–pull amplifier is to deliver a specified amount of output power to a given load with a specified (usually sinusoidal) input signal. With transformer coupling of the load, the reflected load resistance can be adjusted to suit the requirements of the transistor collector circuit and suitable power transistors can then be selected. With the load power P_L specified, and R_L' determined, the collector supply voltage V_{CC} can be determined from Eq. (6–20). The peak collector current can be found by realizing that $V_{CC} = R_{L'}I$. The maximum instantaneous collector dissipation is determined from Eq. (6–24) and this then allows the selection of transistors having suitable voltage, current, and power ratings.

EXAMPLE 6-2

The push–pull amplifier of Fig. 6–6 is required to deliver 10 W of output power to a 4-Ω loudspeaker which is transformer-coupled to the collector circuit. The available collector supply voltage $V_{CC} = 30$ V. Determine the required turns ratio of the output transformer and the minimum voltage, current, and power ratings of the power transistors.

SOLUTION

With $P_L = 10$ W and $V_{CC} = 30$ V, Eq. (6–20) yields the required reflected load resistance R_L' since

$$P_L = \frac{0.5V_{CC}^2}{R_L'}$$

and $R_L' = 0.5(30)^2/10 = 45\ \Omega$.
Since

$$R_L' = \left(\frac{N_1}{N_2}\right)^2 R_L$$

the turns ratio of the output transformer equals

$$a = \sqrt{\frac{45}{4}} \simeq 3.35.$$

Since

$$P_L = 2P_{C(\text{max})}$$

the maximum instantaneous collector dissipation equals

$$P_{C(\text{max})} = \frac{P_L}{2} = 5\text{ W}$$

Hence, each transistor should have a maximum collector dissipation of at least 5 W. The maximum collector current $I_{C(\max)}$ is determined from

$$I_{C(\max)} = \frac{V_{CC}}{R_L'} = \frac{30\text{ V}}{45\ \Omega} = \frac{2}{3}\text{ A}$$

Under maximum signal conditions, each transistor must be able to withstand a collector-to-emitter voltage equal to at least twice the supply voltage. This condition yields

$$V_{CE(\max)} = 2V_{CC} = 60\text{ V}$$

Using the maximum voltage and current ratings of the transistor to calculate the maximum load power which can be delivered and using Eq. (6–22), we obtain

$$P_L = \tfrac{1}{4}\, V_{CE(\max)} I_{C(\max)}$$

or

$$P_L = \tfrac{1}{4} \times 60\text{ V} \times \tfrac{2}{3}\text{ A} = 10\text{ W}$$

Summarizing, the transistor ratings should be

$$V_{CE(\max)} > 60\text{ V}$$

$$I_{C(\max)} > \tfrac{2}{3}\text{ A}$$

$$P_{C(\max)} > 5\text{ W}$$

6–5 CROSSOVER DISTORTION IN PUSH–PULL AMPLIFIERS

The input circuit of the class B amplifier of Fig. 6–6 can be represented as Fig. 6–9. Each half of the input transformer is replaced by its equivalent circuit consisting of voltage source v_s and internal resistance R_s. Base currents i_{b1} into $Q1$ and i_{b2} into $Q2$ depend on the magnitude of input voltage v_s and input resistance R_s and also on the forward-bias characteristics of the base–emitter junctions of $Q1$ and $Q2$. The base current into the device is practically zero until the base–emitter voltage reaches a value of a few tenths of a volt (approximately 0.3 V for germanium and 0.7 V for silicon). This characteristic has a detrimental effect on the performance of the class B push–pull amplifier in terms of its waveform reproduction.

Consider the composite volt–ampere characteristics of the input circuit of Fig. 6–9, as shown in Fig. 6–10. The input characteristics of $Q1$ and $Q2$ are shown "back to back" and it is assumed that the base–emitter voltages of $Q1$ and

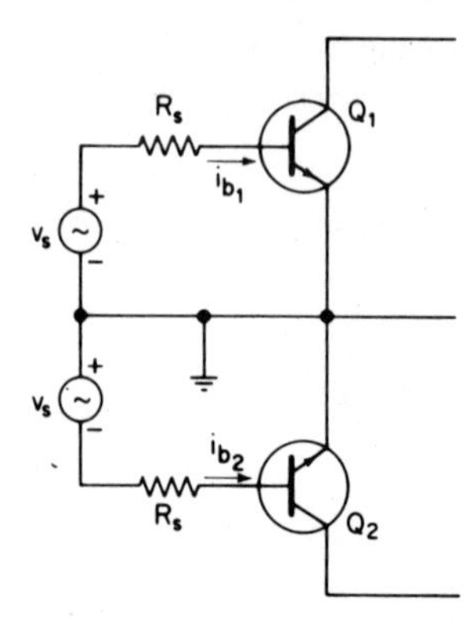

Fig. 6–9 Equivalent circuit for the input circuit of the push-pull amplifier of Figure 6–6. The input transformer is replaced by voltage source v_s and internal resistance R_s.

$Q2$ are equal to the source voltage v_s (neglect R_s). Base current i_{b1} does not have an appreciable value until source voltage v_s has reached a positive value of a few tenths of a volt. Similarly, base current i_{b2} does not have an appreciable value until v_s has reached a negative value of a few tenths of a volt. The area between the knees of the back-to-back characteristics thus represents a region in which the base currents are not directly proportional to the source voltage v_s, and this introduces a form of distortion called *crossover distortion*.

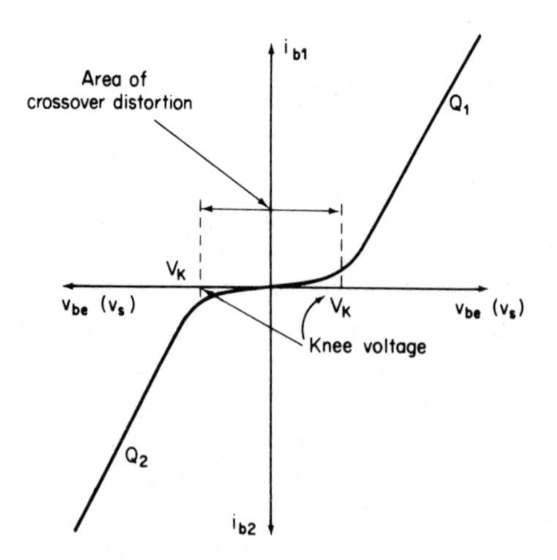

Fig. 6–10 Composite volt-ampere characteristic of the input circuit of the push-pull amplifier of Figure 6–6.

If the source voltage v_s is sinusoidal, as in Fig. 6–11 (a), base currents i_{b1} and i_{b2} have waveforms as indicated in Fig. 6–11 (b). The composite collector current, or load current, is also nonlinear, with a waveform as in Fig. 6–11 (c). The nonlinearity is introduced when the action of the circuit transfers or "crosses over" from one transistor to the other.

If the exact waveform of the input signal must be preserved in the output circuit of the amplifier, as in audio signal reproduction, the cause of crossover distortion must be eliminated or at least sharply reduced. In a practical circuit, this is often achieved by applying a small forward bias to the base–emitter junctions of $Q1$ and $Q2$, thereby placing both transistors at the threshold of strong conduction. The back-to-back input characteristics of $Q1$ and $Q2$ are then shifted along the voltage axis in the manner indicated in Fig. 6–12. Notice that in this case the composite volt–ampere characteristic more nearly approaches a straight line. A commonly used bias circuit is shown in Fig. 6–13, where diode D provides just sufficient bias to place both $Q1$ and $Q2$ at the knee of their input volt–ampere characteristic. This circuit arrangement is popular because diode D stabilizes the Q points of the transistors against temperature variations. A resistive bias network can also be used, simply by replacing the diode by a

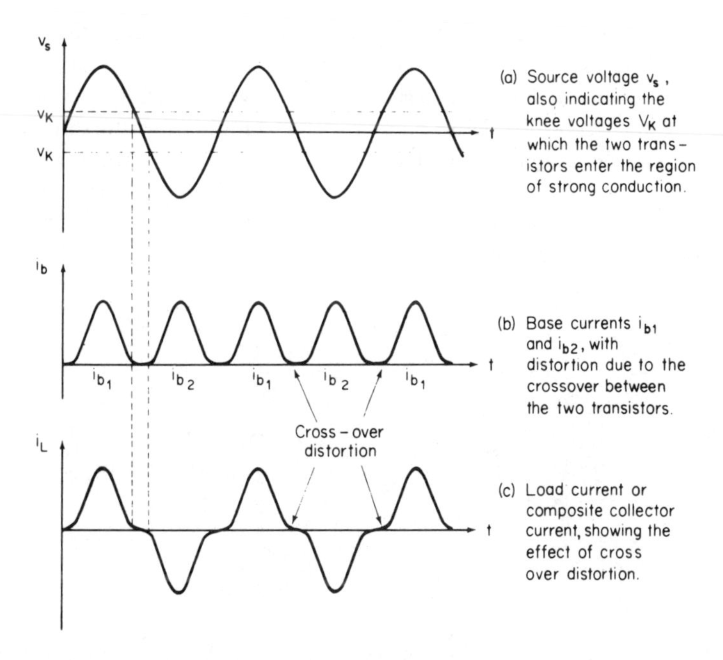

Fig. 6–11 Waveforms for the push-pull amplifier, showing the effects of crossover distortion on the load current.

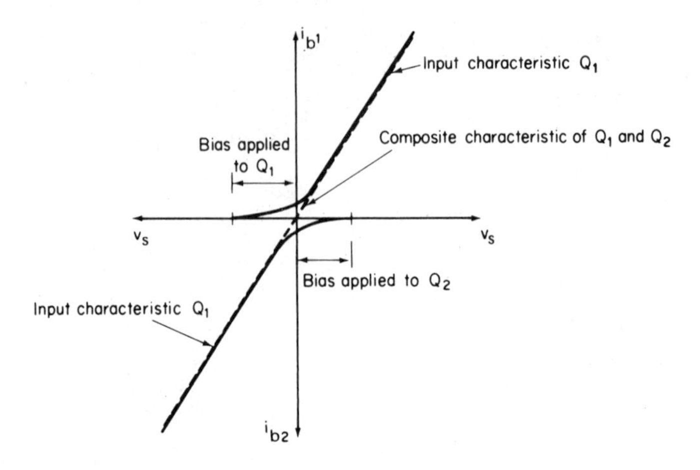

Fig. 6–12 Composite input characteristic of the push-pull amplifier. Fixed bias is applied to both transistors, placing them in their region of strong conduction, even at zero input signal.

resistor of suitable value. The advantage of Q point stabilization, however, is then sacrificed.

In addition to the problems of crossover distortion, another difficulty with the push–pull amplifier is that it requires two transistors of the same polarity and with almost identical characteristics. If the two transistors are not equal, nonlinear distortion results. Most manufacturers supply transistors in "matched pairs," either *npn* or *pnp*. A matched pair consists of two transistors whose characteristics are very similar and they are specially supplied for use in push–pull amplifiers. In addition, push–pull amplifiers generally use a liberal amount of negative feedback to reduce distortion.

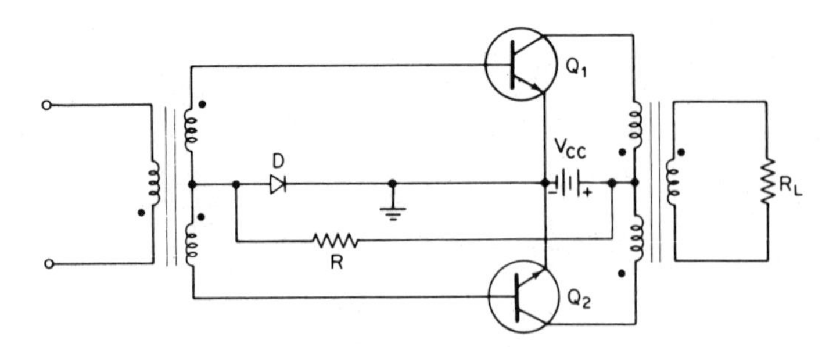

Fig. 6–13 Crossover distortion in a push-pull amplifier can be minimized by providing a small "turn-on" bias for both transistors. In this case the turn-on bias is supplied by diode *D*.

6–6 COMPLEMENTARY SYMMETRY AMPLIFIER

The conventional transformer-coupled push–pull output stage of Fig. 6–6, with transistors of the same polarity, requires two input signals of opposite phase. In Fig. 6–6 this *phase-splitting* function is performed by the input transformer with its center-tapped secondary. The disadvantages of using a transformer for phase splitting are many, particularly in terms of frequency response and distortion. In the audio spectrum, the upper end of the frequency band is affected by the stray inductance and capacitance of the transformer; the lower end of the frequency band is affected by the primary self-inductance. Nonlinear behavior of the magnetic core material of the transformer introduces distortion. These remarks also apply to the output transformer. The use of both input and output transformers therefore tends to degrade the performance of the circuit even further.

The most economical solution to eliminate transformers altogether is found in the *complementary symmetry* circuit, which uses an *npn* and a *pnp* transistor in the configuration of Fig. 6–14. For the sake of simplicity, the components required for dc adjustment have been omitted from the circuit. The transistors are assumed to operate in class B. Note that in this simplified circuit two separate collector supplies are used, instead of the single supply in the normal push–pull circuit. Input signal source v_s delivers signal current i_s to transistors $Q1$ and $Q2$.

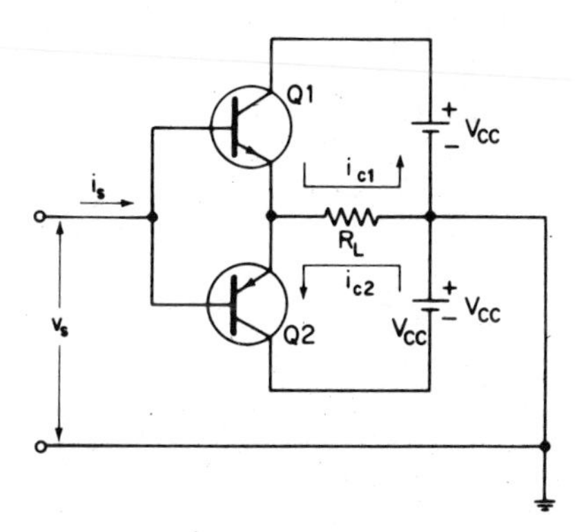

Fig. 6–14 Basic circuit of a complementary-symmetry output stage. The components for dc biasing are omitted for the sake of clarity.

On the positive half-cycle of the input voltage, *npn* transistor $Q1$ becomes forward-biased and conducts, while *pnp* transistor $Q2$ is driven into cutoff. Collector current i_{c1} flows through the load resistor R_L in the direction indicated in Fig. 6–14. On the negative half-cycle of the input voltage, $Q2$ conducts and $Q1$

is cut off, so that collector current i_{c2} flows through the load in the opposite direction. Hence, $Q1$ amplifies the positive portions of the input signal and $Q2$ amplifies the negative portions. The two transistors act as emitter followers, alternately delivering ac output current to the load.

The analysis of the complementary symmetry circuit is rather similar to that of the conventional push–pull circuit and need not be repeated here. A little thought convinces us that all the voltage, current, and power relations derived for the push–pull circuit of Fig. 6–6 apply equally to the complementary symmetry circuit of Fig. 6–14 if R_L' in those relations is replaced by R_L. Hence, the delivered output power to the load equals twice the maximum collector dissipation per transistor and

$$P_L = 2P_{C(\max)} \tag{6–27}$$

The maximum theoretical efficiency of the complementary symmetry amplifier is equal to that of the push–pull amplifier and

$$\eta = 78.5\% \tag{6–28}$$

The basic circuit of Fig. 6–14 is not very practical, since it uses two power supplies and does not contain the required components for biasing and stabilization. Figure 6–15 shows a practical arrangement of the complementary symmetry amplifier using a *single* power source. The circuit includes a driver stage and the necessary components for biasing and feedback.

Load impedance R_L is ac-coupled by capacitor $C3$ to the complementary output transistors $Q2$ and $Q3$. Emitter resistors $R6$ and $R7$ stabilize the output stage $Q2$–$Q3$ against temperature variations. In addition, resistor $R1$ stabilizes driver transistor $Q1$ against temperature variations by applying feedback from the emitters of the output transistors to the base of $Q1$. The dc voltage at the midpoint between the emitters of $Q2$ and $Q3$ is approximately one-half the supply voltage $-V_{CC}$.

To minimize crossover distortion, the output transistors must be biased to a quiescent emitter current of a few milliamperes. If both transistors had the same polarity, this quiescent current could be obtained by applying a small negative base voltage with respect to the emitter voltage, in the manner suggested in Fig. 6–13. Since the output transistors have opposite polarity, the base of $Q2$ requires a negative bias with respect to its emitter and the base of $Q3$ requires a positive bias with respect to its emitter. These bias voltages are supplied by resistor $R4$ in Fig. 6–15, which is inserted in the collector circuit of driver transistor $Q1$. This introduces some asymmetry in the circuit and hence $R4$ should have a small resistance. $R4$ is usually a preset potentiometer to allow exact adjustment of the operating points of the output transistors and to prevent clipping of the output signal at maximum signal excursion. Also, $R4$ is

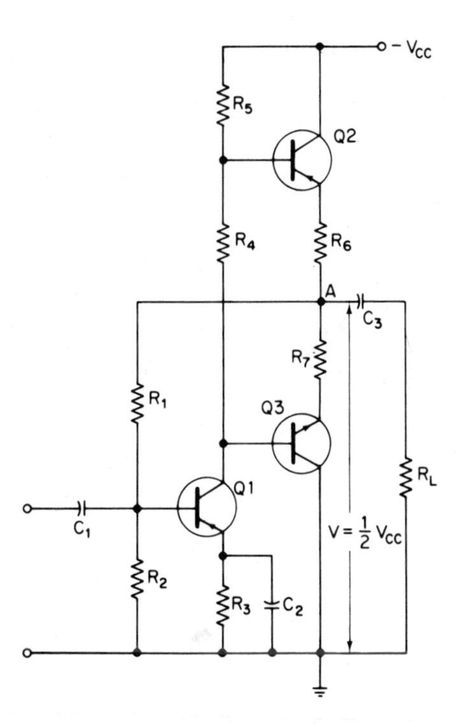

Fig. 6–15 Practical circuit of a complementary output stage.

sometimes paralleled by a thermistor to compensate for temperature dependence of the quiescent currents.

The operation of the circuit is similar to that described for the basic circuit of Fig. 6–14. During the positive half-cycles of the input voltage, applied to the base of driver transistor *Q*1, output transistor *Q*3 conducts and lowers the dc voltage at point *A* (the midpoint between the emitters of the output transistors). During the negative half-cycles of the input voltage, output transistor *Q*2 conducts and raises the dc voltage at point *A*. The variations in the dc voltage at point *A* are transferred to the load via capacitor *C*3.

An example of an economical 1-W stereo pickup amplifier, with a complementary output stage, is shown in Fig. 6–16. The design uses an absolute minimum of components. The simple power supply, designed for an output voltage of 9 V, consists of a 6.3-V 1-A secondary winding with a single rectifier diode and a large storage capacitor. The bias circuits for the input transistors of each channel are decoupled by a common *RC* filter (33 kΩ and 80 μF). The complementary output stage is almost identical to the one described in Fig. 6–15. Since this is a low-cost circuit, no temperature-compensating emitter resistors are used in the output stage. Ganged tone controls and separate or dual-concentric volume controls for each channel are recommended. In each case these

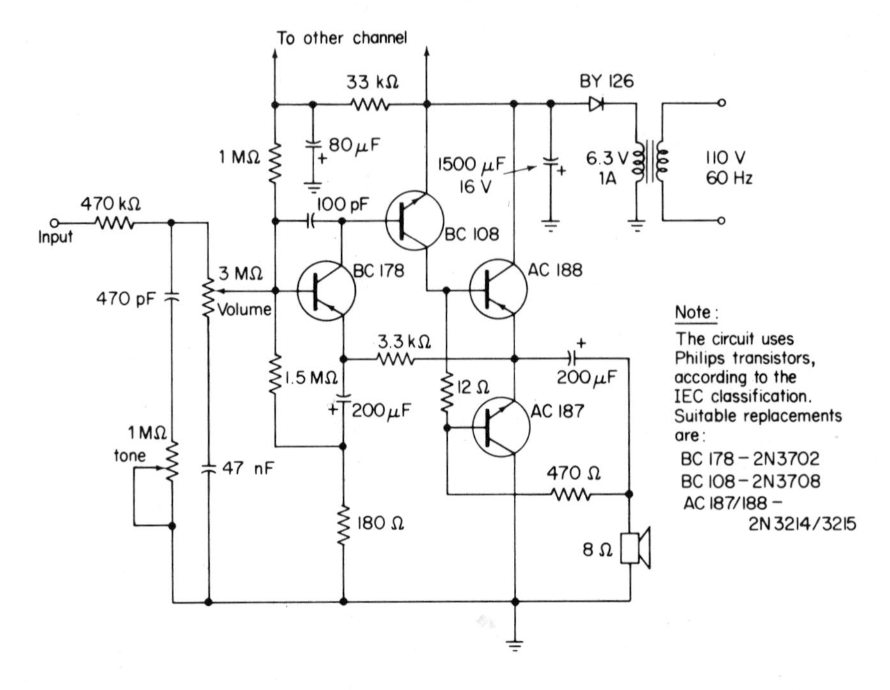

Fig. 6–16 Stereo pickup amplifier with complementary output stage (Courtesy N.V. Philips Gloeilampenfabrieken, Eindhoven, the Netherlands).

may be of the linear type. Feedback is used to achieve a high input impedance, and a good low-frequency response is obtained with a capacitive source such as a ceramic pickup.

The performance specifications per channel are

nominal power output	900 mW into 8-Ω load
sensitivity (1000 Hz) for $P_o = 900$ mW from 1000-pF source	600 mV
frequency response (−3 dB)	110 to 11,500 Hz
tone control	−12 dB at 10,000 Hz
nominal supply voltage	9 V
current consumption at $P_o = 900$ mW	160 mA

6–7 QUASI-COMPLEMENTARY AMPLIFIER

The complementary symmetry output stage of Fig. 6–15 requires transistors with practically identical characteristics. For large power outputs (say, in excess of 10 W) it is rather difficult to manufacture complementary symmetry pairs economically. To retain the advantages of the complementary circuit (single-ended input and no output transformer), complementary transistors are often used in the driver stage of the high-powered amplifier and are then followed in the output stage by two power transistors of the same polarity in a push–pull configuration. This circuit arrangement is known as a *quasi-complementary* amplifier and is shown in Fig. 6–17. Notice that here again two separate power supplies are used and that the load is connected to the midpoint of the supplies instead of to ground. Also, the components for feedback and circuit stabilization are omitted for the sake of simplicity.

Input transistor *Q*1 is biased in the conventional manner by *R*1. The collector load of *Q*1 consists of *R*2 and *R*3, with *R*2 providing the correct bias voltages for the complementary transistors *Q*2 and *Q*3. This bias arrangement ensures that *Q*2 and *Q*3 draw a small quiescent current to avoid crossover distortion. The complementary pair *Q*2–*Q*3 provides the phase-splitting function required for the class B push–pull output stage *Q*4–*Q*5. The current requirements, and hence the power dissipation, in the complementary pair are relatively small since the only power required is that used to provide the base drive for the power transistors. The complementary pair need not be closely matched because their *Q* points can be set adequately by making *R*2 variable.

A qualitative analysis of the circuit of Fig. 6–17 proceeds as follows: When the input signal goes positive, *Q*1 collector current decreases, forcing *Q*2 into heavier conduction and *Q*3 into cutoff. Hence, the base current into power transistor *Q*4 increases while the base current into *Q*5 decreases or becomes zero.

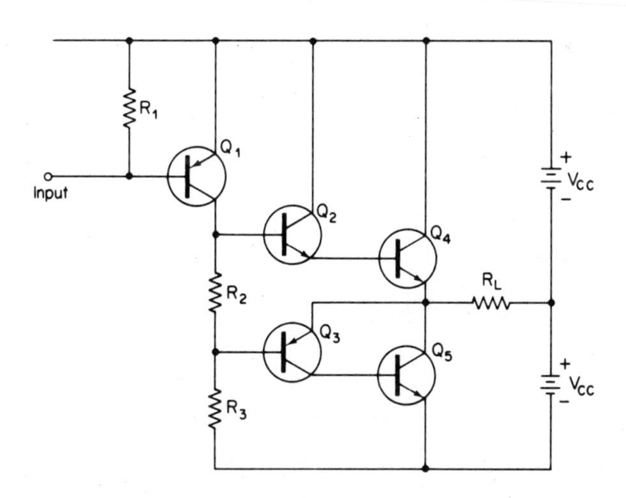

Fig. 6–17 Basic quasi-complementary amplifier, including class-A input stage *Q*1, complementary drivers *Q*2 and *Q*3, and push-pull output stage *Q*4 and *Q*5. The amplifier uses two power sources and a "floating" load.

When the input signal goes negative, the reverse action takes place. Hence, the output current delivered to the load is supplied alternately by each power transistor in a standard push–pull action. Transistor combination $Q2$–$Q4$ can be regarded as a Darlington pair, while transistor combination $Q3$–$Q5$ is connected as a complementary Darlington pair.

The circuit is not very practical since it requires two power supplies and a "floating" load. A practical circuit arrangement, using a single power source, is shown in Fig. 6–18. The load R_L is capacitively coupled to the midpoint of the output stage in a manner identical to that described for the complementary symmetry circuit of Fig. 6–15. Emitter resistors $R4$ and $R5$ provide stabilization of the output stage against temperature variations. Bias resistor $R2$ is usually shunted by a thermistor R_t to stabilize the complementary pair $Q2$–$Q3$. The operation of the circuit is virtually identical to that of the basic circuit of Fig. 6–16 and requires no further elaboration.

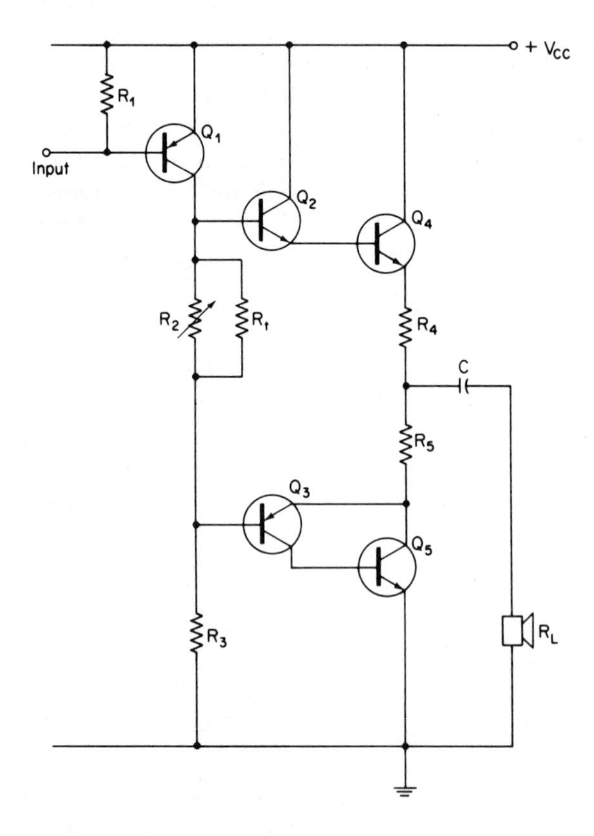

Fig. 6–18 Practical quasi-complementary amplifier with single power source. The complementary pair is temperature compensated by thermistor R_t, and the output stage is stabilized by emitter resistors R_4 and R_5.

QUESTIONS

6–1 For each class of amplifier operation (A, B, and C) indicate the position of the dc operating point on the $I_C - V_C$ collector characteristics. Briefly explain how the collector current in each case varies in response to the input signal.

6–2 In what circuit configurations is a class B amplifier generally used?

6–3 State the theoretical maximum efficiency of a class A amplifier and a class B amplifier.

6–4 What is the quiescent power dissipation (at zero input signal) of a class B amplifier?

6–5 Assume that a load resistor is transformer-coupled to a class A output stage. What can happen to the transistor if the load resistor is disconnected?

6–6 Name the advantages and the disadvantages of transformer coupling.

6–7 Name the various types of distortion that can be present in a power amplifier and in each case explain how this distortion is generated.

6–8 What is meant by crossover distortion? How can crossover distortion be minimized?

6–9 Draw the circuit diagram of a transformer-coupled push–pull stage and explain how the two class B transistors produce the output signal.

6–10 Why is it necessary to use a pair of matched transistors in a push–pull stage?

6–11 What are some of the advantages of a complementary symmetry amplifier over a conventional push–pull amplifier?

6–12 Explain the function of each component in the 1-W stereo amplifier of Fig. 6–16.

6–13 What is the major advantage of a quasi-complementary pair over the complementary symmetry pair?

PROBLEMS

6–1 The transformer-coupled class A power amplifier of Fig. 6–3 operates under the following conditions: $V_{CC} = 48$ V, $R_A = 600\ \Omega$, $R_B = 120\ \Omega$, $R_E = 8\ \Omega$, $C_E = 100\ \mu$F, and $V_{BE} = 0.8$ V. Determine the quiescent collector current and the quiescent collector dissipation.

6–2 In the amplifier of Prob. 6–1 a 4-Ω load is connected to the secondary of the output transformer. This load is reflected as a 20-Ω ac collector load. Determine the maximum power dissipation of the 4-Ω load.

6–3 Find the voltage gain, current gain, and power gain of the amplifier described in Probs. 6–1 and 6–2.

6–4 A class A transformer-coupled power transistor is operated from a 24-V supply. An 8-Ω load is connected to the secondary of the output transformer and it dissipates 4 W of power. Specify the quiescent collector current, the quiescent collector dissipation, and the turns ratio of the output transformer.

6–5 A silicon power transistor, operating class A, is required to deliver a maximum of 10 W to a 6-Ω load. If the collector supply is 48 V, determine the transformer turns ratio, the maximum quiescent collector current, and the collector circuit efficiency.

6–6 Two power transistors, operating in the class B push–pull circuit of Fig. 6–6, are required to deliver 20 W of output power to a 6-Ω transformer-coupled load. The collector supply voltage is 20 V. Determine the maximum ratings of the transistors.

Chapter 7

Rectifiers and Filters

7–1 ELEMENTS OF A REGULATED POWER SUPPLY

In some solid-state equipment the dc voltages necessary to operate the circuitry are generally low and the current consumption small, so dry-cell batteries can be used to supply dc power. This is satisfactory for small power consumers, but when the current drain on the batteries becomes larger, the necessary dc power is usually derived from the ac power line by an *ac-to-dc power supply.*

A *regulated* power supply generally consists of the four sections shown in Fig. 7–1. The *power transformer* converts the ac line voltage into a higher or lower (or sometimes equal) ac voltage, depending on the application, and isolates the remainder of the circuitry from the power line. The transformer output is applied to a *rectifier,* which converts the ac voltage into a dc voltage. The conversion from ac into dc is not perfect, and a certain amount of ac residue or "ripple" remains in the output of the rectifier. The rectifier is usually followed

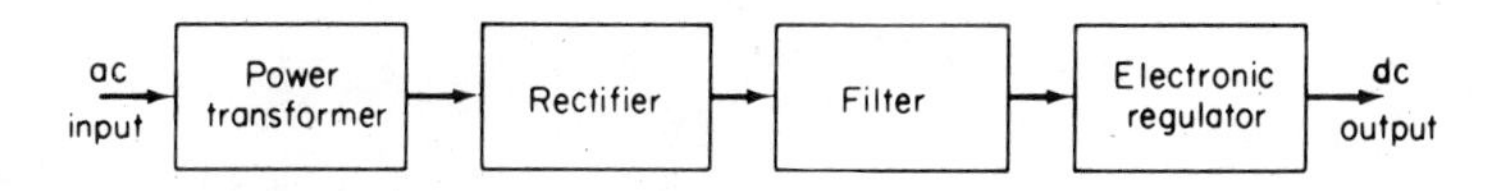

Fig. 7–1 Elements of a regulated ac-dc power supply

by a *filter section*, which reduces this residual ac component to produce a more nearly pure dc voltage. The filtered dc voltage is then applied to an electronic feedback or control circuit, called a *regulator*, which maintains the dc output voltage essentially constant for changes in load conditions or variations in line voltage.

Details of transformer construction and operation, and the magnetic properties governing transformer action, are treated in detail in introductory circuits texts.*

Chapter 7 analyzes the remaining elements of the *unregulated* power supply (rectifier circuits and filter sections), while Chapter 8 describes the various regulating circuits generally used as the final element of the *regulated* power supply.

7-2 SINGLE-PHASE HALF-WAVE RECTIFIER

In the single-phase half-wave rectifier of Fig. 7-2 the secondary winding of the power transformer supplies a sinusoidal voltage $e = E_m \sin\omega t$ to the series combination of rectifier diode D and load resistor R_L. During the positive half-cycle of the applied voltage the diode is forward-biased and acts as a closed switch. Neglecting the diode forward resistance, voltage e is placed across load resistor R_L and causes a load current $i = e/R_L$. During the negative half-cycle of the applied input voltage the diode is reverse-biased and acts as an open switch. The load resistor is then removed from source voltage e and the load current is zero.

For each succeeding cycle of the applied voltage the process repeats and the load resistor carries current only during positive half-cycles. Figure 7-3 shows the voltage and current waveforms of the half-wave rectifier. The load current is *pulsating* but *unidirectional*, with the polarity as indicated in Fig. 7-2.

A dc ammeter in series with R_L reads the *average* or dc value of the load current, I_{dc}. Referring to Fig. 7-3(b), I_{dc} is the average value.

$$I_{dc} = \frac{I_m}{\pi} = 0.318 I_m \tag{7-1}$$

By Ohm's law, the dc load voltage equals

$$E_{dc} = I_{dc} R_L = \frac{E_m}{\pi} = 0.318 E_m \tag{7-2}$$

*See, for example, H. W. Jackson, *Introduction to Electric Circuits,* 3rd ed., Englewood Cliffs, N.J.: Prentice-Hall, Inc., 1970.

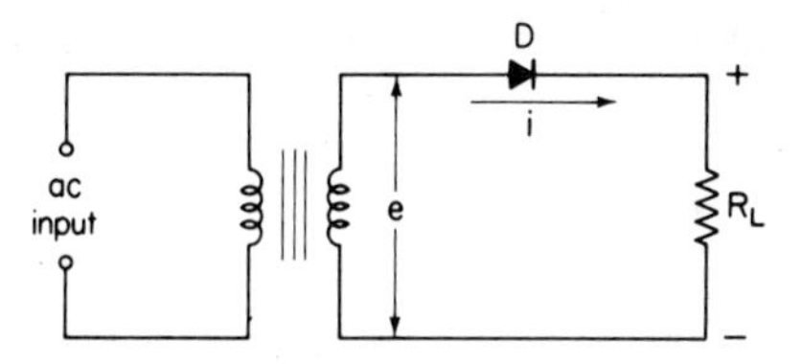

Fig. 7–2 Single-phase half-wave rectifier.

The *effective* or rms value of the total load current of Fig. 7–3(b) can be calculated and this yields

$$I_{\text{rms}} = \frac{I_m}{2} \tag{7-3}$$

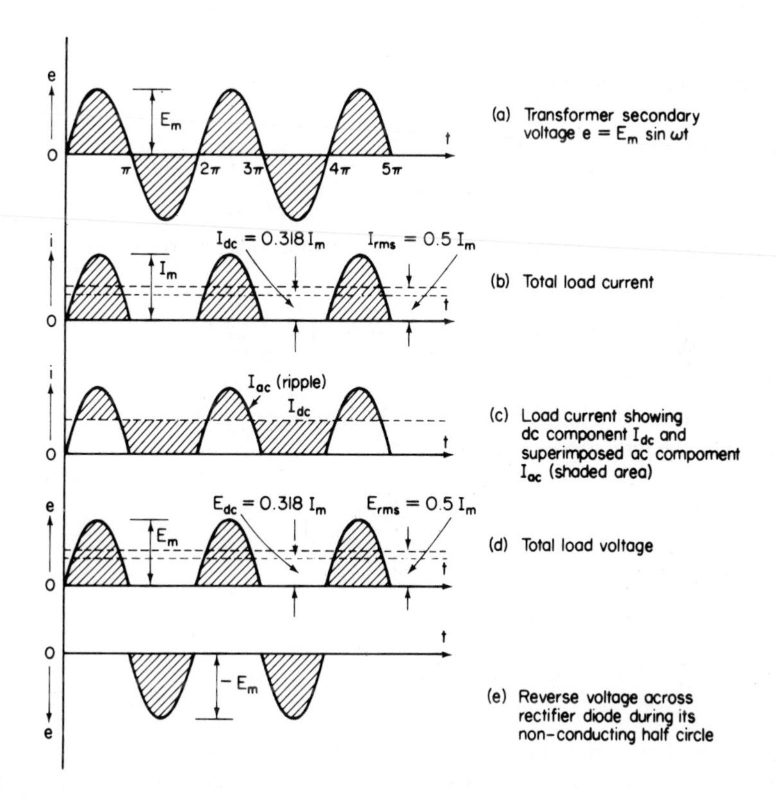

Fig. 7–3 Current and voltage waveforms for the single-phase half-wave rectifier.

This total load current is the vector sum of the dc component I_{dc} and an ac component I_{ac}, so

$$I_{rms} = \sqrt{I_{dc}^2 + I_{ac}^2} \tag{7-4}$$

where I_{ac} = rms value of the ac component. Hence, the ac component of the load current equals

$$I_{ac} = \sqrt{I_{rms}^2 - I_{dc}^2} \tag{7-5}$$

This residual ac component, superimposed on the dc load current, is an unwanted byproduct of the rectification process. It causes an ac variation of the load voltage, called the *ripple* voltage.

The effectiveness of rectification is often expressed in terms of the *ripple factor* r, defined as

$$r = \frac{\text{rms value of the ac components of the output wave}}{\text{average or dc value of the output wave}}$$

or, expressed as a percentage,

$$\%\ \text{ripple} = \frac{\text{rms value of the ac components}}{\text{average or dc component}} \times 100\%$$

Using Eqs. (7–1) and (7–5), the ripple factor can also be expressed as

$$r = \frac{I_{ac}}{I_{dc}} = \frac{\sqrt{I_{rms}^2 - I_{dc}^2}}{I_{dc}} = \sqrt{\left(\frac{I_{rms}}{I_{dc}}\right)^2 - 1} \tag{7-6}$$

(Note that this result is not restricted to the half-wave rectifier.)

For the half-wave rectifier we find

$$\frac{I_{rms}}{I_{dc}} = \frac{I_m/2}{I_m/\pi} = \frac{\pi}{2} = 1.57$$

and

$$r = \sqrt{1.57^2 - 1} = 1.21 \text{ or } 121\% \tag{7-7}$$

This result shows that the ac ripple current or voltage exceeds the dc output current or voltage. The single-phase half-wave rectifier is therefore used only with heavy filtering.

Referring to Fig. 7-3(c), the *frequency* of the ripple component is the same as the frequency of the input voltage. Hence, when the power transformer is connected to the 110-V 60-Hz power line, the ripple frequency is 60 Hz.

During the nonconducting half-cycle of the diode, the peak voltage of the transformer secondary appears across the diode [Fig. 7-3(e)]. When the components for a half-wave rectifier are selected, care should be taken to choose a diode with a *peak inverse voltage* rating (PIV), which is at least as high as the peak value of the transformer secondary voltage (E_m). In practice, a voltage derating of 20-30 percent is allowed for conservative design.

EXAMPLE 7-1

The single-phase half-wave rectifier of Fig. 7-2 uses an ideal diode (zero forward resistance and infinite reverse resistance) to supply power to a 200-Ω load from a power transformer whose secondary voltage is 24 V rms.

Calculate (a) The dc load voltage.
(b) The dc load current.
(c) The ripple voltage.
(d) The PIV rating of the diode.

SOLUTION

(a) Dc load voltage: $E_{dc} = E_m/\pi = 24\sqrt{2}/\pi = 10.8$ V.

(b) Dc load current: $I_{dc} = E_{dc}/R_L = 10.8\text{ V}/200\ \Omega = 54$ mA.

(c) Rms ripple voltage: $E_r = 1.21 \times E_{dc} = 1.21 \times 10.8\text{ V} = 13.1$ V.

(d) Peak transformer secondary voltage $= E_m = 24\sqrt{2} = 34$ V. Using a derating factor of 30 percent, the PIV rating of the diode is 34 V/0.7 = approximately 50 V.

In addition to a very high ripple content, the half-wave rectifier has the disadvantage of being rather inefficient. *Efficiency of rectification* is defined as the ratio of the dc power delivered to the load to the ac power supplied to the circuit. By definition, the dc power delivered to the load is the product of the dc load voltage and the dc load current. Thus,

$$P_{dc} = E_{dc} I_{dc} = I_{dc}^2 R_L = \left(\frac{I_m}{\pi}\right)^2 R_L \tag{7-8}$$

The ac power delivered to the circuit is the product of the rms value of the applied voltage and the rms value of the total current supplied to the circuit.

Note that ac power is supplied to the circuit only during the positive alternations of the transformer secondary voltage. Hence, the ac power is the product of the total load current of Fig. 7–3(b) and the total load voltage of Fig. 7–3(d). Neglecting the forward diode resistance,

$$P_{ac} = E_{rms} I_{rms} = I_{rms}^2 R_L$$

According to Eq. (7–3), $I_{rms} = I_m/2$ and therefore

$$P_{ac} = I_{rms}^2 R_L = \left(\frac{I_m}{2}\right)^2 R_L \tag{7–9}$$

Hence, the efficiency of rectification is

$$\frac{P_{dc}}{P_{ac}} = \frac{(I_m/\pi)^2 R_L}{(I_m/2)^2 R_L} = \frac{4}{\pi^2} = 0.406 \text{ or } 40.6\% \tag{7–10}$$

This result shows that the maximum theoretical efficiency of the half-wave rectifier is only 40.6%. In practice this figure is even lower, since under forward bias conditions the diode has a small voltage drop across it and therefore dissipates some power in the form of heat. In addition, the transformer is subject to copper and iron losses, which further reduce the overall efficiency.

EXAMPLE 7-2

A single-phase half-wave rectifier is required to deliver 10 mA dc at 18 V to a certain load. The power transformer is connected to the 110-V 60-Hz power line. Assuming that the forward voltage drop across the diode is 1.3 V dc under the given load conditions, determine

(a) The full-load transformer secondary voltage.

(b) The no-load transformer secondary voltage (allowing 5 percent for transformer losses).

(c) The transformer turns ratio.

(d) The primary and secondary transformer currents.

(e) The efficiency of rectification.

SOLUTION

(a) The diode voltage drop must be added to the dc load voltage to find the total dc voltage under consideration, so

$$E_{dc} = 18\text{ V} + 1.3\text{ V} = 19.3\text{ V}$$

The peak voltage of the transformer secondary then equals

$$E_m = \pi E_{dc} = 3.14 \times 19.3 \text{ V} = 60.6 \text{ V}$$

Since the transformer secondary produces a full sinusoidal waveform, the rms value of the secondary voltage is

$$E_s = \frac{60.6}{\sqrt{2}} = 43 \text{ V}$$

(b) Allowing 5 percent for internal transformer losses, the no-load secondary voltage E_s' must be

$$E_s' = \frac{43 \text{ V}}{0.95} \simeq 45.3 \text{ V}$$

(c) The primary transformer voltage is 110 V. Hence, the turns ratio is

$$\frac{E_p}{E_{s'}} = \frac{110}{45.3} \simeq 2.44$$

(d) The total secondary current is

$$I_{rms} = \frac{I_m}{2} = \frac{\pi I_{dc}}{2} = 15.7 \text{ mA}$$

The primary transformer current then equals

$$I_p = \frac{15.7 \text{ mA}}{2.43} = 6.46 \text{ mA rms}$$

(Note that the waveforms of the secondary and primary currents are half sine waves, as indicated in Fig. 7-3.)
(e) The efficiency of rectification is P_{dc}/P_{ac}, where

$$P_{dc} = E_{dc} I_{dc} = 18 \text{ V} \times 10 \text{ mA} = 180 \text{ mW}$$

$$P_{ac} = E_{rms} I_{rms}$$

where $E_{rms} = \frac{E_m}{2} = 30.3$ V, so

$$P_{ac} = 30.3 \text{ V} \times 15.7 \text{ mA} = 476 \text{ mW}$$

The efficiency of rectification then is

$$\frac{108 \text{ mW}}{476 \text{ mW}} = 0.378 \text{ or } 37.8\%$$

Finally, the dc load current passes through the transformer winding and may cause dc core saturation, which reduces the transformer output voltage. At high load currents transformer core saturation is a definite possibility. This factor should be taken into account when a half-wave rectifier is being considered.

The half-wave rectifier has several *disadvantages*:

1. A very high ripple content (121 percent).
2. A low efficiency of rectification (40.6 percent maximum).
3. A definite possibility of dc saturation of the transformer core.

The circuit has the advantage of very simple construction at low cost, but it is only used in applications where these advantages outweigh the stated disadvantages.

7–3 SINGLE-PHASE FULL-WAVE RECTIFIER

7–3.1 Basic Circuit

The single-phase full-wave rectifier uses a center-tapped transformer and two rectifier diodes in the configuration of Fig. 7–4. The circuit essentially consists of two half-wave rectifier circuits, each supplied by one half of the power transformer. The secondary voltage of the transformer is measured from *a* to *c*

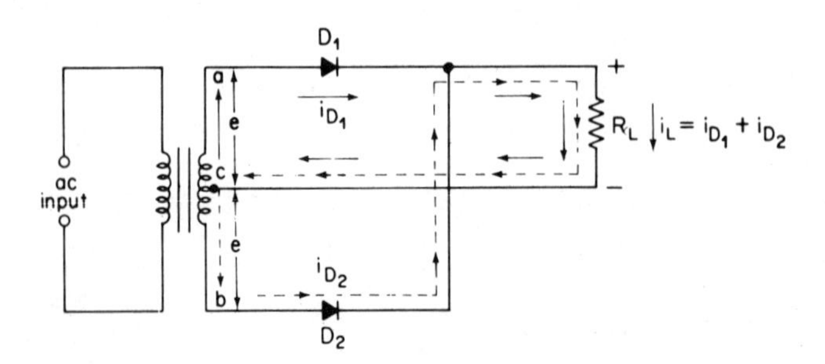

Fig. 7–4 Single-phase full-wave rectifier.

and from b to c, where c is the *center tap* or common point of the system. The secondary voltage is usually specified as, for example, 50–0–50 V, which means that the transformer produces 50 V rms on either side of the center tap and 100 V rms across the total secondary winding.

During one half of the power cycle, when point a is positive with respect to point c (center tap), diode $D1$ is forward-biased and conducts. This completes the top half of the circuit, consisting of transformer winding a-c, diode $D1$, and load resistor R_L, and produces a current through the load. At the same time, point b is negative with respect to point c, so diode $D2$ is reverse-biased, opening the lower half of the circuit.

During the next half of the power cycle, point a is negative with respect to point c, so diode $D1$ is reverse-biased. At the same time, point b is positive with respect to point c and diode $D2$ is forward-biased. This completes the lower half of the circuit, consisting of transformer winding b-c, diode $D2$, and load resistor R_L, and again produces a current through the load. The direction of the load current is the same for both half-cycles of the operation. This is indicated by the solid arrow for diode $D1$ operation, and by the broken arrow for diode $D2$ operation. *Both* halves of the input waveform are used. Figure 7–5 shows the current and voltage waveforms for the full-wave rectifier.

The average or dc value of the load current is simply twice that of the half-wave circuit, so

$$I_{dc} = \frac{2I_m}{\pi} = 0.636I_m \qquad (7\text{–}11)$$

The dc load voltage is also twice that of the half-wave circuit and

$$E_{dc} = I_{dc}R_L = \frac{2E_m}{\pi} = 0.636E_m \qquad (7\text{–}12)$$

The load current again consists of a dc component and an ac component. The dc component is simply I_{dc}, as given by Eq. (7–12). The ac component can be found by calculating the rms value of the total load current of Fig. 7–5(d) and subtracting the dc component.

$$I_{rms} = \frac{I_m}{\sqrt{2}} \qquad (7\text{–}13)$$

so

$$\frac{I_{rms}}{I_{dc}} = \frac{I_m/\sqrt{2}}{2I_m/\pi} = 1.11$$

The ripple factor equals

$$r = \sqrt{1.11^2 - 1} = 0.482 \text{ or } 48.2\% \tag{7-14}$$

The dc power delivered to the load is, by definition,

$$P_{dc} = E_{dc}I_{dc} = I_{dc}^2 R_L = \left(\frac{2I_m}{\pi}\right)^2 R_L \tag{7-15}$$

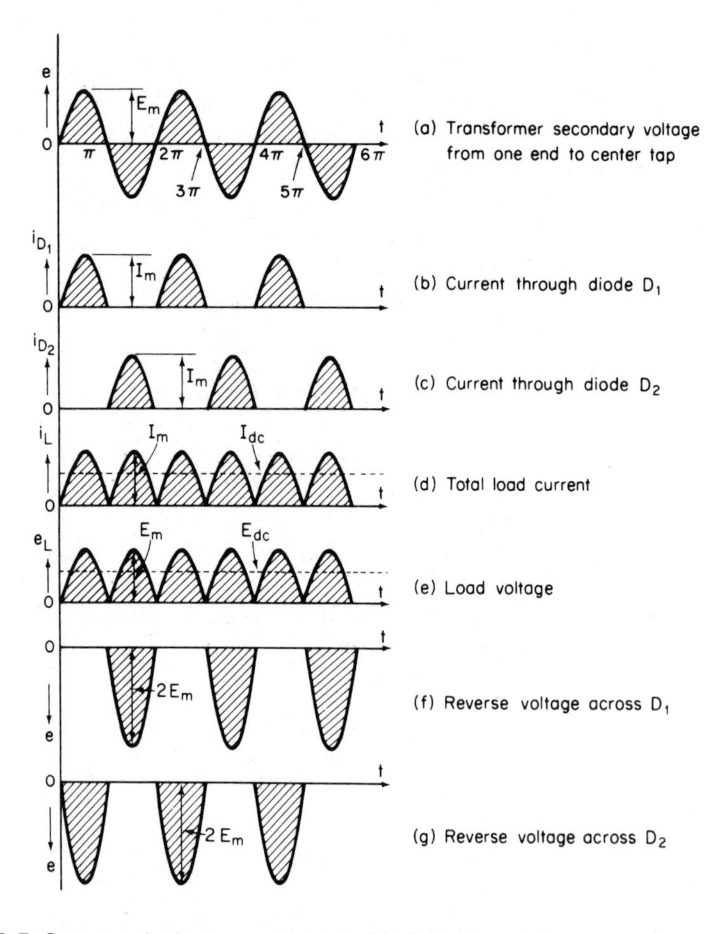

Fig. 7–5 Current and voltage waveforms for the single-phase full-wave rectifier.

The ac input power to the circuit is, by definition,

$$P_{ac} = I^2_{rms}R_L = \left(\frac{I_m}{\sqrt{2}}\right)^2 R_L \tag{7–16}$$

The efficiency of rectification is then

$$\frac{P_{dc}}{P_{ac}} = \frac{(2I_m/\pi)^2R_L}{(I_m/\sqrt{2})^2R_L} = \frac{8}{\pi^2} = 0.812 \text{ or } 81.2\% \tag{7–17}$$

Comparing the results of these calculations with those of the half-wave rectifier indicates that the full-wave rectifier produces considerably less ripple and is twice as efficient as the half-wave circuit.

Consider the waveforms of Fig. 7–5. The frequency of the ripple component is twice the frequency of the input waveform. For a full-wave circuit connected to the 110-V 60-Hz power line, the ripple frequency is 120 Hz.

Consider Fig. 7–4 from the point of view of peak inverse voltages. When diode $D1$ conducts, the peak voltage between the transformer center tap and the cathode of $D2$ equals E_m. This is also the voltage which appears across the load. At the same time, the peak voltage between the center tap and the anode of $D2$ is also E_m, so the total peak voltage across diode $D2$ in its nonconducting state is $2E_m$. The PIV rating of the diodes in a full-wave circuit must therefore be twice s high as for the diode in a half-wave circuit.

EXAMPLE 7-3

The single-phase full-wave rectifier of Fig. 7–4 uses a power transformer with a secondary winding of 100–0–100 V. The load resistor is 1 kΩ. Neglect the transformer losses and the forward voltage drops of the diodes and determine

(a) The dc output voltage.
(b) The dc output current.
(c) The ripple voltage.
(d) The PIV rating of the diodes.

SOLUTION

(a) Dc output voltage: $E_{dc} = 0.636E_m = 0.636 \times 100\sqrt{2}$ V = 90 V.

(b) Dc output current: $I_{dc} = E_{dc}/R_L = 90$ V/1 kΩ = 90 mA.

(c) Ripple voltage: $E_{ac} = 0.482 \times 90$ V = 43.4 V rms.

(d) Peak reverse voltage across the diode is $2E_m = 282$ V. The PIV rating of the diodes, allowing a derating of 30 percent, is 282 V/0.7 = approximately 400 V.

7–3.2 Regulation

So far we have ignored the forward resistance of the diode and the effect this may have on the operation of the circuit. However, a semiconductor diode has a resistance of several ohms. The diode voltage drop should also be taken into account when the dynamic performance of the circuit is considered.

For every half-cycle of the input waveform one diode with a resistance r_B is in series with the load resistor R_L, as in Fig. 7–6. The maximum value of the load current is determined by the total resistance in the circuit, so

$$I_m = \frac{E_m}{r_B + R_L} \tag{7–18}$$

Since, by Eq. (7–11), $I_{dc} = 2I_m/\pi$, the dc current in this case is

$$I_{dc} = \frac{2E_m/\pi}{r_B + R_L} \tag{7–19}$$

Rearranging Eq. (7–19) and substituting in Eq. (7–12) yields

$$E_{dc} = I_{dc}R_L = \frac{2E_m/\pi}{r_B + R_L}R_L = \frac{2E_m}{\pi} - I_{dc}r_B \tag{7–20}$$

This result shows that E_{dc} equals $2E_m/\pi$ at no load, but decreases linearly with an increase in dc load current. The larger the resistance r_B of the diode, the greater this effect becomes. Germanium diodes, with their relatively large resistance, contribute particularly to this undesirable situation.

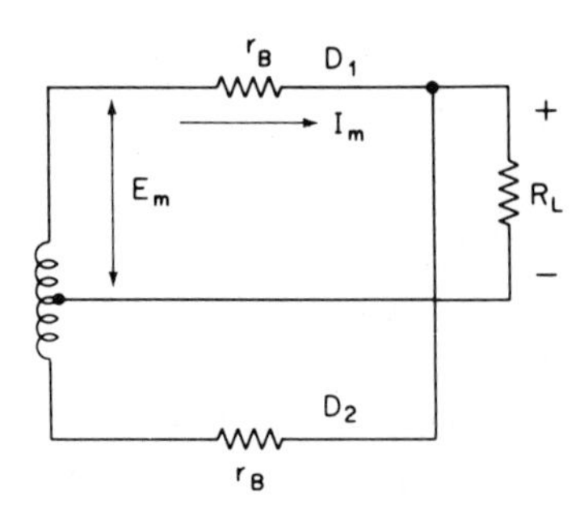

Fig. 7–6 Diode resistance affects the load regulation.

The variation of the dc output voltage as a function of the dc load current is known as *load regulation*. The percentage load regulation is defined as

$$\% \text{ load regulation} = \frac{E_{\text{no load}} - E_{\text{full load}}}{E_{\text{full load}}} \times 100\%$$

If the dc output voltage does not vary with the load current, the percentage regulation is zero. This represents the ideal case.

EXAMPLE 7-4

Calculate the load regulation for the single-phase full-wave rectifier of Fig. 7–4. The transformer secondary voltage is 20–0–20 V rms and the rated output current is 50 mA dc. The diodes used in the circuit are general-purpose low-power germanium diodes with an average bulk resistance of 50 Ω at rated current.

SOLUTION

Taking the diode resistance into consideration, the general expression for the dc output voltage is given by Eq. (7–20) and

$$E_{\text{dc}} = \frac{2E_m}{\pi} - I_{\text{dc}} r_B$$

so

$$E_{\text{dc(no load)}} = 0.636 E_m = 0.636 \times 20\sqrt{2} = 18 \text{ V}$$

and

$$E_{\text{dc(full load)}} = 18 \text{ V} - (50 \text{ mA} \times 50 \ \Omega) = 15.5 \text{ V}$$

The load regulation, expressed as a percentage, equals

$$\% \text{ load regulation} = \frac{18 \text{ V} - 15.5 \text{ V}}{15.5 \text{ V}} \times 100\% \simeq 16\%$$

Referring again to Fig.7–4, we note that each half of the transformer secondary winding carries the direct current, but *in opposite directions*. Hence, the net effect of the dc flux in the secondary winding cancels over the complete power cycle and there is no core saturation of the transformer.

In summary, the full-wave rectifier has *advantages* over the half-wave circuit: a lower ripple factor (48.2 percent), a higher efficiency of rectification (81.2 percent maximum), and no dc saturation of the transformer core. The major *disadvantage* of the full-wave rectifier is that is requires more costly components for construction, such as a center-tapped power transformer and two diodes with a PIV rating of at least $2E_m$. In many applications, however, the advantages of improved circuit performance outweigh the disadvantage of increased cost and the circuit is widely used.

7-4 SINGLE-PHASE FULL-WAVE BRIDGE RECTIFIERS

In the single-phase full-wave bridge rectifier of Fig. 7-7 the secondary winding of the power transformer supplies four rectifier diodes in a bridge configuration. During one-half of the input cycle, when the transformer polarity is as indicated in Fig. 7-7, diodes $D1$ and $D3$ are in the conducting state and current flows in load resistor R_L in the direction indicated by the solid arrows. During the next half of the input cycle, when the polarity of the transformer is reversed, diodes $D2$ and $D4$ are in the conducting state and again current flows in R_L now indicated by the broken arrows. Note that *the current direction is the same* for

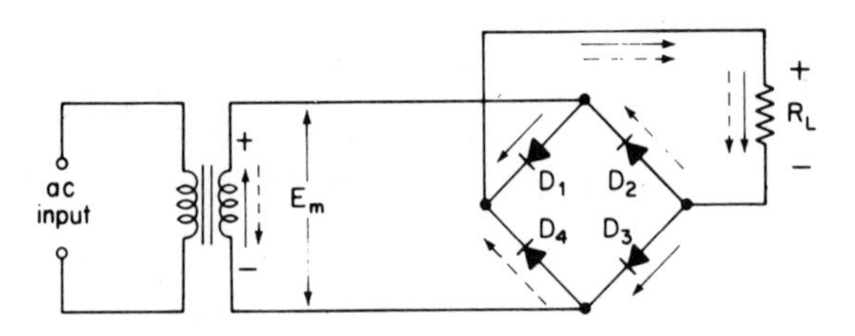

Fig. 7-7 Single-phase full-wave bridge rectifier.

both half-cycles of the input waveform. For one complete cycle of input voltage, therefore, the output current is a pulsating dc with a waveform identical to that of the full-wave circuit shown in Fig. 7-5(d).

Since the waveforms of load current and load voltage for the bridge rectifier are the same as for the full-wave rectifier,

$$I_{dc} = \frac{2I_m}{\pi} = 0.636I_m \tag{7-21}$$

and

$$E_{dc} = \frac{2E_m}{\pi} = 0.636E_m \tag{7-22}$$

Calculation of the ripple component in the output voltage and current yields the same result as for the full-wave circuit.

The major advantage of a full-wave bridge circuit is that a transformer without center tap is used. When the cost of the transformer is the main consideration, as in large rectifiers with low voltage and high current ratings, the bridge circuit is definitely preferred.

The reverse voltage across the non-conducting diodes is equal to the peak value E_m of the applied transformer voltage and diodes with a relatively low PIV rating can therefore be used.

The transformer windings carry purely sinusoidal currents over the entire cycle. There is no unidirectional current in the secondary winding and therefore no dc saturation of the magnetic circuit.

Summarizing, the bridge rectifier is found to be the superior circuit because it produces a smaller rms transformer current for a given load current than both the half-wave and the full-wave circuits. If the load is fixed, therefore, a smaller transformer may be used. However, four diodes are required instead of two or one for the other configurations. The rectifier diodes in the bridge circuit need only withstand half the PIV rating as compared to the full-wave center-tap circuit, so the bridge frequently is the most economical circuit.

Table 7-1 summarizes the main characteristics of the three basic rectifier circuits.

7-5 VOLTAGE DOUBLERS

Figure 7-8 shows the circuit of a full-wave *voltage doubler*, which supplies a dc load voltage of approximately twice the peak transformer voltage. During the positive half-cycle of the ac input voltage, when the top of the transformer secondary is positive, diode $D1$ charges capacitor $C1$ to the peak voltage E_m. During the negative half of the ac input cycle, diode $D2$ charges capacitor $C2$ to the peak voltage E_m. The capacitor voltages are series-aiding, with the polarities as indicated in Fig. 7-8, and the dc output voltage is therefore $2E_m$.

When load resistor R_L is connected, the capacitors discharge through R_L and continually supply a direct load current. During each succeeding half of the ac cycle, however, the capacitors are recharged to their peak voltages, replenishing the charge lost in the form of load current. The actual dc load voltage approaches $2E_m$ for small load current demands but reduces sharply when the

TABLE 7–1 A COMPARISON OF THE THREE BASIC RECTIFIER CIRCUITS.

Schematic	Output waveform	Number of diodes	PIV	Type of transformer	dc output voltage	Ripple factor	Ripple frequency	Maximum efficiency
Half-wave rectifier		1	E_m	Single winding	$0.318\ E_m$	0.21	f	40.6%
Full-wave rectifier		2	$2\ E_m$	Centre tap	$0.636\ E_m$	0.482	2f	81.2%
Full-wave bridge rectifier		4	E_m	Single winding	$0.636\ E_m$	0.482	2f	81.2%

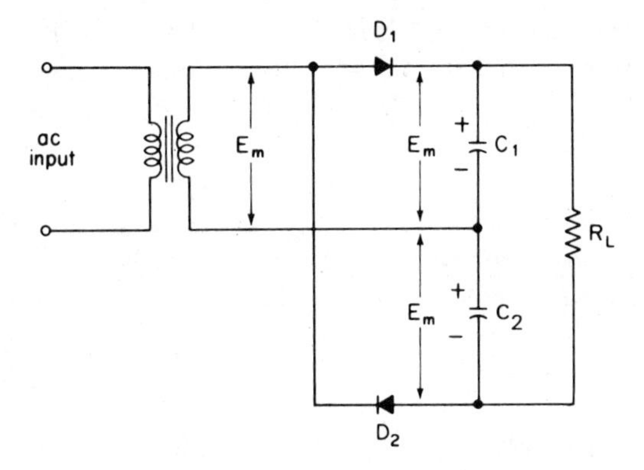

Fig. 7–8 Full-wave voltage doubler.

load current increases significantly, accompanied by an increase in the ac ripple voltage. The frequency of the ripple component is twice the line frequency.

The voltage doubler is primarily used to provide high dc voltages, often directly from the ac power line without the use of a transformer. When the circuit is connected to the 110-V 60-Hz power line, without a transformer, the no-load dc output voltage equals $2Em = 2\sqrt{2} \times 110\ \text{V} \simeq 310\ \text{V}$.

An alternative arrangement of two diodes and two capacitors in a voltage doubler circuit is shown in the *cascade voltage doubler* of Fig. 7–9. During the negative half of the ac input cycle, when the top of the transformer is negative, diode $D1$ is forward-biased and capacitor $C1$ charges to the peak voltage E_m in the polarity indicated. During the next half-cycle when the top of the

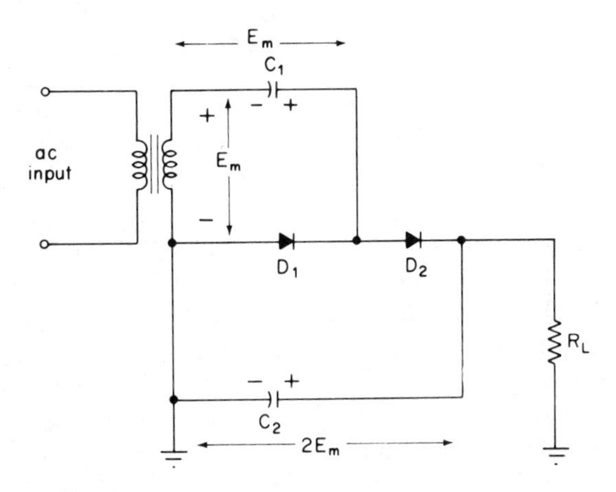

Fig. 7–9 Cascade voltage doubler.

transformer is positive, diode $D1$ is reverse-biased and acts like an open circuit. The voltage across $C1$ and the transformer winding are now series-aiding, so that capacitor $C2$ charges through diode $D2$ to a peak voltage of $2E_m$. The dc output voltage of the circuit is therefore twice the peak value of the applied ac input voltage.

When the load resistor is connected $C2$ discharges through R_L to supply the direct load current. During each succeeding cycle, capacitor $C2$ is recharged to a peak voltage $2E_m$, replenishing the charge lost in the form of direct load current.

7-6 CAPACITOR-INPUT FILTER

7-6.1 Basic Circuit

Filtering is concerned with the process of reducing the ripple component in the dc output voltage of a rectifier. A commonly used filter consists of a capacitor placed across the output terminals of the rectifier, as shown in the bridge circuit of Fig. 7-10. This simple *capacitor-input* filter is very widely used in applications which do not require a high degree of regulation (small load current). Its main features include a high dc output voltage for a given ac input voltage, a relatively low ripple content, and fairly good output voltage regulation at light loads.

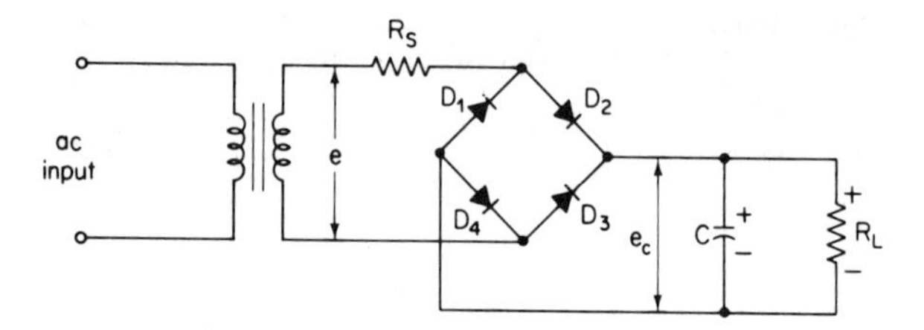

Fig. 7-10 Full-wave bridge with capacitor-input filter.

Consider ideal diodes and assume that initially load resistor R_L is disconnected from the output. During time interval t_0–t_1 on the voltage waveform of Fig. 7-10, diodes $D2$ and $D4$ conduct and capacitor C charges to the peak voltage E_m of the transformer secondary. When the applied voltage falls below the capacitor voltage, the diodes become reverse-biased and the capacitor charge current ceases. Since R_L is disconnected, the capacitor cannot discharge, and the dc output voltage of the circuit equals E_m.

Assume now that R_L is connected to the circuit at instant t_1, when the capacitor voltage e_c has just reached its peak value E_m. When the supply voltage falls below this peak value, the capacitor discharges through the load and supplies a load current $i_L = e_c/R_L$. During its time of discharge, from t_1 to t_2, the capacitor voltage decreases *exponentially* with a time constant CR_L. At $t = t_2$, the supply voltage exceeds the capacitor voltage and the capacitor charges again to the peak voltage E_m through diodes $D1$ and $D3$. Recharging occurs during the relatively short time interval t_2–t_3. This process repeats for every half-cycle of the input waveform, with the capacitor charging in short-duration pulses (t_2–t_3) and discharging through the load resistor during the longer time intervals (t_3–t_4).

The waveform of Fig. 7–11(a) shows that a capacitor-input filter produces an average or dc load voltage which is considerably higher than $0.636E_m$, derived for the full-wave bridge circuit, while the amplitude of the ripple voltage is greatly reduced.

When C is large, and the load current small, the capacitor discharges only very slightly and the dc output voltage approaches E_m with negligible ripple. When the load current increases, the capacitor discharges more rapidly, and hence the output voltage decreases, while the ripple increases. Therefore, an unregulated power supply consisting of a full-wave bridge configuration with a capacitor-input filter is an excellent choice when the load is fixed (no load-voltage regulation problems) and the current demand is small (small ripple voltage).

The most serious weakness of the capacitor-input filter is that the capacitor charges up in short-duration high-amplitude current pulses, as shown

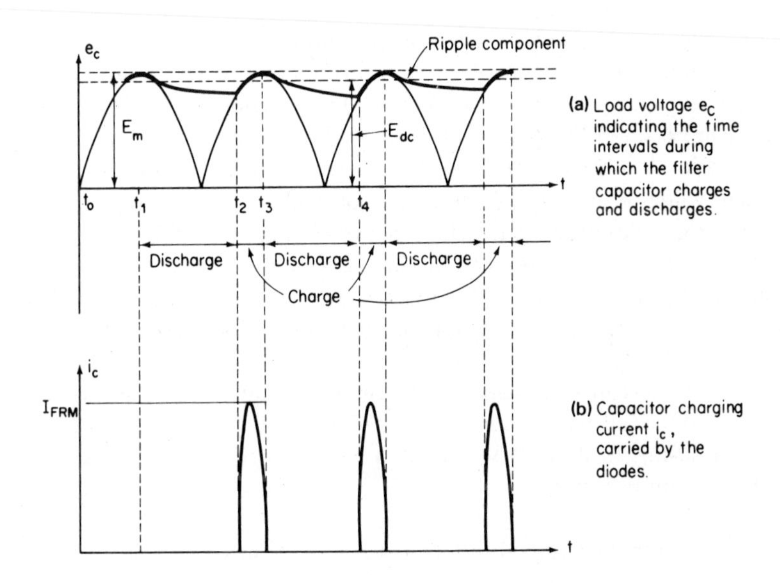
Ripple component
e_c
E_m
E_{dc}
t_0
t_1
t_2 t_3
t_4
t
Discharge
Discharge
Discharge
Charge
i_c
I_{FRM}
t
(a) Load voltage e_C indicating the time intervals during which the filter capacitor charges and discharges.
(b) Capacitor charging current i_c, carried by the diodes.

Fig. 7–11 Capacitor-input filter.

in Fig. 7–11(b). This can place a severe strain on the diodes, because these current pulses can easily exceed the rated peak diode current. In a practical circuit, the repetitive peak forward current (I_{FRM}) through the diodes may be as high as 10 times the average forward diode current $I_{F(\mathrm{AV})}$.

Another weakness of the capacitor-input filter is the possibility of a very high surge current (I_{FSM}) when the supply is first switched on. Filter capacitor C may be completely discharged at the instant that ac linepower is first applied. If it happens that the power is applied when the ac line voltage is at its peak, the diodes carry a very large *surge current* (30–100 times the average current). This *turn-on* surge current is frequently the determining factor in the choice of rectifier diodes. The turn-on surge current (called the nonrepetitive peak forward current, I_{FSM}) may be reduced by placing a small resistor R_s in series with the diodes, as indicated in Fig. 7–10. It should be noted that the addition of R_s degrades the voltage regulation of the circuit. In general, R_s is chosen to produce a turn-on surge current of approximately 40 times the load current. The duration of the surge is normally taken at the time constant CR_s, which is a parameter which determines, in part, the dc output voltage regulation and the percentage ripple.

The summary of filter characteristics, contained in Table 7–2, lists the mathetmatical expressions for dc load voltage and ripple factor for the capacitor-input filter.

7–6.2 Graphical Solution

Most circuit designers use a *graphical procedure* in designing filter circuits, which gives reasonably accurate results with minimum effort. For a typical graphical approach to the solution of a capacitor filter problem, reference is made to Figs. 7–12 to 7–15. The graphs relate the surge time constant (ωCR_s) to the ratio of surge current (I_{FSM}) and average forward current ($I_{F(\mathrm{AV})}$), for different values of output voltage regulation and percentage ripple. It is found that the largest dc output voltage from a given ac input is obtained when the time constant CR_s falls between certain limits, and that the best performance is obtained when $0.2 < \omega CR_s < 0.4$. For 60-Hz operation, this means that a given rectifier circuit will produce its best voltage regulation when $0.5 \text{ ms} < CR_s < 1.0 \text{ ms}$.

The procedure for selecting the correct circuit components is given with reference to the full-wave bridge rectifier of Fig. 7–10. This circuit is required to supply a 20-mA load at 12 V dc with a voltage regulation of 10 percent. The relevant circuit parameters are

$$\text{dc output current at full load} = I_{\mathrm{dc-fl}} = 20 \text{ mA}$$

$$\text{dc output voltage at full load} = E_{\mathrm{dc-fl}} = 12 \text{ V}$$

TABLE 7-2 SUMMARY OF FILTER SECTIONS.

	No filter	Capacitor-input	Choke-input	π-section	RC-section
Circuit	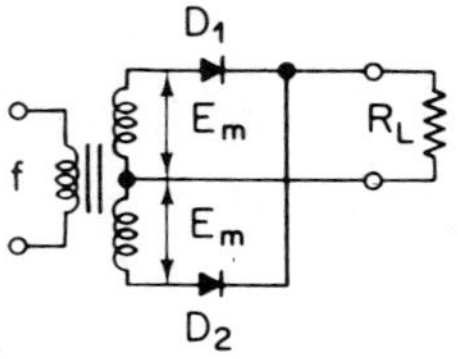 f, D_1, D_2, E_m, E_m, R_L	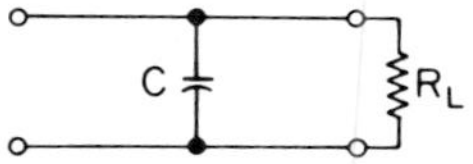 C, R_L	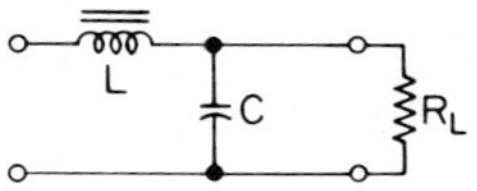 L, C, R_L	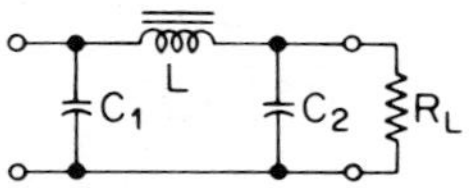 C_1, L, C_2, R_L	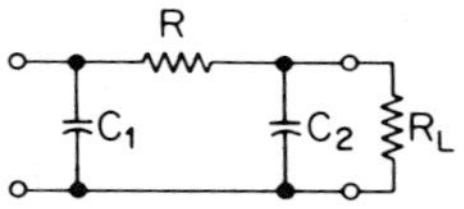 C_1, R, C_2, R_L
E_{dc} (no-load)	$2\,E_m/\pi$ or $0.636\,E_m$	E_m	E_m	E_m	E_m
E_{dc} (full-load)	$2\,E_m/\pi$ or $0.636\,E_m$	$\dfrac{E_m}{1+\dfrac{1}{4fR_LC}}$	$2\,E_m/\pi$	$\dfrac{E_m}{1+\dfrac{1}{4fR_LC_1}}$	$\dfrac{E_m}{1+\dfrac{1}{4fR_LC_1}}$
Ripple factor	0.48	$\dfrac{1}{4\sqrt{3}\,fR_LC}$	$\dfrac{\sqrt{2}}{3}\,\dfrac{X_C}{X_L}$	$\sqrt{2}\,\dfrac{X_{C_1}X_{C_2}}{R_LX_L}$	$\sqrt{2}\,\dfrac{X_{C_1}X_{C_2}}{R_LR}$

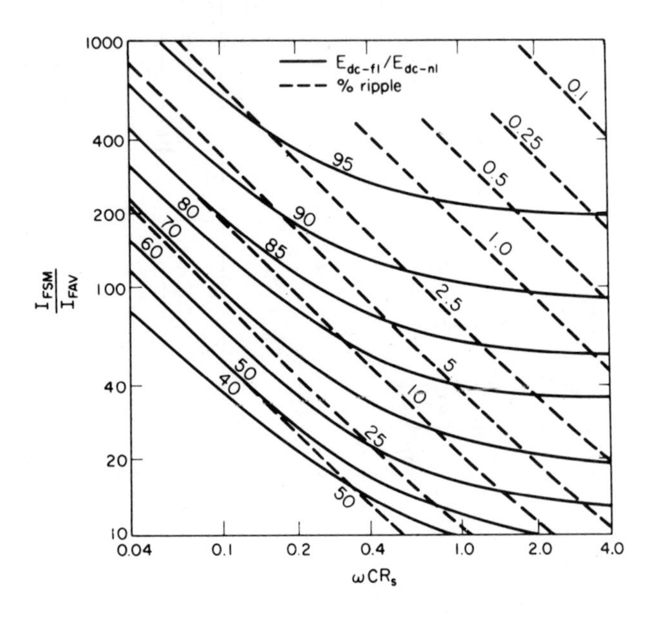

Fig. 7–12 Performance graph for half-wave rectifier with capacitor-input filter.

$$\text{line frequency} = f = 60 \text{ Hz}$$

$$\% \text{ voltage regulation} = \frac{E_{\text{dc-nl}} - E_{\text{dc-fl}}}{E_{\text{dc-fl}}} \times 100\% = 10\%$$

The circuit analysis proceeds as follows:

1. The no-load dc output is calculated from the percentage regulation and we find that $E_{\text{dc-nl}} = 1.1 \times 12 \text{ V} = 13.2 \text{ V}$. This voltage, incidentally, is the peak capacitor voltage.
2. The ratio of full-load to no-load dc output voltage is $12/13.2 = 0.91$ or 91 percent. In practice, when the percentage regulation is low, this ratio is simply 100 percent minus the percentage regulation. In this case, $100 - 10 = 90$ percent.
3. From the graph of Fig. 7–13 we find the ratio of turn-on surge current I_{FSM} and average forward current $I_{F(\text{AV})}$. Assuming the optimum value of $\omega CR_s = 0.3$, we find $I_{FSM}/I_{(\text{FAV})} = 70$. Since $I_{F(\text{AV})} = 20$ mA, the nonrepetitive peak forward current of the diode $I_{FSM} = 70 \times 20 \text{ mA} = 1400 \text{ mA}$ or 1.4 A. The rectifier diode to be selected must be able to carry this current over a time interval equal to the time constant CR_s. For $\omega CR_s = 0.3$, this time interval is

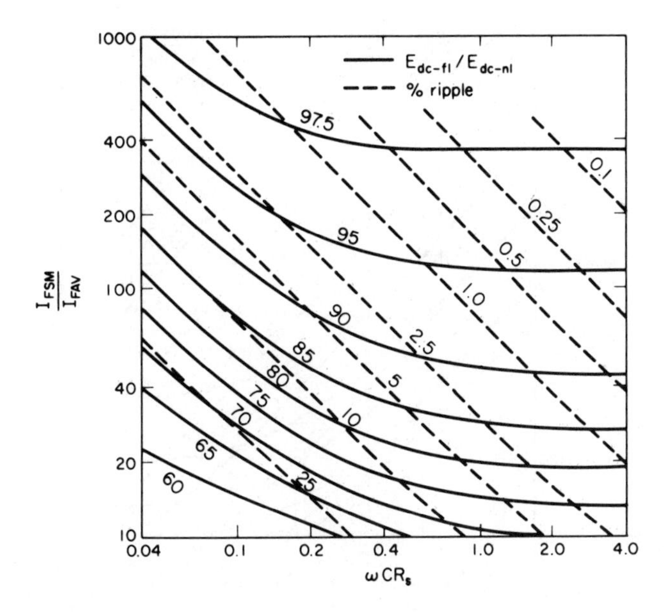

Fig. 7–13 Performance graph for full wave rectifier with capacitor-input filter.

0.8 ms. If we allow a safety margin of 100 percent, the turn-on surge current of 1.4 A extends over a time interval of 1.6 ms.

4. Allowing a forward voltage drop $V_F = 0.3$ V across the diode, the open-circuit voltage of the transformer secondary must be $E_m = 13.2\text{ V} + 0.3\text{ V} = 13.5$ V. Using a derating factor of 30 percent, the PIV rating of the diodes must be 13.5 V/0.7 = approximately 20 V.
5. The surge limiting resistor $R_s = 13.5\text{ V}/1.4\text{ A}$ = approximately 10 Ω. Note that R_s includes the winding resistance of the transformer secondary, the reflected primary resistance, and the dynamic resistance of the diode. The smallest possible transformer results when the windings provide the entire value of R_s minus the dynamic resistance of the diode.
6. The filter capacitance equals $C = 0.3/\omega R_s = 800/R_s = 80\ \mu$F.
7. The percentage ripple is found from the graph of Fig. 7–13 as 4 percent. The rms ripple voltage is $E_r = 0.04 \times 12\text{ V} = 480$ mV.
8. The rms capacitor current I_C is found from the ripple voltage. For the full-wave bridge, $I_C = 2E_r\omega C \cong 30$ mA. The filter capacitor is now specified as follows:

$$\text{capacitance, } C = 80\ \mu\text{F}$$

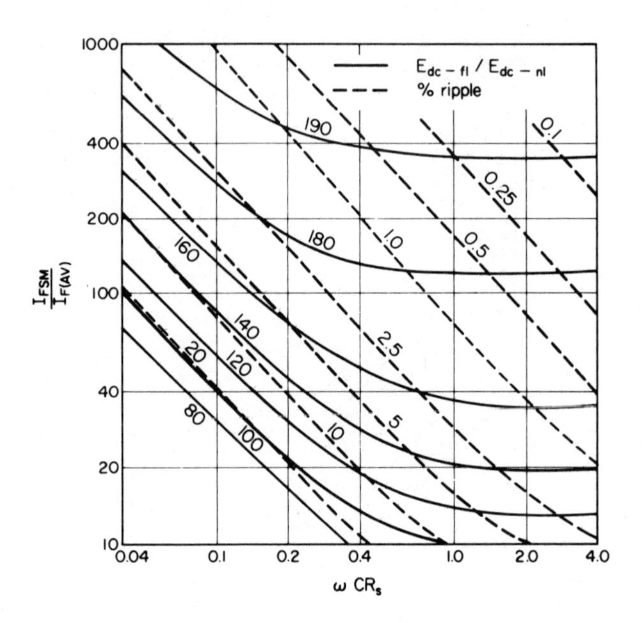

Fig. 7–14 Performance graph for voltage doubler with capacitor input filter.

$$\text{dc working voltage} = 20 \text{ V (derated by 30\%)}$$
$$\text{rms current, } I_C = 30 \text{ mA}$$

9. Figure 7–15 relates the ratio of the recurrent peak rectifier current I_{FRM} to the average dc rectifier current $I_{F(AV)}$. In this graph,

$$k = n \frac{I_{FSM}}{I_{F(AV)}} \frac{E_{dc-fl}}{E_{dc-nl}}$$

where $n = 1$ for a half-wave rectifier

$n = 2$ for a full-wave or bridge rectifier

$n = 0.5$ for a voltage doubler

For the case in question we find the following graph parameters:

$$k = 2 \times 70 \times 0.9 = 125$$

$$k\omega CR_s = 125 \times 0.3 = 37.5$$

$$\frac{100}{k} = \frac{100}{125} = 0.8$$

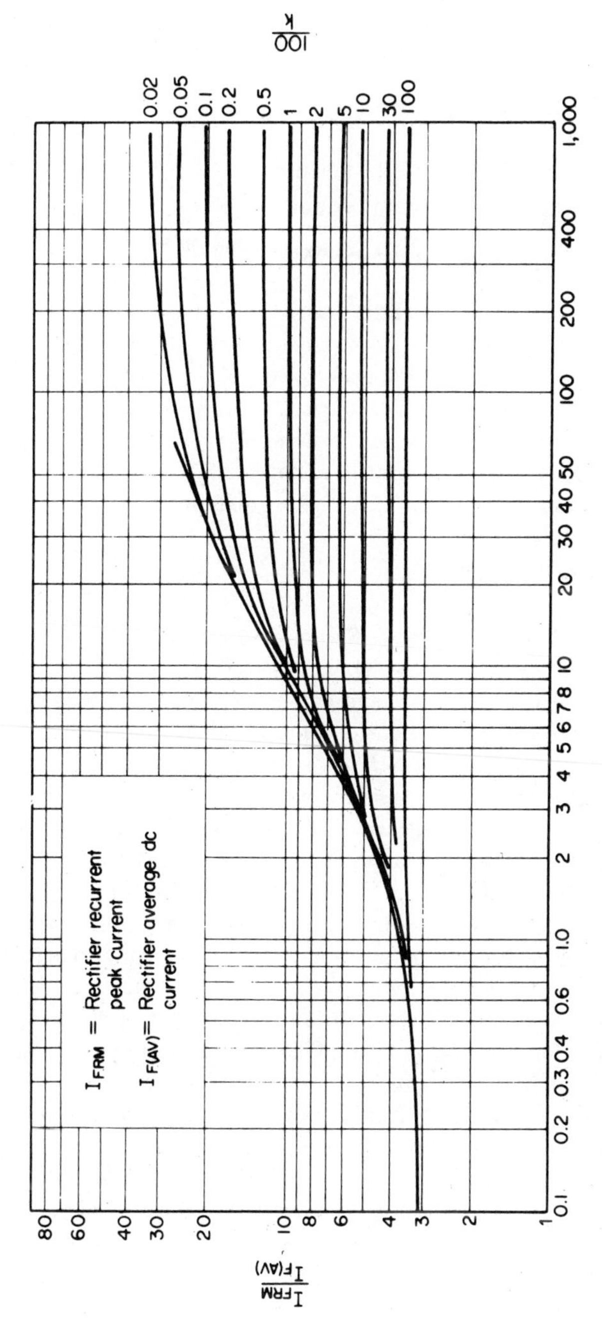

Fig. 7–15 Ratio of recurrent peak rectifier current to dc rectifier current.

Applying these parameters to Fig. 7–15, we find that $I_{FRM}/I_{F(\mathrm{AV})} \cong 10$. The recurrent peak rectifier current then is $I_{FRM} = 10 \times 20$ mA $= 200$ mA. The diode is now fully specified as follows:

$$\text{average forward current, } I_{F(\mathrm{AV})} = 20 \text{ mA}$$

$$\text{repetitive peak forward current, } I_{FRM} = 200 \text{ mA}$$

$$\text{nonrepetitive peak forward current, } I_{FSM} = 1.4 \text{ A}$$

$$\text{peak inverse voltage, PIV} = 20 \text{ V}$$

This procedure illustrates a straightforward graphical approach to a mathematically complex situation. The method does not produce the lowest possible ripple content in the dc output voltage, but does give a practical compromise in terms of component selection. Ripple may, of course, be reduced by increasing the size of the filter capacitor (increase in time constant CR_s). However, an increase in C must be accompanied by an increase in R_s to reduce the turn-on surge current, and this further degrades the output voltage regulation.

7–7 CHOKE-INPUT FILTER

The simple capacitor-input filter of Section 7–6 can be designed to produce a very small ripple voltage simply by increasing the capacitance of the filter capacitor. The ripple component can be reduced much more effectively, however, by filter networks consisting of inductors and capacitors. One such arrangement, called a *choke-input* or *L-section filter*, is shown in Fig. 7–16(a). When the design requirements of a filter are severe enough to require an LC network, it is generally desirable to use a full-wave or bridge rectifier because their ripple component is inherently lower than that of the half-wave circuit.

Series inductor L in Fig. 7–16(a) has the fundamental property of opposing any change in current through it and therefore tends to reduce the ac variations in the load current. Shunt capacitor C opposes any change in voltage across itself, and hence tends to reduce the ac variations in the load voltage.

Input voltage e_{in} to the filter section is the output of the bridge rectifier and has the waveform of Fig. 7–16(b). This waveform can be represented by a Fourier series. Analysis of this series shows that e_{in} consists of a dc component $E_{\text{dc}} = 2E_m/\pi$ and an infinite number of ac components with frequencies $2f$, $4f$, $6f$, etc., called "harmonics." The ac component with frequency $2f$ (second harmonic) is the largest component of this ac waveform with an rms value of $(4/3\pi\sqrt{2})E_m$. It is justifiable to neglect all harmonics higher than the second, which then means that e_{in} can be represented by a dc component

$$E_{\text{dc}} = \frac{2}{\pi} E_m \tag{7–23}$$

and an ac component

$$E_{ac} = \frac{4}{3\pi\sqrt{2}} E_m = \frac{\sqrt{2}}{3} E_{dc} \tag{7-24}$$

Voltage e_{in} is applied to the filter network of Fig. 7-16(c). Neglecting the small resistance of the inductor, the dc component E_{dc} appears across the capacitor and hence across the output terminals of the filter. The ac component is attenuated by the reactive voltage divider consisting of L and C. The values of L and C are usually chosen so that $X_L \gg X_C$. In addition, $X_C \ll R_L$, so R_L is effectively removed from the circuit as far as ac components are concerned.

The ripple voltage developed across the capacitor (filter output) is determined by voltage division of X_C and X_L and has an rms value

$$E_{rms} \cong E_{ac} \frac{X_C}{X_L} = E_{dc} \frac{\sqrt{2}}{3} \frac{X_C}{X_L} \tag{7-25}$$

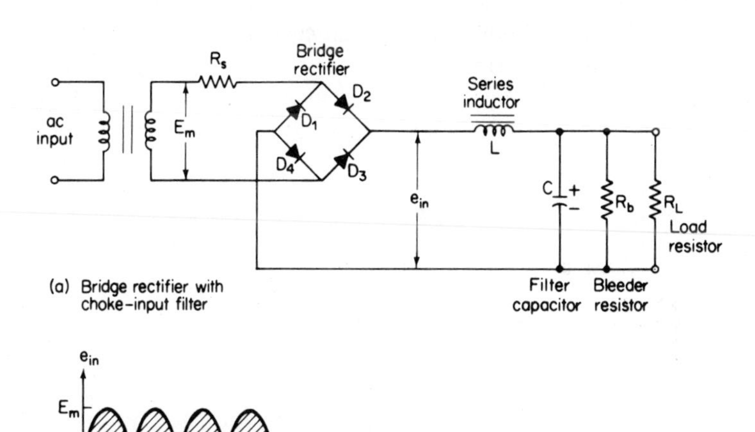

(a) Bridge rectifier with choke-input filter

e_{in} E_m 0 t

(b) Input waveform to filter section

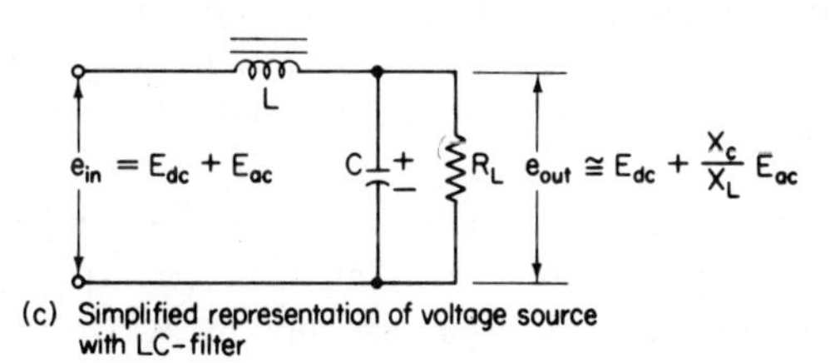

(c) Simplified representation of voltage source with LC-filter

Fig. 7-16 Choke-input filter.

The ripple factor, by definition, equals

$$r = \frac{E_{\text{rms}}}{E_{\text{dc}}} = \frac{\sqrt{2}}{3}\frac{X_C}{X_L} = \frac{\sqrt{2}}{3}\frac{1}{(2\pi f)^2 LC} \tag{7-26}$$

When the rectifier circuit is connected to the 110-V 60-Hz power line, the ripple frequency is 120 Hz. Substituting $f = 120$ into Eq. (7–26), we find that

$$\text{percentage ripple} = \frac{\sqrt{2}}{3} \times \frac{100\%}{(2\pi f)^2 LC} = \frac{83}{LC}\,\% \tag{7-27}$$

where L is expressed in henrys and C in microfarads. Equation (7–27) expresses the percentage ripple only in terms of the circuit components L and C; the percentage ripple is therefore independent of the dc load conditions.

The foregoing analysis is valid only when the diodes supply current at all times. We recall from the discussion of the capacitor-input filter (Section 7–6) that the diodes actually supply current in short-duration pulses to recharge the capacitor. If a small series inductor is inserted into the circuit, as in Fig. 7–16(a), these current pulses are lengthened, although current "cutout" may still occur. As the series inductance is increased, a value is reached where the diodes supply current to the circuit continuously and no cutout occurs. This value of inductance is called the *critical inductance* L_c. It can be shown that the critical inductance

$$L_c = \frac{R_L}{3\omega} \tag{7-28}$$

For a line frequency of 60 Hz, the critical inductance becomes

$$L_c = \frac{R_L}{1130} \tag{7-29}$$

where R_L is expressed in ohms and L_c in henrys.

Equation (7–29) shows that the critical inductance becomes very large if R_L is very large (small dc load current). To limit the value of L_c, a *bleeder resistor* R_b is usually connected across the capacitor. The bleeder resistor causes a minimum amount of dc current at all times and this then effectively determines the maximum size of L_c.

The bleeder resistor also serves another function. If a high-quality capacitor is used, it can hold its charge for a very long time, particularly when the load is disconnected entirely or when the power supply is switched off. The bleeder resistor provides a discharge path for the capacitor and eliminates a possible serious shock hazard.

Instead of using a bleeder resistor to limit the value of inductance, the inductor is sometimes replaced by a *swinging choke*, whose inductance varies with the dc current through it. This type of choke has little or no airgap in its magnetic circuit, so its inductance is high at low dc currents. At light loads, therefore, the inductance rises, increasing the ac reactance and reducing the value of current at which discontinuous current flow begins. At heavy load currents the inductance is small so that any load current surges can be supplied directly from the diode and transformer and not from the capacitor. This feature improves the voltage regulation of the circuit.

EXAMPLE 7-5

A single-phase full-wave rectifier with choke-input filter is to supply a 40-mA load at 24 V with a ripple that must be less than 1 percent. Specify the components of a circuit which produces the required results.

SOLUTION

The load resistance is $R_L = 24\ \text{V}/40\ \text{mA} = 600\ \Omega$. The rms value of the ripple voltage is $E_{\text{rms}} = 0.01 \times 24\ \text{V} = 0.24\ \text{V}$. The critical inductance is $L_c = R_L/3\omega = 600/1130 = 0.53\ \text{H}$. The product LC must be at least as large as $LC = 0.83/0.01 = 83$. The minimum values of L and C are now specified. The actual values are determined by commercially available components. For example, the manufacturer's catalogues list a choke with the following parameters: $L = 2\ \text{H}$ at $I_{\text{dc}} = 50$ mA. The winding resistance of this choke is 50 Ω. Using this choke, the minimum capacitance required is $C = 83/L = 41.5\ \mu\text{F}$. A commercially available capacitor is an electrolytic capacitor with $C = 50\ \mu\text{F}$ at a dc working v of 50 V. Allowing a transformer winding resistance of 30 Ω, a choke resistance of 50 Ω, and an average forward diode resistance of 20 Ω, the total circuit resistance equals 100 Ω. Therefore, the peak transformer voltage is $E_m = (\pi/2)(E_{\text{dc}} + I_{\text{dc}}R_t) = (\pi/2)(24\ \text{V} + 4\ \text{V}) = 44\ \text{V}$. A stock transformer would be chosen with a center-tap voltage of 48–0–48 V and a current rating of at least 40 mA. The PIV rating of the diode equals at least $2E_m$ or 88 V.

7–8 π-SECTION FILTER

In Fig. 7–17 an additional capacitor is added to the input of the L-section filter of the previous Section. This arrangement is called a *π-section filter*, because it resembles the Greek letter pi. It produces a considerable reduction in the ripple components, at the expense of the dc output voltage regulation. The filter is used when the design requirements demand a very low ripple and it is therefore almost exclusively used in conjunction with full-wave rectifiers and voltage doublers.

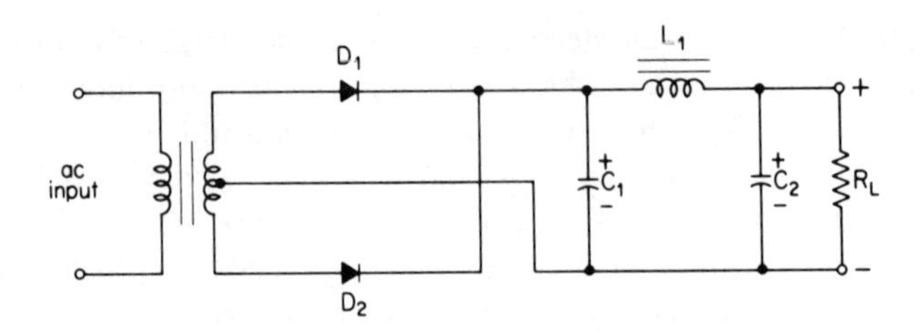

Fig. 7–17 Single-phase full-wave rectifier with π-section filter.

The action of the π filter can best be understood by considering inductor L_1 and capacitor C_2 as an L-section filter which acts upon the output voltage of capacitor-input filter C_1.

The dc load voltage is approximately equal to the dc voltage across C_1, reduced by the dc voltage drop across the inductor. The dc voltage regulation of the circuit, already inherently poor in a capacitor-input filter, is therefore even poorer in a π-section filter.

The ac component appearing across C_1 is further reduced by the L-section filter L_1C_2 which follows it. The reactance of L is very much larger than the reactance of C_2, so to close approximation the reduction in the ac component is given by the voltage division provided by L_1 and C_2, and we can write

$$\text{ripple factor, } r = \sqrt{2}\,\frac{X_{C1}}{R_L}\,\frac{X_{C2}}{X_{L1}} \tag{7–30}$$

where the reactances are calculated at the frequency of the ripple component (120 Hz for 60-Hz power-line frequency).

If a π-section filter is followed by another L section consisting of inductor L_2 and capacitor C_3, a further reduction in ripple occurs by a ratio X_{C3}/X_{L2} and the ripple factor becomes

$$r = \sqrt{2}\,\frac{X_{C1}}{R_L}\,\frac{X_{C2}}{X_{L1}}\,\frac{X_{C3}}{X_{L2}} \tag{7–31}$$

7–9 RC FILTER

An inductor is a relatively expensive, bulky, and heavy component. Especially in solid-state applications, when rigorous filtering requires the action of a π-section filter, the inductor is often replaced by a resistor to form an *RC filter*, as in Fig. 7–18.

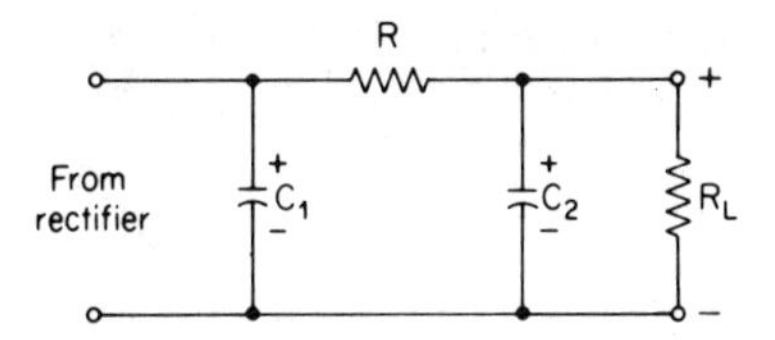

Fig. 7–18 *RC* filter

The dc output voltage is again derived from the dc voltage across capacitor C_1, reduced by the dc voltage drop across the resistor. As in the case of the π filter, the ripple component is reduced by the action of the voltage divider R-C_2. This reduction in ripple voltage equals approximately X_{C2}/R, so the ripple factor is

$$r = \sqrt{2}\,\frac{X_{C1}}{R_L}\,\frac{X_{C2}}{R} \qquad (7\text{–}32)$$

If the resistance of R is chosen equal to the reactance of the choke of the π-section filter, the ripple remains unchanged. In practice, however, R would then become rather large and would degrade the output voltage regulation.

Generally, the *RC* filter is considerably cheaper and less bulky than the π filter, but its effectiveness in reducing the ripple component may be considerably smaller.

7–10 SUMMARY OF FILTER SECTIONS

Table 7–2 summarizes the main characteristics of various filter sections, when used with a single-phase, full-wave, or bridge rectifier. In the listing of the full-load output voltages, the resistances of rectifier diodes and filter elements are neglected. Appropriate dc voltage losses in these elements should be subtracted from the values given in the table. The frequency of the input voltage applied to the rectifier circuit is indicated by f. In the majority of cases, $f = 60$ Hz, but there are occasions when this frequency is different ($f = 400$ Hz is used in some applications). In calculating the ripple factor, it should be remembered that the reactances of inductors and capacitors must be calculated at the frequency of the ripple voltage and not the line frequency. For example, in a full-wave or bridge circuit the reactance of the series inductor is $X_L = \omega L = 4\pi fL$, since the ripple frequency equals $2f$.

7–11 RECTIFIER DIODES IN SERIES AND IN PARALLEL

7–11.1 Rectifier Diodes in Parallel

Rectifier diodes can be connected in parallel to increase the power-handling capacity and reliability of a rectifier system. The circuit layout and the connections should be made as symmetrical as possible, but even then differences in the diode volt–ampere characteristics may cause unequal current sharing and result in overheating and possible destruction of one or more rectifier diodes.

Several precautions can be taken to avoid the risk of damage to the diodes. A widely used precaution is *current derating*, which simply is a reduction in the nominal possible rating of the diodes. In most cases, satisfactory current derating is obtained when

$$d = 0.8 + \frac{0.2}{n} \tag{7–33}$$

where d = current derating factor
n = number of diodes in parallel

As alternatives to current derating the following measures can also be taken:

1. Use of specially selected diodes with almost identical volt–ampere characteristics. Selected diodes with minimum current spread are available from most manufacturers at some extra cost.
2. Derating of the operating temperature of the diodes by using larger heatsinks (see Section 7–12). These larger heatsinks should have a thermal resistance no greater than 0.75 of the nominal value specified (or calculated).
3. Use of a resistor in series with each diode, thereby forcing equalization of the diode currents (Fig. 7–19). The resistance of the series resistor should be chosen such that the voltage drops across the resistor and the diodes are equal at the peak forward current. The disadvantage of this system is that the series resistors dissipate power which does not contribute to the power delivered to the load circuit. Also, the regulation of the circuit is adversely affected by the additional resistance in the diode circuit.
4. Use of balancing transformers to equalize the diode currents, as in Fig. 7–20. This system is mainly used in high-power rectifier circuits. The inductance of each transformer winding should then be

$$L = \frac{20}{\omega I_{FRM}} \tag{7–34}$$

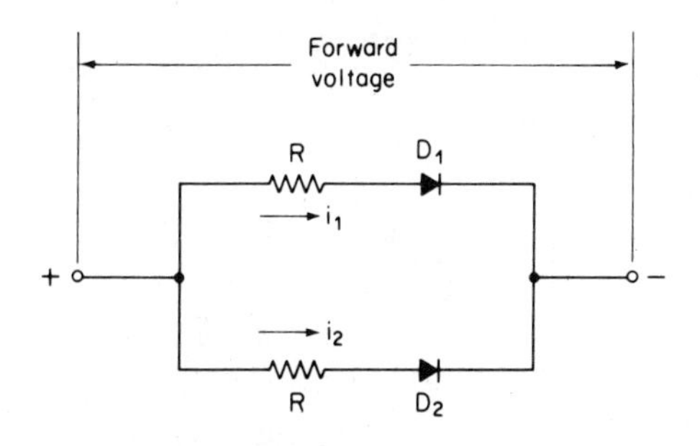

Fig. 7–19 Parallel-connected diodes with series resistors for current equalization.

where I_{FRM} is the nominal peak forward current through the diode and L is expressed in henrys.

7–11.2 Rectifier Diodes in Series

Rectifier diodes must be connected in series when the peak reverse value V_{RRM} of the applied voltage exceeds the PIV rating of a single diode. Since the leakage resistances of nonavalanche diodes may differ considerably, poor voltage sharing between the diodes in a series string may result, with the possibility of damage to one of the diodes.

A common method of equalizing the voltages across the diodes is shown in Fig. 7-21. Resistors R swamp the spread in diode leakage resistance and provide approximately equal distribution of the reverse voltages. The value of R is controlled by the requirement that the reverse voltage across the diode with the highest leakage resistance must be lower than the maximum rated PIV of the diode. Therefore, resistance R is taken lower than the minimum expected diode leakage resistance. Resistor R is often paralleled by a capacitor C to ensure adequate sharing of reverse recovery voltage transients and input transients.

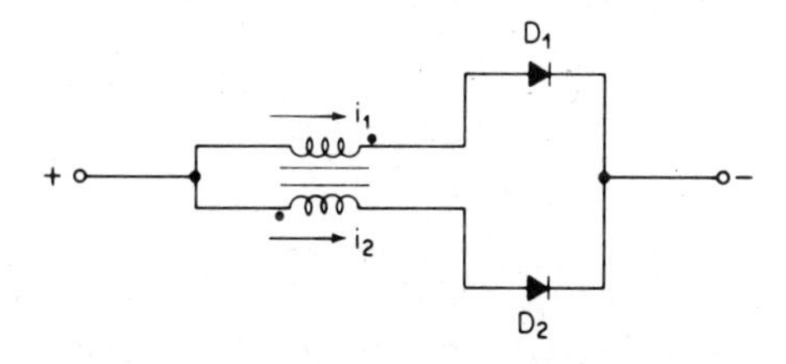

Fig. 7–20 Parallel-connected diodes with balancing transformer for current equalization.

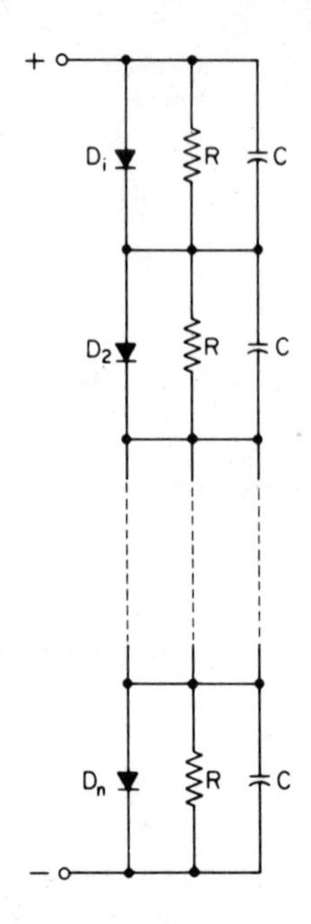

Fig. 7–21 Series-connected diodes with voltage sharing network.

In a practical circuit, the swamping resistor R is approximately one-fourth the minimum expected diode leakage resistance. For resistors with 10 percent tolerance, R is given by

$$R < 0.16 \frac{V_{RRM}}{I_{RRM}} \tag{7–35}$$

where V_{RRM} = maximum rated PIV of the single diode
I_{RRM} = maximum reverse current, specified at V_{RRM} and at maximum rated junction temperature T_j

The number of diodes required in a series string is given by

$$n > 1 + 1.5 \frac{V_{R(tot)} - V_{RRM}}{V_{RRM}} \tag{7–36}$$

where n = number of diodes in the series string

$V_{R(tot)}$ = continuous total reverse voltage across the string

V_{RRM} = maximum PIV rating of a single diode

Controlled avalanche diodes (Zener diodes) under reverse bias conditions do not require parallel resistors since proper voltage sharing is automatically ensured by reverse breakdown.

7–12 HEATSINK EVALUATION

7–12.1 Introduction

One of the important ratings of semiconductor power devices is the maximum permissible junction temperature, T_j. The life expectancy and reliability of the device depend on the junction temperature and it is therefore important that this temperature be kept as low as space and economic considerations permit.

Figure 7–22 shows a diagrammatic representation of a semiconductor device. The crystal with its heat-generating junction is in close thermal contact with its metal encapsulation or case. Most of the heat generated inside the device is transferred to the relatively heavy mounting base, from which it is dissipated to the surrounding environment. In power devices such as rectifier diodes, the mounting base is not large enough to dissipate the generated heat adequately and the surface area of the case can then be extended by attaching the device to a thermally conductive metal plate, called a *heatsink*.

The heatsink may be a simple flat piece of aluminum or steel. In many cases the surface area of the heatsink is increased by folding or by finned extrusions.

7–12.2 Thermal Resistance

With the semiconductor mounted on a heatsink, the flow of heat takes place from the heat-generating junction via the mounting base of the case to the

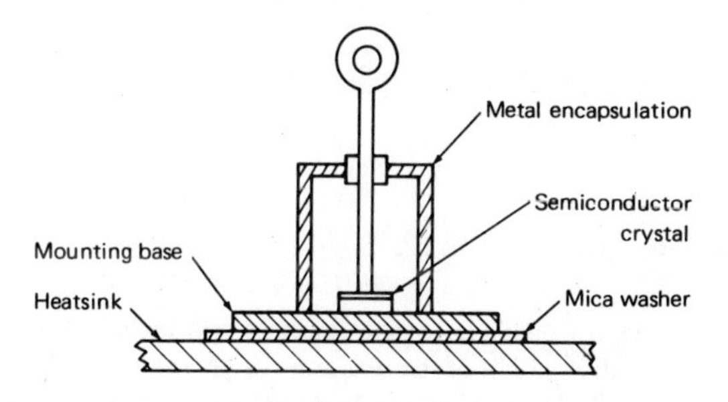

Fig. 7–22 Sketch of a semiconductor device mounted on a heatsink.

heatsink, and then by conduction, convection, and radiation to the surrounding environment. We recognize three distinct parts in the heat-flow system, each exhibiting a certain resistance to the flow of heat. The total resistance to heat flow is called the *thermal resistance*, R_{th}, which is related to the power dissipation of the device, and to the difference in temperature between the heat-generating junction and the operating environment (ambience). This relation is expressed in the form of a *thermal law*, similar to Ohm's law for electric circuits, which states that

$$\Delta T = PR_{th(j-a)} \qquad (7\text{–}37)$$

where $\Delta T_j = T_j - T_a$ (junction temperature minus ambient temperature, in °C)

P = maximum power dissipation of the device, in watts

$R_{th(j-a)}$ = thermal resistance from junction to ambience, in °C/W

In Eq. (7–37), ΔT_j is known, since T_j is stated in the data sheet of the device, and T_a is the ambient temperature which can be measured. P represents the actual maximum power dissipation of the device and is known from its circuit configuration. The total thermal resistance can then readily be calculated.

In an actual case $R_{th(j-a)}$ consists of the sum of the thermal resistances from junction to ambience, so that

$$R_{th(j-a)} = R_{th(j-c)} + R_{th(c-s)} + R_{th(s-a)} \qquad (7\text{–}38)$$

where $R_{th(j-c)}$ = thermal resistance from junction to case

$R_{th(c-s)}$ = thermal resistance from case to heatsink

$R_{th(s-a)}$ = thermal resistance from heatsink to ambience

The situation is illustrated graphically in Fig. 7–23. By rearranging Eq. (7–38) we can calculate the thermal resistance of the heatsink.

EXAMPLE 7-6

A 2N4347 power transistor is used as a series regulator in a power supply. The maximum power dissipation of the device is 10 W and the ambient temperature is not expected to exceed 50°C. Calculate the required thermal resistance of the heatsink.

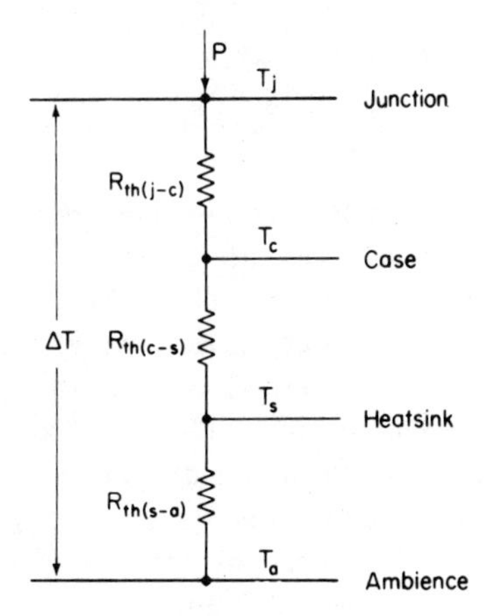

Fig. 7–23 Thermal resistances and their relation to power dissipation and temperature differential.

SOLUTION

From the diode manual we find that the maximum permissible junction temperature of the 2N4347 is $T_j = 200°C$. The thermal resistance from junction to case is listed as $R_{th(j-c)} = 1.5°C/W$. The thermal resistance from case to heatsink is not given in the transistor manual, but experience indicates that $R_{th(c-s)} = 0.5°C/W$ is a reasonable value. Using the thermal law, we find the total thermal resistance from junction to ambience:

$$R_{th(j-a)} = \frac{\Delta T_j}{P} = \frac{200°C - 50°C}{10\ W} = 15°C/W$$

Subtracting the known parts of the thermal resistance path, the thermal resistance of the heatsink is

$$R_{th(s-a)} = 15°C/W - 2°C/W = 13°C/W$$

Example 7-6 indicates that it is a fairly simple procedure to calculate the required thermal resistance of a heatsink, using the device ratings and external circuit conditions.

The next step is to determine the heatsink material and to calculate required heatsink size for adequate dissipation.

7-12.3 Physical Characteristics

Various heatsink materials have different thermal conductivity properties.

Copper has a very high thermal conductivity and is used for heatsinks requiring low thermal resistance with maximum surface area.

Aluminum also has a high thermal conductivity and has the advantage of light weight and low cost. Aluminum is the obvious choice for most applications.

Steel has a moderate thermal conductivity, but is very low in cost compared to copper and aluminum. The most common application of steel heatsinks is where the device is mounted directly onto the equipment chassis, with the chassis serving as the heatsink.

The *surface finish* of the heatsink has an effect on its emissivity. Bright or polished metal surfaces have poor emission properties, while matte black painted or black anodized surfaces have the best emission properties. A matte paint finish of any color is considerably better than a bright surface.

The *mounting attitude* of the heatsink is also of importance. For the best transfer of heat, vertical mounting with both sides of the heatsink exposed to convection is best. Frequently the space available to mount the heatsink is limited, and the total surface area may then be extended by folding or the use of multiple fins.

In some cases, natural convection cooling is not sufficient, especially for thermal resistances below 2°C/W, and forced air cooling is then required.

7-12.4 Calculation of Heatsink Dimensions

The total thermal resistance of the heatsink consists of two parts, namely a part responsible for *conduction* of heat within the metal itself, and a part responsible for *radiation* from the metal surface into the environment.

The *conduction* properties of the heatsink depend on the physical characteristics of the metal, such as thickness and type of material. Table 7-3 lists the thermal conduction properties for various metals of different thicknesses. The table shows that the thermal resistance of steel is approximately twice that of aluminum. The *radiation* properties of the metals listed in Table 7-3 depend on several factors such as the total surface area of the heatsink, the surface finish, and the position (horizontally or vertically placed). Figure 7-24 relates these parameters in a nomogram. Its use is best demonstrated by an example.

EXAMPLE 7-7

A heatsink with a thermal resistance $R_{th(s-a)} = 13°C/W$ is required to dissipate the heat generated by a 2N4347 power transistor, operated at 10-W dissipation (see also Ex. 7-6). A flat piece of 20-gage aluminum is available, which will be

TABLE 7-3 THERMAL CONDUCTION PROPERTIES OF HEATSINK MATERIALS.

Gage (SWG)		24	22	20	18	16	14	12
Thickness	inches	0.022	0.028	0.036	0.048	0.064	0.080	0.104
	mm	0.56	0.71	0.92	1.22	1.63	2.04	2.54
$R_{th\,(radiation)}$ in °C/W	Copper	2.04	1.81	1.59	1.38	1.20	1.07	0.94
	Aluminum	2.74	2.43	2.14	1.86	1.61	1.44	1.26
	Steel	5.87	5.19	4.80	3.97	3.44	3.08	2.70

mounted vertically to allow radiation from both sides. Calculate the size of this heatsink.

SOLUTION

From Table 7–3 we determine the thermal resistance of 20-gage aluminum, and we find that $R_{th(conduction)} = 2.14°C/W$. This indicates that the heatsink should have radiation properties corresponding to

$$R_{th(radiation)} = 13°C/W - 2.14°C/W = 10.86°C/W$$

Use the nomogram of Fig. 7–24. Align a straightedge from the point $k = 0.85$ (vertical mounting of bright aluminum) on the k scale to the point $R_{th(radiation)} = 10.86°C/W$ on the R_{th} scale. The A scale then indicates the required area and we read that $A = 51.5\ cm^2$. Since the most effective shape of a heatsink is approximately square, the dimensions of the bright aluminum sheet should be approximately 7 by 8 cm. If the surface of the aluminum sheet is blackened, the required area is only 27 cm^2.

For heatsinks of finned construction, appropriate nomograms are available from the manufacturers but the procedure for calculating type and size of heatsink is essentially the same. Also, forced-air cooling considerably reduces the required heatsink area and again nomograms are available which relate the size of the heatsink to the total power dissipation and the velocity of the forced air.

PROBLEMS

7–1 A half-wave rectifier circuit has a dc output voltage of 24 V. What is the rms value of the ripple voltage?

7–2 A full-wave rectifier circuit has a no-load voltage of 30 V dc and a full-load voltage of 28 V dc. Calculate the percentage regulation.

7–3 The secondary voltage of the power transformer in a full-wave rectifier has an rms value of 60 V, measured from the center tap to the ends. Determine the PIV rating of the diodes, allowing 30 percent voltage derating.

7–4 One of the diodes in a full-wave rectifier circuit develops a short circuit. What is the effect on the dc output voltage and current?

7–5 Draw the circuit diagram of a full-wave rectifier with capacitor-input filter to deliver a negative output voltage.

7–6 A silicon rectifier diode with a forward voltage drop of 0.5 V is used in the half-wave circuit of Fig. 7–2. This simple power supply delivers a dc current of 0.5 A to a 48-Ω load.

Determine (a) The secondary transformer voltage.
(b) The ripple voltage.

7–7 A half-wave rectifier uses a germanium power rectifier with an average forward resistance of 200 Ω. The load voltage is 400 V dc across a resistor of 5 kΩ.

Calculate (a) The secondary transformer voltage.
(b) The dc diode current.
(c) The rms value of the total load current.
(d) The rms value of the ripple voltage.

7–8 Two silicon diodes are used in the full-wave rectifier circuit of Fig. 7–4. The supply delivers a dc load current of 1 A at 12 V. Neglect the forward voltage drop across the diodes and *calculate*

(a) The rms current in the transformer secondary.
(b) The rms voltage of the transformer secondary.
(c) The ripple voltage.

7–9 A bridge rectifier using silicon diodes is connected across a 75-V rms transformer winding. The load resistance is 100 Ω.

Calculate (a) The dc load voltage.
(b) The dc load current.
(c) The PIV rating of the diodes, allowing 50 percent derating.
(d) The ripple voltage.

7–10 A single-phase half-wave rectifier uses a 40-μF capacitor across a resistive load. The dc load voltage is 60 V and the dc load current is 100 mA.
Calculate (a) The secondary transformer voltage.
(b) The ripple voltage.

7–11 The full-wave rectifier of Fig. 7–4 uses silicon diodes with an average forward resistance of 20 Ω. The transformer voltage, measured across the entire secondary winding, is 60 V rms. The dc load current can vary from 0 to 400 mA.
Calculate (a) The no-load dc voltage.
(b) The full-load dc voltage.
(c) The percentage load regulation.

7–12 Assume that one of the diodes in the full-wave rectifier circuit of Prob. 7–11 is open-circuited.
Determine (a) The no-load voltage.
(b) The full-load voltage.
(c) The percentage load regulation.

7–13 The bridge rectifier of Fig. 7–7 supplies 50 mA dc to a 3-kΩ load. Neglect the forward voltage drops across the diodes and *determine*
(a) The transformer secondary voltage.
(b) The ripple voltage.
(c) The PIV ratings of the diodes.

7–14 A full-wave rectifier uses an L-section filter with L = 10 H and C = 20 μF. The transformer secondary is rated as 100–0–100 V rms. Assume that the inductance is the critical value and *calculate*
(a) The dc load voltage.
(b) The dc load current.
(c) The ripple voltage.

7–15 A certain rectifier diode has the following ratings: P_{tot} = 20 W, T_j = 240°C, and $R_{th(j-s)}$ = 3°C/W. The ambient temperature is T_{amb} = 20°C. Calculate the required thermal resistance of the heatsink.

7–16 A certain rectifier diode requires a heatsink with a thermal resistance of $R_{th(s-a)}$ = 20°C/W. This diode is to be mounted directly onto the steel chassis of the power supply. It can be assumed that heat dissipation takes place from the upper surface of the chassis only. Use the nomogram of Fig. 7–24 and determine the minimum area required for mounting the diode.

7–17 A number of rectifier diodes, each with a current rating of $I_{F(AV)}$ = 2 A, are placed in parallel to supply a 10-A dc load current. Determine the current derating factor and the required number of parallel diodes.

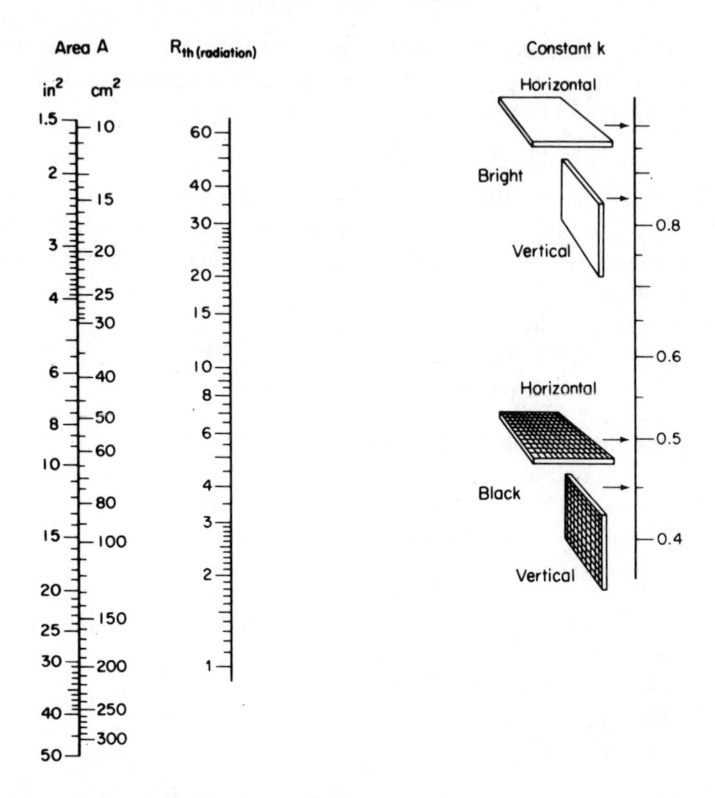

Fig. 7–24 Nomogram to determine the thermal resistance $R_{th(radiation)}$ of flat heatsinks for different attitudes and surface finishes.

7–18 Three identical rectifier diodes, each with a peak reverse voltage rating of $V_{RRM} = 40$ V, are placed in series. Determine the maximum reverse voltage that can be applied to this series string.

Chapter 8

Power Supply Regulators

8-1 ZENER DIODE REGULATORS

8-1.1 Introduction

The characteristics of the Zener diode in the breakdown region were described in Section 2-5. It was shown that at a given reverse voltage, called the *Zener voltage* V_Z, the current through the diode increases sharply while the voltage drop across the diode remains essentially constant. This suggests that the Zener diode is useful as a control element in applications where a constant voltage is required, and we therefore find the Zener diode as a *voltage regulator* in power supply circuits.

The volt-ampere characteristic of a Zener diode in its normal mode of operation (under reverse-bias conditions) is shown in Fig. 8-1.

The *Zener voltage* V_Z is defined as the reverse voltage just below the knee of the curve, where the current increases very sharply. V_Z is not really constant but increases slightly with an increase in reverse current, indicating that the Zener diode in the breakdown region has a small internal resistance. This internal resistance is given by the slope of the curve and is called the *dynamic slope* resistance r_Z, where

$$r_Z = \frac{\Delta V_Z}{\Delta I_Z} \tag{8-1}$$

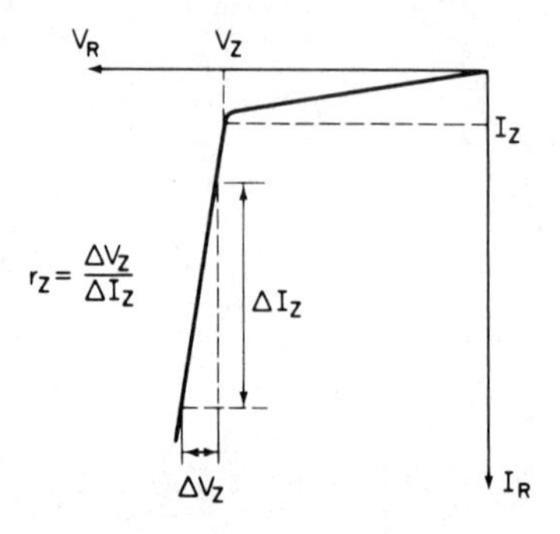

Fig. 8–1 Conduction characteristics of a Zener diode.

The dynamic slope resistance generally increases from a few ohms for diodes with low Zener voltages to several hundred ohms for the high-voltage devices. Also, diodes with high power ratings have much smaller dynamic resistances than those in the low-power range. The dynamic slope resistance is given in the data sheet; this parameter plays a role in determining the actual Zener voltage of an operating circuit.

Another important rating of the Zener diode is the total *power* dissipation P_t, which is approximately equal to the product of the Zener voltage V_Z and the reverse current through the device. The total power dissipation determines the maximum allowable reverse current. Care must be taken to limit this current to prevent damage or complete destruction through overheating.

Finally, the ratings of semiconductor devices are usually given at an ambient temperature of 25°C. The Zener diode generally has a positive *temperature coefficient* of operating voltage, which means that the Zener voltage rises as the temperature increases. The temperature coefficient is given in the data sheet as S_Z, where S_Z is expressed in mV/°C.

8–1.2 Basic Zener Circuit

Figure 8–2 shows a basic Zener diode voltage regulator. The Zener diode is connected in series with current limiting resistor R_S across the dc voltage source, which could be the output terminals of a full-wave rectifier with capacitor filter. To cause Zener action, the supply voltage V must be greater than the Zener voltage V_Z. In general, the supply voltage should be approximately 1.2 to 1.5 times the Zener voltage.

Load resistor R_L can be connected across the diode by closing switch S. We first assume a no-load condition, with switch S open and R_L removed from the circuit.

Maximum power dissipation P_t and Zener voltage V_Z are given in the data sheet; they determine the maximum current I_t which can be allowed in the diode as

$$I_t = \frac{P_t}{V_Z} \tag{8-2}$$

Since the Zener diode conducts and has a constant voltage V_Z across it, voltage drop V_R *across current limiting resistor* R_S must also be constant, and equals

$$V_R = V - V_Z \tag{8-3}$$

The minimum resistance of R_S required to limit the diode current to its maximum value I_t equals

$$R_S = \frac{V - V_Z}{I_t} \tag{8-4}$$

Switch S is now closed, load resistor R_L is connected across the diode, and the load current is

$$I_L = \frac{V_Z}{R_L} \tag{8-5}$$

Since the current in the circuit is controlled by R_S and is constant, the current through the Zener diode must decrease, and

$$I_Z = I_t - I_L \tag{8-6}$$

As long as the demand for load current does not exceed the available circuit current I_t, the Zener diode remains in the breakdown region and the output voltage of the system remains constant. If the load current demand exceeds the available current I_t, the diode stops conducting and behaves like an ordinary

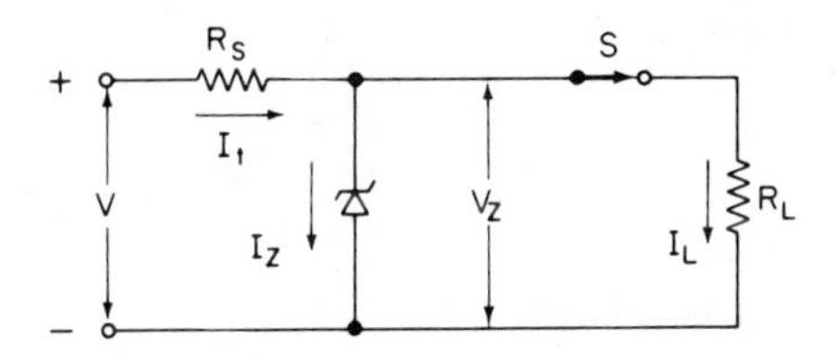

Fig. 8–2 Simple Zener diode voltage regulator.

reverse-biased diode (an open switch). In this case the output voltage follows Ohm's law and voltage regulation no longer exists.

EXAMPLE 8-1

The voltage regulator of Fig. 8–2 uses a Zener diode with the following ratings: $V_Z = 10$ V at $I_Z = 100$ mA, $P_t = 1$ W, and $r_Z = 10\ \Omega$. The dc supply voltage $V = 15$ V and load resistor $R_L = 200\ \Omega$.

Calculate (a) The minimum resistance of current limiting resistor R_S.
(b) The percentage regulation of the circuit.

SOLUTION

(a) The maximum current allowed in the circuit is

$$I_t = \frac{P_t}{V_Z} = \frac{1\text{ W}}{10\text{ V}} = 100\text{ mA}$$

The voltage drop across R_S equals

$$V_R = V - V_Z = 15\text{ V} - 10\text{ V} = 5\text{ V}$$

The minimum resistance of the limiting resistor is

$$R_S = \frac{V_R}{I_t} = \frac{5\text{ V}}{100\text{ mA}} = 50\ \Omega$$

(b) At no-load, the Zener current equals the maximum current and

$$I_{Z(\text{max})} = 100\text{ mA}$$

Since the internal resistance of the diode $r_Z = 10\ \Omega$, the no-load output voltage is

$$V_{\text{no load}} = V_Z + I_{Z(\text{max})} r_Z = 10\text{ V} + 1\text{ V} = 11\text{ V}$$

When the load is connected, the load current is

$$I_L = \frac{V_Z}{R_L} = \frac{10\text{ V}}{200\ \Omega} = 50\text{ mA}$$

The current through the Zener diode is therefore

$$I_Z = I_t - I_L = 50\text{ mA}$$

The full-load output voltage is

$$V_{\text{full load}} = V_Z + I_Z r_Z = 10 \text{ V} + 0.5 \text{ V} = 10.5 \text{ V}$$

The percentage voltage regulation is defined as

$$\% \text{ regulation} = \frac{V_{\text{no load}} - V_{\text{full load}}}{V_{\text{full load}}} \times 100\%$$

and equals

$$\% \text{ regulation} = \frac{11 \text{ V} - 10.5 \text{ V}}{10.5 \text{ V}} \times 100\% = 4.76\%$$

This type of voltage regulation, where the Zener diode attempts to keep the output voltage constant under varying load conditions, is called *load regulation*.

8-1.3 Simple Zener Regulator

Example 8-1 describes the simplest case in which we are concerned with a *constant supply voltage* and a *fixed load.* The arrangement is not very efficient because the Zener current under full load is relatively large and does not contribute to the power delivered to the load. In a case like this, the supply voltage should not be more than a few volts above the Zener voltage for efficient operation.

Consider now the case where the load is constant but the input voltage to the regulator circuit varies. For convenience, assume that the supply voltage in Ex. 8-1 is reduced to 12 V, while the remaining circuit parameters are unchanged.

At $V = 12$ V, the voltage drop across the current-limiting resistor is $V_R = 12 \text{ V} - 10 \text{ V} = 2$ V. The total current in the circuit is then $I_t = 2\text{V}/50\ \Omega = 40$ mA. The input current of 40 mA is not sufficient to supply the load demand of 50 mA, necessary to maintain the output voltage at 10 V. The load voltage drops below the Zener voltage, stopping Zener conduction. The regulator is reduced to a simple series combination of R_S and R_L, Fig. 8-3. The current, $I_t = 12 \text{ V}/250\ \Omega = 48$ mA, and the load voltage, $V_L = 48 \text{ mA} \times 200\ \Omega = 9.6 \text{ V} < 10$ V, hence regulation has ceased.

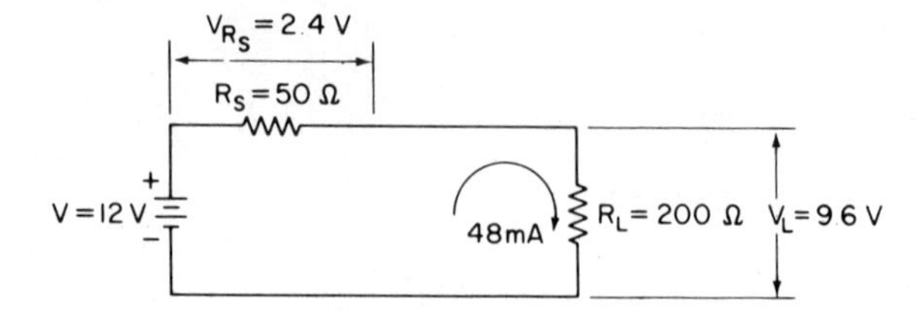

Fig. 8–3 The Zener diode is inoperative because the supply voltage is too low to sustain the Zener voltage.

EXAMPLE 8-2

Refer to Fig. 8-4 where a 20-V 5-W Zener diode is used to regulate a 0.2-A load. A 20-Ω series resistor protects the Zener diode.

Calculate (a) The maximum permissible input voltage.
(b) The input voltage at which the regulator becomes ineffective.

SOLUTION

(a) From the power dissipation rating we find the maximum Zener current is

$$I_{Z(\max)} = \frac{P_t}{V_Z} = \frac{5\text{ W}}{20\text{ V}} = 0.25\text{ A}$$

The maximum circuit current is

$$I_t = I_{Z(\max)} + I_L = 0.45\text{ A}$$

Hence, the voltage drop across the 20-Ω limiting resistor is

$$V_R = 0.45\text{ A} \times 20\ \Omega = 9\text{V}$$

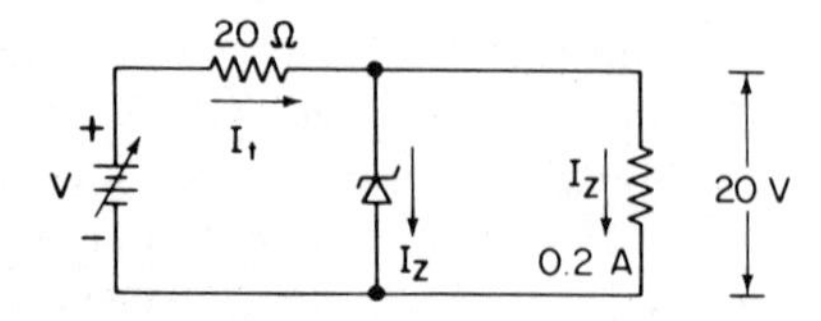

Fig. 8–4 Zener diode circuit for Example 8-2.

The maximum input voltage then is

$$V_{max} = V_R + V_Z = 29 \text{ V}$$

If the input voltage is greater than 29 V, the rated power dissipation of the Zener diode is exceeded, and the diode may be destroyed.

(b) The regulator becomes ineffective when the total circuit equals the load current. In this case, the voltage drop across R_S is

$$V_R = 0.2 \text{ A} \times 20\ \Omega = 4 \text{ V}$$

The input voltage at which the regulator becomes ineffective is

$$V_{min} = V_R + V_Z = 24 \text{ V}$$

The major disadvantage of the simple Zener diode regulator is that the Zener current at maximum input voltage may be several times larger than the load current, an inefficient operation, and the circuit is therefore mainly used in low-power applications, not exceeding a few watts.

8-2 BASIC TRANSISTOR REGULATORS

8-2.1 Shunt Regulator

There are two basic types of voltage regulators: the *shunt* regulator and the *series* regulator. In the shunt regulator, the regulating element, represented by variable resistor R_r, is placed in parallel with the load as in Fig. 8-5.

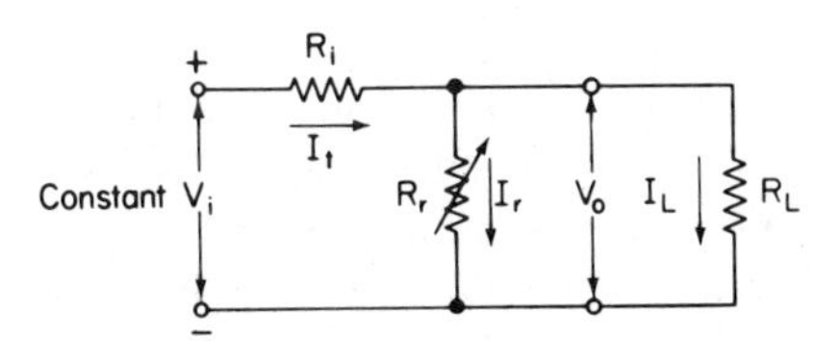

Fig. 8-5 Basic shunt regulator.

To maintain the output voltage V_o constant under varying load conditions, the total current drawn from the source (V_i) must be kept constant. This can be accomplished by variable resistor R_r across the output terminals. If, for example, the demand for load current decreases (R_L becomes larger), the resistance of R_r can be reduced so that it draws more current. This interaction of R_L and R_r must be coordinated in such a way that the total current $I_t = I_r + I_L$ is kept constant. In the simple schemes, R_r is a Zener diode which absorbs variations in I_L over a limited range, as discussed in Section 8–1. In the more sophisticated arrangements, R_r is a suitably controlled transistor, as in Fig. 8–6.

Silicon power transistor $Q1$ is the *regulating element.* Zener diode $CR1$ controls the base current of $Q1$ and it is called the *reference element.* Power transistor $Q1$ conducts only when the Zener diode conducts, in which case output voltage V_o is approximately 0.7 V above the Zener voltage, *and constant.* If V_o tends to increase due to a decrease in load current, the Zener diode is made to conduct more and this increases the base current into $Q1$. The collector current of $Q1$ then increases accordingly and the total current drawn from the source tends to remain constant. It is obvious that the Zener diode can be a low-power device, because it only supplies the relatively small base current for $Q1$. The control transistor is the high-power device, absorbing the current changes required to compensate for the loade voltage variations.

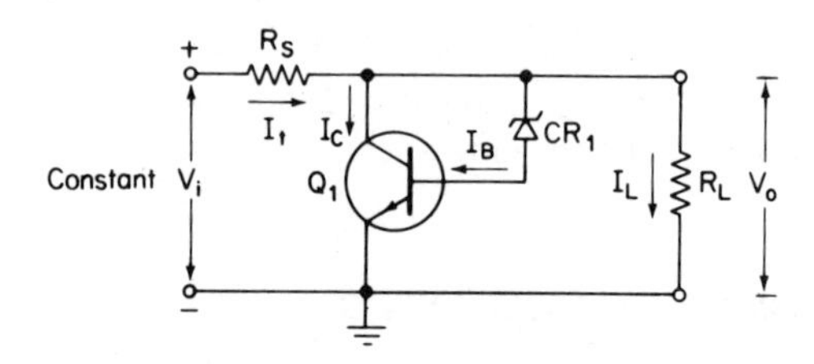

Fig. 8–6 Transistor shunt regulator.

EXAMPLE 8-3

The transistor shunt regulator of Fig. 8–6 is to supply a load current of 0-500 mA at an output voltage of 10 V dc. The input voltage to the regulator, supplied by a rectifier-filter circuit, is 15 V. Making reasonable approximations, *calculate*

(a) The resistance of series resistor R_s.
(b) The maximum power dissipation of power transistor $Q1$.
(c) The maximum power dissipation of Zener diode $CR1$.

SOLUTION

(a) The collector current of $Q1$ must compensate for load current variations such than when $I_L = 500$ mA, $I_C = 0$ mA, and when $I_L = 0$ mA, $I_C = 500$ mA. Neglecting the relatively small Zener current ($Q1$ base current), the current through R_s remains practically constant at 500 mA. Hence,

$$R_s = \frac{V_o - V_i}{I_L + I_C} = \frac{5\text{ V}}{500\text{ mA}} = 10\ \Omega$$

(b) In the worst condition, the load current is zero and the shunt regulator carries the total current of 500 mA. Hence, $I_L = 0$ mA and $I_C = 500$ mA. The maximum power dissipation of $Q1$ is

$$P_c = V_{CE}I_C = 10\text{ V} \times 500\text{ mA} = 5\text{W}$$

(c) Assume that the current gain of $Q1$, is approximately 20 (a reasonable assumption). The Zener current then equals

$$I_Z = I_B = \frac{I_C}{\beta} = \frac{500\text{ mA}}{20} = 25\text{ mA}$$

The maximum power dissipation of $CR1$ is

$$P_Z = V_Z I_Z = 10\text{ V} \times 25\text{ mA} = 0.25\text{ W}$$

Example 8–3 indicates that at full load the shunt current is zero, but as the load current decreases the current through the shunt regulator increases until at no-load the entire current is carried by the shunt transistor. The circuit carries full current at all times and the shunt regulator therefore has low efficiency at small loads.

The outstanding advantage of the shunt regulator is that it is short-circuit-proof. A "short" across the load terminals forces the output voltage, and hence the voltage across the transistor, to zero. The power dissipation in the transistor then drops to practically zero and the transistor escapes damage.

8–2.2 Series Regulator

In the series regulator of Fig. 8–7 the control or regulating element is represented by variable resistor R_r. The internal resistance of the source is R_i; if load current decreases, output voltage V_o rises. If, as I_L decreases, variable resistor R_r increases, the voltage drop across R_r increases and compensates for the initial increase in V_o.

In a practical circuit, R_r is a power transitor controlled by a Zener diode voltage reference, as in Fig. 8–8. Silicon power transistor $Q1$ is the series-

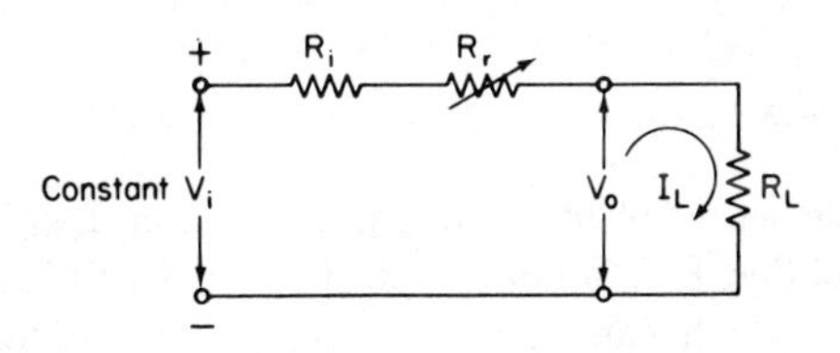

Fig. 8–7 Basic series regulator.

regulating element and carries the entire load current at all times. Since the base current into $Q1$ is relatively small compared to the total current through series resistor R_S, the Zener diode current is practically constant and the Zener reference voltage is also constant. The base of $Q1$ is therefore held at a constant potential. The emitter-to-base voltage V_{BE} of $Q1$ is the difference between reference voltage V_R and output voltage V_o. If the output voltage V_o tends to increase as the result of a change in load conditions, V_{BE} tends to decrease. A lower base–emitter voltage reduces the conduction in $Q1$ and hence raises the voltage drop across $Q1$, compensating for the initial increase in V_o.

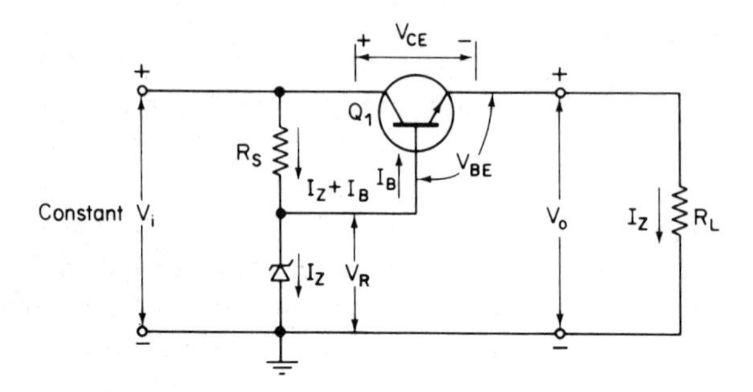

Fig. 8–8 Transistor series regulator.

The series regulator is much more widely used than the shunt regulator. In the series regulator the load current is carried by the regulating transistor and no current is wasted through a shunt path, as in the shunt regulator. Full-load efficiency of the series regulator is approximately the same as for the shunt regulator, but at small loads the series circuit is definitely the preferred circuit.

The simple series regulator of Fig. 8–8 does not have inherent overload protection and the regulating transistor can be destroyed by a momentary short circuit across the output terminals, unless suitable protection is used (Section 8–6).

8–3 FEEDBACK REGULATORS

8–3.1 Basic Feedback Circuit

To achieve a higher degree of line and load regulation than is possible with the simple regulators of Sec. 8–2, a more sophisticated circuit is required, such as a *feedback regulator,* which can have different configurations. But since these circuits operate on the same principle, any form of feedback regulator can be represented by the block diagram of Fig. 8–9. The output voltage V_o is sensed by a *voltage monitor*, producing voltage KV_o, proportional to V_o, which is continuously compared to *reference* voltage V_R, produced by a Zener diode. The output of the voltage *comparator* is amplified by an *error amplifier* to provide sufficient current drive to control the conduction of the *series regulator*, the action of which is such that it compensates for any changes V_o in the the manner described for the simple circuit of Fig. 8–8.

The output *voltage monitor* usually consists of a resistive voltage divider across the output terminals of the regulator, as in Fig. 8–10. Clearly, the output voltage of the divider is given by

$$V_s = \frac{R_2}{V_1 + R_2} V_o = KV_o \tag{8–7}$$

Usually, a potentiometer is inserted between the two fixed elements R_1 and R_2, so the sampled voltage KV_o is variable between certain limits.

The source impedance of the voltage monitor, looking into the V_s terminals, is

$$Z_s = \frac{R_1 R_2}{R_1 + R_2}$$

If this source impedance is too high, one of the resistors may be replaced by a Zener diode, which has a low dynamic resistance, and the source impedance then approaches r_Z.

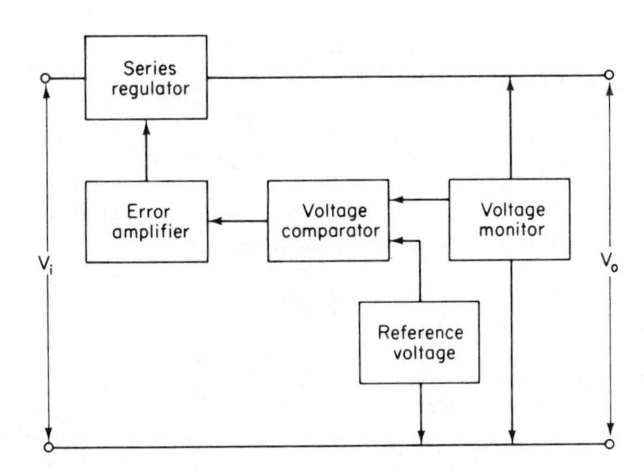

Fig. 8–9 Block diagram of a feedback voltage regulator.

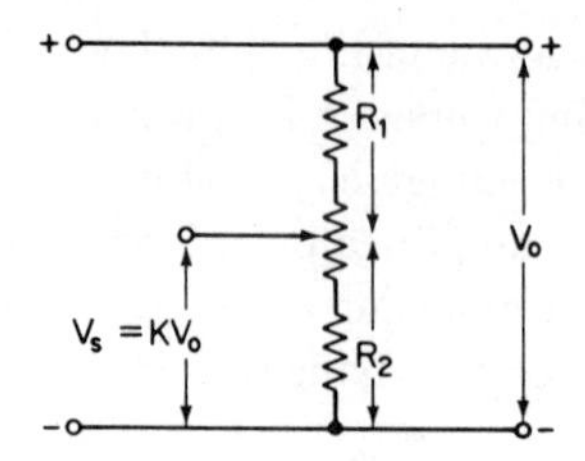

Fig. 8–10 Resistive voltage divider to monitor the output voltage.

The *voltage reference* is usually a Zener diode. It is connected to the regulated dc output voltage, as in Fig. 8–11. The output voltage is regulated, and the Zener breakdown is given by

$$I_Z = \frac{V_o - V_R}{R_3} \tag{8–8}$$

8–3.2 Error Amplifiers

The voltage comparator can be one of several configurations. A popular circuit is shown in Fig. 8–12, where a single transistor is connected as both *error detector* and *error amplifier*. Resistors R_1 and R_2 form the output voltage monitor. If the small base current of $Q1$ is neglected, the voltage monitor produces an output voltage $KV_o = [R_2/(R_2 + R_2)]V_o$ which is applied to the base of $Q1$. Resistor R_3 and Zener diode $CR1$ provide the reference voltage V_R, which is applied to the emitter of $Q1$. Assuming that the emitter current of $Q1$ is very small compared to the Zener current of $CR1$, the reference voltage V_R is practically constant.

With the emitter of $Q1$ maintained at V_R, and the base of $Q1$ at KV_o, the emitter-to-base voltage of $Q1$ is

$$V_{BE} = KV_o - V_R \tag{8–9}$$

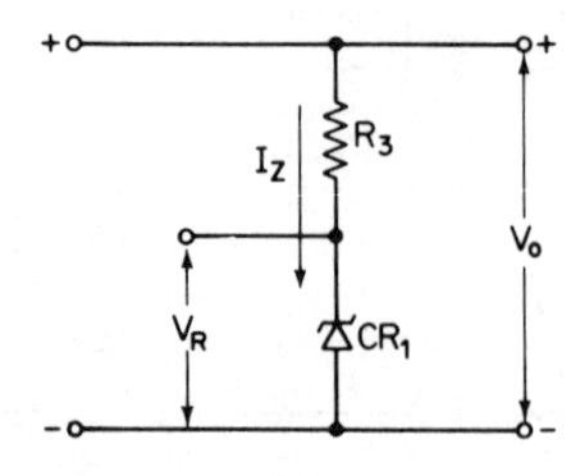

Fig. 8–11 Zener diode provides the reference voltage.

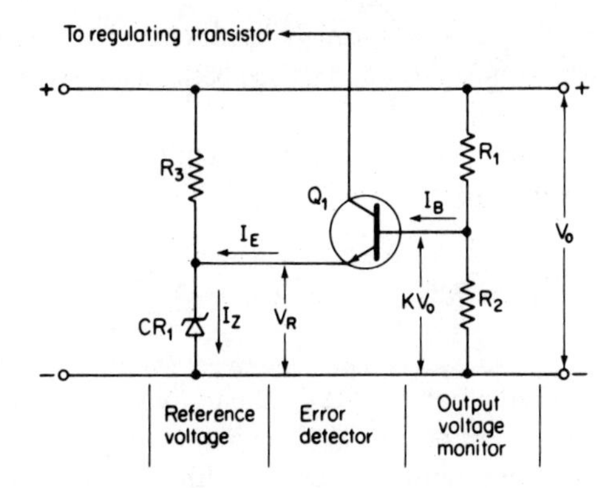

Fig. 8–12 Single transistor *Q*1 functions as a voltage comparator and error amplifier in a feedback regulator.

If V_o changes by some small amount ΔV_o, there is a corresponding change ΔV_{BE} in the emitter-to-base voltage of *Q*1 and a resulting variation ΔI_C in collector current. The collector current of *Q*1 controls the conduction of the series regulating transistor directly, or via an additional amplifier stage. Although the regulating transistor is not shown in Fig. 8–12, its connection and operation are the same as for Fig. 8–8.

Another voltage comparator circuit is the *differential amplifier* (see also Sec. 5–6). The differential amplifier produces an output voltage proportional to the difference between two input voltages. A typical circuit is given in Fig. 8–13, where *Q*1 and *Q*2 form the differential amplifier. The two voltages to be compared are the reference voltage V_R and the sampled output voltage KV_o. As

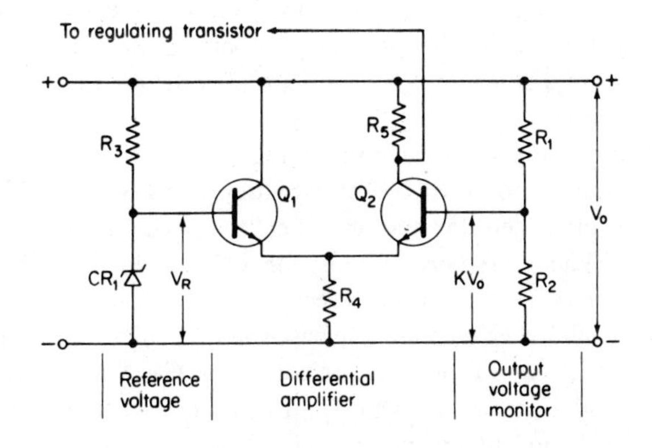

Fig. 8–13 A differential amplifier is connected as the voltage comparator in a feedback regulator.

in the previous case, the currents through R_3–$CR1$ and R_1–R_2 should be much larger than the base currents of $Q1$ and $Q2$, so both bases essentially see constant voltage sources.

For identical transistors in a balanced configuation, the base voltages of $Q1$ and $Q2$ must be equal, so

$$V_R = KV_0 \qquad \text{where } K = \frac{R_2}{R_1 + R_2}$$

If V_o changes by some small amount ΔV_o, the output voltage at the collector of $Q2$ changes in direct proportion to KV_o. The output of the differential amplifier is applied to the series regulating transistor, with or without additional driver stages.

The voltage gain of the error amplifier can be estimated by considering the characteristics of the series regulating transistor. For example, if the regulating transistor requires a bias change of 0.5 V to produce a given change in load current through it, corresponding to an output voltage change of, say, 500 μV, then the error amplifier must provide a voltage gain of 0.5 V/500 μV = 1000. This voltage gain includes the attenuation of the monitoring voltage divider, so the amplifier itself may have to provide as much as 30 times the gain calculated, depending on the attenuation of the divider.

In addition, there is also a current gain requirement because the base current of the series regulating transistor must be changed by some amount. It is therefore often necessary to provide for additional current amplification between the error amplifier and the regulating transistor.

8–3.3 Feedback Regulator

The final element in the feedback regulator is the regulating device itself. Using the general scheme of the block diagram of Fig. 8–9, the series regulator generally consists of a silicon power transistor of sufficient capacity to carry the entire load current.

Combining the regulator circuit of Fig. 8–12 with the series regulating transistor results in the complete feedback regulator of Fig. 8–14. Transistor $Q1$ again acts as the error detector and amplifier, comparing the attenuated output voltage to the Zener reference. Transistor $Q2$ provides additional current amplification to drive the base of series regulating transistor $Q3$.

If output voltage V_o increase, the base voltage of $Q1$ rises, which results in a corresponding increase in the collector current of $Q1$. Since resistor R_4 is the relatively large collector resistor for $Q1$, an increase in collector current causes the collector of $Q1$, and hence the base of $Q2$, to drop in potential. The base current of $Q2$ therefore decreases and the base current of $Q3$ decreases. (Note that $Q2$ and $Q3$ form a Darlington pair.) A decrease in $Q3$ base current increases

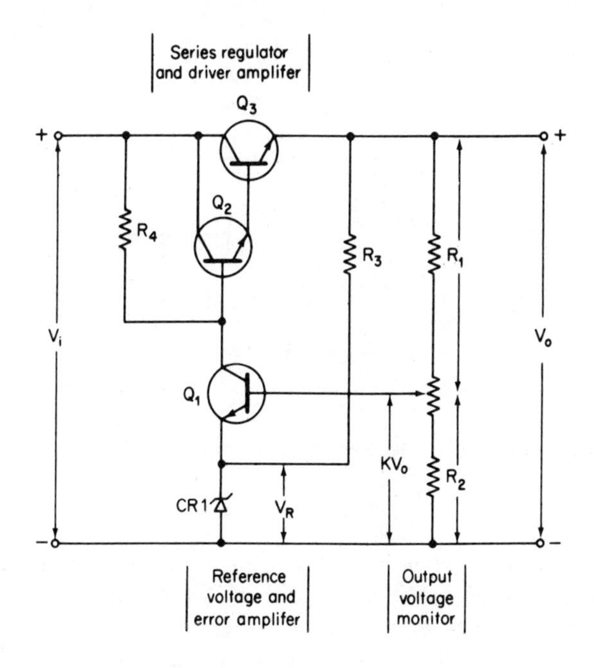

Fig. 8–14 A complete series of feedback regulator.

the apparent resistance of the series regulator and causes a greater voltage drop across $Q3$. The increased drop across $Q3$ compensates for the rise in output voltage V_o which initiated the action of the feedback regulator. The regulated output voltage V_o in Fig. 8–14 can be adjusted by varying the R_1-R_2 ratio.

8–4 REGULATORS WITH ADJUSTABLE OUTPUT VOLTAGE

The circuit of Fig. 8–15 uses a differential amplifier as the feedback element to control series regulator $Q3$. The base of $Q1$ is connected to reference voltage V_R developed by Zener diode $CR1$. The base of $Q2$ senses the output voltage through voltage divider R_1–R_2. The feedback system adjusts itself so that the base voltage of $Q1$ equals the base voltage of $Q2$.

If the output voltage is to be set at 0 V, the circuit of Fig. 8–15 can be modified by grounding the base of $Q1$, thus providing zero reference voltage. However, to maintain $Q1$ and $Q2$ in the active region, emitter resistor R_3 must be returned to a negative supply, as in Fig. 8–16. In the circuit of Fig. 8–16 the output is continuously variable between maximum ($V_o \simeq V_i$) and zero volts by the single output potentiometer which replaces resistors R_1 and R_2 of the

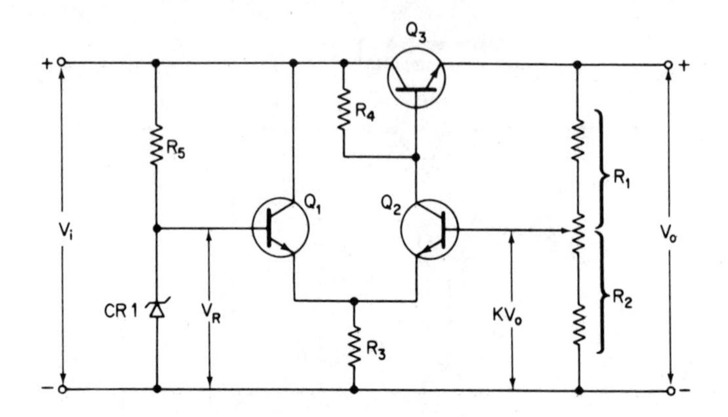

Fig. 8–15 Differential amplifier *Q*1-*Q*2 controls series regulator *Q*3.

previous circuit. The differential amplifier must always be in the balanced or null condition, i.e., $V_{B2} = 0$. This is entirely possible, since somewhere along voltage divider R_1–R_2 there must be a point where $V_{B2} = 0$.

Expressing V_{B2} in terms of the circuit constants given in Fig. 8–16,

$$V_{B2} = -V_a + (V_o + V_a)\frac{R_2}{R_1 + R_2} = 0 \tag{8–10}$$

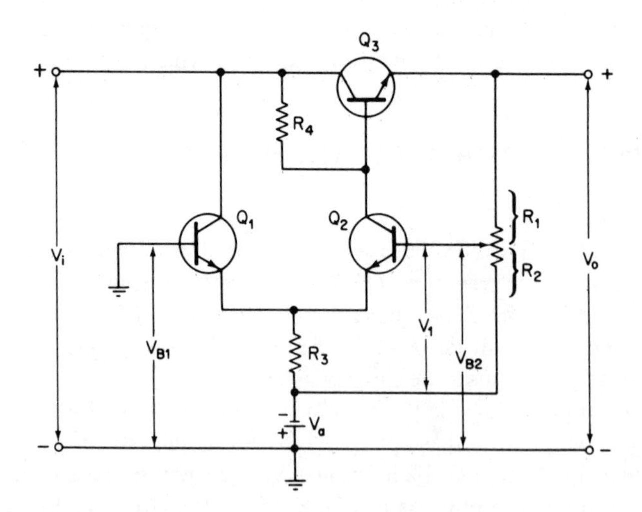

Fig. 8–16 Auxiliary negative supply $-V_a$ allows adjustment of the output voltage between maximum and zero volts.

Setting $K = R_2/[R_1 + R_2]$, as before, Eq. (8-10) simplifies to

$$V_{B2} = -V_a + (V_o + V_a)K = 0 \tag{8-11}$$

Rearranging and solving for V_o,

$$V_o = V_a \frac{1 - K}{K} \tag{8-12}$$

Equation (8-12) essentially indicates how V_o is related to V_a and the setting of the potentiometer. For the output voltage to be zero volts, the potentiometer must be moved up completely, so that $R_1 = 0$ and $K = 1$. The circuit may be refined by adding buffer stages between $Q2$ and series regulator $Q3$.

8-5 PRACTICAL SERIES REGULATOR

If the feedback regulator is required to supply an output voltage continuously variable between zero volts and some positive value, a constant current source can be used effectively as the reference medium. An example is shown in Fig. 8-17. Transistor $Q4$ represents the constant current source, operated from an auxiliary supply providing a negative voltage $-V_a$. The voltage drop across the

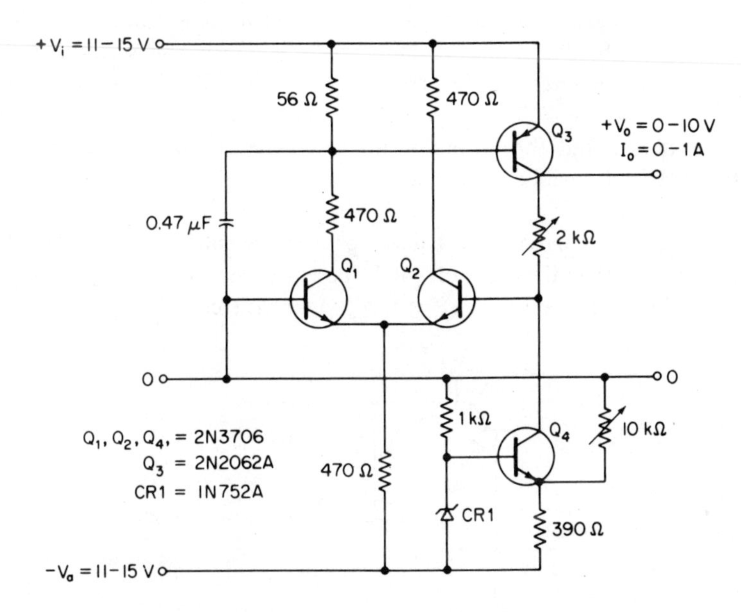

Fig. 8-17 Series regulator with adjustable output voltage from 0 to 10 volts.

390-Ω emitter resistor of $Q4$ is compared to the reference voltage provided by Zener diode $CR1$ (−5.6 V). $Q4$ attempts to pass a current which keeps the voltage across its emitter resistor closely equal to the Zener reference voltage. Since the Zener voltage is constant, the emitter current of $Q4$ (and hence its collector current) is also constant. The 10-kΩ potentiometer in the $Q4$ emitter leg provides additional current stabilization and is part of a balanced-bridge circuit with the base-emitter of $Q4$ in the center of the bridge.

Auxiliary supply $-V_a$ has a common ground connection with main supply $+V_i$. The constant collector current of $Q4$ passes through the 2-kΩ potentiometer in the $Q3$–$Q4$ circuit and provides continuous adjustment of the output voltage between 0 V and +10 V. Difference amplifier $Q1$–$Q2$ controls series regulating transistor $Q3$. If the output voltage increases (due to a change in load conditions or a change in input voltage $+V_i$), the current in $Q2$ increases and in $Q1$ decreases. Regulating transistor $Q3$ then receives less base current and the output voltage decreases toward its original value.

8–6 OVERLOAD AND SHORT-CIRCUIT PROTECTION

An accidental short circuit across the output terminals of a regulated supply may destroy the series regulating transistor. Fuses or circuit breakers are usually not fast-acting enough and the series transistor is destroyed before the fuse has reacted to an overload.

A basic scheme for limiting the load current to some maximum value is indicated in Fig. 8–18. In this circuit transistor $Q3$ is the series regulator receiving its base current from the error amplifier, possibly via a current driver stage (see also Fig. 8–14). A current-sensing resistor R_{SC} is connected between the emitter of $Q3$ and the output terminal of the regulator. An additional transistor $Q4$ with (possibly) a diode $D1$ is added to the circuit to form the output current limiting circuitry.

The leakage current of $Q4$ develops a small voltage across $D1$ which holds $Q4$ cut off. When the load current I_L increases, the voltage drop across R_{SC} increases. When $I_L R_{SC}$ exceeds $(V_{D1} + V_{BE4})$, transistor $Q4$ switches on and

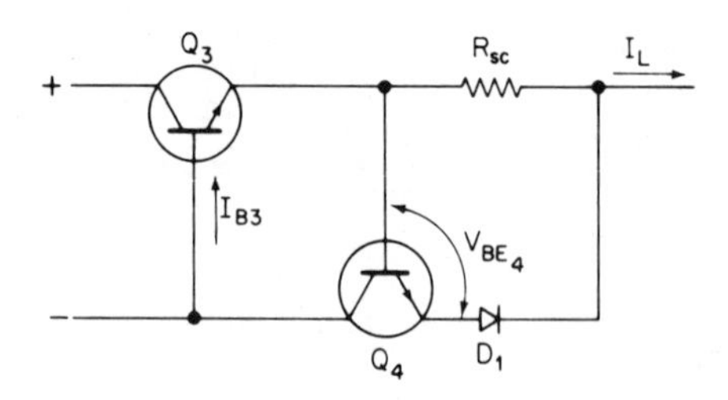

Fig. 8–18 A basic method of output current limiting.

establishes a current path between the output terminal of the circuit and the base of *Q*3. The collector current of *Q*4 subtracts from the base current of *Q*3, and this causes a decrease in the current through *Q*3. The load current is therefore limited to a certain maximum value.

The resistance of R_{SC} is usually small. E.g., if the normal load current delivered by the regulator is 200 mA and an overload of 50 percent can be tolerated, the output current must be limited to 300 mA. Suppose the voltage drop across D1 is 0.6V and that *Q*4 requires a base-emitter voltage of 0.6 V to turn on. The minimum resistance of R_{SC} then is

$$R_{SC} = \frac{V_{D1} + V_{BE4}}{I_{L(\max)}} = \frac{0.6\text{ V} + 0.6\text{ V}}{300\text{ mA}} = 4\ \Omega$$

A different value of R_{SC} sets the current limit to another value.

Figure 8–19(a) to (d) shows several building blocks which can be connected together as a voltage regulator with protective circuitry for overload and short-circuit protection. Figure 8–19(a) represents a simple *series regulator* whose circuit is identical to the basic feedback regulator of Fig. 8–14. The voltage for the 10-kΩ collector resistor of error amplifier *Q*1 is derived from a 24-V secondary supply and is regulated with a 15-V Zener diode.

If the circuit of Fig. 8–19(b) is added to the series regulator, the circuit becomes a *current* limiter, adjustable by the 100-Ω potentiometer in the return lead of the supply. Current limiting occurs when the voltage drop across the 100-Ω potentiometer, caused by the load current, exceeds the sum of the voltage drops across the stabilizer diode and the base-emitter of *Q*4. *Q*4 then carries current and the voltage at the base of *Q*2, and hence the output voltage, decreases. The circuit of Fig. 8–19(b) can only be used when series regulator transistor *Q*3 can continuously carry the additional power. If this is not the case, the output stage of the regulator should be switched off when the load current exceeds a certain maximum value.

The output stage of the regulator can be switched off by replacing the circuit of Fig. 8–19(b) by that of Fig. 8–19(c). In constrast to the previous circuit, the emitter voltage of *Q*4 is derived from the regulator output and is approximately 2 V. With increasing load current, the voltage drop across the 5-Ω potentiometer, and hence the base-emitter voltage of *Q*4, increases. When *Q*4 starts conducting, the base voltage of *Q*2 decreases and the output voltage decreases. This in turn reduces the emitter voltage of *Q*4 and the base-emitter voltage of *Q*4 increases. This feedback arrangement quickly cuts the output stage off. When the circuit reacts to overload or short-circuit conditions, the final stage remains cut off and does not automatically return to its normal state when the overload is removed. In fact, the power needs to be removed momentarily to reset the circuit.

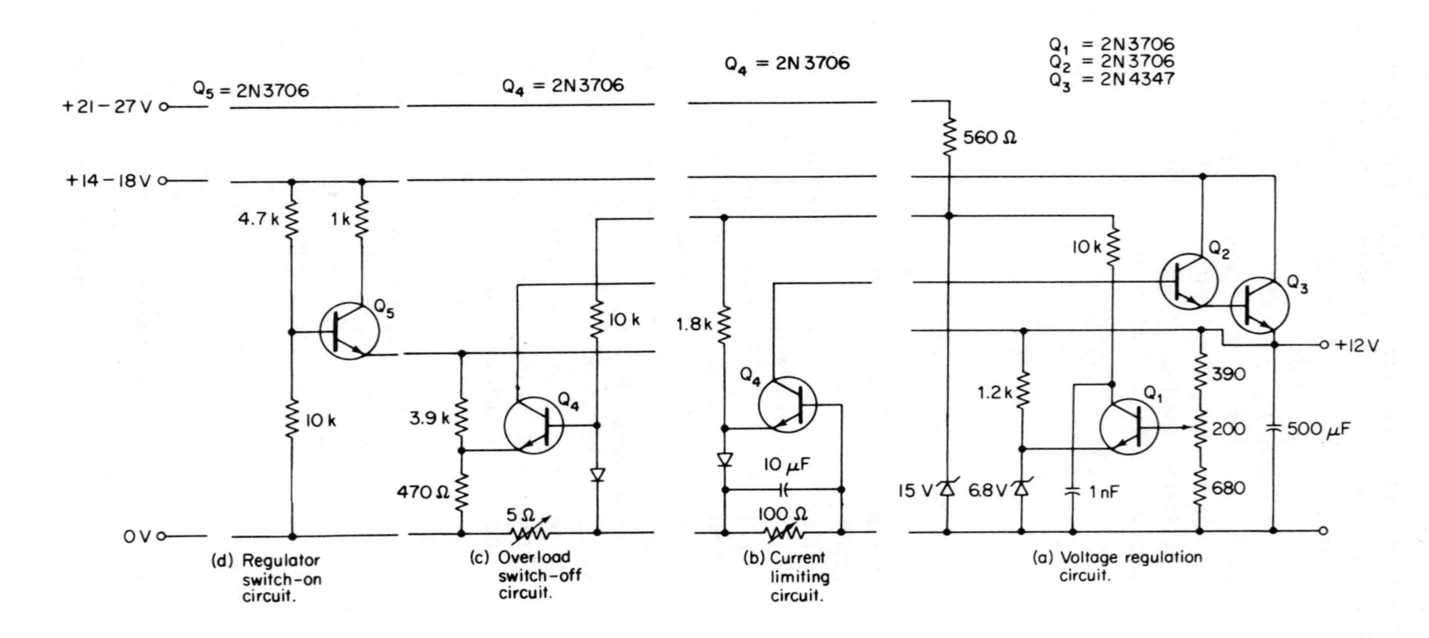

Fig. 8–19 A series voltage regulator with various circuits for overload protection.

This problem may be overcome by adding an extra stage to the circuit of Fig. 8–19(c). This extra stage is shown in Fig. 8–19(d), where *Q*5 is normally a cutoff transistor. The voltage divider, consisting of the 4.7-kΩ and the 10-kΩ resistors, develops a base voltage for *Q*5 which is negative with respect to its emitter voltage. Only when the output falls below the nominal value, or becomes zero due to a short circuit, the base-emitter voltage of *Q*5 reverses. *Q*5 then starts conducting and provides a current through the external short, controlled by the 1-kΩ collector resistor. If the short is removed, or when the load resistance increases again, the *Q*5 emitter current provides a voltage drop across the load, which flips the *Q*4 stage back.

8–7 MONOLITHIC VOLTAGE REGULATOR

Integrated circuits play an important role in circuits such as voltage regulators. The TBA 281 or μA 723C is a monolithic voltage regulator comprising a temperature-compensated reference amplifier, an error amplifier, a series-regulating power transistor, and current-limiting circuitry. The circuit, Fig. 8–20, features high ripple rejection, low temperature drift, and low standby

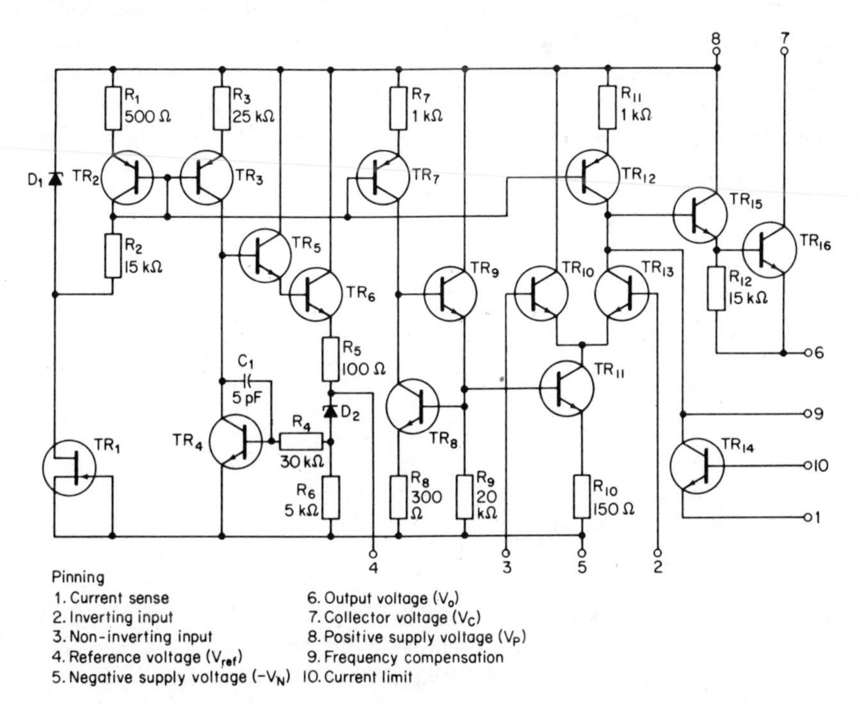

Fig. 8–20 Monolithic voltage regulator TBA 281 or μA 723C. This circuit is contained in a standard 10-pin TO-74 package.

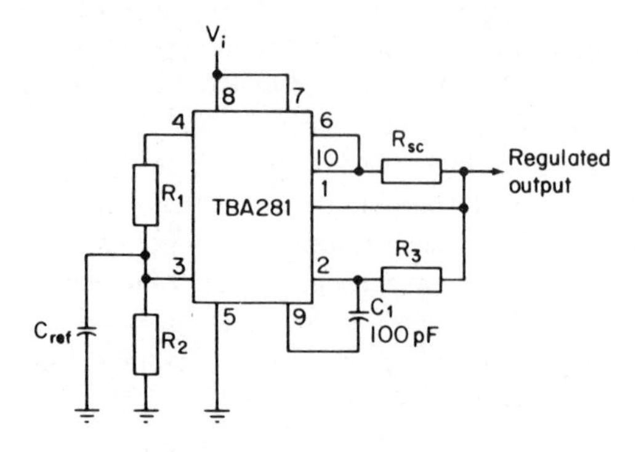

Fig. 8–21 Low-voltage regulator (V_o = 2 to 7 volts).

current, while external power transistors can be added if the load current exceeds the maximum current limit of the basic package.

The IC is packaged in a standard 10-pin TO-74 can (8.5 mm diameter and 6.6 mm height). A few external components are added to determine the operating function of the device.

As a *low-voltage regulator,* the circuit of Fig. 8–21 is used. Resistors R_1 and R_2 act as a voltage divider for the internally generated Zener reference voltage. For a fixed positive output voltage of 5 V, resistor R_1 = 2,150 Ω and R_2 = 5 kΩ. The output voltage is adjustable within the limits of 2 to 7 V, by replacing R_1 and R_2 by the circuit of Fig. 8–22, where R_1 = 750 Ω, R_2 = 2.2 kΩ, and R_v = 500 Ω.

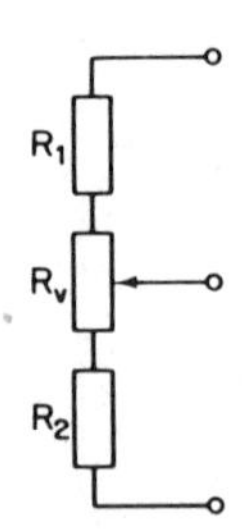

Fig. 8–22 Output voltage adjustment.

As a *high-voltage regulator* where the output voltage is adjustable from 7 to 37 V, the circuit of Fig. 8–23 is used. Note that here resistors R_1 and R_2 are moved to the output terminals of the device, where they form the output voltage divider. The internally generated reference voltage is connected directly to the error amplifier.

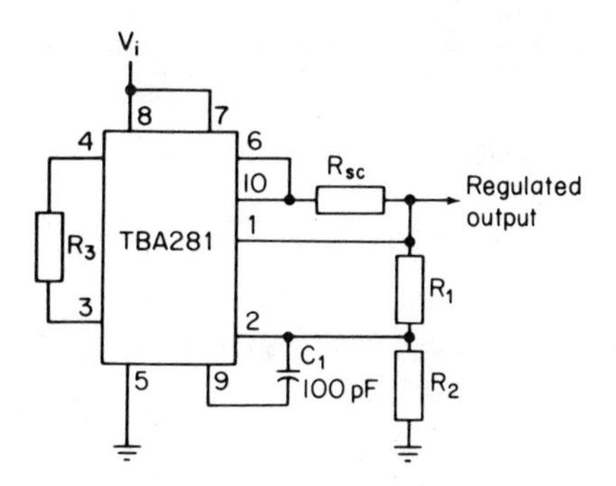

Fig. 8–23 High-voltage regulator (V_o = 7 to 37 volts).

Short-circuit protection is provided by connecting an external resistor R_{SC} (from 5 to 10 Ω) between the output pin (pin 6) and the current sense connection on TR14 (pin 1).

The performance characteristics of the IC are high. Line regulation for an input voltage range of 12 to 40 V is typically 0.1 percent of the output voltage. Load regulation for a load current range of 1 to 50 mA is typically 0.03 percent of the output voltage. The input voltage range is from 9.5 to 40 V; the output voltage range is from 2 to 37 V.

It is clear that the economy of the integrated circuit in terms of size and cost, and the considerable time saving in using a ready-made circuit over a specially built circuit, makes the IC voltage regulator attractive.

QUESTIONS

8–1 Under what conditions is the shunt regulator preferred over the series regulator?

8–2 Explain why the temperature sensitivity of the base–emitter voltage of the series transistor in the regulator of Fig. 8–8 does not cause a significant change in the regulated output voltage.

8–3 Explain the function of each component in the feedback regulator of Fig. 8–14.

8–4 Explain the action of the differential amplifier in the series regulator of Fig. 8–17 when the positive supply voltage decreases.

8–5 What is the effect on circuit operation if the Zener diode in Fig. 8–17 accidentally overheats and becomes an open circuit?

8–6 Is it necessary for the two transistors in the difference amplifier of Fig. 8–15 to have identical characteristics? Explain.

8–7 Draw a block diagram of a series feedback regulator and explain the function of each block.

8–8 What is the function of transistor *Q*2 in Fig. 8–14?

8–9 Explain how the regulated output voltage in the series regulator circuit of Fig. 8–14 changes when the potentiometer in the output voltage monitor is moved down (small R_2 and large R_1).

8–10 Explain how the auxiliary negative supply in Fig. 8–16 allows the output voltage to be adjusted to zero volts. Is it possible to produce a negative output voltage? Explain.

8–11 Redraw the circuit of Fig. 8–16 to produce a negative output voltage.

8–12 Complete the circuit of Fig. 8–12 by adding the series regulating transistor.

8–13 Under what load conditions does the shunt regulating transistor in Fig. 8–6 carry the largest current? Explain.

8–14 Assume that the output terminals of the series regulator of Fig. 8–8 are accidentally short-circuited. What effect does this have on the transistor? On the Zener diode?

8–15 What is the effect on the output voltage of the series regulator of Fig. 8–8 when the ambient temperature increases from its normal value of 25°C to 100°C? Explain.

8–16 Explain the function of diode *D*1 in the current limiting circuit of Fig. 8–18.

8–17 What is the reason for the Darlington pair in the output circuit of the voltage regulator of Fig. 8–19(a)?

8–18 What is the function of the 500-μF capacitor across the output terminals of the regulator of Fig. 8–19(a)? What happens if this capacitor becomes open-circuited?

8–19 Explain the operation of the current-limiting circuit of Fig. 8–19(b).

8–20 Suppose that the 100-Ω variable resistor in the return lead of Fig. 8–19(b) is set to a smaller value. Does the circuit trip with a smaller or a larger load current? Explain.

8–21 Explain how short-circuit protection is achieved in the IC voltage regulator of Fig. 8–20.

8–22 How can the power-handling capability of the IC voltage regulator of Fig. 8–20 be increased?

8–23 Explain how the reference voltage in the comparator circuit of the IC regulator of Fig. 8–20 is generated.

PROBLEMS

8–1 A 5-W 8.6-V Zener diode is used as a medium-power voltage regulator for a 0.5-A load in the circuit of Fig. 8–24. A 10–Ω series resistor protects the Zener diode.

Calculate (a) The maximum input voltage without exceeding the power rating of the Zener diode.

(b) The minimum input voltage at which the regulation becomes ineffective.

8–2 A certain Zener diode has the following characteristics at $\mathcal{T}_{amb} = 25°C$:

$$V_Z = 9 \text{ V at } I_Z = 1 \text{ mA}$$

$$r_Z = 12\ \Omega$$

$$S_Z = 6 \text{ mV/°C}$$

Calculate (a) The Zener voltage when $IZ = 10$ mA.

(b) The Zener voltage when $\mathcal{T}_{amb} = 100°C$.

8–3 In the basic Zener regulator circuit of Fig. 8–24 the diode has the following characteristics at $\mathcal{T}_{amb}$ 25°C:

$$V_Z = 8.4 \text{ V at } I_Z = 20 \text{ mA}$$

$$r_Z = 1.8\ \Omega$$

$$S_Z = 5 \text{ mV/°C}$$

The load current may vary from 0–100 mA. The input voltage may vary from 12 to 18 V. Assume that the minimum Zener current is 10 percent of the maximum possible Zener current.

Calculate (a) The required resistance of series resistor R_s.

(b) The maximum power dissipation in the Zener diode.

(c) The output impedance of the circuit.

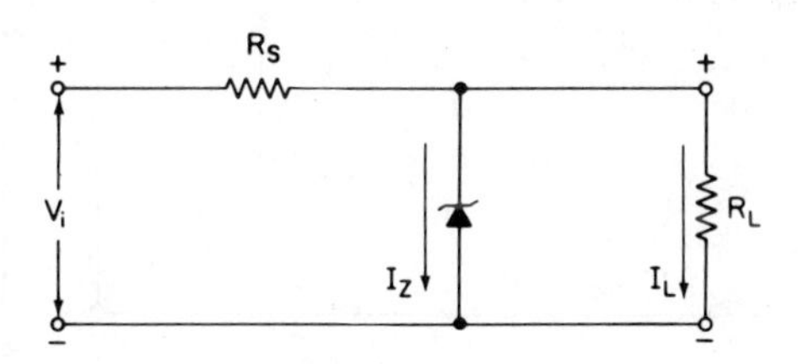

Fig. 8–24 Simple Zener diode voltage regulator for Problems 8–1, 8–3, and 8–4.

8–4 The voltage regulator of Fig. 8–24 uses a silicon power Zener diode with the following ratings and characteristics:

$V_Z = 15$ V at $I_Z = 1$ A $\quad T_j = 25°C$

$r_Z = 0.4\ \Omega$ $\quad P_{tot} = 20$ W

$S_Z = 7$ mV/°C $\quad I_{ZRM} = 20$ A

The supply voltage is constant at $V_i = 25$ V. The load current varies between 0 and 1.2 A. Assuming that the junction temperature of the Zener diode is kept at 25°C, *calculate*

(a) The resistance of protective resistor R_s.
(b) The percentage load regulation of the circuit.

8–5 The shunt regulator of Fig. 8–25 uses a 9-V 400-mW low-power Zener diode and a silicon power transistor of sufficient capacity to provide an output voltage $V_o = 10$ V to a 1-A load. The dc supply voltage varies from 15 to 20 V.

Calculate (a) The required resistance of protective resistor R_s.
(b) The dissipation of the silicon power transistor when the load is in place and the input voltage is maximum.
(c) The dissipation of the transistor when the load is removed and the input voltage is minimum.

8–6 Assume that the silicon power transistor in Fig. 8–25 has a current gain $h_{FE} = 40$. Estimate the power dissipation of the Zener diode under the most unfavorable conditions of input voltage and load current.

8–7 The series regulator of Fig. 8–26 uses a 10.9-V 400-mW Zener diode and a medium-power silicon transistor with a current gain $h_{FE} = 50$. The output voltage of the regulator $V_o = 10$ V, implying that the base-emitter voltage drop of the silicon transistor $V_{BE} = 0.9$ V. Assume that V_{BE} is constant. To

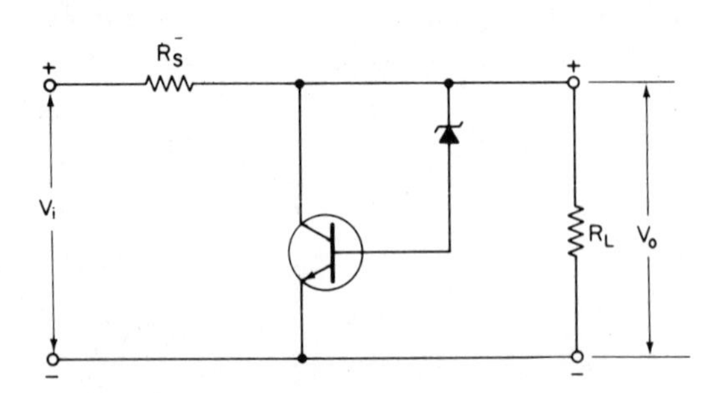

Fig. 8–25 Simple shunt regulator for Problems 8–5 and 8–6.

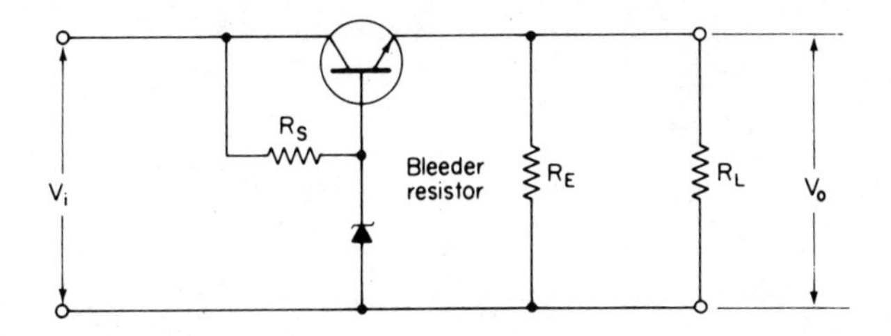

Fig. 8–26 Series regulator for Problem 8–7.

avoid the possibility of transistor cutoff, a bleeder resistor R_E is connected across the output terminals providing a minimum emitter current $I_{E(\min)}$ = 10 mA, even when the load is disconnected. The input voltage may vary from 15 to 20 V and the load current may vary from 0 to 500 mA.

Calculate (a) The maximum power dissipation in the transistor.
(b) The resistance of protective resistor R_s.
(c) The maximum power dissipation in the Zener diode.

Chapter 9

Oscillators

9–1 GENERAL CONSIDERATIONS

9–1.1 Criteria for Oscillation

An oscillator is a circuit that converts input energy from a dc source into ac output energy of some periodic waveform, at a specific frequency and a known amplitude. The characteristic feature of the oscillator is that it maintains its ac output signal without an external input signal.

A *sinusoidal oscillator* is essentially a form of feedback amplifier (see also Section 5–5), where special requirements are placed on the voltage gain A_v and the feedback network $\boldsymbol{\beta}$. Consider the feedback amplifier of Fig. 9–1, where the feedback voltage $v_f = \boldsymbol{\beta} v_o$ supplies the entire input voltage v_i, so

$$v_i = v_f = \boldsymbol{\beta} v_o = A_v \boldsymbol{\beta} v_i \tag{9–1}$$

Solving Eq. (9–1) for v_i we obtain

$$v = A_v \boldsymbol{\beta} v_i \quad \text{or} \quad (1 - A_v \boldsymbol{\beta}) v_i = 0 \tag{9–2}$$

If an output voltage is to be produced, the input voltage cannot be zero. Hence, for v_i to exist, Eq. (9–2) requires that

$$1 - A_v \boldsymbol{\beta} = 0 \quad \text{or} \quad A_v \boldsymbol{\beta} = 1 \tag{9–3}$$

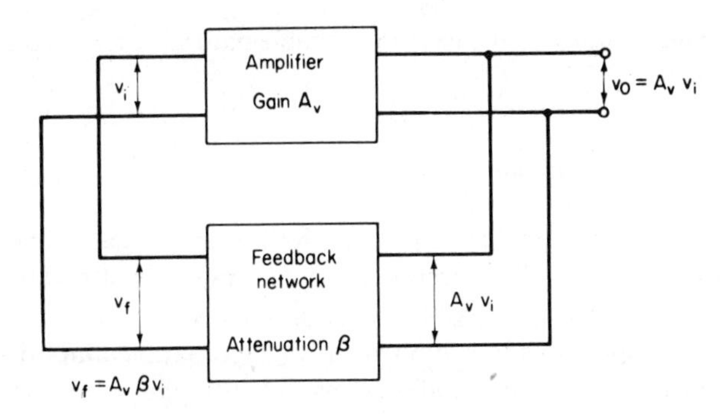

Fig. 9–1 Amplifier and feedback network connected to form a closed-loop system.

Equation (9–3) is the *Barkhausen criterion,* which defines two basic requirements for oscillation:

1. The voltage gain around the amplifier and feedback loop, called the *loop gain,* must be unity, or $A_v\beta = 1$.
2. The phase shift between v_i and v_f, called the *loop phase shift,* must be zero.

If these two conditions are satisfied, the feedback amplifier of Fig. 9–1 generates a sinusoidal output waveform.

The amplifier, or the feedback network, or perhaps both, contains reactive elements which affect the frequency and phase characteristics of the waveform passing through the system. Hence, Eqs. (9-1) and (9-3) can only be satisfied at one frequency.

The Barkhausen criterion states that the magnitude of the feedback voltage must be identical to the magnitude of the input voltage. The circuit only functions as an oscillator if the product of the voltage gain A_v and the feedback attenuation β equals 1. If the loop gain is less than 1, the circuit cannot sustain the output voltage and cannot oscillate. If the loop gain is greater than 1, the oscillator output tends to increase to the point where the amplifier operates nonlinearly and waveform distortion sets in. This, of course, is an undesirable situation, and should be avoided. In a practical oscillator, the loop gain is chosen slightly larger than 1 (say 5 to 10 percent) to ensure that oscillation does not cease when the amplifier gain drops marginally.

In general, an amplifier which is well stabilized against temperature variations and power supply fluctuations is a prerequisite. Amplitude limiting is often accomplished by additional circuitry such as automatic gain control (AGC). Frequency stability can be closely controlled by high-Q resonant circuits in the feedback path. In crystal-controlled oscillators, for example, the frequency of oscillation is determined by the extremely stable natural resonant frequency of a quartz crystal.

9–1.2 Basic Circuit Requirements

Oscillators in the radio frequency (RF) range usually operate with resonant circuit loads. This has several advantages in terms of purity of output waveform and frequency stability. If the loop gain is greater than unity, the amplitude of the output voltage is limited by the fact that the amplifier tends to operate in the saturation and cutoff regions. This implies a distorted output waveform.

If the amplifier load is a high-Q resonant circuit, the load current at the fundamental or resonant frequency f_r predominates. Load current components at any other frequency are deemphasized. A resonant load, therefore, has an inherent filtering action, and the purity of the output waveform is a function of the circuit Q (sharpness of the response curve). The feedback voltage βv_o, derived from the resonant load voltage (or current), is maximum at the resonant frequency, and the Barkhausen criterion for unity loop gain can be satisfied only at this frequency. The loop gain at the off-resonance frequencies is smaller than unity, hence these components do not contribute to the oscillation.

In addition, the phase shift per cycle of frequency variation is determined by the Q of the resonant circuit. A high-Q circuit with a sharp response curve (see Fig. 9–2) provides a very rapid change of phase at its resonant frequency, and this improves the frequency stability of the oscillator.

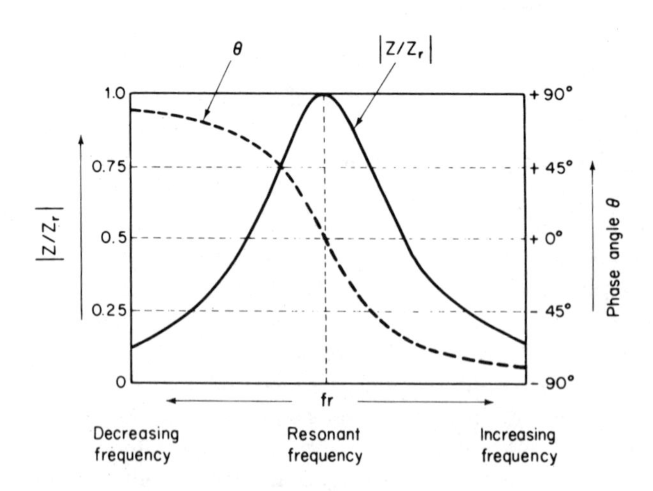

Fig. 9–2 General circuit for a resonant-load oscillator.

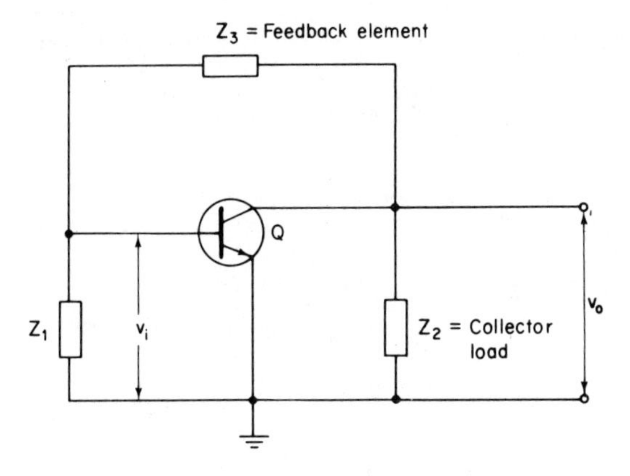

Fig. 9–3 Impedance and phase angle as a function of frequency in a parallel-resonant circuit.

Figure 9–3 shows the general form of a feedback oscillator, using a BJT in the common-emitter connection as the voltage amplifier. The dc supply and bias components are omitted for the sake of clarity. Signal feedback from the collector to base is provided by the Z_3 impedance. The voltage gain of the amplifier is given by the general expression

$$A_v = -A_i \frac{Z_L}{Z_{\text{in}}} \tag{9–4}$$

The minus sign indicates the inherent phase reversal between output and input voltage or current, and A_i represents the *CE* current gain h_{fe} of the transistor.

Neglecting the high output resistance of the BJT itself, the collector load impedance Z_L equals

$$Z_L = Z_2/\!/(Z_1 + Z_3) = \frac{Z_2(Z_1 + Z_3)}{Z_1 + Z_2 + Z_3} \tag{9–5}$$

The base of the transistor looks into a source impedance of

$$Z_s = Z_1/\!/(Z_2 + Z_3) = \frac{Z_1(Z_2 + Z_3)}{Z_1 + Z_2 + Z_3} \tag{9–6}$$

The total input impedance of the amplifier is

$$Z_{\text{in}} = Z_s + r_{ie} \tag{9–7}$$

where r_{ie} is the input resistance of the *CE* transitor, defined as $r_{ie} = r_b + (h_{fe} + 1)r_e \simeq h_{ie}$.

Substituting in Eq. (9–4), the amplifier gain *without feedback* is

$$A_v \cong -h_{fe}\frac{Z_L}{Z_s + h_{ie}} = -h_{fe}\frac{Z_2(Z_1 + Z_3)}{(Z_1 + Z_2 + Z_3)h_{ie} + Z_1(Z_2 + Z_3)} \tag{9–8}$$

Feedback is provided by Z_3, and the feedback factor

$$\beta = \frac{Z_1}{Z_1 + Z_3} \tag{9–9}$$

The loop gain of the amplifier with feedback is

$$A_v\beta = -h_e\frac{Z_1Z_2}{(Z_1 + Z_2 + Z_3)h_{ie} + Z_1(Z_2 + Z_3)} \tag{9–10}$$

If the general impedances are replaced by pure reactances, so that $Z_1 = jX_1$, $Z_2 = jX_2$, and $Z_3 = jX_3$, Eq. (9–10) becomes

$$A_v\beta = -h_{fe}\frac{-X_1X_2}{j(X_1 + X_2 + X_3)h_{ie} - X_1(X_2 + X_3)} \tag{9–11}$$

For the circuit to oscillate, the Barkhausen criterion must be satisfied, and the loop gain must be *real* and *positive* (zero phase shift) and at least *unity*. If the gain is to be real, the imaginary term in the denominator of Eq. (9–11) must be zero, so

$$X_1 + X_2 + X_3 = 0 \tag{9–12}$$

Equation (9–12) implies that the circuit oscillates at the resonant frequency of the reactive circuit consisting of X_1, X_2, and X_3. The expression for the loop gain [Eq. (9–11)] then simplifies to

$$A_v\beta = -h_{fe}\frac{X_2}{X_2 + X_3} \tag{9–13}$$

Since, by Eq. (9–12), $X_2 + X_3 = -X_1$, Eq. (9–13) further simplifies to

$$A_v\beta = h_{fe}\frac{X_2}{X_1} \tag{9–14}$$

Equation (9–14) states that, if the loop gain is to be positive, reactances X_1 and X_2 must be of equal sign, and they must therefore be the *same kind of reactance* (two capacitors or two inductors). It then also follows [Eq. (9–12)], that X_3 must be of opposite sign, and will be an inductor if X_1 and X_2 are capacitors, and vice versa.

The final condition for oscillation requires that the loop gain must be at least unity. This then means, by Eq. (9–14), that the current gain of the transistor must have a minimum value of

$$h_{fe} \geq \frac{X_1}{X_2} \tag{9–15}$$

These are the conditions for resonant circuit oscillators.

9–2 TYPICAL RF OSCILLATOR CIRCUITS

9–2.1 Hartley Oscillator

The Hartley oscillator of Fig. 9–4 is one of the most popular RF circuits, often used as the local oscillator in a superheterodyne broadcast receiver. The bipolar junction transistor in the common-emitter connection is the voltage amplifier and is biased by a universal bias circuit consisting of R_A, R_B, and R_E. Emitter bypass capacitor C_E increases the voltage gain of this single transistor stage.

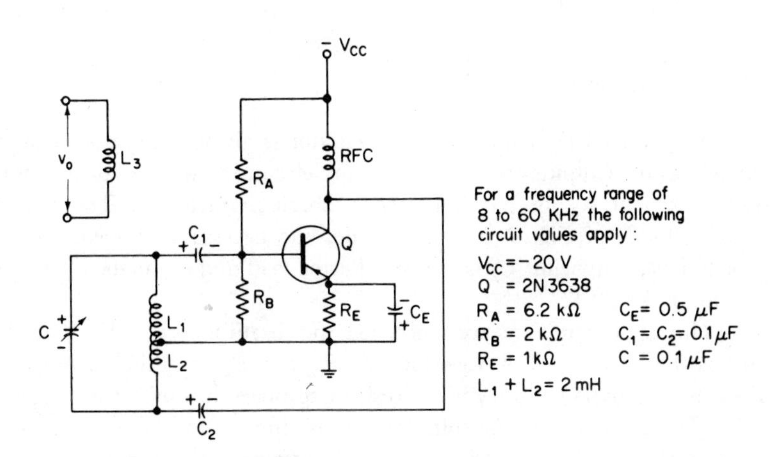

Fig. 9–4 Hartley oscillator.

The radio frequency choke (RFC) in the collector circuit acts as an open circuit at the RF frequency and prevents RF energy from entering the power supply. The reactive elements Z_1, Z_2, and Z_3 of the general feedback circuit of Fig. 9–3 are contained in a parallel-resonant circuit consisting of tapped inductor $(L_1 + L_2)$ and variable capacitor (C). The variable capacitor corresponds to Z_3 of Fig. 9–3 and provides ac signal feedback from collector to base. The lower half L_2 of the coil corresponds to Z_2 of Fig. 9–3 and is the ac collector load. The upper half L_1 of the coil corresponds to Z_1 of Fig. 9–3.

Provided that the loop gain of the feedback amplifier is 1, the circuit oscillates at the resonant frequency of the LC tank circuit. This resonant frequency is

$$f_r = \frac{1}{2\pi \sqrt{(L_1 + L_2)C}} \tag{9–16}$$

For the circuit values given in Fig. 9–4, the frequency of oscillation can be varied from approximately 8 to 60 kHz. To obtain a good sinusoidal output waveform, the resonant circuit should have a reasonably sharp response curve with a Q of approximately 20.

The output signal can be taken from the collector by capacitive coupling, provided that the load is large and the frequency of oscillation is not affected. In the usual case, however, a third coil L_3, closely coupled to the resonant coil, is used to transfer energy to the load. The circuit of Fig. 9–4 shows this type of output connection.

9–2.2 Colpitts Oscillator

The counterpart of the Hartley oscillator is given in the circuit of Fig. 9–5, known as the Colpitts oscillator. The tank circuit in this case consists of two capacitors C_1 and C_2, in parallel with a single inductor L. The inductor corresponds to Z_3 of Fig. 9–3, and provides the necessary ac feedback from collector to base. Capacitor C_2 is the ac collector load and capacitor C_1 is in the base circuit of the BJT.

As far as dc conditions are concerned, the circuit is identical to that of the Hartley oscillator of Fig. 9–4. Resistors R_A, R_B, and R_E form the universal bias circuit, while C_E bypasses the emitter resistor to place the emitter at ac ground potential. The RF choke in the collector supply line presents an open circuit to RF frequencies and prevents RF energy from entering the power supply.

The two variable capacitors are mounted on a common shaft and the frequency of oscillation is varied by changing these capacitances. Provided that

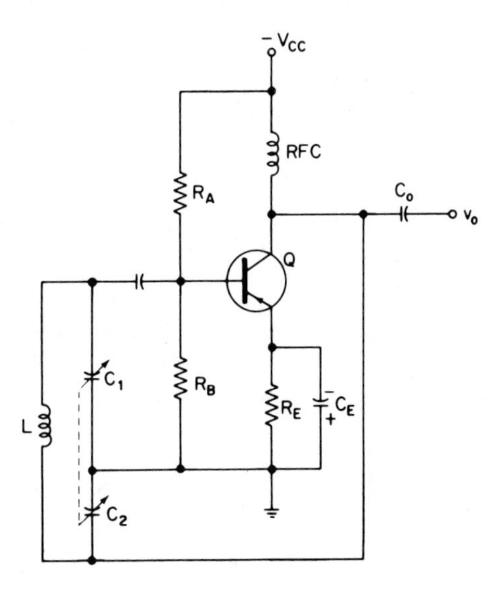

Fig. 9–5 Colpitts oscillator.

the loop gain of the amplifier is at least unity, the circuit oscillates at the resonant frequency of the LC tank circuit, which equals

$$f_r = \frac{1}{2\pi \sqrt{L(C_1 + C_2)}} \tag{9–17}$$

The output signal is taken directly from the collector via output capacitor C_o. The load impedance must be high enough not to affect the resonant frequency of the tank circuit.

9–2.3 Tuned-Collector Oscillator

Figure 9–6 shows another type of resonant-circuit feedback oscillator, known as the tuned-collector oscillator. The common-emitter transistor is biased in the conventional manner by a universal bias circuit consisting of R_A, R_B, and R_E. R_A and R_E are ac-bypassed by C_A and C_E, respectively. The ac collector load consists of a parallel-resonant circuit consisting of a fixed coil and a variable capacitor. The closely coupled coil L_2 is so phased (see the polarity dots) that positive signal feedback is applied from collector to base. The amplifier gain is adjusted so that adequate feedback of signal energy occurs at the resonant frequency of the LC tank circuit; for the off-resonance signal components the feedback is insufficient to obtain the required loop gain of 1.

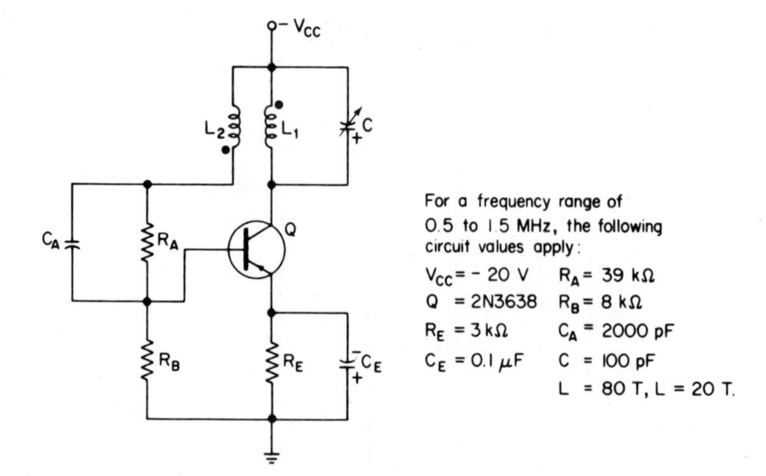

Fig. 9–6 Tuned-collector oscillator.

The frequency stability of these oscillator circuits depends to a large extent on the Q of the resonant circuit. The resonant frequency can easily be affected by an externally connected load and, especially if the load varies, the frequency stability of the circuit may be in jeopardy. To minimize the effect of load variations on frequency stability, it may be necessary to isolate the load from the oscillator circuit by means of a class A buffer amplifier. Variations in temperature, which affect both transistor parameters and circuit elements, also contribute to frequency instability.

9–3 CRYSTAL OSCILLATORS

9–3.1 Characteristics of the Quartz Crystal

Exceptional frequency stability can be achieved by using a *piezoelectric crystal* in place of the *LC* tank circuit of the resonant circuit oscillators of Section 9–2.

The piezoelectric crystal used in electronic oscillators is a small, thin slice of *quartz,* cut from a raw, natural quartz crystal, and placed between two metal electrodes. The method in which this slice is cut from the original crystal, and the dimensions of the slice, determine the electrical properties of the crystal. The piezoelectric effect, exhibited by the quartz crystal, occurs when mechanical stress is applied to opposite faces of the crystal, so a small voltage is developed across the crystal faces. Conversely, a minute deformation of the crystal takes place when a voltage is applied to its opposite faces.

On application of an ac voltage, the crystal is excited into mechanical vibration, which reaches a maximum at the *natural* or *resonant* frequency of the crystal. This natural frequency is characteristic for each crystal and is determined by the type of cut, the lateral dimensions and the thickness of the slice, and the mode of vibration. Resonant frequencies as low as a few kilohertz are obtained with the thicker crystals, while thin crystals can have a resonant frequency of 20 MHz or higher. A crystal for a frequency of 20 MHz typically has a thickness of approximately 80 μm; to obtain a frequency tolerance of ± 5 parts per million the thickness tolerance is 4×10^{-4} μm. After the crystal is cut and ground down to its required thickness, it is provided with electrodes by depositing metal on opposite faces or by clamping the crystal between metal plates. The crystal with its electrodes is then fitted into a moisture-repellent glass envelope, which is sealed by a high-frequency heating process.

The resonant frequency of the quartz crystal is extremely stable, although it may change slightly with variations in temperature. For ultimate frequency stability, the crystal is usually placed in a temperature-controlled oven.

Figure 9–7(a) shows the schematic symbol of a quartz crystal with its metal electrodes. Its electrical equivalent is shown in Fig. 9–7(b). The series combination of L, R, and C_1 describes the electrical properties of the crystal itself, while parallel capacitance C_2 represents the capacitance between the two metal plates, with the crystal serving as the dielectric. The inductance of the quartz crystal is very high, with typical values ranging from 0.1 to 100 H. The internal resistance R of the crystal is generally low, so the Q ($Q = X_L/R$) is exceptionally high, with typical values up to 100,000. The internal series capacitance C_1 is very small (on the order of 1 pF); the parallel capacitance C_2 is on the order of 4 to 40 pF.

When an ac voltage is applied across the crystal, it is excited into a mechanical vibration. At a given frequency of excitation, the series reactances of L and C_1 are equal but of opposite phase, and the crystal is in *series resonance.* Slightly above this resonant frequency, the reactance of the L-C_1 series combination becomes inductive and equal to X_{C2}, and the crystal now behaves like a parallel resonant circuit. The series and parallel resonant frequencies are very

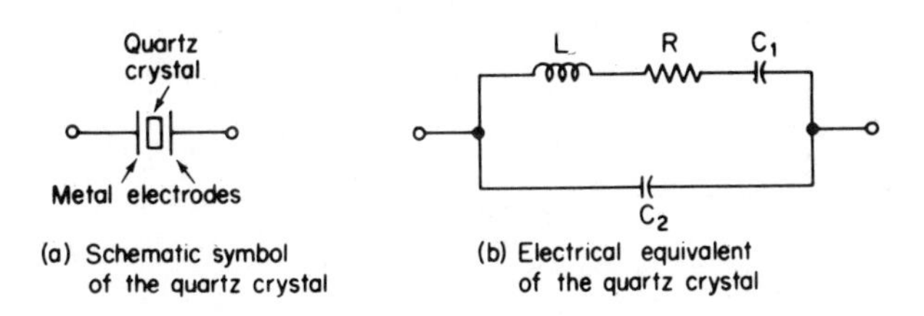

Fig. 9–7 Symbol and equivalent circuit of a quartz crystal.

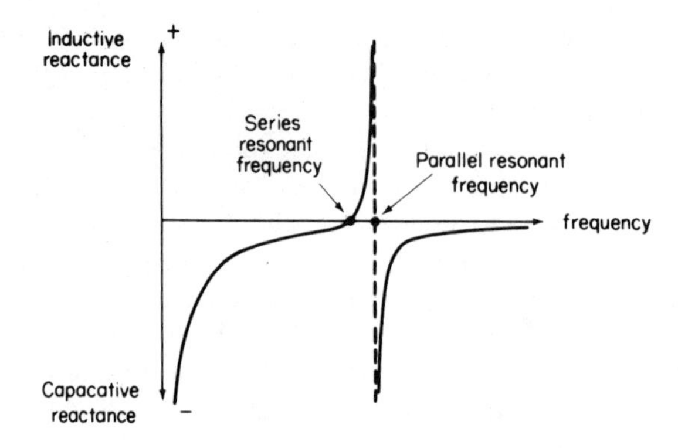

Fig. 9–8 Series and parallel reactive function of a quartz crystal.

close together (usually within 1 percent), and the reactance of the circuit changes very rapidly for a small change in frequency near resonance. This effect is shown in Fig. 9–8. The very high Q and the close proximity of the resonant frequencies are responsible for the excellent frequency stability of the crystal oscillator.

9–3.2 Crystal Oscillator Circuits

Figure 9–9 shows a variation of the Colpitts circuit, known as the *Pierce crystal oscillator*. The crystal in the feedback path acts like an inductive reactance at the operating frequency, and essentially replaces the inductance in the Colpitts

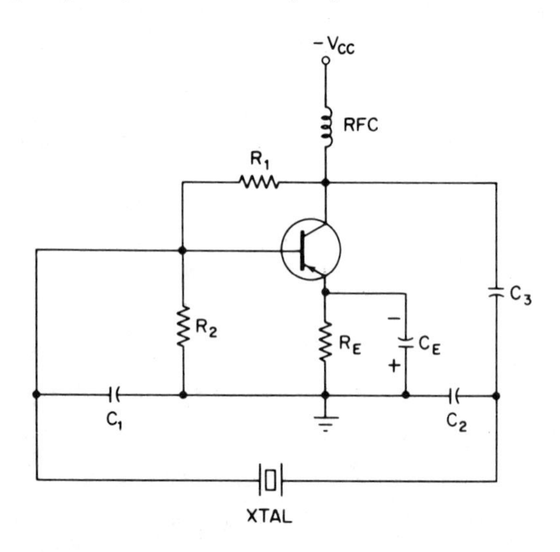

Fig. 9–9 The Pierce crystal oscillator is basically a Colpitts circuit with the crystal acting like an inductive reactance.

circuit of Fig. 9–5(b). The required 180-degree phase shift is provided by the crystal and capacitors C_1 and C_2. The frequency of oscillation is entirely determined by the natural resonant frequency of the crystal. The circuit has a very high degree of frequency stability due to the exceptionally high Q of the quartz crystal. A change in frequency is only possible by replacing the crystal with another one of a different resonant frequency, and the Pierce oscillator is therefore essentially a *fixed-frequency* circuit.

Another version of the Pierce crystal oscillator, this time using a FET as the active element, is shown in Fig. 9–10. The use of a FET has several advantages. In the first place, the high-impedance gate circuit of the FET does not "load" the crystal, so the resonant feedback path maintains the advantage of its high Q. Second, the drain voltage of the FET is generally so low that the crystal is protected from damage by overexcitation. In the circuit of Fig. 9–10 the crystal again acts like an inductive reactance. The two capacitors, required to complete the resonant feedback path, are provided by the internal gate and drain capacitances of the FET. It may be necessary to add a small capacitance across the drain (as indicated) to counteract the effect of the inductance of the RF choke in the drain lead.

An entirely different type of low-frequency crystal oscillator is shown in Fig. 9–11, where a conventional two-stage BJT amplifier is provided with positive feedback via a high-Q crystal feedback element. Phase reversal in the feedback path is now not required, and the crystal simply acts as a *tuned feedback circuit,* resonant at the natural frequency of the crystal. This circuit is also designed for fixed-frequency operation.

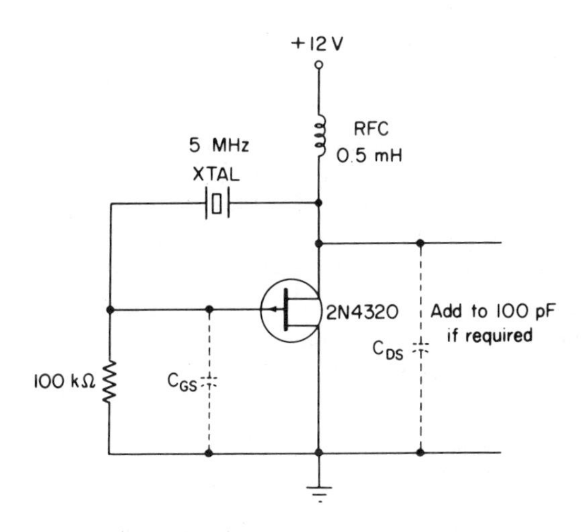

Fig. 9–10 FET Pierce crystal oscillator.

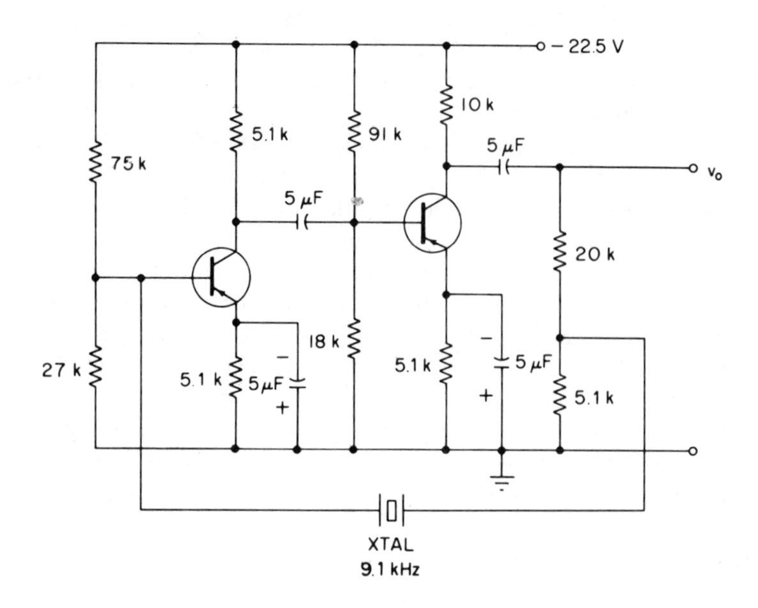

Fig. 9–11 Two-stage crystal oscillator with long-term frequency stability.

9–4 RC OR PHASE-SHIFT OSCILLATORS

9–4.1 Principles of Operation

The phase-shift oscillator of Fig. 9–12 consists of a conventional single *CE* amplifier stage with a universal bias circuit, followed by three cascaded stages of identical series capacitors and shunt resistors (shunt-*R* phase-shift network). Feedback is provided from the output of the *RC* network back to the amplifier input. The *CE* amplifier stage provides an initial 180-degree phase shift from the base to the collector, and the *RC* network introduces an additional amount of phase shift. At a specific frequency, the phase shift introduced by the network is exactly 180 degrees, so the loop phase shift is 360 degrees and the feedback voltage is in phase with the input voltage. Provided that the voltage gain of the amplifier is sufficient, the circuit oscillates at that particular frequency.

The feedback factor β of the *RC* network can be calculated by conventional circuit analysis, which yields

$$\beta = \frac{v_o}{v_i} = \frac{1}{1 - 5\alpha^2 - j(6\alpha - \alpha^3)} \tag{9–18}$$

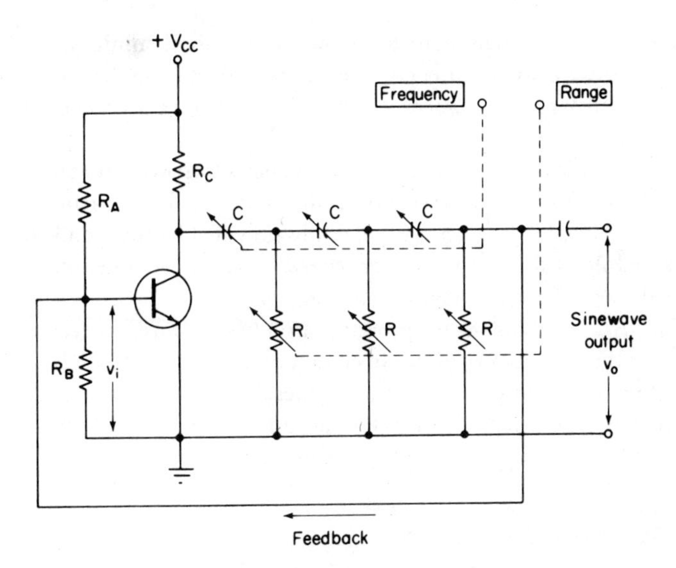

Fig. 9–12 Phase-shift oscillator with three cascaded *RC* sections to provide the necessary 180-degree phase shift between collector output and base input. The three ganged capacitors provide continuous frequency adjustment.

where $\alpha = 1/\omega RC$. To satisfy the Barkhausen criterion for oscillation, the phase shift produced by the *RC* network must be 180 degrees. This occurs when the imaginary term in the denominator of Eq. (9–18) is zero, or when $\alpha^2 = 6$, so the frequency of oscillation is found to be

$$f = \frac{1}{2\pi RC\sqrt{6}} \qquad (9\text{–}19)$$

At this frequency of oscillation the feedback factor $\beta = -1/29$.

To satisfy the condition that the loop gain $A_v\beta$ equals at least unity, the voltage gain of the amplifier must be at least 29. The transistor therefore must have a sufficiently high current gain to produce the required voltage gain.

The minimum number of *RC* stages in the feedback network is three, with each section providing 60 degrees of phase shift. If only two sections were to be used, each section would have to provide 90 degrees of phase shift and this would result in infinite attenuation. On the other hand, there is no clear advantage in using more than three sections.

The *RC* oscillator is ideally suited to the range of *audio frequencies*, From a few cycles to approximately 100 kHz. At the higher frequencies the network impedance becomes so low that it may seriously load the amplifier, thereby reducing its voltage gain to below the required minimum value, and oscillations cease. At low frequencies the loading effect is not usually a problem and the required large resistance and capacitance values are generally readily available.

With identical RC sections, the frequency of oscillation can be changed by simultaneously varying all resistors or all capacitors. In a practical oscillator, a continuously *variable frequency* is generally provided by three ganged variable capacitors, while different *frequency ranges* can be obtained by switching different values of shunt resistors into the network.

The RC oscillator can produce an almost pure sine-wave output, provided that the amplifier operates in class A and the voltage gain is maintained just above the minimum requirement. In general, the oscillator is best suited for fixed-frequency operation. It has the advantage of simple circuitry, good frequency stability, and reasonably pure waveform.

9–4.2 MOST Phase-Shift Oscillator

Some phase-shift oscillators use field effect devices such as FETs or MOSTs. Consider the circuit of Fig. 9–13, where amplifier $Q1$ is followed by a three-stage phase-shifting network and output buffer amplifier Q2. The phase-shifting network consists of series resistors and shunt capacitors (a shunt-C network). The necessary dc gate biase for the enhancement MOSTs is derived from drain supply V_{DD} through the 51-kΩ drain resistor of $Q1$ and the series resistors of the RC phase-shift network. The voltage gain of the MOST amplifier is generally lower than that of the conventional BJT stage, and it is therefore essential to minimize the voltage loss in the RC network. In the oscillator circuit of Fig. 9–13, each section of the RC network has a progressively

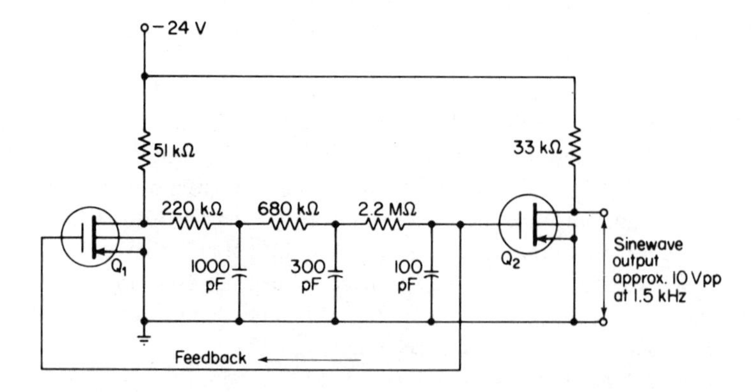

Fig. 9–13 MOST low-distortion phase-shift oscillator. The shunt-*C* feedback network contributes to the purity of waveform.

higher impedance level to minimize the loading effect on the previous section. For the values given, the circuit oscillates if amplifier $Q1$ has a voltage gain of approximately 18.

Provided that the three RC sections are identical, the shunt-C network produces the required 180-degree phase shift at a frequency of approximately

$$f \cong \frac{1}{3RC} \tag{9-20}$$

If the sections are not identical, as in Fig. 9–13, the frequency of oscillation can be calculated by classical network analysis. This analysis, however, is beyond the scope of this text.

Because of the loading effect of the final RC stage on the amplifier, the calculated frequency may be in error by as much as a factor of 2. The required frequency, however, can easily be obtained by adjusting one or two of the resistors.

Additional voltage gain, plus isolation between load and oscillator circuit is provided by MOST output stage $Q2$. Because the RC network is essentially a low-pass filter, the output waveform is remarkably free of harmonic distortion, and distortion figures of a few percent are easily obtained.

9–5 WIEN BRIDGE OSCILLATOR

9–5.1 Wien Bridge

The Wien bridge is a frequency-selective network consisting of resistive and capacitive elements only. It is often used as the *feedback network* in a sinusoidal oscillator, and it then determines the frequency of oscillation. In this application, the input voltage v_i to the bridge is the amplifier output voltage, and the output voltage v_o of the bridge is fed back to the amplifier input.

Consider the fundamental circuit of the Wien bridge of Fig. 9–14. The bridge has a series RC combination in one arm and a parallel RC combination in the adjoining arm. The other arms are purely resistive. The equations for bridge balance yield

$$\mathbf{Z}_1\mathbf{Z}_4 = \mathbf{Z}_2\mathbf{Z}_3 \quad \text{or} \quad \mathbf{Z}_3 = \mathbf{Z}_1\mathbf{Z}_4\mathbf{Y}_2 \tag{9-21}$$

where $\mathbf{Z}_1 = R_1 - j/\omega C_1$, $\mathbf{Y}_2 = 1/R_2 + j\omega C_2$, $\mathbf{Z}_3 = R_3$, and $\mathbf{Z}_4 = R_4$. Substituting the appropriate values into Eq. (9–21) we obtain

$$R_3 = \left(R_1 - \frac{j}{\omega C_1}\right) R_4 \left(\frac{1}{R_2} + j\omega C_2\right) \tag{9-22}$$

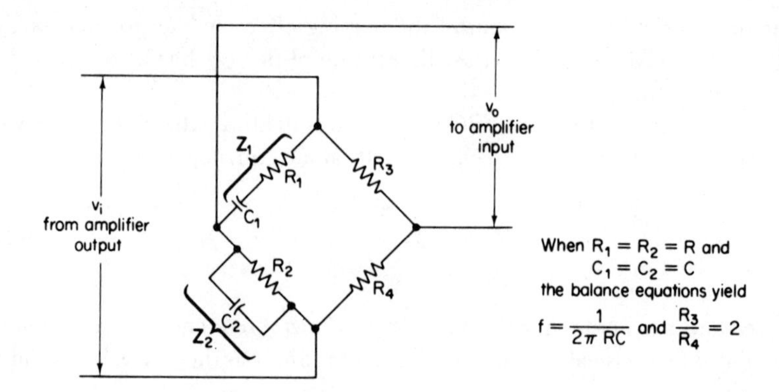

Fig. 9–14 Wien bridge.

and expanding

$$R_3 = \frac{R_1 R_4}{R_2} + j\omega C_2 R_1 R_4 - \frac{jR_4}{\omega C_1 R_2} + \frac{R_4 C_2}{C_1} \tag{9–23}$$

At bridge balance, both the real terms and the imaginary terms must equate. Separating them in Eq. (9–23) yields, for the real terms,

$$\frac{R_3}{R_4} = \frac{R_1}{R_2} + \frac{C_2}{C_1} \tag{9–24}$$

and for the imaginary terms

$$\omega C_2 R_1 = \frac{1}{\omega C_1 R_2} \tag{9–25}$$

where $\omega = 2\pi f$. Equation (9–25) can be solved to obtain an expression for the frequency of the input or excitation voltage and

$$f = \frac{1}{2\pi \sqrt{C_1 C_2 R_1 R_2}} \tag{9-26}$$

In the usual case, the bridge components are selected so that $R_1 = R_2 = R$ and $C_1 = C_2 = C$, and Eq. (9–24) then yields

$$\frac{R_3}{R_4} = 2 \tag{9–27}$$

while the balancing or resonant bridge frequency is

$$f = \frac{1}{2\pi RC} \tag{9-28}$$

In other words, the bride is *balanced* (zero output voltage) when the resistance ratio of the nonreactive arms satisfies Eq. (9–27) and when the excitation voltage has a frequency given by Eq. (9–28). The resonant frequency is determined only by the values of R and C in the reactive arms and not by the resistive arms of R_3 and R_4.

When the Wien bridge is used as the feedback network in an oscillator, the circuit must be modified slightly.

9-5.2 Wien Bridge Oscillator

The Wien bridge oscillator of Fig. 9–15 consists of a conventional two-stage RC-coupled amplifier, using two CE transistors with universal bias circuits. The output voltage of the two-stage CE amplifier is in phase with its input voltage. The Wien bridge feedback network, which provides zero phase shift at its resonant frequency, consists of the series and parallel RC arms and resistors R_1 and R_2 associated with transistor Q_1.

The elements of the Wien bridge, identified in Fig. 9–15, are reproduced in Fig. 9–16. The impedances of the reactive arms at the resonant frequency ($f = 1/2\pi RC$) can be written as

$$\mathbf{Z}_1 = R - \frac{j}{\omega C} = (1 - j)R \tag{9-29}$$

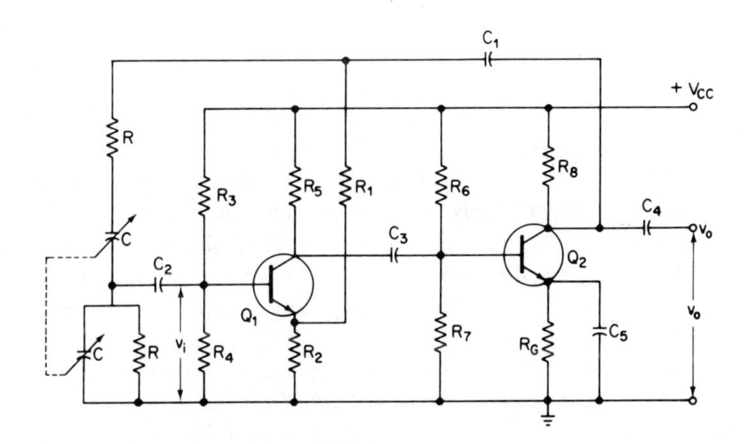

Fig. 9–15 Two-stage Wien bridge oscillator.

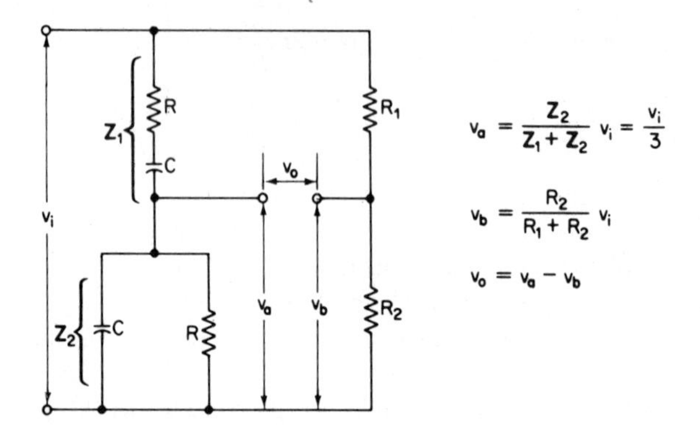

Fig. 9–16 Voltages in the Wien bridge.

and

$$\mathbf{Z}_2 = \frac{1}{1/R + j\omega C} = \frac{(1 - j)R}{2} \tag{9–30}$$

Hence, the voltage drop v_a across $\mathbf{Z}_2$

$$v_a = \frac{\mathbf{Z}_2}{\mathbf{Z}_1 + \mathbf{Z}_2} v_i = \frac{v_i}{3} \tag{9–31}$$

The voltage drop v_b across R_2 can be written as

$$v_b = \frac{R_2}{R_1 + R_2} v_i \tag{9–32}$$

In the Wien bridge oscillator of Fig. 9–15 feedback is applied from the collector of $Q2$ through coupling capacitor C_1 to the top of the bridge. The reactance of C_1 must be low at the frequency of oscillation so that no phase shift will be introduced by it. Resistor R_2 serves the dual purpose of $Q1$ emitter resistor and bridge element. Continuous variation of the frequency can be achieved by changing the two identical ganged air capacitors. Different frequency ranges can be provided by switching in other values for the two identical Wien bridge resistors.

A Wien bridge oscillator with automatic gain control (AGC) is shown in Fig. 9–17. This oscillator uses a high-gain differential amplifier, whose differential input is connected to the bridge output. The amplifier output is fed back to the top of the bridge.

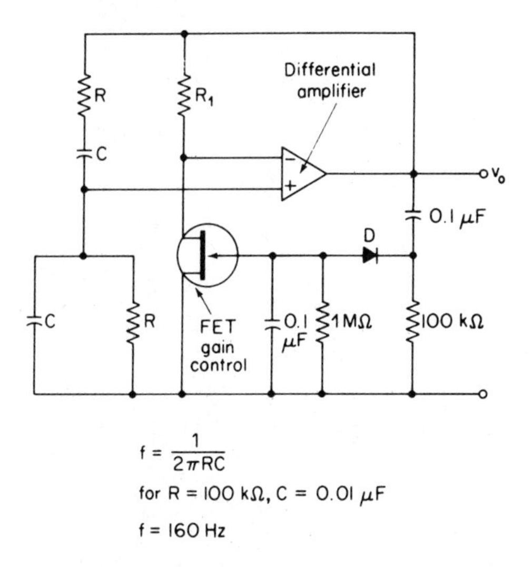

Fig. 9-17 Wien bridge oscillator with IC differential amplifier and automatic gain control (AGC) circuit. The dc supply connections to the IC amplifier are not shown.

The left side of the bridge contains the usual *RC* elements. The right side of the bridge consists of fixed resistor R_1 and a FET connected as a voltage controlled resistor. The voltage which controls the FET resistance is obtained by rectifying and filtering part of the amplifier output. With zero input signal, the FET resistance is low and the bridge is off balance, so the input to the differential amplifier is large. When the amplifier output builds up to the desired level, the AGC circuit biases the FET toward pinchoff and this reduces the input to the differential amplifier. Because the differential amplifier has a high gain, the bridge is almost balanced at the resonant frequency. The AGC circuit maintains just enough unbalance in the circuit to maintain the loop gain slightly higher than 1.

The frequency of oscillation is again $f = 1/2\pi RC$, and it can be continuously varied over a 10:1 frequency range by the ganged air capacitors *C*. Additional decade frequency ranges can be provided by a 10:1 change in bridge resistors *R*.

PROBLEMS

9-1 Assume that the amplifier of Fig. 9-18 has infinite input impedance and zero output impedance. Determine the closed-loop gain at $\omega = 0$.

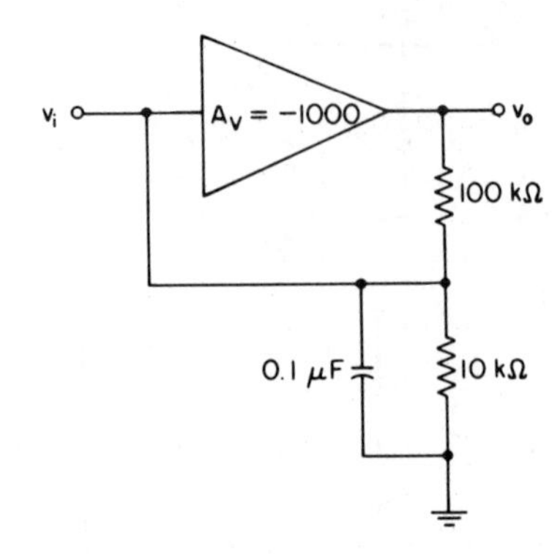

Fig. 9–18 Feedback amplifier for Problems 9–1 and 9–12.

9–2 Refer to Fig. 9–18 and determine v_o in both magnitude and phase if v_i = 10 mV at ω = 1000 rad/s.

9–3 Derive the expression for the frequency of the Hartley oscillator of Fig. 9–4, assuming no loading effects.

9–4 Refer to the Colpitts circuit of Fig. 9–5. Calculate the frequency of oscillation if $L = 50\ \mu$H, $C_1 = C_2 = 300$ pF. Assume that the transistor operates class A and neglect any loading effects.

9–5 Derive an expression for the frequency of oscillation of the Pierce crystal oscillator of Fig. 9–9, using the equivalent circuit of Fig. 9–7(b) for the crystal. What does the frequency of oscillation become as R approaches zero?

9–6 Derive an expression for the frequency of oscillation of a three-stage shunt-C phase shift oscillator.

9–7 Calculate the frequency of oscillation for the shunt-C phase-shift oscillator of Fig. 9–13.

9–8 Verify Eq. (9–18) for the feedback factor of the phase-shift network of Fig. 9–12, assuming that the network does not load the amplifier. Prove that the phase shift is 180 degrees for $\alpha^2 = 6$ and that $\beta = -1/29$ at this frequency.

9–9 Refer to the feedback network of Fig. 9–19.
(a) Calculate the feedback factor $\beta = v_o/v_i$.
(b) Sketch the circuit of a phase-shift oscillator using this network.
(c) Find the frequency of oscillation, assuming that the network does not load the amplifier.
(d) Find the required minimum amplifier gain for oscillation.

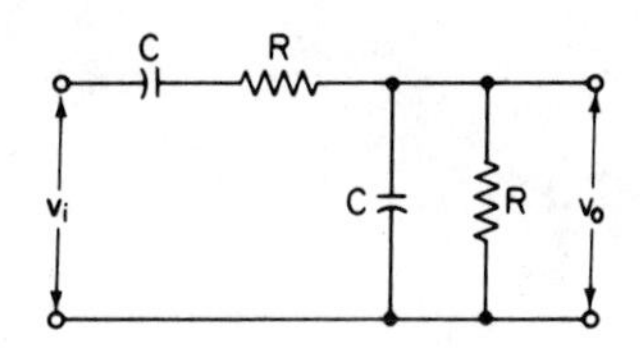

Fig. 9–19 Circuit for problem 9–9.

9–10 A certain quartz crystal has the following parameters (see Fig. 9–7): $L = 10\ H$, $R = 120\ \Omega$, $C_1 = 0.1$ pF, and $C_2 = 20$ pF.
Calculate (a) The series resonant frequency.
(b) The parallel resonant frequency.
(c) The Q of the crystal.

9–11 Assume that the two-stage amplifier in the Wien bridge oscillator of Fig. 9–15 has a voltage gain of 40 dB. If R_2 has a resistance of 470 Ω, calculate the minimum required resistance of R_1.

9–12 A four-section phase-shift oscillator uses four 10-kΩ shunt resistors and four 0.1-μF series capacitors. Assume class A operation and neglect the loading effects.
(a) Calculate the approximate frequency of oscillation.
(b) What is the frequency of oscillation if the 10-kΩ resistors are placed in series and the 0.1-μF capacitors in parallel?

Chapter 10

Special Solid-State Devices and Applications

10–1 INTRODUCTION

There are a number of solid-state devices which do not exhibit the conventional characteristics associated with the *pn* junction diode or the bipolar transistor. These special devices include certain types of diodes, whose conduction properties deviate from the usual one-way current flow of the rectifier diode.

The unusual properties of the devices discussed in this chapter are based on design and manufacturing techniques, where special attention is given to the levels of impurity concentration, the properties of the depletion regions, and the use of additional electrodes to control conduction across the *pn* junctions. Only the properties and applications of the more common solid-state devices on the market can be given. Table 10–1 is a quick summary of a number of special solid-state devices, their electrical properties and the likely application.

10–2 TUNNEL DIODE

10–2.1 Tunnel Diode Operation

The *tunnel diode* is basically a low-power *pn* junction with special and heavy doping of the *p*-type and *n*-type semiconductor materials. This heavy doping

TABLE 10–1 SUMMARY OF SPECIAL SOLID-STATE DEVICES.

Device	Schematic symbol	Electrical characteristics	Areas of application
Tunnel diode		Negative resistance at small forward bias voltage	Amplifiers, oscillators and pulse generators in vhf/uhf circuits. High–speed switching in logic and timing circuits
Backward diode	BD	Negative avalanche current at zero bias	Low–voltage signal rectifier or detector in vhf/uhf circuits
Varactor diode		Voltage–variable junction capacitance	Tuned resonant circuits and frequency multiplication at vhf frequencies
Silicon controlled rectifier (SCR)	G	Unilateral conduction, initiated by external control pulse	Control and switching in medium and high–power ac and dc circuits. Typical circuits include: rectifiers, converters, inverters, motor control, heat control, and other industrial applications
Triac	G	Bidirectional conduction, initiated by external control pulse	
Diac		Bidirectional conduction, initiated by avalanche breakdown	
Silicon controlled switch (SCS)	A, G_A, G_C, C	Unilateral conduction, initiated and terminated by external control pulses	Low–power switching element in logic circuits, timing circuits, counters, lamp–drivers
Unijunction transistor (UJT)	E, B_2, B_1	Negative resistance at low forward bias voltage	Oscillators, comparators, trigger circuits and timing circuits at vhf frequencies

results in an extremely thin depletion region and large numbers of available charge carriers, accompanied by a favorable repositioning of the energy levels in the p and n regions. As a result, when the applied forward or reverse bias is only slightly larger than zero volts, electrons from the valence band of the n region can pass directly, and very easily, across the narrow depletion region into the conduction band of the p region. This new current mechanism, caused by quantum mechanical action, is called *tunneling*. Under forward-bias conditions, tunneling occurs at very small voltages, well below the knee voltage for conventional diodes. Depending on the type of semiconductor material and the impurity concentration, the tunnel effect occurs at forward voltages ranging from a few millivolts to several hundred millivolts.

A typical *voltage–current characteristic* for a germanium tunnel diode is shown in Fig. 10–1. The tunneling action makes the diode highly conductive for all reverse-bias voltages. Under conditions of forward bias, electrons from the n region tunnel directly across the junction into the p region, and the forward current increases rapidly to its *peak-point* value I_P at peak-point voltage V_P. A further increase in forward bias beyond V_P reduces the forward current to a minimum value I_V at *valley-point* voltage V_V. Beyond V_V the diode current again increases, but now in exactly the same manner as for the conventional diode (the diffusion mechanism).

The region between peak point and valley point, where the tunnel diode current decreases with increasing forward bias, gives the diode a *negative resistance* characteristic. This suggests that the tunnel diode, when operated in the negative resistance region, can be used as an amplifier, an oscillator, or a switching element.

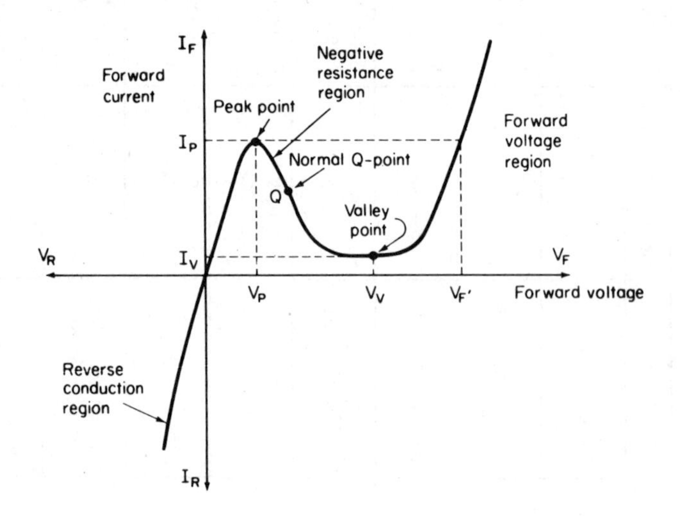

Fig. 10–1 Typical voltage–current characteristic of a tunnel diode.

10–2.2 Ratings and Characteristics

The tunnel diode is represented by one of the *schematic symbols* of Fig. 10–2. Ratings and device characteristics are given by the manufacturer's data sheet, which describes the static and dynamic behavior of the diode in its normal (negative resistance) mode of operation. A sample sheet is shown in the Appendix.

Fig. 10–2 Commonly used schematic symbols for the tunnel diode.

In addition to the usual maximum ratings for current, power dissipation, and temperature, the data sheet provides static and dynamic device characteristics. The static parameters are related to the typical voltage-current characteristics of Fig. 10–1 and define the peak point and valley point voltages and currents. The dynamic parameters are related to the ac *equivalent circuit* of Fig. 10–3 and describe the characteristics of the tunnel diode in its negative resistance region.

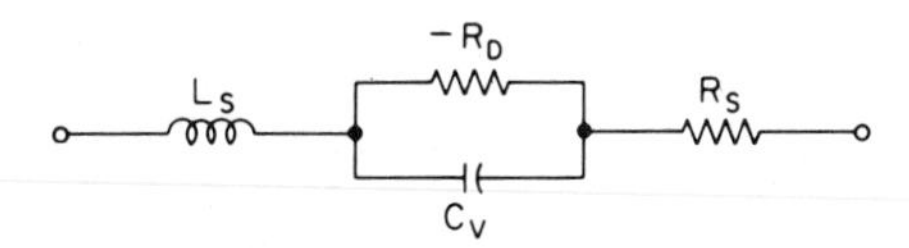

Fig. 10–3 Equivalent circuit for the tunnel diode.

In this equivalent circuit, the total series inductance L_S (which largely controls the high-frequency response of the diode) includes the internal lead inductance. Total series resistance R_S includes the terminal resistance and the internal lead resistance. The negative device resistance, represented by $-R_D$, is measured at the inflection point of the negative resistance characteristic of Fig. 10–1 and is typically on the order of 5 to 100 Ω. Shunting capacitance C_V of the equivalent circuit is called the *valley-point capacitance* and includes the internal junction capacitance as well as the case capacitance. Since the reactive components of the ac equivalent circuit are extremely small, the high-frequency performance of the tunnel diode is excellent and extends well into the gigahertz range.

In amplifier or oscillator operation, the tunnel diode is biased approximately halfway on the negative resistance portion of the voltage-current characteristic, as indicated by the location of the Q point on the curve of Fig. 10–1.

Since both V_P and V_V are in the millivolt range (typical values are $V_P = 60$ mV and $V_V = 350$ mV), the location of the Q point must be established and maintained accurately. The tunnel diode voltage–current characteristic is relatively independent of temperature. The stability of the operating point may be affected, however, by temperature dependence of external circuit components and, in this case, circuit compensation (such as negative feedback) may be required.

10–2.3 Tunnel Diode Oscillators

Although tunnel diodes are low-power devices (typically below 100 mW dissipation), their low-noise characteristics, good frequency stability, and low internal capacitance make them ideally suited in high-frequency applications such as oscillators or switching circuits.

Figure 10–4(a) shows the tunnel diode (TD) in a basic series–parallel sine-wave oscillator. The tunnel diode is biased by resistors R_1 and R_2, so the operating point is located approximately in the center of the negative resistance region of the voltage–current characteristic. Figure 10–4(b) shows that this allows equal positive and negative excursions around the Q point to obtain maximum signal swing.

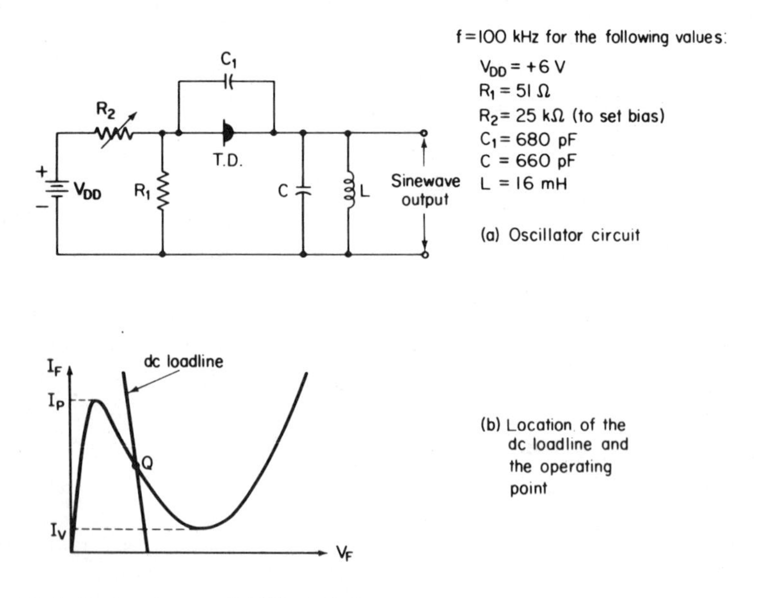

Fig. 10–4 Basic series-parallel TD oscillator.

The frequency of oscillation is determined mainly by the resonant frequency of the LC tank circuit and the design equations are expressed as follows:

$$C + \frac{C_1}{1 - R_T g_d} = \frac{1}{L\omega^2} \tag{10–1}$$

where R_T is the total dc resistance of the circuit, equal to

$$R_T = \frac{R_1 R_2}{R_1 + R_2} + R_S + R_{\text{dc(coil)}} \tag{10–2}$$

and

$$C_1 = \sqrt{\frac{g_d(1 - R_T g_d)}{R_T \omega^2}} \tag{10–3}$$

where g_d = negative conductance of the TD
R_S = total series resistance of the TD

For a 100-kHz TD oscillator, using a suitable tunnel diode, the circuit values are as tabulated in Fig. 10–4(a).

Figure 10–5 shows a crystal-controlled TD oscillator operating in the citizens' band at 27.255 MHz. The oscillator frequency operates within the tolerance range of the quartz crystal and a bias range of 110 to 150 mV.

10–2.4 Tunnel Diode Switching Circuits

The tunnel diode (TD) acts ideally as a monostable or bistable switching element, producing an output voltage excursion when triggered by an input pulse. Consider the *monostable switching circuit* of Fig. 10–6(a).

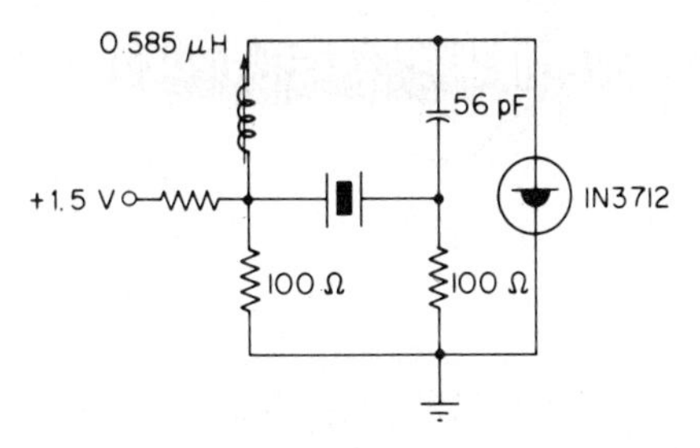

Fig. 10–5 Citizens band 27.255 MHz crystal-controlled TD oscillator. (*Courtesy General Electric Company.*)

The quiescent operating point of the TD is determined by supply voltage V_{DD} and anode resistor R_A. The dc loadline intersects the voltage–current characteristic to the left of the peak point, as indicated in Fig. 10–6(b). We observe that the quiescent anode voltage V_Q is only very slightly smaller than the peak point voltage V_P.

A positive trigger pulse of short duration momentarily raises the anode voltage above the peak point value V_P and forces the TD to enter its negative resistance region. Inductor L opposes the sudden decrease in anode current and produces an induced emf which switches the TD rapidly to the high-voltage state, indicated by the point market $V_{(high)}$ on the characteristic curve. The TD attempts to return to its stable Q-point position, and both current and voltage

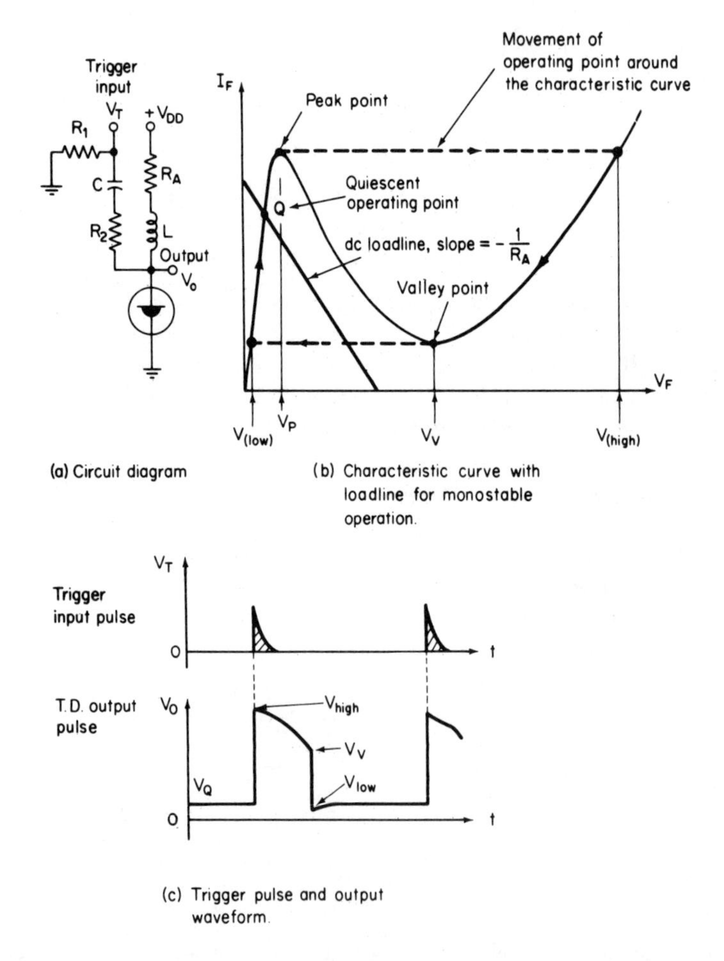

Fig. 10–6 The tunnel diode as a monostable switching element.

decrease along the curve to the valley point. As the anode voltage drops below V_V, the current suddenly increases as the TD enters the negative resistance region, an action which is opposed by the inductor. The resulting induced emf rapidly switches the TD to the low-voltage state, indicated by the point marked $V_{(low)}$ on the curve. The device then returns to the quiescent Q point along the positive resistance portion of the curve and rests in this stable condition.

The application of another positive trigger pulse causes the process just described to repeat. The voltage excursion at the TD anode terminal produces an output waveform as shown in the diagram of Fig. 10–6(c).

The switching circuit of Fig. 10–6(a) operates as a *bistable switching circuit* when the supply voltage and the load resistor are adjusted so that the loadline intersects the voltage–current characteristic at *two* points, marked A and B in Fig. 10–7(a). Both points are located on the positive resistance portions of the curve and represent *stable* operating points.

We assume that the tunnel diode initially rests at quiescent operating point A. A *positive trigger pulse* at the anode switches the TD to the high-voltage state and it reaches the second stable operating condition at point B, where it rests. The output voltage at the anode has changed from the low-voltage quiescent value V_A to the high-voltage quiescent value V_B, as indicated in the waveform diagram of Fig. 10–7(b).

To return the TD to its original operating position, a *negative trigger pulse* is required. A pulse of sufficient amplitude to reduce the anode voltage below the valley-point voltage V_V switches the TD back to the low-voltage state and it returns to the quiescent point at A, where it rests. The output voltage at the TD anode has now changed from the high-voltage quiescent value V_B to the low-voltage quiescent value V_A. Subsequent positive and negative trigger pulses cause the TD to switch between the two stable states.

10–3 TUNNEL RECTIFIER OR BACK DIODE

The tunneling effect of a normal tunnel diode is obtained by careful control of the doping concentrations of its junctions. The *peak-to-valley ratio* I_P/I_V is used to describe the extent of the negative resistance portion of its forward characteristic. It is possible, by a careful reduction of the doping levels, to modify the forward characteristic so that the TD exhibits a very low peak-to-valley ratio but to retain the reverse characteristic of very low resistance (high conduction). When the tunnel diode is so manufactured that its reverse conduction characteristics are emphasized rather than its negative resistance characteristics, we speak of a *tunnel rectifier or back diode*.

Figure 10–8 compares the voltage-current characteristics of a conventional rectifier diode and a tunnel rectifier. In the conventional rectifier diode there is substantial conduction in the forward direction (above the knee voltage),

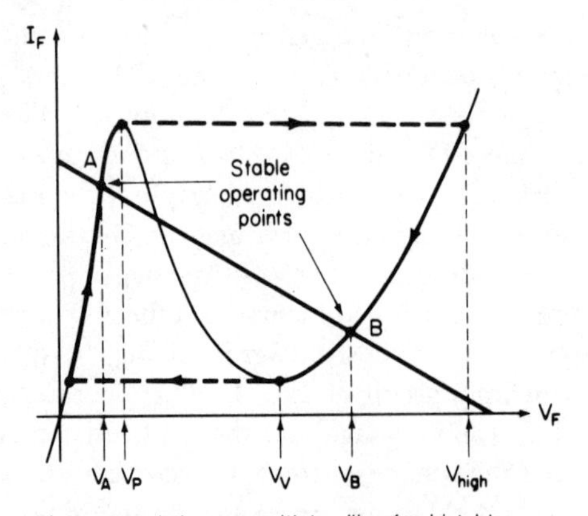

(a) Characteristic curve with loadline for bistable operation.

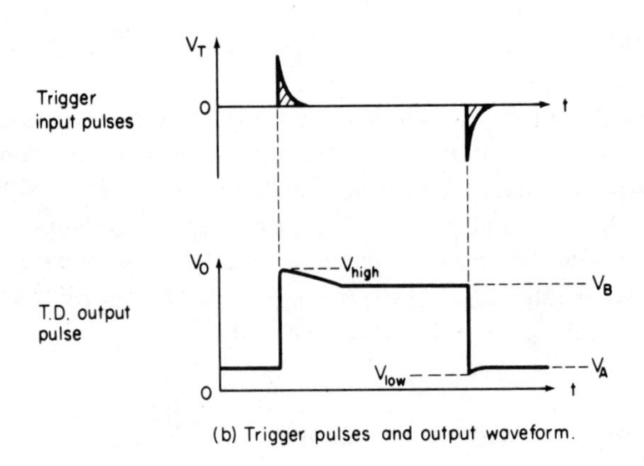

(b) Trigger pulses and output waveform.

Fig. 10–7 **The tunnel diode as a bistable switching element.**

and practically no conduction in the reverse direction. In the tunnel rectifier a substantial amount of reverse current flows at extremely small reverse voltages, while the forward current is relatively small. The tunnel rectifier can therefore effectively be used as a low-voltage rectifier or detector, keeping in mind, of course, that the polarity of the rectified signal is opposite to that for the conventional rectifier. This is the reason the tunnel rectifier is often referred to as a "back diode."

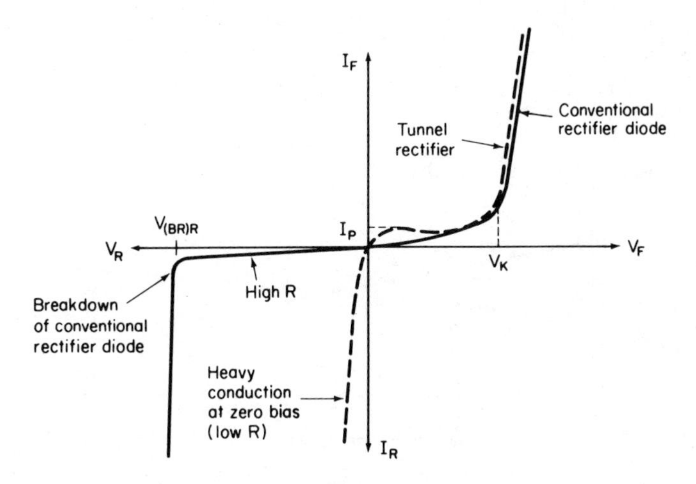

Fig. 10–8 Comparison of the voltage-current characteristics of a tunnel rectifier and a conventional rectifier. There is a marked similarity in the forward region, but in the reverse region the tunnel rectifier conducts heavily at zero bias, whereas the rectifier diode presents an almost infinite resistance.

Since the back diode has the same excellent high-frequency response characteristics as the tunnel diode, it finds application in very high frequency and microwave circuits.

10–4 VARACTOR DIODE

10–4.1 Operation and Characteristics

In the conventional junction diode, the depletion region forms a barrier which separates the positive and negative charges on each side of the junction. These opposite charges are analogous to the charges on opposite plates of a capacitor, with the depletion region itself serving as the dielectric medium between the plates. Any *pn* junction therefore possesses *junction capacitance.*

The depletion region widens when the junction is reverse-biased, and the junction capacitance therefore decreases. In a conventional *pn* junction, where the impurity concentration is approximately uniform throughout the *p*-type and *n*-type semiconductor materials, this capacitance versus reverse-voltage relationship is essentially nonlinear, as shown in the typical silicon diode $C–V_R$ characteristic of Fig. 10–9.

The varactor diode is a specially manufactured *pn* junction with a variable impurity concentration throughout its *p* and *n* regions. The doping concentration of this *graded junction* decreases toward the junction and is very

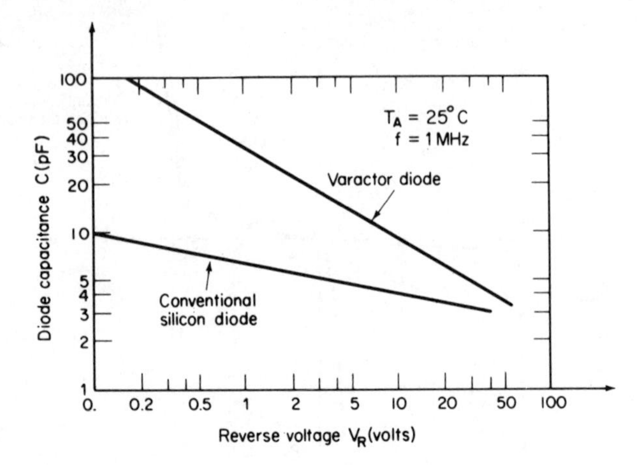

Fig. 10–9 Comparison of a diode capacitance as a function of reverse voltage for a conventional silicon diode and a varactor diode.

light in the regions adjacent to the junction. The graded impurity profile of the varactor diode produces a much steeper C–V_R curve, so that relatively large capacitance changes can be obtained with rather small voltage variations. Figure 10–9 shows the difference between the capacitance curves for the conventional diode and the varactor diode.

It then follows that the varactor diode can be used as a *voltage-sensitive variable capacitance,* and as such it finds application in high-frequency tuning circuits, oscillators, and frequency multipliers.

Figure 10–10 shows the schematic symbol and the simplified equivalent circuit of a varactor diode. This basic circuit, consisting of voltage-variable junction capacitance C_J and series resistance R_S, adequately represents the varactor diode at the lower end of the frequency spectrum. For applications at microwave frequencies, it may be desirable to use the conventional equivalent

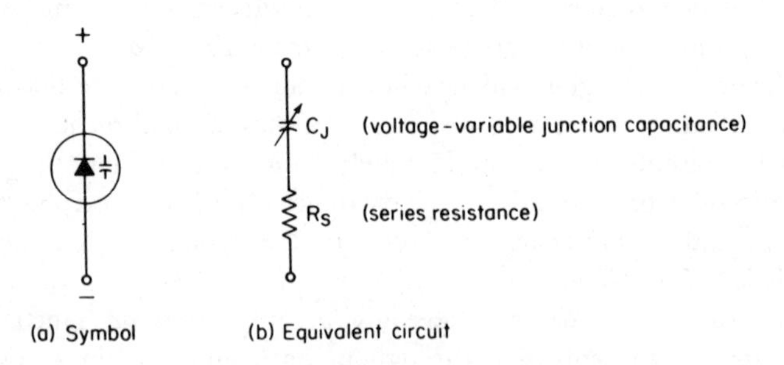

Fig. 10–10 Schematic symbol and basic equivalent circuit for the varactor diode.

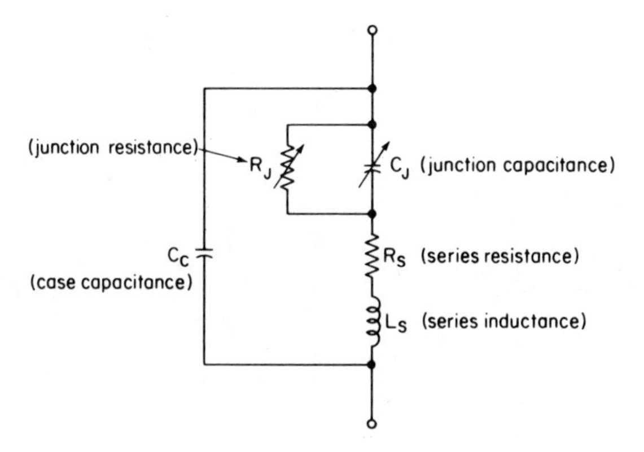

Fig. 10–11 Conventional equivalent circuit of the varactor diode.

circuit of Fig. 10–11, which more accurately describes the varactor properties in the very high frequency/ultra high frequency range.

In this equivalent circuit, diode junction capacitance C_J is the voltage-variable capacitance at the depletion layer. It is one of the electrical characteristics of the device and is stated at a specific operating frequency and a given reverse voltage. Case capacitance C_C and series inductance L_S are constants that detract from performance. Another characteristic is the figure of merit Q, resulting from the reactive character of the varactor diode and defining the efficiency of the diode response to an input frequency (bandwidth). Related to Q is the cutoff frequency f_{co}, defined as that frequency at which the reactance of the diode equals its total resistance, or $Q = 1$.

10–4.2 Applications

The varactor diode, with its excellent high-frequency stability, is often used in very high frequency circuits such as oscillators and amplifiers, frequency multipliers, and microwave circuits.

In the conventional *Colpitts oscillator* of Fig. 10–12, two varactor diodes and a coil form the tank circuit of a 25–35 MHz variable-frequency oscillator. The frequency of oscillation is varied by an external dc control voltage, which applies reverse bias to the two varactor diodes. Their junction capacitances are varied over the range defined by the $C–V_R$ characteristics, and this changes the resonant frequency of the tank circuit. To ensure good frequency stability, particular attention should be paid to the temperature coefficient of the diode capacitance.

Varactor diodes with small series resistance and particularly sharp, nonlinear capacitance characteristics are often used in *frequency-multiplier*

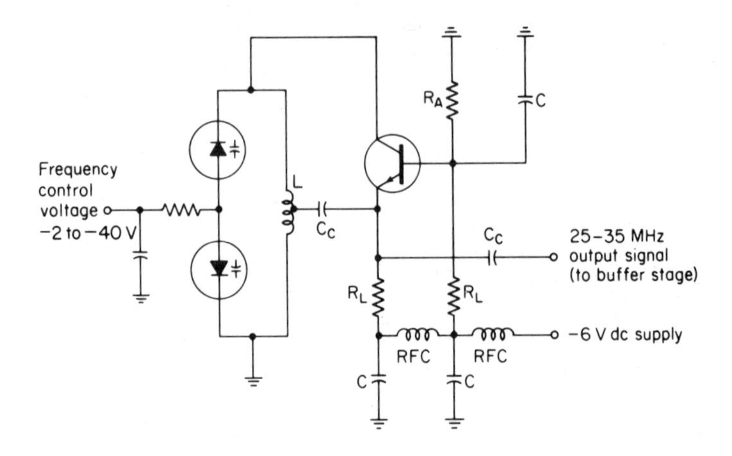

Fig. 10–12 Hf/vhf oscillator with varactor frequency control.

applications. When such a varactor diode is heavily driven by a sine-wave input voltage, it generates harmonic voltages with great efficiency. Consider the simplified *frequency doubler* of Fig. 10–13, which is driven by a sinusoidal voltage v_i at the fundamental frequency f. A resonant filter in the input loop, tuned to the fundamental frequency, only allows the fundamental current component i_f in the input loop. A second-harmonic current i_{2f}, generated by the varactor diode, flows toward the load Z_L. A filter tuned to the second-harmonic frequency is placed in the output loop and it blocks the fundamental frequency component of the total current. The output voltage at frequency $2f$ is usually buffered and amplified.

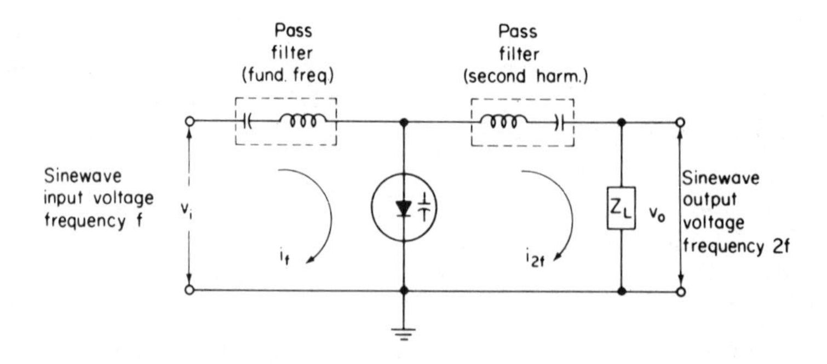

Fig. 10–13 Varactor diode frequency doubler.

10–5 THYRISTORS

10–5.1 Types of Thyristors

Thyristors are bistable solid-state devices that have two or more junctions, and that can be switched between conducting states (from OFF to ON, or from ON to OFF). Thyristors generally are heavy-current devices (to several hundred amperes) and are extensively used in applications involving power rectification, control, and switching.

1. The *silicon-controlled rectifier* (SCR) is a four-layer device with three electrodes: an anode, a cathode, and a control gate. The basic junction diagram and schematic symbol are shown in Fig. 10–14(a). The SCR is a *unidirectional* reverse-blocking thyristor, providing current conduction in only one direction. It can be switched from the OFF state to the ON state by a positive trigger pulse applied to the gate.
2. The *triac* is a four-layer device with three electrodes: terminal 1, terminal 2, and a control gate. The basic junction diagram and schematic symbol of a triac are shown in Fig. 10–14(b). It is a *bidirectional* device which can block voltages of either polarity and conduct current in either direction. It can be switched from the OFF state to the ON state, in either direction, by a positive or a negative gate pulse. The triac is often considered analogous to two SCRs connected in parallel but oriented in opposite directions (inverse-parallel connection).
3. The *diac* is a two-electrode, three-layer *bidirectional* avalanche diode, which can be switched from the OFF state to the ON state in either direction. The basic junction diagram and schematic symbol of a diac are shown in Fig. 10–14(c). It does not have a control gate. It is switched ON, for either polarity of applied voltage, simply by exceeding the avalanche breakdown voltage.

10–5.2 SCR Operation

Figure 10–15 shows the typical voltage–current characteristic of an SCR. In the forward-bias region (anode positive with respect to cathode) the SCR has two distinct operating states. As the foward bias is initially increased from zero volts, the SCR allows only a small forward current and exhibits a high forward resistance. This region of operation is called the *forward blocking region,* or *OFF state.* As the forward bias is further increased, the OFF-state current increases very slowly until the *breakover voltage* $V_{(BO)}$ is reached. At $V_{(BO)}$ the SCR suddenly converts to the *high conductance region,* or *ON state,* where the anode current is limited mainly by the resistance in the external circuit.

The breakover voltage can be varied by applying a current pulse of the right character to the gate electrode, causing the SCR to switch from the OFF state to the ON state at a lower forward voltage, as indicated in Fig. 10–16. As the amplitude of the gate pulse is increased, the breakover point moves toward the origin until the curve resembles that of a conventional rectifier diode. Normally, the SCR is operated well below $V_{(BO)}$ and is then made to switch ON by a gate signal of sufficient amplitude. This assures that the SCR starts forward conduction at exactly the right instant.

After the SCR is switched ON, a certain minimum anode current, called the *holding current* I_H, is required to maintain the device in the ON state. If the anode current drops below I_H, the SCR reverts to the OFF state. In fact, to switch the SCR from ON to OFF, the anode current must be reduced to a value below I_H.

In the reverse-bias region (anode negative with respect to cathode), the SCR presents a very high impedance and allows only a small reverse current.

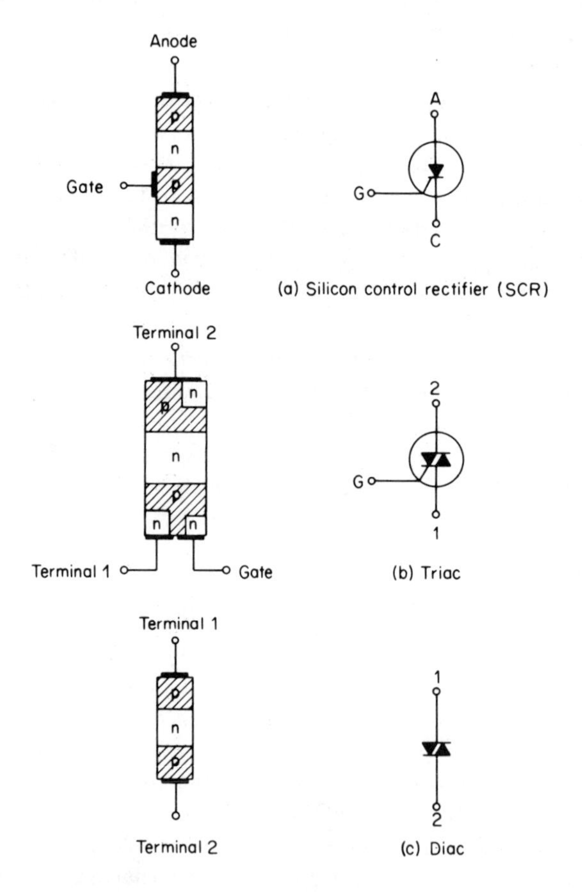

Fig. 10–14 Junction arrangements and schematic symbols for the SCR, the triac, and the diac.

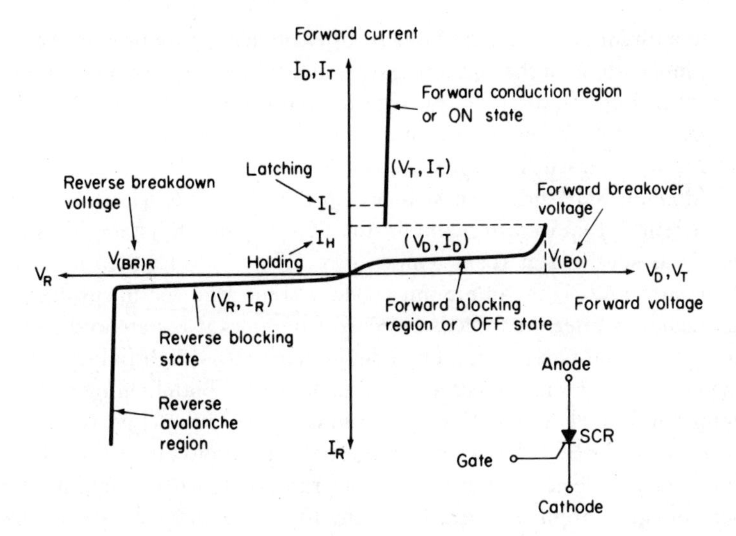

Fig. 10–15 SCR voltage-current characteristic.

This is the *reverse-blocking region.* The SCR remains blocked until the reverse bias exceeds the reverse breakdown voltage $V_{(BR)R}$. At this avalanche breakdown point the SCR conducts heavily and is subject to the usual problems of thermal runaway. Thermal runaway means destruction of the device, and $V_{(BR)R}$ should therefore never be exceeded.

10–5.3 SCR Ratings and Characteristics

Like all solid-state devices, SCRs must be operated within the maximum ratings specified by the manufacturer. These ratings, as presented in the typical data sheet of the 2N3228 SCR in the Appendix, define limiting values.

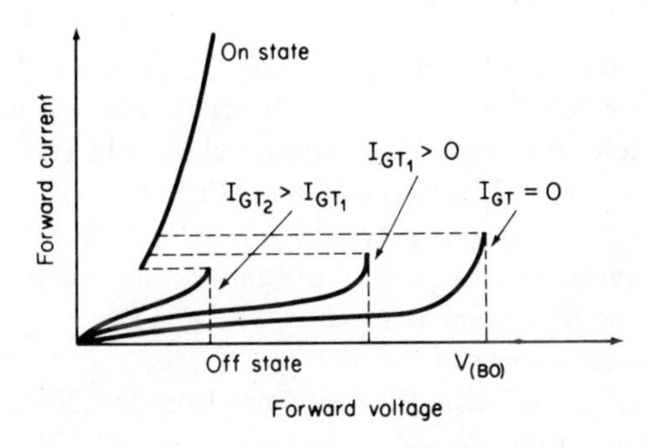

Fig. 10–16 SCR breakover voltage decreases with increasing values of gate current.

In addition to the usual power rectifier nomenclature used in the *ratings,* detailed information on the dynamic gating and switching parameters is given in the electrical *characteristics.* Gate-triggering requirements are supplied in terms of the *gate-to-cathode voltage* V_{GT} and the *gate current* I_{GT}; typical values are specified in the characteristics.

After an SCR has been switched to the ON state, a certain minimum anode current is required to maintain the SCR in the ON state. This *holding current* I_H is specified at room temperature ($T = 25°C$) with the gate open. *Latching current* I_L is the minimum anode current required to sustain conduction immediately after the SCR is switched from the OFF state to the ON state and the gate signal is removed. The latching current is slightly larger than the holding current, but for most SCRs less than twice I_H. Both holding and latching current are indicated on the voltage–current characteristic of Fig. 10–15.

Because the SCR has a certain internal capacitance, the forward blocking capability of the device is sensitive to the rate at which the forward anode–cathode voltage is applied. A quickly rising forward voltage causes a capacitive charge current ($i = CdV_D/dt$) which may become large enough to trigger the SCR into conduction. The SCR data sheet of the Appendix includes the parameter dV_D/dt, called the *rate of rise of the OFF-state voltage,* which the SCR can withstand before triggering occurs.

Turn-on time t_{on} refers to the switching characteristics of the SCR. The turn-on time is defined as the time interval between the initiation of a gate signal and the time when the anode current reaches 90 percent of its maximum value. The turn-on time is dependent on the magnitude of the gate trigger pulse, decreasing with increasing gate current magnitude.

The commutated *turn-off time* t_q of an SCR is the time interval between the cessation of forward current and the reapplication of forward blocking voltage. The turnoff time depends on several circuit parameters, in particular the SCR junction temperature and the ON-state current prior to switching.

10–5.4 Control of the Phase Angle

Many ac power control applications require that the thyristor be switched full ON or full OFF, an action similar to that of a relay. In cases like this, the SCR handles the heavy load current, while only a small gate current is required to trigger the SCR into conduction. Figure 10–17(a) shows a simple method of controlling the ac load power by gating the SCR into the ON state.

With maximum resistance R in the gate circuit, the SCR is in the OFF state. As R is reduced, a point is reached where sufficient gate current I_{GT} is provided at the positive peak of the ac input voltage waveform (90 degrees) to trigger the SCR ON. The SCR then conducts from the 90-degree point to the 180-degree point (where the anode current drops to zero), for a total *conduction angle* of 90 degrees. This conduction angle can be increased to almost 180

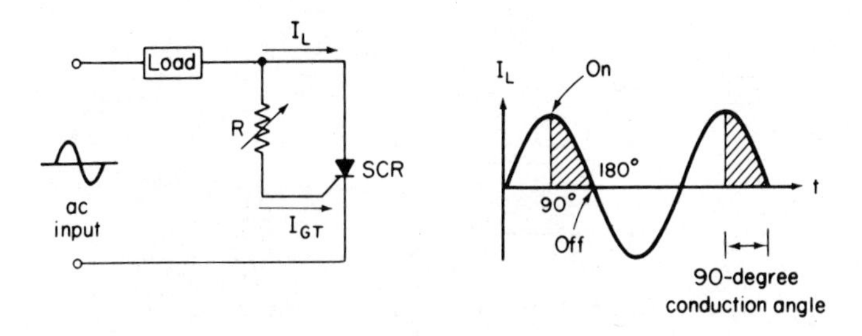

(a) SCR gating circuit, with resistor R controlling gate current I_{GT}

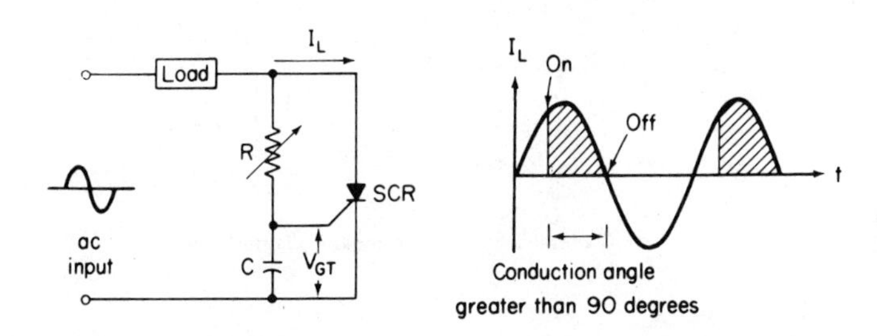

(b) RC network in SCR gating circuit, controlling gate voltage V_{GT}

Fig. 10–17 Basic gate trigger circuits, controlling the conduction angle of the SCR.

degrees by reducing the resistance in the gate circuit and thereby increasing I_{GT}. In the half-wave circuit of Fig. 10–17(a), the ac power delivered to the load can be controlled over a considerable range simply by adjusting the conduction angle of the SCR.

A different method of obtaining conduction angle control greater than 90 degrees for half-wave operation is shown in the *RC* trigger network of Fig. 10–17(b). Here the SCR is in series with the ac load and in parallel with the *RC* network. At the start of each positive half-cycle, the SCR is in the OFF state (zero anode current) and the rising ac input voltage drives current through *R* and *C* to charge the capacitor. When the capacitor voltage reaches the SCR breakover voltage, the capacitor discharges through the gate terminal and turns the SCR ON. Full ac power is then transferred to the load for the remainder of the half-cycle. When the potentiometer resistance is decreased, the capacitor charges more rapidly and the SCR breakover voltage is reached earlier in the cycle. In this case the ac power delivered to the load is increased.

Many triggering circuits can be used to meet the needs of the SCR. In the typical ac power control circuits of Fig. 10–18, the trigger circuits are labeled CONTROL.

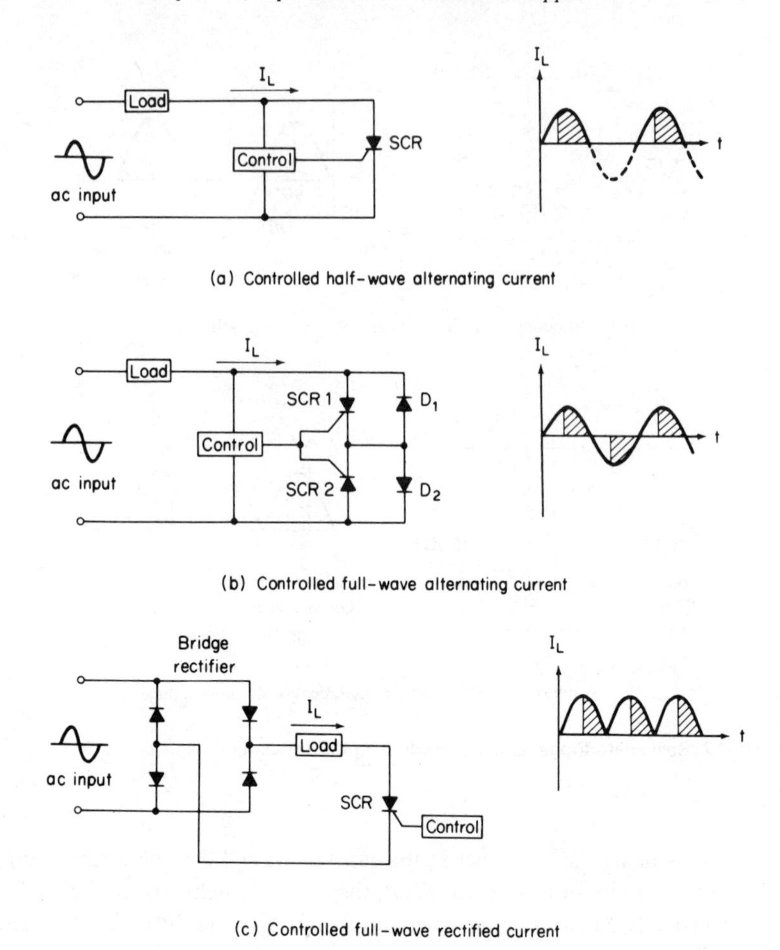

Fig. 10–18 SCR ac phase control in power applications.

10–5.5 SCR Applications

SCR *inverters* provide an efficient and economical way of converting direct current or voltage into alternating current or voltage. In this application the SCR serves as a controlled switch, alternately opening and closing a dc circuit. Consider the basic inverter circuit of Fig. 10–19(a), where an ac output is generated by alternately closing and opening switches S_1 and S_2. Replacing the mechanical switches with SCRs, whose gates are triggered by an external pulse generator, we obtain the practical SCR inverter of Fig. 10–19(b).

It should be realized that the SCR is basically a *latching* device, which means that the gate loses control as soon as the SCR conducts, and anode current continues to flow in spite of any gate signal that may be applied. To have each SCR perform the ON-OFF switching function at the desired instant, special *commutating* circuitry must be added. In the practical circuit of Fig. 10-19(b) capacitor C is connected across the anodes of the SCRs to provide this commutation, as the following discussion explains:

When conduction is initiated by applying a positive trigger pulse to SCR1 (SCR2 is assumed to be OFF), the voltage across SCR1 decreases rapidly as the current through it increases. Capacitor C charges through SCR1 in the polarity shown. The load current flows from the dc supply through inductor L, one-half of the transformer winding and SCR1.

When a firing pulse is now applied to the gate of SCR2, this SCR turns ON and conducts. At this instant, capacitor C begins to discharge through SCR1 and SCR2. This discharge current flows through SCR1 in a *reverse* direction. Conduction of SCR1 ceases as soon as the charge stored in the SCR has been removed by the reverse recovery current. At this time, with SCR1 turned OFF, the capacitor voltage (approximately $-2E$) appears across SCR1 as a reverse voltage, long enough to allow this SCR to recover for forward blocking.

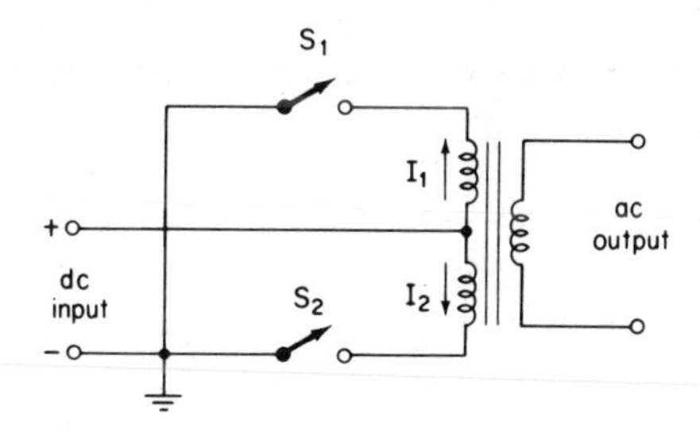

(a) Basic inverter circuit using mechanical switches

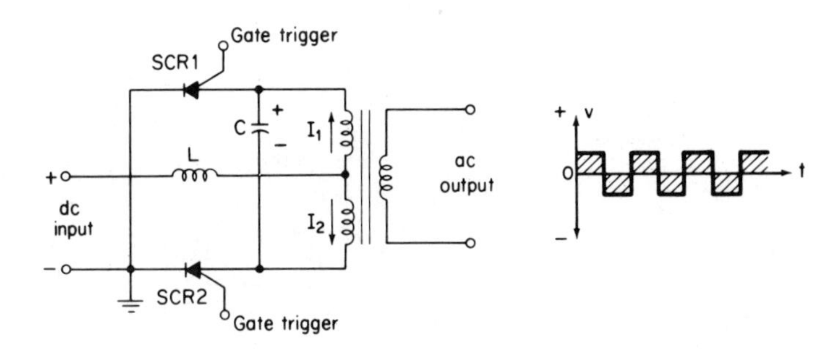

(b) Practical SCR inverter with commutating capacitor

Fig. 10-19 Inverter circuits.

Simultaneously during this interval, conducting SCR2 allows the capacitor to discharge through the transformer winding and inductor L. The function of L is to control the discharge rate of C to allow sufficient time for SCR1 to turn OFF. Capacitor C discharges rapidly from $-2E$ to zero and then charges up in the opposite direction to $+2E$. The load current is now carried through the second half of the transformer winding and SCR2.

When the trigger current is applied again to the gate of SCR1, this device conducts and the process just described repeats.

SCRs can be used effectively to control the application of power to ac and dc motors, and hence change their operating characteristics to obtain different speed and torque curves. An example of a basic *dc motor control circuit* is shown in Fig. 10–20. The unfiltered or "raw" output from a full-wave bridge rectifier is applied to the field and armature windings of the dc motor. These windings are placed in series with an SCR whose conduction angle is controlled by the R_1–R_2 trigger circuit in the conventional manner.

10–5.6 Triac: Operation and Applications

Figure 10–21 shows the diagrammatic junction arrangement of a triac. Both main terminals make contact with an n-type emitter as well as a p-type emitter. The n-type emitter at terminal 2 is located directly opposite the p-type emitter at terminal 1, and the p-type emitter at terminal 2 is located opposite the n-type emitter at terminal 1. This suggests that the triac consists of two four-layer devices (SCRs) in parallel, but oriented in opposite directions. The triac can therefore *block or conduct* current between the two main terminals *in either direction.* The gate terminal, which makes contact with both p-type and n-type materials, allows the application of either positive or negative trigger currents.

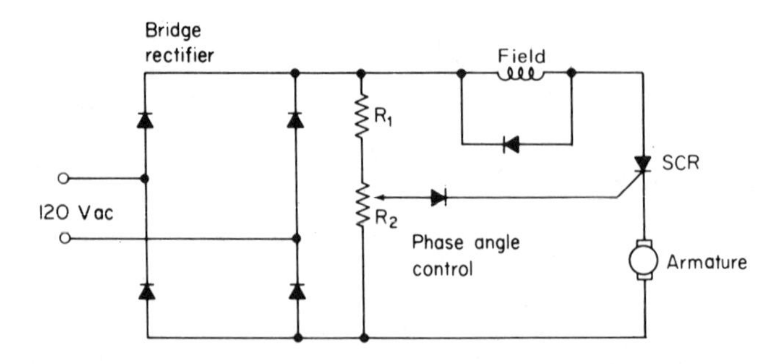

Fig. 10–20 Full-wave dc motor control.

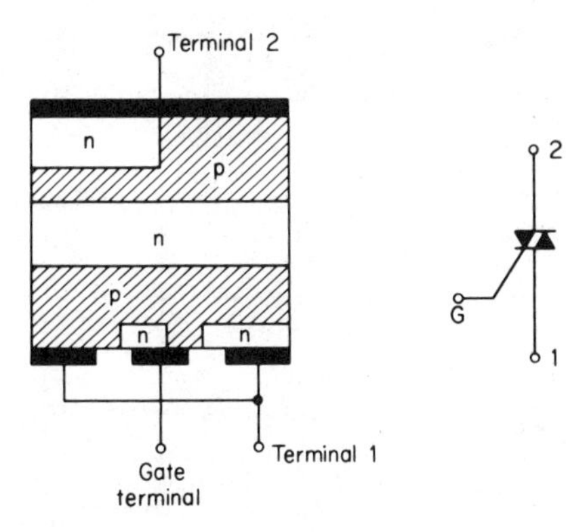

Fig. 10-21 Junction arrangement and schematic symbol of the triac.

The voltage-current characteristic of the triac is shown in Fig. 10-22. We observe that the triac has the same OFF-state and ON-state regions as the SCR, but now for either polarity of voltage applied across the main terminals. When main terminal 2 is positive with respect to main terminal 1, the triac is *positively biased* and operates in the first quadrant of its characteristic. When main terminal 2 is negative with respect to main terminal 1, the triac is *negatively biased* and operates in the third quadrant of the characteristic. Like an SCR, the triac switches from the OFF state to the ON state at the breakover voltage $V_{(BO)}$. The breakover point can be reached at lower main terminal voltages by applying a positive or a negative trigger pulse to the gate.

Gate triggering occurs in any of the four operating modes summarized in Table 10-2. The gate-triggering requirements are different for each operating

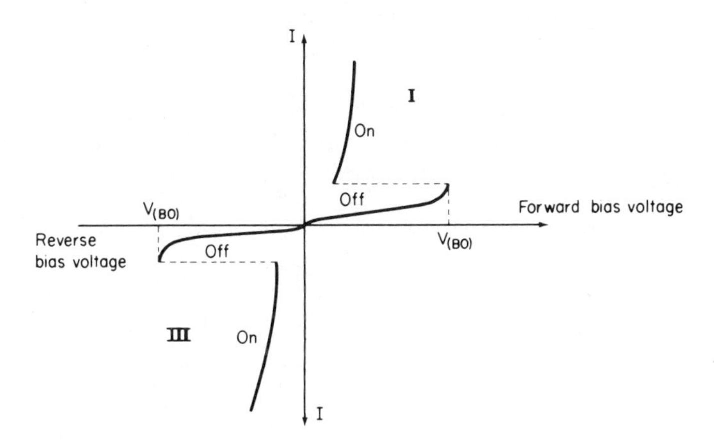

Fig. 10-22 Voltage-current characteristic of the triac.

TABLE 10–2 TRIAC TRIGGERING MODES.

Triac triggering modes		
Main terminal 2 to main terminal 1 voltage	Gate to main terminal 1 voltage	Operating quadrant*
Positive	Positive	I(+)
Positive	Negative	I(−)
Negative	Positive	III(+)
Negative	Negative	III(−)

*Note: Plus (+) or minus (−) sign in the quadrant designation indicates the polarity of the gate trigger voltage or current

mode. The most sensitive trigger mode is usually mode *I*(+), where the triac is forward-biased and a positive gate trigger is applied. Mode *I*(+) is identical to conventional SCR operation.

The maximum trigger current rating published in the triac rating sheet is the largest value of gate current required to trigger the triac in any operating mode.

The method of triggering a triac is very similar to that for the SCR. Figure 10–23 shows a basic circuit where the variable resistance in the gate circuit controls the conduction angle of the triac. As the resistance is decreased, the gate current increases until the triac turns ON at both the peak positive (90 degrees) and peak negative (270 degrees) points on the voltage waveform. The conduction angle can be increased by appropriate reduction of *R*. Nearly 360 degrees can be obtained by moving the firing point back from 90 degrees to zero on the positive half-cycle, and from 270 degrees to 180 degrees on the negative half-cycle.

Triacs are used extensively in full-wave power control applications, to supply either fixed or variable power to the load. Fixed load power is delivered when the triac is used as a static OFF–ON switch in the fixed-phase firing mode,

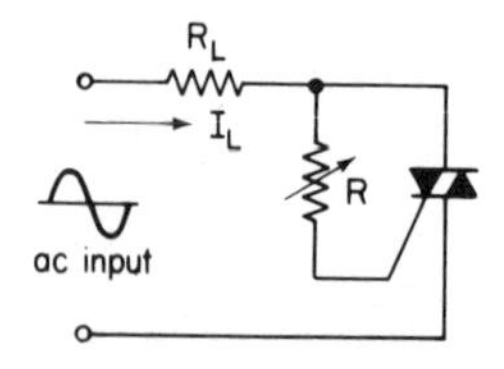

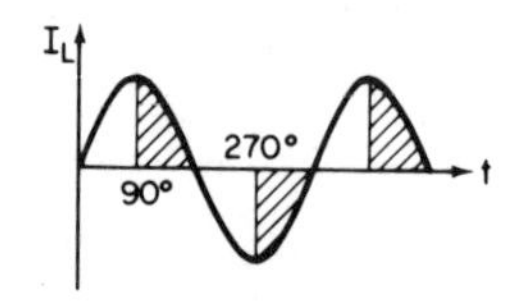

Fig. 10–23 Basic triac gate trigger circuits.

which supplies any desired portion of the input voltage to the load. A simple *fixed-phase control* circuit is shown in Fig. 10–24. In this circuit, the triac is fired at fixed points on the positive and negative half-cycles of the ac input waveform, as determined by the time constant of the *RC* network in the gate circuit. Commutation of the triac, taking place at the zero crossing of each half-cycle, can pose problems with inductive loads, where the transition from the ON state to the OFF state can produce large voltage transients which tend to drive the triac into premature conduction.

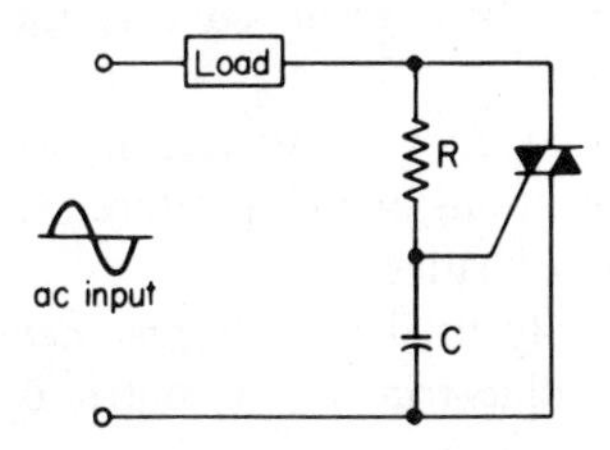

Fig. 10–24 *RC* network in fixed-phase control of triac conduction.

Variable or *proportional control* of load power can be obtained with the circuit of Fig. 10–25, which shows a low-current triac in series with the load. The gate trigger circuit consists of a two-section *RC* network and a neon-bulb threshold device. The time constant of the *RC* network can be varied by R_1 and allows triggering of the triac over a large range, from very small conduction angles to almost the complete 360-degree cycle. The neon bulb, requiring a very precise and stable voltage to ignite, serves as a threshold device to the triac gate and removes any dependence of the trigger circuit on variations in the gate trigger characteristics. In the usual application, however, a diac is the preferred threshold device.

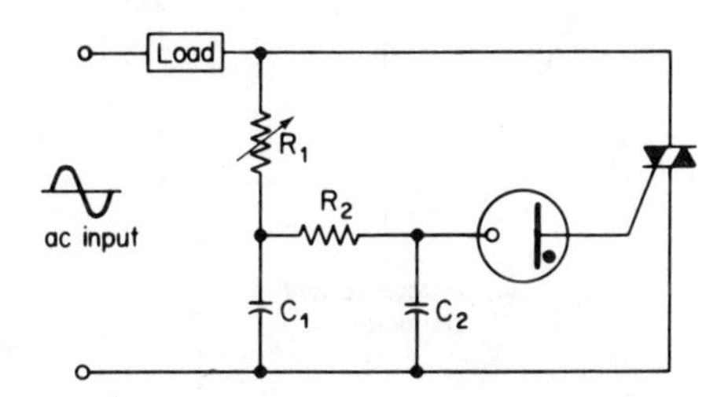

Fig. 10–25 Triac proportional control circuit using a neon bulb threshold device.

10-5.7 Diac: Operation and Applications

The diac is a two-terminal, three-layer *pnp* device, which can be made to conduct in either direction. Figure 10-26 shows the basic junction arrangement, the schematic symbol, and the voltage-current characteristic.

Voltage of either polarity between the diac terminals initially causes a small leakage current across the *pn* junction which is reverse-biased. When the applied voltage exceeds the breakdown voltage of that reverse-biased junction, the junction enters the avalanche breakdown region and the diac current suddenly increases sharply. In this ON-state condition, the voltage across the diac decreases with increasing current, so that the diac exhibits a negative resistance characteristic.

Unlike the bipolar junction transistor, the two *pn* junctions of the diac have equal doping concentrations, which produces symmetrical switching characteristics for both voltage polarities.

The diac is primarily used as a *trigger device* in triac power control circuits. A basic circuit for diac-triac phase control is shown in Fig. 10-27. The

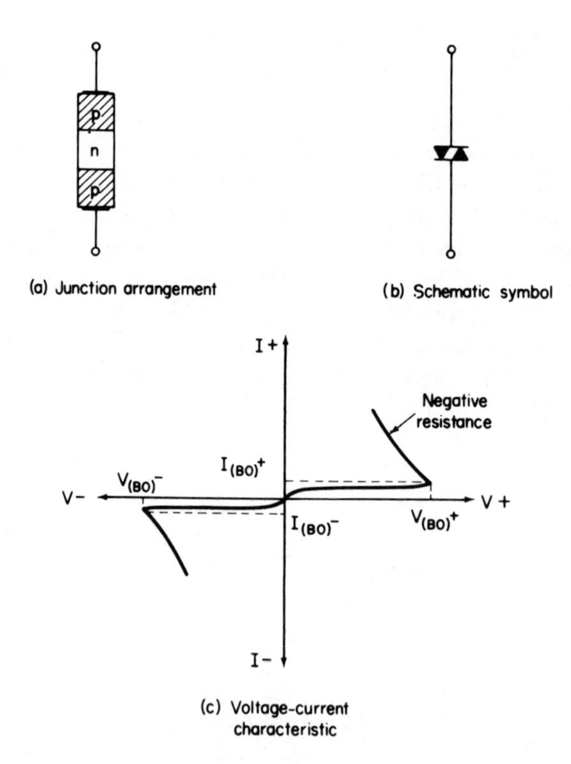

Fig. 10-26 Junction diagram, schematic symbol, and voltage-current characteristic of a diac.

diac switches ON when the capacitor voltage reaches the diac breakover voltage $V_{(BO)}$ and the diac then supplies gate current to the triac. The conduction angle of the triac is controlled by the time constant of the RC network in the diac circuit.

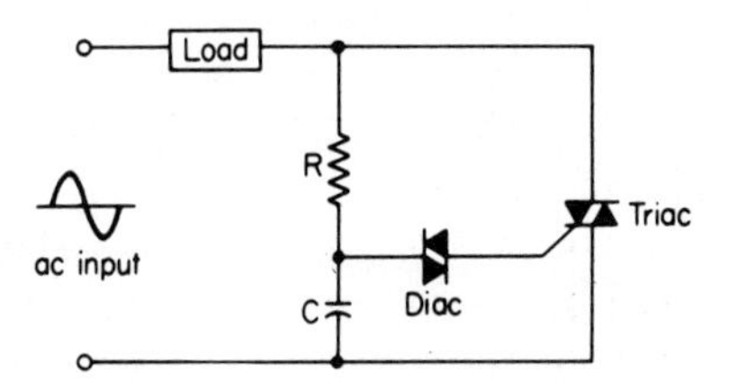

Fig. 10–27 General circuit of diac-triac phase control.

In the typical application of Fig. 10–28, the diac-triac combination is used as a *proportional speed control* for a universal motor. This single time constant circuit is designed to control motor speed over a limited range and is useful in applications such as fan and blower motor controls, where a small change in motor speed causes a relatively large change in air velocity. The R_1–C_1 time constant controls the point on the input voltage waveform at which the diac fires and supplies the gate current necessary to switch the triac ON. The triac is switched OFF again when the load current is reduced to zero.

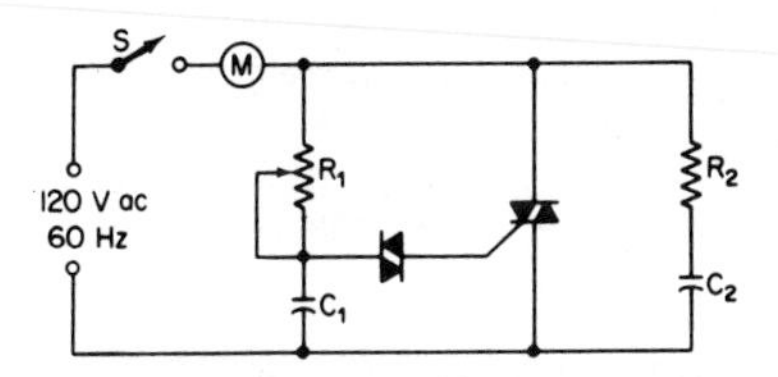

Fig. 10–28 Proportional speed control circuit for a universal motor.

Since motors are basically inductive loads, the applied input voltage is not in phase with the load current and the triac current therefore goes through its zero crossing when the applied voltage has not yet reached zero. This *commutating* voltage, appearing across the triac terminals at the moment that its current is zero, may have a rate of rise (dv/dt) which is high enough to retrigger the triac. The R_2–C_2 network, connected across the triac, limits the commutating dv/dt and tends to prevent unwanted triac triggering.

10–6 SILICON-CONTROLLED SWITCH

10–6.1 Operation

The silicon-controlled switch (SCS) is a unilateral four-layer silicon device, with electrodes connected to the four terminals, as shown in the basic junction diagram of Fig. 10–29(a). The four electrodes are known as anode (4), anode gate (3), cathode gate (2), and cathode (1).

When the SCS is manufactured by the planar technique, it resembles an *npn* transistor surrounded by a *p*-diffused ring which forms a *pnp* transistor across the surface, as shown in the sectional view of Fig. 10–29(b). The planar technique makes all four terminals readily accessible, and the device enjoys the inherent parameter stability of all planar silicon structures. Figure 10–29(c) shows the schematic symbol of an SCS.

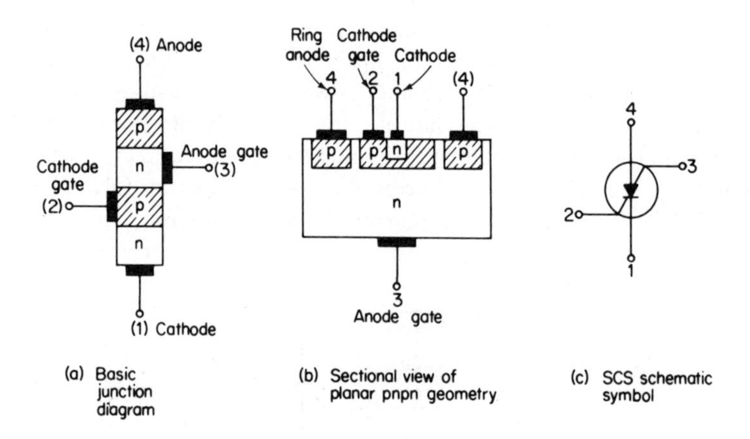

Fig. 10–29 Junction arrangement and schematic symbol of the SCS.

The basic junction diagram of Fig. 10–29(a) suggests the equivalent circuit of Fig. 10–30(a), which shows an *npn* transistor with a diode in the collector circuit. Normal transistor action requires that the collector (3) is positive with respect to the emitter (1). To allow SCS current, the diode must be forward-biased, and the anode terminal (4) must therefore be positive with respect to the collector terminal (3). This reasoning leads to the SCS bias conditions and we recognize three possible operating modes, which are summarized in Fig. 10–30.

In the *forward-blocking mode* of Fig. 10–30(b), the anode (4) is positive, but the reverse-biased base–emitter junction (2) keeps the transistor cut off. (In lieu of negative V_{BE}, the base can be left disconnected or short-circuited to the

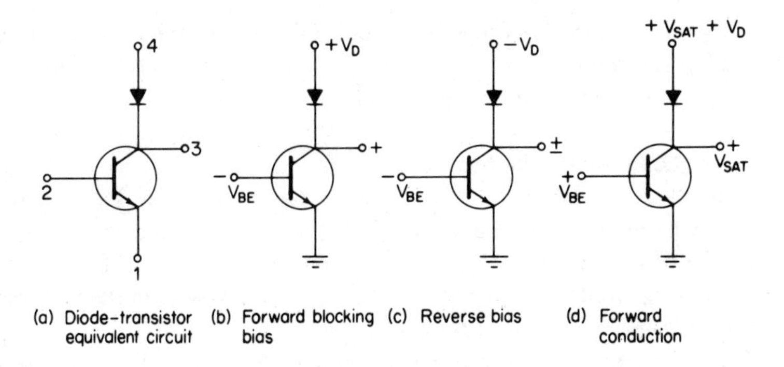

Fig. 10–30 Biasing a *pnpn* device.

emitter.) The maximum blocking voltage, defined as the maximum permissible positive anode voltage, is determined by the collector-to-base breakdown voltage.

In the *reverse-bias mode* of Fig. 10–30(c), the anode is negative with respect to the emitter, the diode is reverse-biased, and there is no anode current. The maximum reverse voltage, defined as the maximum permissible negative anode voltage, is determined by the diode breakdown voltage plus the emitter breakdown voltage of the transistor.

In the *forward-conduction mode* of Fig. 10–30(d), the anode is positive and the SCS is made to conduct by the positive bias on the base. The anode voltage and collector voltage differ only by the forward voltage drop across the conducting diode.

A different view of the SCS is obtained when we consider the two-transistor equivalent circuit of Fig. 10–31(b), which is derived from the basic junction diagram of Fig. 10–31(a). This equivalent circuit considers the SCS as a *complementary pair with regenerative feedback,* which switches the device ON when the product of the transistor gains is greater than unity. Base current I_B into the *npn* transistor is multiplied by the *npn* current gain β_1 and becomes base

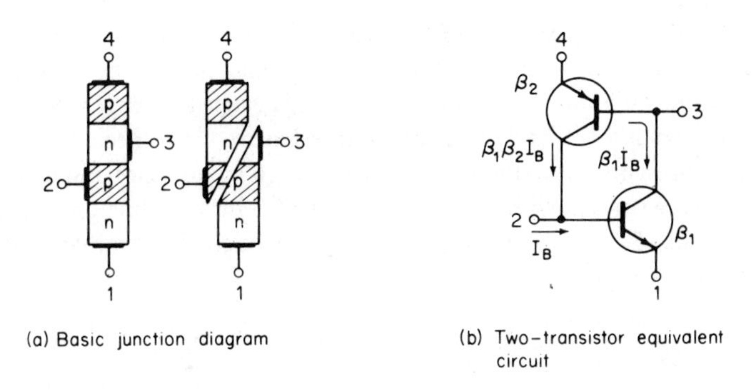

Fig. 10–31 Development of the two-transistor SCS equivalent circuit.

current $\beta_1 I_B$ for the *pnp* transistor. After multiplication by the *pnp* current gain β_2, it becomes the reinforcing base current $\beta_1\beta_2 I_B$ for the *npn* transistor. If this reinforcing base current exceeds the initial base current ($\beta_1\beta_2 \geq 1$), regeneration takes place and both transistors are driven into saturation. This condition is known as the ON state for the *pnpn* device.

The product of the two transistor gains is the critical factor which determines whether the SCS switches ON. Since the β of a transistor is a function of collector current I_C, collector voltage V_{CE}, base bias V_{BE}, and temperature T, it should be realized that the ON or OFF state of the SCS can be affected by changes in any one of these parameters.

For example, the typical forward characteristic of Fig. 10–32 shows that a reduction in anode current switches the SCS OFF. This is due to the fact that at low anode currents the transistor betas decrease, so their product becomes less than 1 and regeneration ceases.

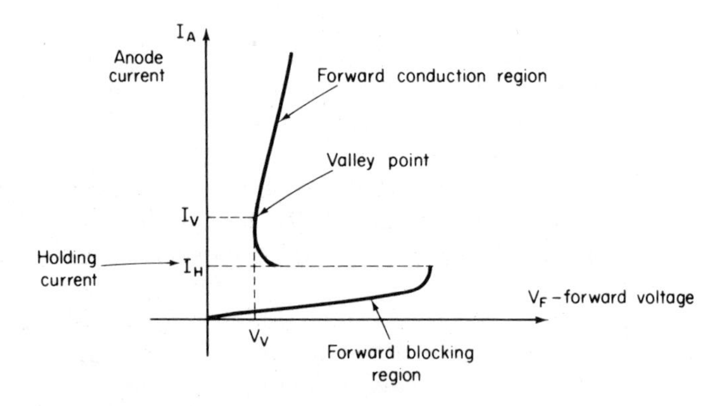

Fig. 10–32 Typical forward characteristic of an SCS, indicating that the device switches OFF when the anode current drops below the holding current I_H.

The SCS can be switched ON accidentally if the anode voltage is applied suddenly, or if the SCS is subjected to high-frequency transients. This phenomenon is known as the *rate effect.* The rate effect is caused by internal capacitance between electrodes (2) and (3), called interbase capacitance. The interbase capacitance presents a low-impedance path at high frequencies and can result in substantial base current which may trigger the SCS into conduction. If the circuit is such that the rate effect causes device triggering, it can be eliminated by proper circuit design.

10–6.2 Trigger Circuits

The equivalent circuits of Section 10–6.1 point out that the SCS can be used as a conventional transistor, simply by ignoring the anode connection.

However, because of its inherent regenerative properties and high triggering sensitivity, the SCS is mainly used as a switching device which can be turned ON or OFF at will.

The SCS turns ON readily, simply by satisfying the stated voltage requirements for conduction. Figure 10–33 summarizes some typical circuits which can be used to trigger the SCS ON.

In Fig. 10–33(a), the cathode gate is left open and the voltage at the anode gate is set by the R_1–R_2 voltage divider to a triggering threshold of $[R_1/(R_1 + R_2)]V$ volts. When the anode voltage rises above this threshold the SCS will trigger ON.

Figure 10–33(b) is a variation of Fig. 10–33(a), using conventional transistor bias stabilization. When the anode voltage exceeds the stabilized collector voltage, the transistor saturates.

In Fig. 10–33(c) the Zener diode in the cathode circuit sets the threshold voltage. A positive input to the cathode gate, exceeding the Zener voltage, triggers the SCS ON.

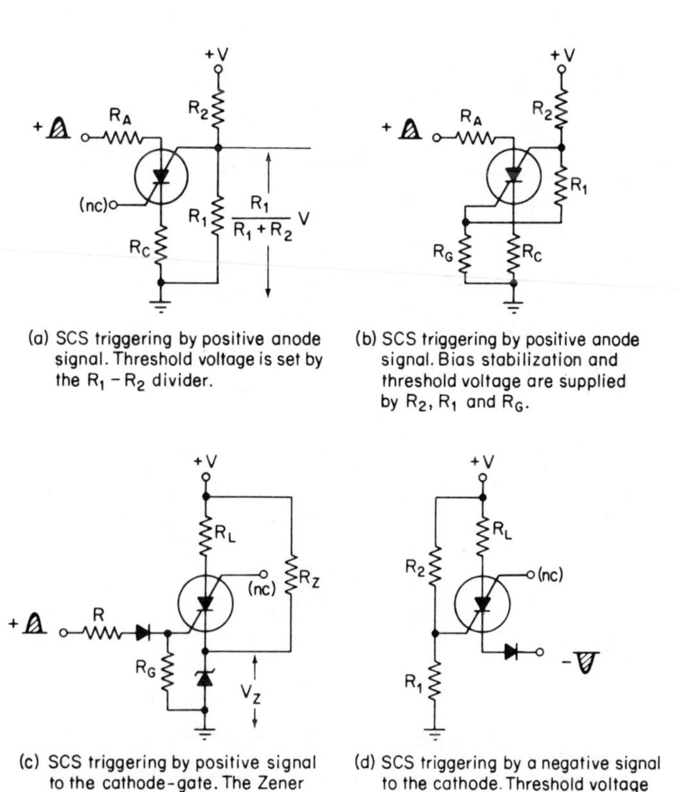

(a) SCS triggering by positive anode signal. Threshold voltage is set by the $R_1 - R_2$ divider.

(b) SCS triggering by positive anode signal. Bias stabilization and threshold voltage are supplied by R_2, R_1 and R_G.

(c) SCS triggering by positive signal to the cathode-gate. The Zener diode provides the threshold voltage.

(d) SCS triggering by a negative signal to the cathode. Threshold voltage to the cathode-gate is supplied by R_1 and R_2.

Fig. 10–33 Various methods of switching the SCS to the ON state.

Figure 10–33(d) shows that the SCS can be triggered by applying a negative waveform of sufficient amplitude to the cathode terminal.

Silicon-controlled switches turn readily ON, but they are much more difficult to switch OFF again. Figure 10–34 summarizes a number of turnoff methods. The simplest turnoff is achieved by operating from ac, as in Fig. 10–34(a). The *pnpn* device turns OFF when the anode becomes negative. When operating from a dc supply, the anode can be opened by a switch, as in Fig. 10–34(b). Removal of the anode current stops transistor action and turns the SCS OFF. Resistor R_{RE} suppresses the rate effect when the switch is closed again.

The gates can also be used to turn the device OFF. In Fig. 10–34(c), a negative cathode-gate input is used, while in Fig. 10–34(d) a positive anode-gate pulse is applied. In both instances, the gate pulse applies reverse bias to the *pn* junction and stops transistor action, so the saturated transistor turns off as the current carriers recombine. A positive anode-gate pulse in Fig. 10–34(c), where the load is in the anode circuit, represents a poor method of turning the SCS off,

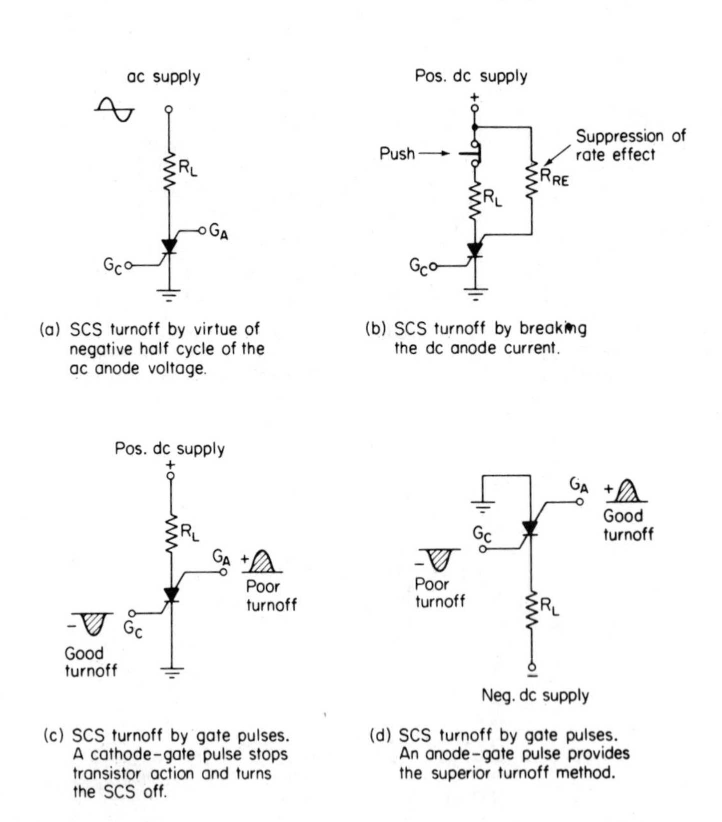
ac supply
R_L
G_A
G_C
(a) SCS turnoff by virtue of negative half cycle of the ac anode voltage.
Pos. dc supply
+
Push
Suppression of rate effect
R_RE
R_L
G_C
(b) SCS turnoff by breaking the dc anode current.
Pos. dc supply
+
R_L
G_A
+
Poor turnoff
-
G_C
Good turnoff
(c) SCS turnoff by gate pulses. A cathode-gate pulse stops transistor action and turns the SCS off.
G_A
+
Good turnoff
G_C
-
Poor turnoff
R_L
Neg. dc supply
(d) SCS turnoff by gate pulses. An anode-gate pulse provides the superior turnoff method.

Fig. 10–34 SCS turnoff methods.

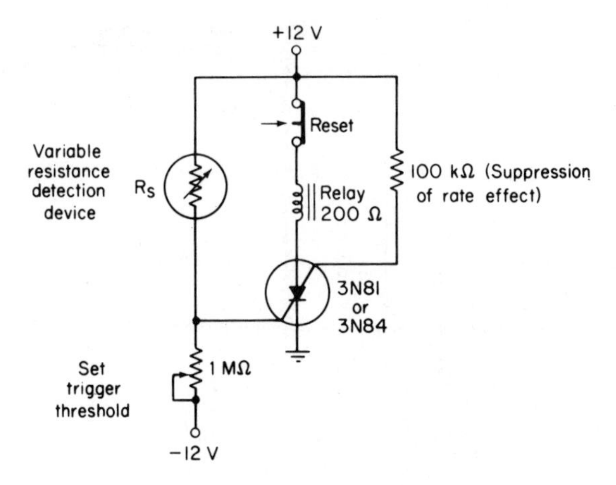

Fig. 10–35 Alarm circuit, where the SCS is triggered by sensing the change in voltage produced by a decrease in the resistance of R_S. (*Courtesy General Electric Company.*)

since it does not reverse bias the anode junction. Similarly, a negative cathode-gate pulse in Fig. 10–34(d), where the load is in the cathode circuit, does not reverse bias the cathode junction and poor turnoff results can be expected.

10–6.3 Applications

The number of circuit applications is limited only by the designer's imagination and resourcefulness. The following circuits provide an insight into SCS usage and show how the device is switched from the ON state to the OFF state.

In the *alarm circuit* of Fig. 10–35 R_S represents a temperature-, light-, or radiation-sensitive resistor which, together with the potentiometer, sets the voltage at the cathode gate. When R_S decreases sufficiently to forward bias the SCS, the alarm relay is energized. The circuit is reset by pushing the reset button momentarily, breaking the anode current path. The 100-kΩ resistor across the reset button suppresses the rate effect. The circuit can readily be modified to accept increasing values of R_S for SCS triggering, by interchanging R_S and the potentiometer.

In the *neon-lamp driver* circuit of Fig. 10–36, the *pnpn* device operates purely as an SCR, which is fired by a positive trigger input to the cathode gate. Notice that the anode gate is not connected (*nc*). With a dc supply source, the SCS latches in the ON condition but can be reset by applying a −1-V reset pulse to the cathode gate via a diode. When an ac supply is used, the circuit is nonlatching since the SCS turns OFF when the anode current is reduced to zero.

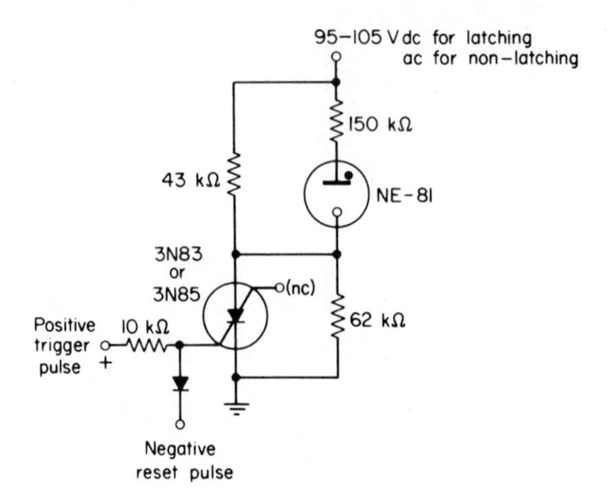

Fig. 10–36 Neon lamp driver circuit, where a positive gate pulse triggers the SCS and a negative gate pulse resets the circuit. (*Courtesy General Electric Company.*)

The SCS also finds application in logic circuits, such as the *pulse coincidence detector* of Fig. 10–37. When positive inputs A and B (2 to 3 V amplitude) occur simultaneously, both SCSs fire and an output voltage is generated across R_L. The SCS triggering is so sensitive that less than 1 μs overlap of pulse inputs A and B is required to trigger the devices.

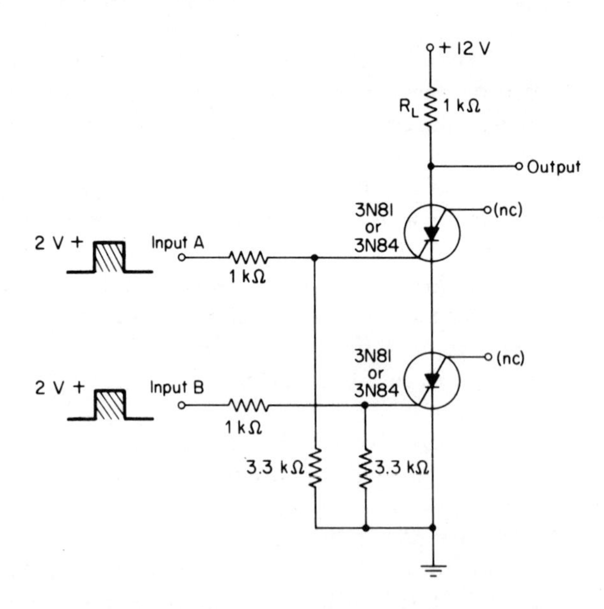

Fig. 10–37 Pulse coincidence detector. (*Courtesy General Electric Company.*)

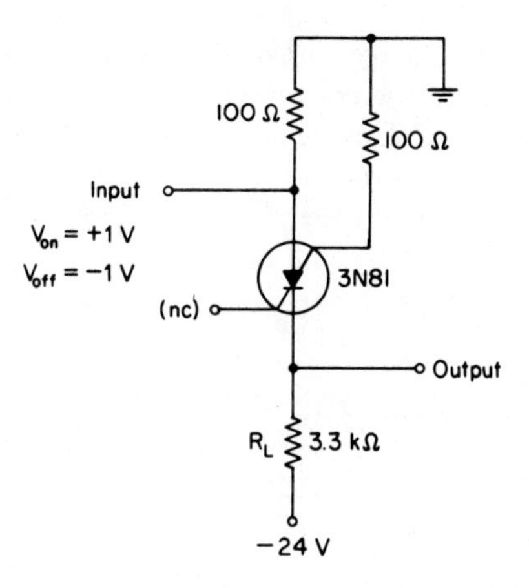

Fig. 10–38 Bistable memory element. (*Courtesy General Electric Company.*)

Figure 10–38 shows the SCS as a *bistable memory cell.* The circuit operates from a negative supply, with the load in the cathode circuit. A positive input pulse to the anode latches the SCS in the ON state and provides a positive output. A negative input pulse reverse biases the anode junction and the device switches OFF, producing a negative output.

Figure 10–39 shows the SCS in an application as a *relaxation oscillator.* When the switch is first closed, capacitor C charges through inductor L to a

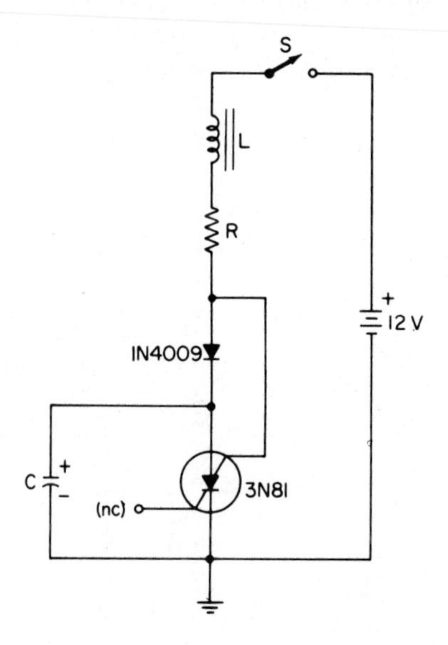

Fig. 10–39 *RLC* oscillator. (*Courtesy General Electric Company.*)

voltage above the supply voltage, provided that the Q of the RLC circuit is sufficiently high. When the current reverses, the diode blocks so that the SCS anode is at a higher potential than the anode gate. This triggers the SCS and the capacitor starts to discharge through the SCS. During the discharge cycle, the anode gate approaches ground potential and deprives the anode of holding current. This turns the SCS OFF and the capacitor charges again to repeat the cycle.

10–7 UNIJUNCTION TRANSISTOR

10–7.1 Operation and Characteristics

The unijunction transistor (UJT) is a three-terminal device whose characteristics are quite different from those of a conventional bipolar transistor. The junction diagram of Fig. 10–40(a) shows that the UJT consists of an n-type silicon type bar with ohmic contacts at either end and a pn junction along one side of the bar. The end terminals are called base one ($B1$) and base two ($B2$), and the pn junction is called the emitter (E).

A cross-sectional view of an actual bar-type UJT structure is shown in Fig. 10–40(b). A ceramic disc with the same thermal expansion coefficient as silicon is used as a mounting platform. A single crystal of n–type silicon, with typical dimensions of $8 \times 10 \times 60$ mils, is attached to this ceramic disc. The two base contacts are attached to the silicon bar as indicated. A 3-mil aluminum wire, alloyed to the top of the bar nearest the $B2$ terminal, forms the pn emitter junction. Leads are attached to the three elements, and the entire structure is passivated and encapsulated.

The schematic symbol of the UJT and a simplified equivalent circuit are shown in Fig. 10–41. The ohmic resistance of the silicon bar between the $B1$ and

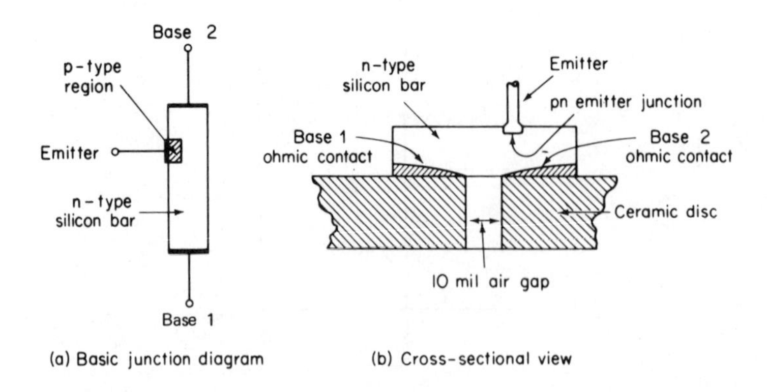

Fig. 10–40 Junction diagram and cross-sectional view of a bar-type unijunction transistor.

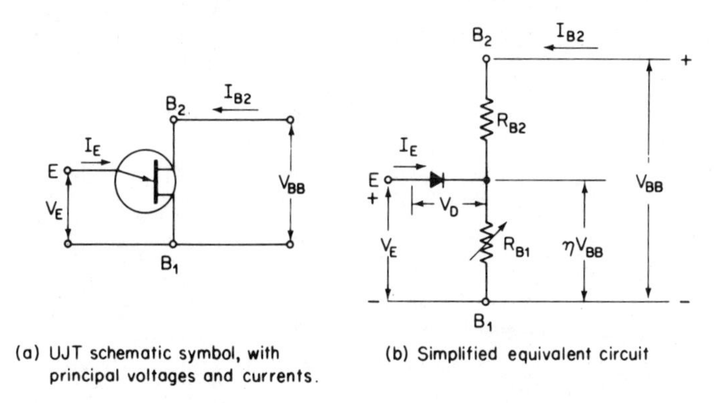

Fig. 10–41 Schematic symbol and equivalent circuit for the UJT, including the principal voltages and currents.

$B2$ terminals is called the *interbase resistance* R_{BB} and is represented by the sum of resistances R_{B1} and R_{B2}. The interbase resistance of a UJT is typically on the order of 5 to 10 kΩ. The emitter is represented by the diode connected to the R_{B1}–R_{B2} junction.

In the normal method of operation, a positive supply voltage V_{BB} is applied to the $B2$ terminal, while the $B1$ terminal is connected to ground. With the emitter terminal open, the silicon bar acts like a resistive voltage divider which develops a fraction η of the supply voltage at the emitter junction. If the emitter voltage V_E is less than ηV_{BB}, the emitter diode is reverse-biased and only a small reverse current flows across the junction. If V_E is greater than ηV_{BB}, the emitter diode is forward-biased and current flows between the emitter and base one. This emitter current consists of holes which are attracted to the $B1$ terminal. With holes flowing toward $B1$, charge-compensating electrons are supplied by V_{BB}. These excess electrons lower the resistance of R_{B1} and this causes a reduction in the voltage across R_{B1}. Therefore, an increase in emitter current is accompanied by a decrease in emitter voltage, which results in the negative resistance characteristic of Fig. 10–42.

The *negative resistance region* is delineated by the peak point and the valley point. To the left of the peak point is the *cutoff region,* where V_E is smaller than ηV_{BB} and the device supports only the small reverse current. To the right of the valley point is the *saturation region,* where the E-$B1$ section acts like a normal forward-biased diode and I_E increases steeply.

The UJT specification sheet lists the absolute maximum ratings of power dissipation, currents and voltages, and temperature ranges. The electrical characteristics include the following important parameters:

Interbase resistance R_{BB} is the resistance measured between the $B1$ and $B2$ terminals, with the emitter open. R_{BB} is on the order of 5 to 10 kΩ and increases with temperature at the rate of approximately 0.8 percent/°C.

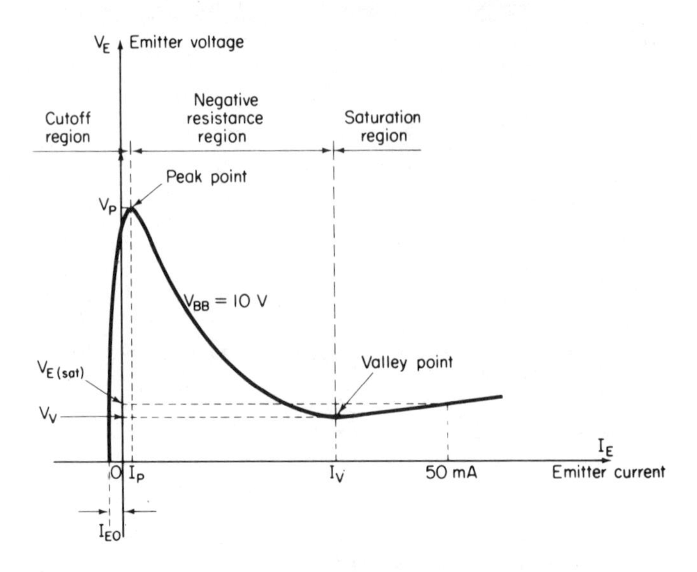

Fig. 10–42 Emitter characteristic curve of a typical UJT, showing some of the important parameters.

Intrinsic standoff ratio η is defined in terms of the peak-point voltage V_P by the equation $V_P = \eta V_{BB} + V_D$, where V_D is the forward voltage drop across the emitter diode. η is essentially constant with temperature.

Peak-point current I_P is the emitter current at the peak point, required to trigger the UJT into conduction. *Peak-point voltage* V_P is the emitter voltage at the peak point, and depends on the intrinsic standoff ratio and interbase voltage V_{BB}. *Valley-point current* I_V and *valley-point voltage* V_V define the location of the valley point. Both I_V and V_V increase with increasing interbase voltage.

Emitter saturation voltage $V_{E(\text{sat})}$ indicates the forward voltage drop from emitter to base one in the saturation region. *Emitter reverse current* I_{EO} is measured with an applied voltage between base two and emitter, with base one open. This current varies in the same way as the I_{CO} of a bipolar transistor.

10–7.2 Relaxation Oscillator

The UJT relaxation oscillator is the basic circuit for many applications such as sawtooth generators, pulse generators, timing circuits, and triggering circuits.

Figure 10–43 shows the basic configuration of a UJT relaxation oscillator. When initially switch S is closed, capacitor C_T charges through resistor R_T, and emitter voltage V_E rises exponentially toward the supply voltage V_{BB}. When V_E reaches the peak voltage V_P of the UJT, the emitter diode becomes forward-biased and the UJT triggers on, providing a low-resistance path between E and $B1$. The capacitor then discharges through the E-$B1$ path and V_E decreases rapidly. When V_E can no longer sustain the bias required for UJT conduction [$V_{E(\min)}$ is approximately 0.5 $V_{E(\text{sat})}$], the E-$B1$ discharge path is broken and the charging cycle repeats.

The frequency of oscillation is determined by the magnitudes of timing resistor R_T and timing capacitor C_T, as well as by the peak-point voltage of the UJT. For small values of resistance R_1 and R_2 in Fig. 10–43, the frequency of oscillation is given by

$$f \cong \frac{1}{R_T C_T \ln\left(\dfrac{1}{1-\eta}\right)} \tag{10-4}$$

where η is the intrinsic standoff ratio of the UJT.

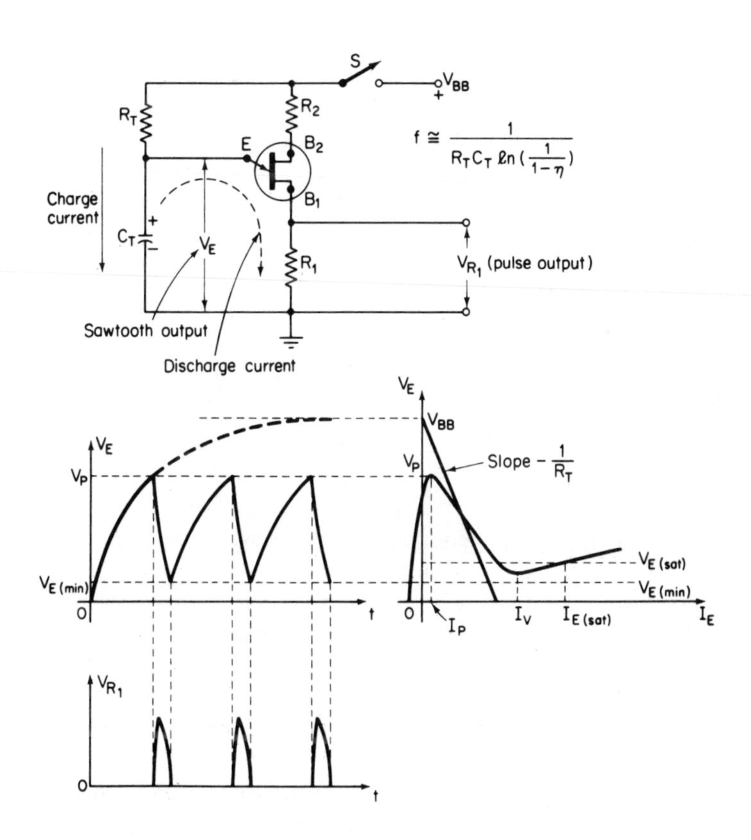

Fig. 10–43 Basic relaxation oscillator with typical waveforms.

The various waveforms in the circuit are included in Fig. 10–43. We observe that the emitter voltage resembles a sawtooth waveform which shows a convex curvature due to the nonlinear capacitor charging current. Base one terminal provides a positive pulse output; base two terminal provides a negative pulse output.

There are two basic conditions which must be satisfied for reliable oscillator operation. The first condition is that the dc loadline, defined by V_{BB} and R_T, must intersect the emitter characteristic to the right of the peak point, as indicated in Fig. 10–43. This ensures that the R_T can supply sufficient current to the emitter to trigger the UJT. This condition is expressed by the relation $(V_{BB} - V_P)/R_T > I_P$. The second condition is that the dc loadline must intersect the characteristic to the left of the valley point. If this condition is not satisfied, the loadline intersects the curve in the saturation region and the UJT may not turn off after the first cycle. This condition is expressed by $(V_{BB} - V_V)/R_T < I_V$.

Although these conditions impose limitations on the magnitude of the timing resistor R_T, they are not severe. With a typical maximum value of $I_P = 12$ μA and a typical minimum value of $I_V = 8$ mA, R_T still has an allowable range of approximately 1000:1.

Finally, when the size of the timing capacitor C_T becomes small (say below 0.01 μF), the amplitude of the emitter voltage waveform decreases to the extent that the frequency stability of the oscillator is affected.

10–7.3 Linear Sawtooth Generator

To improve the linearity of the sawtooth waveform of the basic UJT relaxation oscillator the *bootstrap* method is often used. In a bootstrap circuit the timing capacitor is charged with a *constant current* and the voltage across the capacitor therefore rises linearly.

In the 50-kHz sawtooth generator of Fig. 10–44, a constant voltage is maintained across the charging resistor by the Zener diode and the emitter-follower amplifier *Q*2. The constant voltage across R_T maintains a constant charging current into C_T over the entire cycle.

Transistor *Q*2 performs a double function in that it not only serves as the driver for the bootstrap circuit but also as the output amplifier stage. The 3.9-kΩ load resistor of *Q*2 is returned to the negative supply voltage to prevent clipping at the bottom of the sawtooth waveform.

Transistor *Q*3 acts as a constant-current generator for the Zener diode to improve its linearity and also assists *Q*2 in supplying current to the load.

The frequency of oscillation can be adjusted by the 5-kΩ potentiometer over a 5:1 range and the circuit generates a linear sawtooth waveform up to 50 kHz.

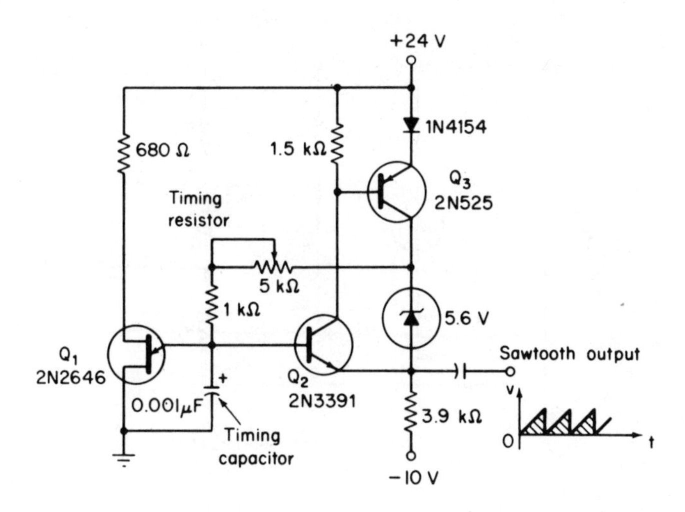

Fig. 10–44 Linear sawtooth generator. (*Courtesy General Electric Company.*)

10–7.4 Synchronization of the Relaxation Oscillator

The UJT relaxation oscillator can be synchronized to an external frequency source by positive pulses at the emitter or negative pulses at the base two terminal. The amplitude of these triggering pulses must be large enough to reduce the peak-point voltage V_P below the instantaneous emitter voltage V_E.

In power control applications, where a UJT can be used to trigger an SCR into conduction, it is often desirable to synchronize the UJT to the line frequency. A basic ac line-synchronized trigger circuit is shown in Fig. 10–45.

The Zener diode is used to regulate the peaks of the full-wave rectified ac, as indicated in the waveform diagram, and provides a practically constant voltage to the R_T–C_T timing circuit. Capacitor C_T charges exponentially to provide the UJT emitter voltage V_E. At the end of each half-cycle, the supply voltage to the UJT drops to zero, and the UJT peak-point voltage $V_P = \eta V_{BB}$ also falls to zero. *Any* charge on the capacitor then forward-biases the emitter diode and the capacitor discharges through R_1 to provide an output trigger pulse. It is important to note that the capacitor is discharged at the end of each half-cycle and the trigger circuit is therefore synchronized to the line frequency.

10–7.5 Regulated AC Power Supply

Figure 10–46(a) shows a regulated ac supply which uses a UJT trigger circuit to control the conduction angles of two SCRs in a full-wave configuration.

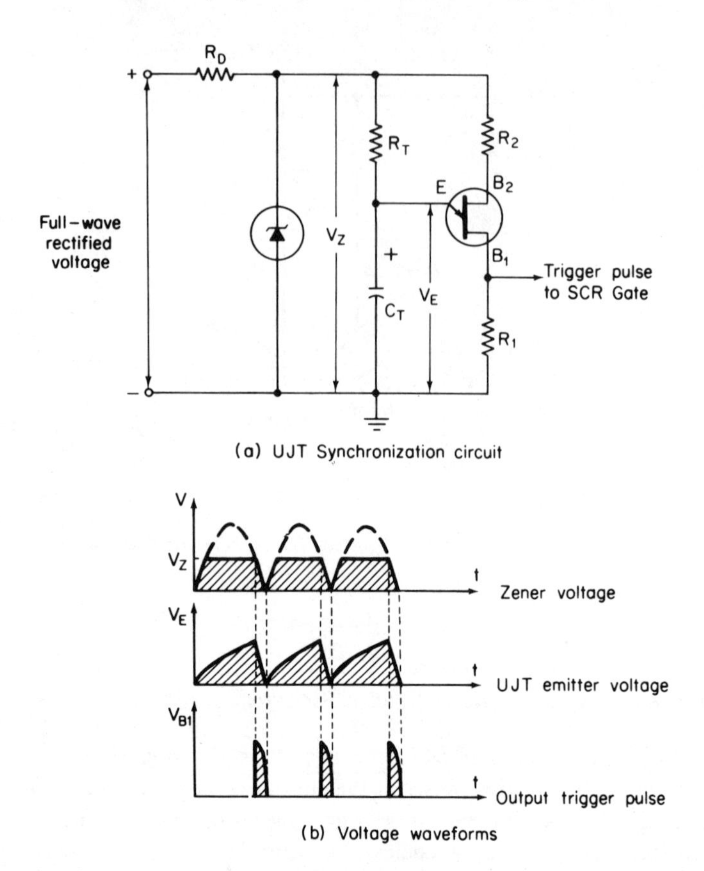

Fig. 10–45 Line synchronization of the UJT relaxation oscillator.

The two SCRs are forward-blocking to the ac line voltage and are made to conduct at the desired phase angle by triggering pulses from the UJT.

As in the synchronized trigger circuit of Section 10–7.4, the charging voltage for timing capacitor C_T is essentially constant and equal to the Zener voltage V_Z. The UJT emitter voltage V_E therefore follows the normal exponential characteristic of the rising capacitor voltage. The UJT triggers when emitter voltage V_E equals the peak-point voltage V_P and then provides a firing pulse of the correct polarity to the appropriate SCR via pulse transformer T. Each SCR in turn conducts for a controlled portion of the ac input waveform and delivers a certain amount of ac power to the load. Figure 10–46(b) shows the various waveforms in the circuit and indicates their time relationship.

The regulation ability of the circuit is derived from the presence of resistor R_2. The interbase voltage V_{BB} of the UJT equals the Zener voltage V_Z plus a fraction of the rectified line voltage as determined by the voltage dividing ratio of R_1 and R_2. If therefore the ac line voltage rises, V_{BB} also increases. This raises peak-point voltage V_P, since V_P is determined by the intrinsic standoff ratio η (a device parameter that is constant) and V_{BB} (a circuit variable). With a higher peak-point voltage, the triggering of the UJT is delayed until V_E equals

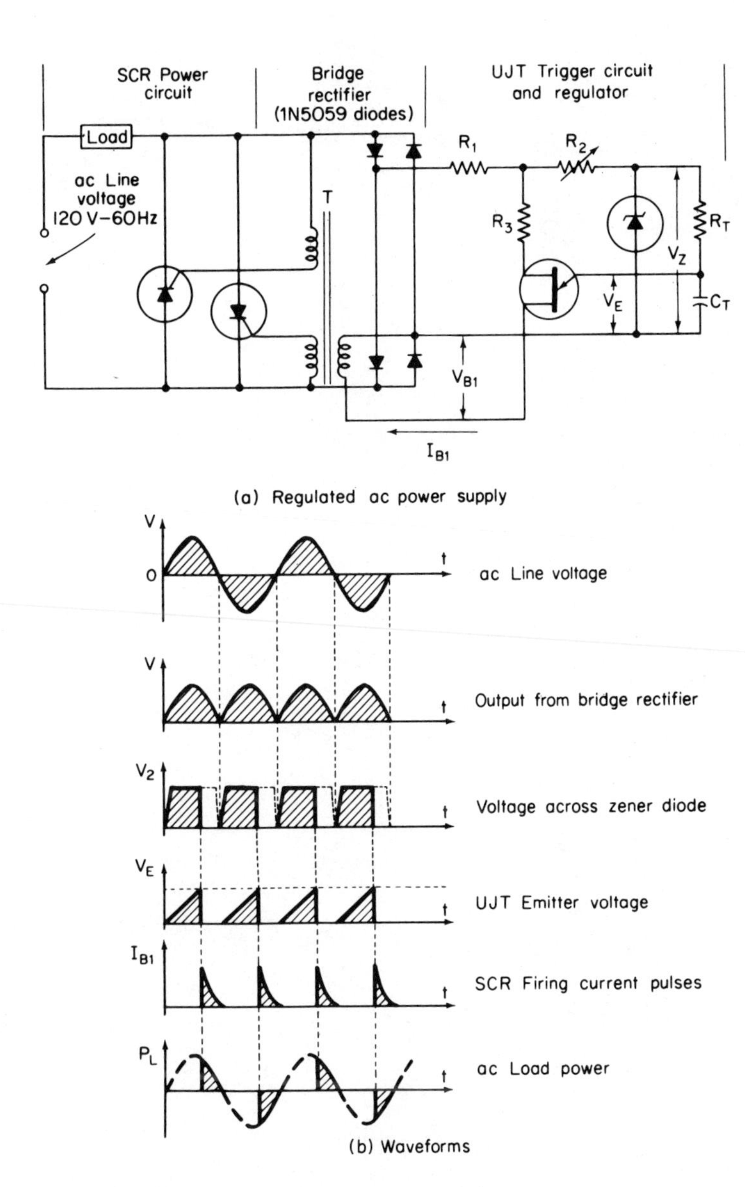

Fig. 10–46 UJT phase control of a regulated power supply.

this new value of V_P. A delay in the SCR trigger pulses decreases their conduction angles and this decreases the ac load power. The reduction in load power compensates for the increase in load power which would have occurred with the higher line voltage. Proper selection of the R_1–R_2 voltage division ratio produces compensation over a limited range of ac line voltage variations.

Chapter 11

Analog and Digital Computers

11–1 INTRODUCTION

Mankind's attempt to escape the drudgery of mathematical computation began millenia ago. An early mechanical aid, the abacus, of ancient Greek origin, relegated in our society to the role of a child's toy, is still being used to advantage in many areas of the world. Skilled operators of the device until fairly recently were able to keep up with our computer operators. The abacus was followed through the centuries by many ingenious mechanical devices, culminating in desktop mechanical calculators, which are now going out of use. There are two types of electronic computers—the analog and the digital. Within these types we also differentiate between two classes of computers: general purpose machines, designed to handle all kinds of computations, and special purpose computers, which are designed to perform only specific computations, with the borderline between these not always easy to define. The analog computer is gradually being totally displaced by the digital machine, but for an adequate understanding of digital computers, at least a brief look at the analog machines is essential.

11–2 ANALOG VERSUS DIGITAL

11–2.1 Analog Representation

The devices and circuits that have been described and analyzed in Chapters 3 through 6 have as their basic purpose the amplification of voltages and currents in such a way as to preserve their shape as nearly as possible. This means that the amplification has to be as linear as possible. If, for example, music is to be reproduced, then the voltages to be amplified represent the tones involved. If a single note consisting of a pure sine wave of pressure versus time is applied to a microphone, there results a sine wave of voltage, which is then handled by the system, and hopefully reproduced at the output as a sine wave of pressure versus time (Fig. 11–1). If the voltage is plotted versus the pressure, however, a straight line results (Fig. 11–2)—i.e., the voltage varies linearly with the pressure. Another way of expressing this relationship is to state that the voltage is the analog of the pressure. In the devices that have been considered, it was assumed that the signal to be amplified was an analog signal, in the sense that the voltage represented the analog of some quantity.

11–2.2 Digital Representation

There is, however, an alternative method of representing the quantity: the digital method. Instead of transmitting a voltage that is directly and linearly proportional to a quantity, it is possible to "digitize" the quantity. It is first determined what value the quantity has at a given time, then that value is converted into digital form, using 1's and 0's.

If the sine wave of Fig. 11–1 is to be transmitted over a digital system, it must be sampled at various instants, such as *a, b, c, d,* etc., and each sample has to be digitally encoded. The digits are transmitted, and at the output they can be decoded to retrieve the value of the voltage, or they can be fed out directly. In an

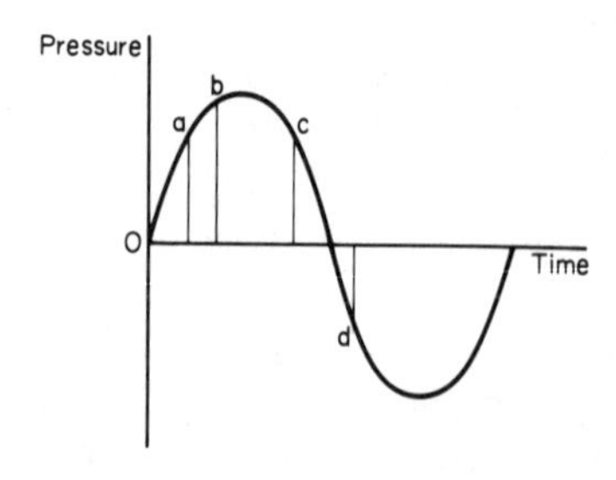

FIG. 11–1 Pressure vs. time.

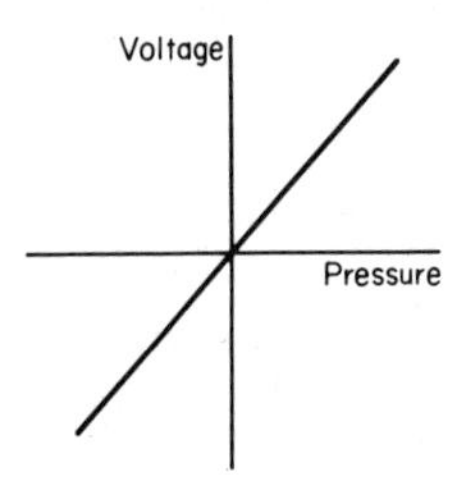

FIG. 11–2 Voltage vs. pressure.

analog system, the voltage is simply transmitted as it stands, an obviously simpler procedure, which has, however, the disadvantage of not providing certainty of accuracy.

When digital computers were first developed, the need arose for representing the numbers 1 and 0. This was originally done with mechanical relays, then with vacuum tubes, later with transistors, and today with integrated circuits. Even though the original computers were slow, bulky, unreliable, and energy consuming, and today's computers are fast, small, reliable, and energy efficient, there are nevertheless certain building blocks, which, though their physical realization has changed drastically, still remain true to their original basic nature.

It seems natural to assume that, when mankind began to count, fingers were used for that purpose, and this was a primitive digital system, as is also emphasized by the derivation of the word. Since there were ten fingers, a system based on 10 naturally resulted. For simple purposes, such a system sufficed. But as the tasks men could perform grew in complexity, this system became inadequate, and it was found that representing quantities by 1, 2, 3, etc., would need to be supplemented by fractions such as 1/2, 1/4, 1/3, etc., and later by decimals. A quantity could then be represented to any degree of accuracy by retaining enough significant figures after the decimal point. The decimal system served mankind well until the modern age, when it turned out that it was inadequate for the purposes then to be accomplished. Although mankind's learning to count with the aid of fingers led to an early digital system, it did not lead to the most advantageous digital system from the electronic viewpoint. A binary system, utilizing only the 1 and the 0, has certain advantages over the decimal system. While more units of information, or bits, are required, the advantage is that there are only two states between which a decision must be reached. Table 11–1 illustrates how numbers can be represented in number systems using various bases. For example, the number 1000 requires only 4 separate symbols in the decimal system, while it requires 10 symbols in the

TABLE 11-1

Decimal (10)	Binary (2)	Ternary (3)	Duodecimal (12)	Hexadecimal (16)
1	1	1	1	1
2	10	2	2	2
3	11	10	3	3
4	100	11	4	4
5	101	12	5	5
6	110	20	6	6
7	111	21	7	7
8	1000	22	8	8
9	1001	100	9	9
10	1010	101	t	t
11	1011	102	e	e
12	1100	110	10	w
100	1 100100	1 021	84	64
1000	1111 101000	1101 001	6e4	3f8

Where t = 10, e = 11, w = 12, f = 14

binary system. But it is far easier to distinguish between just the 1 and the 0 than to discriminate among the 10 different levels that may exist for each digit of the decimal system.

11-2.3 Comparison

When attempting to produce, amplify, transmit, and detect an analog signal, certain difficulties arise. As the signal proceeds through the various stages it must traverse from input to output; it is possible that nonlinearities may appear. That is, the voltage no longer accurately represents the quantity it is supposed to represent because the amplifiers are not perfectly linear. Secondly, there is always some noise in the system (Fig. 11-3). Noise interferes with the signal because it adds random voltages to the signal voltages, so that, again, the voltage no longer accurately represents the quantity. As the signal proceeds through a line or over the air, the signal-to-noise ratio decreases substantially. Thirdly, the detection process, which is, by its very nature, nonlinear, may not correctly reproduce the original quantity. And finally, at the output, it may be

FIG. 11-3 (a) Noise voltage. (b) Signal plus noise.

difficult to measure exactly what was transmitted—e.g., it may be difficult to tell the difference between 72.43 and 72.44.

As long as electronic circuits had only the functions of transmitting music, voice, or television, the requirement for an extremely accurate mode of transmission did not exist. But when the need arose for computation, for transmission of data over vast distances with great accuracy, and for noise-free reproduction, analog systems proved inadequate.

11-2.4 Numerical Analysis

It cannot be said that there were no accurate analog computers because excellent analog machines have been developed. But digital computers were able to outpace them with their accuracy, speed, and versatility. Returning to what was said above, suppose it becomes necessary to differentiate between 72.43 and 72.44. If the number 1 is to be represented by one volt, then it becomes necessary to differentiate between two voltages that differ from one another by only .01 volt, a formidable task. But if the number 72.43 is represented by 7243 pulses, a method has been devised of counting these pulses, giving virtual certainty that 7243 pulses have been counted, not 7244. Electronically, the problem has been greatly simplified since the circuits need only determine the presence or absence of pulses; they need not measure quantities, making the chances of inaccuracies extremely small. The noise immunity is also obtained. If one volt means there is a pulse, and zero volts means there is no pulse, then any noise pulse that is small with respect to one volt cannot affect the result. In an analog system, a noise pulse of .01 volt would produce an error since it would cause a reading of 72.44 or 72.42 instead of 72.43.

The advantages of a digital system over an analog system can be seen from the above numerical example. No advantage comes without an accompanying disadvantage, and in this case the disadvantages are the need of more electronics and the need for wide channels. But in any system where accuracy is important, the advantages outweigh the disadvantages. On the other hand, for systems where the required accuracy can easily be obtained by analog methods, analog circuits continue to find their applications.

11-3 THE ANALOG COMPUTER

11-3.1 General Approach

The analog computer, as its name indicates, employs voltages and currents to represent those quantities that are to be computed and those quantities that are the inputs and outputs—that is, accepting voltages and currents representing the inputs, it produces, by use of electronic analog methods, voltages and currents representing the outputs. To accomplish this purpose, a number of standard circuit configurations have been developed, which can be combined, or "patched," to suit a particular problem, so that not only special purpose analog computers but also general purpose machines can be constructed.

11–3.2 The Potentiometer

The simplest of all analog devices is the ordinary variable resistor, known also as the rheostat or potentiometer (Fig. 11–4). It is shown symbolically as a resistor with an arrow, but actually it is ordinarily either a single-turn or

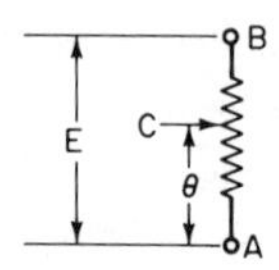

FIG. 11–4 Potentiometer.

multi-turn rotary device comprising a wiper, or arm, that moves along a conductive surface that may consist of a variety of materials such as wires, carbon, or plastic. As the wiper (C) moves from one end (A) to the other (B), the resistance between C and A increases, while the resistance between C and B decreases. If a potential E is applied between B and A, then the potential between C and A varies with the angle of rotation θ of the potentiometer. In most applications, especially in analog computers, the potentiometer is designed to provide a linear relationship between potential and degree of rotation, but other relationships can also be achieved, such as exponential, logarithmic, etc., for special applications. For present purposes, a linear relationship is assumed, so that

$$E_{CA} = k\,E\,\theta \tag{11–1}$$

where E_{CA} = the potential between the arm (C) and lower end (A)
θ = the angle of rotation
k = a constant of proportionality

It can be seen that the potentiometer can perform three important functions:

1. It can provide an input to the computer that is a voltage directly proportional to whatever quantity may be represented by rotation θ.

2. It can provide an output voltage proportional to whatever rotation the computer may have given to the arm.
3. It can accomplish multiplication of two quantities, provided that potential E is proportional to one of those quantities, and rotation θ is proportional to the other. That is, if

$$E = k_1 X,$$

and

$$\theta = k_2 Y,$$

then

$$\begin{aligned} E_{CA} &= k_1 X k_2 Y \\ &= kXY \end{aligned} \tag{11-2}$$

The potentiometer is not, of course, the only element accomplishing these functions.

11-3.3 The Operational Amplifier

A class of devices known as *operational amplifiers* has provided the basic building blocks for analog computers, which could hardly have been constructed without them. These devices are extremely versatile since not only can they multiply, integrate, differentiate, and scale, but also they are useful in adders, can be simply phase inverters, etc. So versatile are these devices that even in the digital era they still perform some functions.

An operational amplifier is a direct-coupled or dc amplifier consisting basically of a differential amplifier as described and analyzed in Sec. 5-6. There are many variations on the basic theme such as additional input and output circuits, constant current elements, etc., as shown in Figs. 5-21 and 5-22. Due to the fact that these devices can be employed for so many different functions, they lend themselves particularly well to integrated-circuit realization, with the standardization conducive to mass production and economy.

As is shown in Sec. 5-6, the gain of an operational amplifier is large, but it is hardly ever operated open loop. Generally it is employed in a negative feedback configuration. When the gain is assumed large, $A \gg 1$. The gain of the amplifier,

$$A' = \frac{A}{1 - A\beta} \tag{5-35}$$

reduces simply to

$$-\frac{1}{\beta} \tag{5–39}$$

so that the gain depends almost entirely on the feedback. For example, if a gain of 10 is desired, one need only connect a 10-K resistor and a 90-K resistor in series, giving

$$\beta = \frac{10\ \text{K}\Omega}{10\ \text{K}\Omega + 90\ \text{K}\Omega} = \frac{10\ \text{K}\Omega}{100\ \text{K}\Omega} = 0.1$$

$$\text{Gain} = -\frac{1}{\beta} = -\frac{1}{0.1} = -10$$

The circuit of either Fig 11–5(a) or 11–5(b) can be used.

The minus sign only indicates that a 180° phase reversal or change of polarity has occurred between output and input. But this is not a limiting factor since the operational amplifier, being a differential amplifier, has two inputs. If the phase reversal is not desired, the input signal can simply be applied to the other input. Feedback, however, is always negative, unless the objective is oscillation (see Chapter 14).

The symbol for the operational amplifier is a triangle, as shown in Fig. 11–6. The two input terminals and the output terminal are indicated. More than this is not really of concern to the user, as long as the condition

$$A\beta \gg 1$$

is maintained. This, in turn, is true unless there is a radical component failure. Any variation in gain, such as that due to temperature variations or component

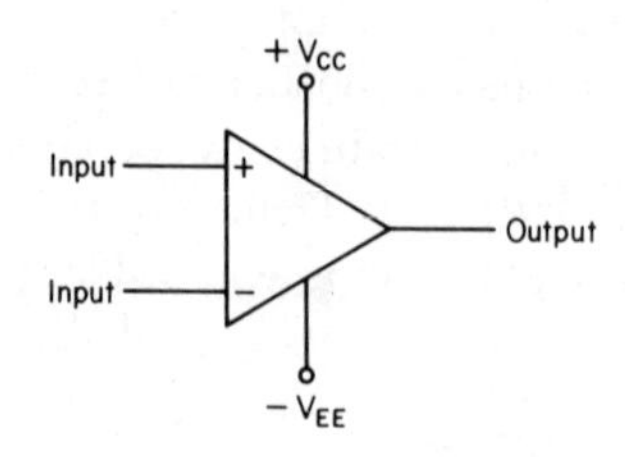

FIG. 11–5 Operational amplifier circuits.

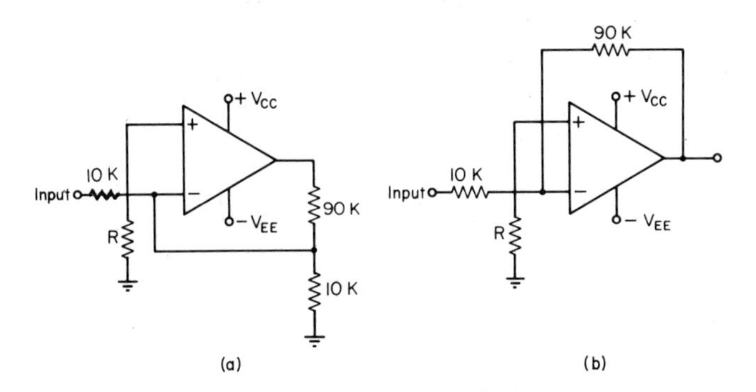

FIG. 11-6 Differential amplifier symbol.

tolerances, does not affect the output until it gets to the point where the above condition no longer holds. Used by itself, as it stands, the operational amplifier can multiply by a constant (scaler) or function as an inverter (gain = −1).

11-3.4 Addition and Subtraction

Addition, or summation, is accomplished in the analog computer by a circuit such as that shown in Fig. 11-7. Due to the operational amplifier's versatility, the same circuit can also be used for subtraction by feeding those quantities to be subtracted to the inverting input while feeding those quantities to be added to the noninverting input. For illustrative purposes, three positive and two negative inputs are chosen, though, of course, other combinations are equally possible. The problem to be solved is

$$E_0 = E_1 + E_2 + E_3 - E_4 - E_5 \tag{11-3}$$

where E_0 is the output voltage and E_1 through E_5 are the input voltages. It is assumed that resistors R_1 through R_6 are all equal and designated R. It is further assumed that the input resistance to the operational amplifier is much greater than R, and that the input current into the amplifier can thus be assumed to be zero. By means of Kirchoff's current law (the algebraic sum of all currents at a point = 0), it can be concluded that

$$i_1 + i_2 + i_3 = i_0$$

in the absence of negative inputs. However, due to the symmetrical construction of the operational amplifier, the negative inputs act as if they were positive inputs with the sign reversed, so we can write for this case that

$$i_1 + i_2 + i_3 - i_4 - i_5 = i_0 \tag{11-4}$$

By Ohm's law,

$$\frac{E_1}{R} + \frac{E_2}{R} + \frac{E_3}{R} - \frac{E_4}{R} - \frac{E_5}{R} = \frac{E_0}{R} \tag{11-5}$$

Hence

$$E_1 + E_2 + E_3 - E_3 - E_4 - E_5 = E_0 \tag{11-3}$$

If, instead of solving the problem considered here, one wished to solve the problem

$$E_0 = aE_1 + bE_2 + cE_3 - dE_4 - eE_5 \tag{11-6}$$

this can be readily accomplished by using unequal resistors R_1 through R_5 and adjusting those resistors in accordance with the coefficients a through e. By letting R_1 through $R_5 = R$ and adjusting R_6, it is possible to compute

$$E_0 = K(E_1 + E_2 + E_3 - E_4 - E_5) \tag{11-7}$$

where K is any constant, and where, therefore, $R_6 = KR$.

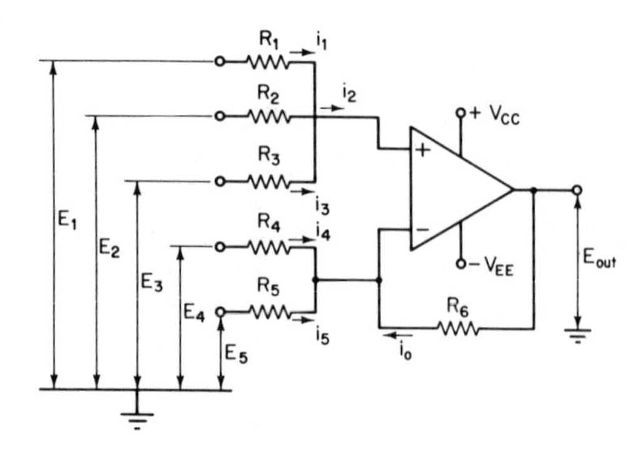

FIG. 11–7 Summing amplifier.

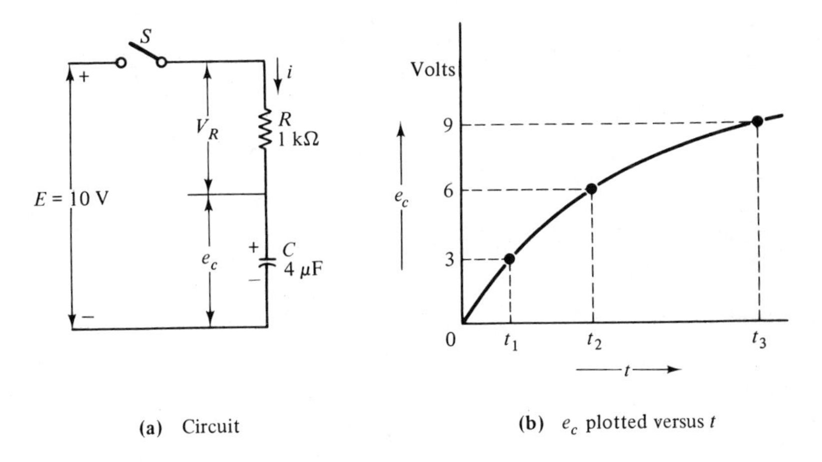

(a) Circuit

(b) e_c plotted versus t

FIG. 11–8 (a) RC circuit. (b) Capacitor voltage vs. time.

11–3.5 Charging and Discharging a Capacitor

Many electronic circuits are based on the charging and discharging of a capacitor. Consider the circuit of Fig. 11–8(a). Assume switch S open and an initial charge of E_0 volts on the capacitor. Then S is closed, and a current begins to flow, slowly charging up the capacitor. The equation of the voltage across the capacitor as it charges can be shown to be

$$e_c = E - (E - E_0)\,\epsilon^{-t/RC} \tag{11–8}$$

where C is the capacitance, R is the resistance, E is the supply voltage, t is the time, e_c is the instantaneous capacitor voltage, and ϵ is the base of natural logarithms ≈ 2.72.

For the special case of $E_0 = 0$, the equation reduces to

$$e_c = E(1 - \epsilon^{-t/RC}) \tag{11–9}$$

If the voltage source is removed and replaced by a short circuit, the capacitor is initially charged to a voltage E_0, and then the switch is closed, the voltage across the capacitor is

$$e_c = E_0\,\epsilon^{-t/RC} \tag{11–10}$$

A plot of these equations is shown in Figs. 11–8(b) and 11–9.

It is also possible to solve the charging equation for the time as follows:

$$(E - E_0)\,\epsilon^{-t/RC} = E - e_c \tag{11–11}$$

$$\epsilon^{-t/RC} = \frac{(E - e_c)}{(E - E_0)} \tag{11–12}$$

$$\epsilon^{t/RC} = \frac{(E - E_0)}{(E - e_c)} \tag{11–13}$$

Taking the natural logarithm of both sides,

$$\frac{t}{RC} = \ln \frac{(E - E_0)}{(E - e_c)} \tag{11–14}$$

$$t = RC \ln \frac{(E - E_0)}{(E - e_c)} \tag{11–15}$$

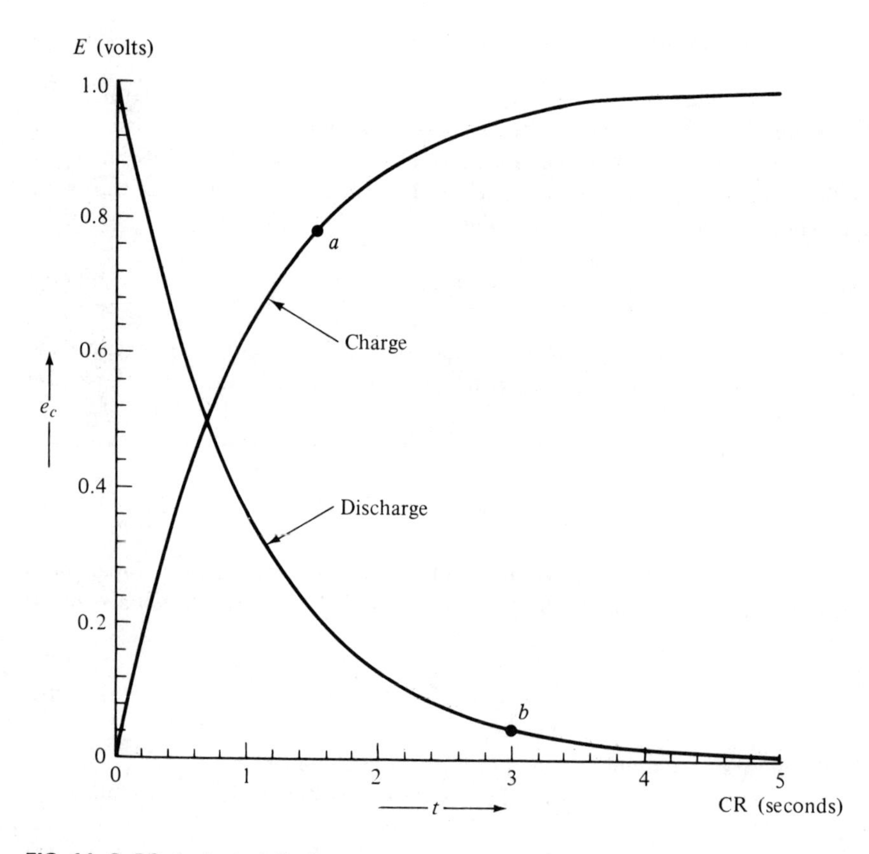

FIG. 11–9 RC charge and discharge.

The time t is a function of the various voltages and of the RC product. Once these values have been chosen, the time is fixed. This makes it possible to use the charging and discharging of a capacitor as a timing element, and this is done in many circuits, some of which are described below. Additionally, the charging process can be used to generate a ramp. By inspecting the curves of Fig. 11-9, it can be seen that if only the first parts of the curves are used, the functions are quite linear. As more and more of each curve is used, they become more and more nonlinear. This consideration shows that as long as the time the circuit functions is considerably less than one RC time constant, the curve is linear. Once a single time constant is reached, the capacitor voltage or charge has reached 63.2% of its final value; at two time constants it has reached 86.5%; at three, 95%; and at four, 98.2%, which can be considered, for practical purposes, to be the final value. That is, after four time constants, the capacitor can be considered to be fully charged. But to produce a linear curve

$$t \ll RC \tag{11-16}$$

11-3.6 Integration

The Miller integrator is so called because it relies on the Miller effect. (A capacitor connected between the output and the input of an amplifier has approximately the effect of a capacitor at the input multiplied by the gain of the amplifier—i.e., $C_{in} \approx G\,C_{fb}$, where G is the gain of the amplifier, C_{in} is the equivalent capacitance across the input, and C_{fb} is the capacitance in the feedback leg.) The integrator is a very simple device; all that is required to construct it is to replace the resistor in the operational amplifier circuit with a capacitor (Fig. 11-10). An integrator can also consist of just the resistor and the capacitor, but

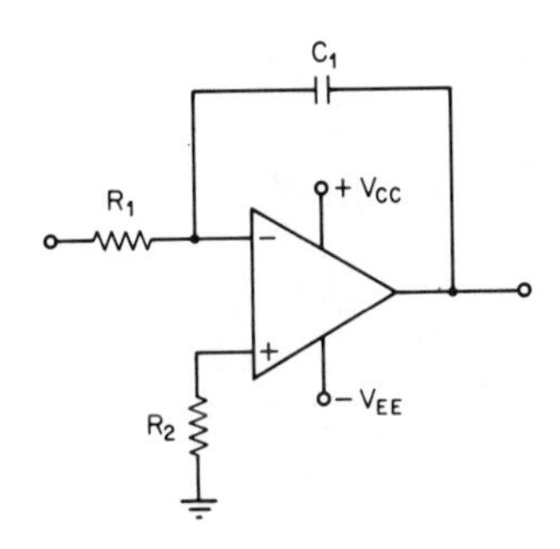

FIG. 11-10 Miller integrator.

this is not very effective. Exactly what integration is can be explained rigorously only by use of the calculus, which is beyond the scope of this discussion. What follows is a nonmathematical explanation of the process of integration: it is the computation of the area under the curve of a function, as shown in Table 11–2 and in Fig. 11–11. Consider first the circuit in Fig. 11–8. If voltage V_i is applied, a current flows into the capacitor C_1. This current is initially equal to V_i/R_1, but as the capacitor charges, the current decreases.

The charge of the capacitor equals the product of voltage and capacitance—i.e.,

$$Q = C\,V \tag{11–17}$$

hence,

$$V = \frac{Q}{C}$$

Charge accumulates as current flows into the capacitor, so that charge is the product of current and time; i.e.,

$$Q = I\,t \tag{11–18}$$

Thus, the charge can be considered the integral of the current. If it were not for the facts that, as the capacitor begins to charge, its voltage rises, and that hence the current drops, then the charge would be the true integral of the current.

TABLE 11–2

Integral	Function	Derivative
$\frac{-\cos \omega t}{\omega}$	$\sin \omega t$	$\omega \cos \omega t$
t	1	pulse
$\frac{kt^2}{2}$	kt	k
$\frac{e^{at}}{a}$	e^{at}	ae^t
triangular wave	square wave	pulses
kt	k	0

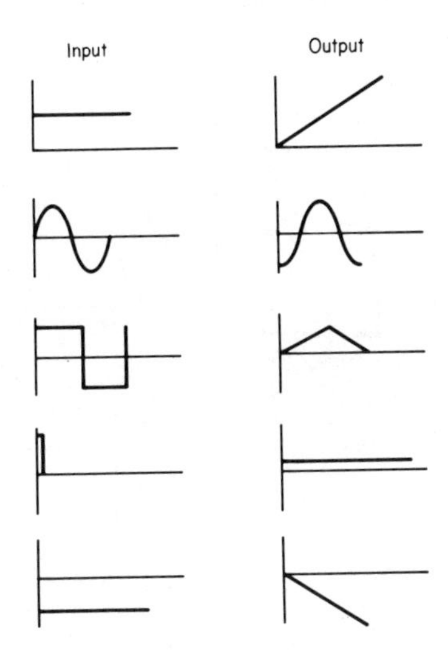

FIG. 11-11 Integrator waveforms.

Thus, the circuit considered does not produce the true integral, but only approximates it, to the extent that C_1 is chosen very large.

This difficulty is overcome by the Miller integrator of Fig. 11-10. Since the input impedance of the operational amplifier is very large, it can still be assumed that all the current flowing through R_1 also flows into C_1. However, the voltage at the output rises as the voltage across the capacitor rises (the voltage at the input of the operational amplifier ≈ 0). Thus, the current remains constant. The voltage across the condenser rises $(A + 1)$ times as much as would be the case in Fig. 11-8; hence, the circuit behaves like a capacitor equal to

$$(A + 1)\, C_1 \tag{11-19}$$

The time constant is $R_1C_1\,(A + 1)$. Since integration of functions is an important computation, many integrators are employed in analog computers. Of course, integrators are also required in digital computers, where they have different configurations, as discussed in Chapter 13.

If a dc voltage is applied to the input of a Miller integrator, a voltage is generated, which varies linearly with time and is called a *ramp*. Hence, the Miller integrator can function as a ramp generator, with provision being made for a pulse to return the ramp periodically to zero, from where the ramp starts over again. An alternative to the Miller ramp generator is the bootstrap circuit of Fig. 11-12. Here an emitter follower Q_2 is the amplifier, providing a gain ≈ 1,

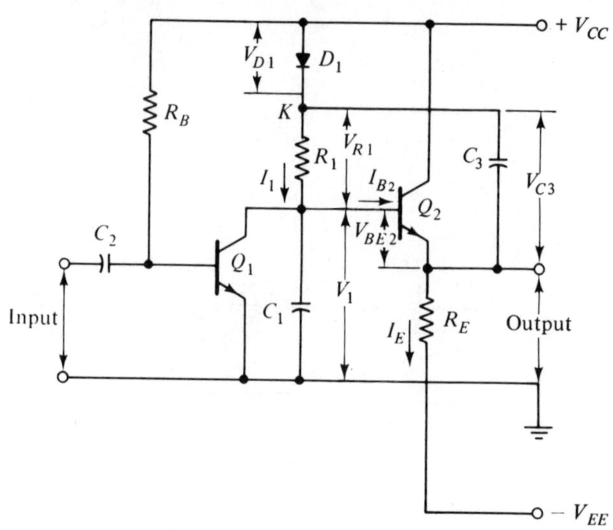

(a) Bootstrap ramp generator

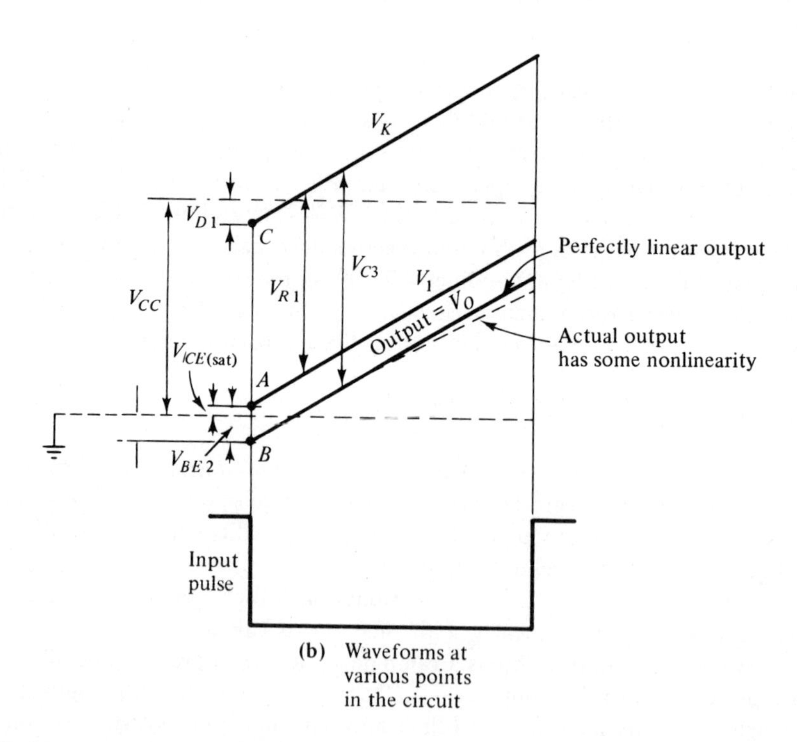

(b) Waveforms at various points in the circuit

FIG. 11–12 Bootstrap ramp generator.

but a low output impedance with a high input impedance. Initially, transistor Q_1 is conducting; hence $V_1 \approx 0.2$ volt. The output is close to zero. Capacitor C_3 has a charge such as to give it voltage V_{C3}, approximately equal to V_{CC} (actually less by the forward drop of diode D_1 and the fact that the output is slightly below zero). Since C_3 is very large, the voltage V_{C3} is approximately maintained throughout the ramp generation. Now a pulse is applied, turning Q_1 OFF and effectively converting that transistor into an open circuit. V_{C3} now causes a constant current to flow through resistor R_1 into capacitor C_1. The voltage of this capacitor is the ramp voltage V_1, fed through the emitter follower to the output. Even though V_1 and the output rise linearly, V_{C3} does not change appreciably. When the period designated for the ramp is over, the pulse disappears from Q_1, turning that element ON, discharging C_1, and allowing any lost charge to be restored to C_3. The ramp is not perfectly linear; a more linear ramp can be generated by employing the IC version of Sec. 14-2.7.

11-3.7 Differentiation

The opposite of integration is differentiation. To make a differentiator, the integrator elements are exchanged; i.e., the resistor R_2 is in the feedback leg and the capacitor C_1 is in the input (Fig. 11-13). The function of R_1 is explained below. To define differentiation rigorously requires the calculus, but it can adequately be explained by considering it to be the computation of the slope of a function, as shown in Table 11-2 and in Fig. 11-15. A rudimentary differentiator can be constructed without the operational amplifier (Fig. 11-14), but the results are inadequate. The question may arise: How does a differentiator differ from an ordinary RC coupling network? The answer lies in the choice of components. For a coupling network the RC product is chosen as large as practical, while for the differentiator the RC product is chosen small with respect to the time interval concerned. Also, in a coupling network it is tacitly assumed that sinusoidal waveforms are to be passed, while a differentiator ordinarily has pulse waveforms applied to it.

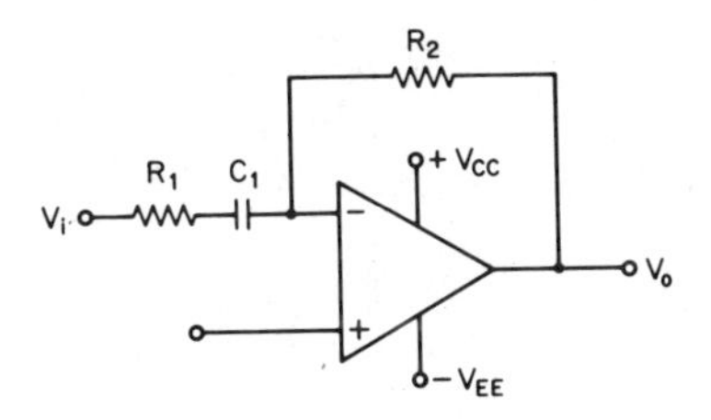

FIG. 11-13 Differentiator circuit.

Ignoring R_1 for the moment, the input current, I_1, is the same as the current in the feedback resistor R_2. This is true since the input current to the amplifier is negligible, the amplifier input impedance being very high. It was shown that the voltage across a capacitor is the integral of the current. This means that the current is the derivative (differential) of the voltage. Since the output voltage is the product of R_2 and I_1, the output voltage must be proportional to the derivative of the input voltage.

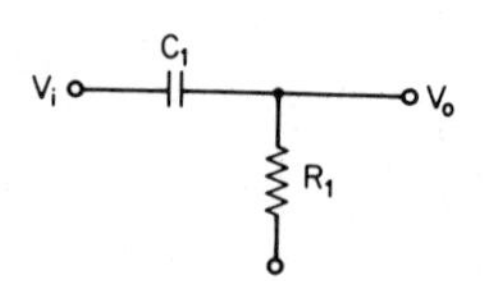

FIG. 11–14 Elementary differentiator circuit.

The process just described works well, except for the fact that at high frequencies the impedance of capacitor C_1 keeps decreasing. This causes the amplifier to have greater and greater gain, until it reaches dangerous levels. To prevent this, resistor R_1 is added, which is small enough not to affect the

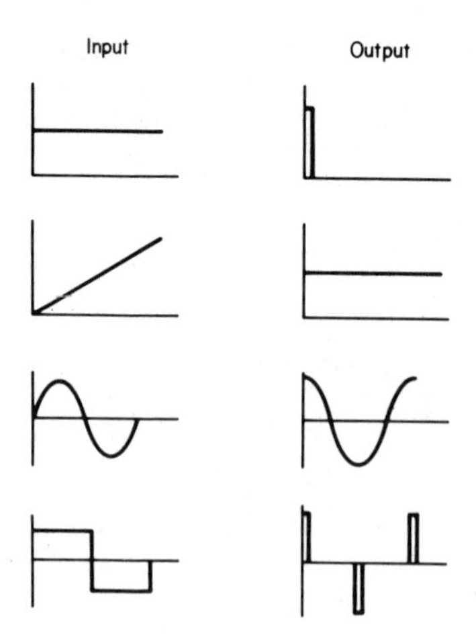

FIG. 11–15 Differentiator waveforms.

differentiation process, yet large enough to prevent excessive gain growth. To satisfy these conditions,

$$R_1 < 0.1X_{C1} \tag{11-20}$$

at the frequencies involved in the application.

11-3.7 Summary

The elements described above, in combination with such devices as servomechanisms, modulators, detectors, operational amplifiers (scalers), and inverters—constitute the analog computer. Since this discussion is illustrative and designed to give some perspective on computation, but is not intended to be exhaustive, there is no need to pursue the analog computer further, in view of the fact that it is gradually being displaced by the digital computer.

11-4 THE DIGITAL COMPUTER

11-4.1 Basis

As was pointed out in Sec. 11-2.4, the virtue of the digital computer lies in the fact that all data and information are immediately encoded in such a way as to involve only two possible states, normally designated 1 and 0, or sometimes called ON and OFF, or "signal" and "no signal." Since there exist only two possible states, any significant information requires a great many of these quantities. The name *bit,* a contraction of the words "binary digit," designates one unit, either 1 or 0. On the other hand, the word *byte* was created to designate a unit of information, such as the number 7 or the letter A (together designated as *alphanumerics*), or a sign, such as +, −, or *. Since the number of possible bytes is large, it is obvious that each byte comprises several bits. For example, the number 7 may require three or four bits, depending on the encoding system used. But all digital computers have one thing in common: they all employ bits representing 1's and 0's. Information is grouped into "words"—i.e., groups of bytes—with each byte containing a number of bits.

11-4.2 Inputs

The digital inputs to the computer can be derived from a variety of sources. Some of these are punched cards, manual operation of buttons on a console, tape (perforated or magnetic), reading magnetic ink, laser reading of black-and-white patterns, and inputs from a class of devices known as analog-to-digital converters. These are employed where the input is a quantity that occurs naturally in analog form, but which must be converted in order to constitute an input to the computer.

11–4.3 Storage

Storage in the computer involves different types of memories. We differentiate among long-term memories, short-term memories, and those that are in between. If, for example, A is to be added to B and the result to C, then the intermediate result $(A + B)$ must be stored long enough for C to be added to it. Since the quantity $(A + B)$ cannot possibly be of interest to any other problem than this particular one, there is no need to retain it. It is thus an example of a quantity stored in a short-term memory. The number π, on the other hand, is involved in numerous problems so it belongs in permanent or long-term memory. Other examples of this type are the number ϵ, the base of natural logarithms = 2.718, or the number 2.3, the conversion factor between natural logarithms and logarithms to the base 10. An example of an intermediate storage might be a constant k, where there are, say, a hundred numbers, all to be multiplied by k, but where k is no longer of interest once the hundred calculations have been carried out. Such information as subscriber lists, bank records, drivers' licenses, etc., must be stored permanently, even through possible power failures, but it need not remain in the computer. This information is thus best stored on magnetic tape or magnetic discs.

In computer terminology, the term *ROM* ("read only memory") is used to describe any information storage permanently in the computer, such as the number π. *RAM* ("random access memory") designates the computer-stored information capacity, where this information is of immediate but not permanent interest. Finally, *mass storage* is used to indicate the type of information that is permanent but need not remain in the computer, such as described in the paragraph above. This kind of information is stored outside the computer and can be reinserted at will.

In addition to the type of memory, the capacity of a computer is also important. The greater the memory capacity, the greater the complexity and variety of problem the computer can handle. Many computers, besides having a certain amount of memory built in, can be expanded by adding additional memory externally. The question of format is intimately related to the memory capacity. If the format is excessively large for the problem to be solved, the capacity is more quickly exhausted by being filled. Of course, if, on the other hand, the format is too small, it cannot solve the problem. Hence, format and information storage capacity have to be attuned to one another.

Depending on whether they are to be used for short-term, long-term, or intermediate purposes, memories have taken the form of anything from ordinary capacitors and flip-flops through magnetic tapes, magnetic disks, bubble memories, memory tubes, etc. New types of memories are constantly under development, where such factors as ease of storage, ease of readout, ease of changing, speed, and information density are being improved. But all memories have one thing in common: they store information that is put in at one time in order to use that information at a later time. Hand in hand with the concept of memory must go the concept of *address*.

Once information has been stored, it can never be found again unless the location where it has been stored is known. Thus, no piece of information is useful without the knowledge of where it has been stored. The location of the information is called the *address*. Any memory contains many addresses, and it is necessary to keep track of what has been stored where. The computer must consider the address when first storing the information and, again, when that information is to be retrieved. *Access* is the word used to describe the process of reaching the proper address. One way of illustrating this entire procedure is to compare it to the process of obtaining the weather from a telephone. If 937-1212 is the number one must dial to get the weather, then anyone knowing that number can obtain the report. Anyone not knowing the number cannot. The telephone company, on the other hand, must be careful to store the weather reports at 937-1212 and not at 837-1212; otherwise no one will obtain the report. In other words, the address is important for storage and retrieval.

11–4.4 Functions

Expressing the functions of the computer most generally, it can accomplish any function that can be spelled out in definite terms. Starting with the elementary functions of adding, subtracting, multiplying, and dividing, the computer capability also includes integration, differentiation, taking roots, finding logarithms, looking up trigonometric functions, etc. But it can do much more than that. It can perform logic—i.e., it can determine whether among a group of quantities all are present (AND), one is present (OR), or none is present (NOR). It can make comparisons—i.e., decide whether one quantity is greater, smaller, or equal to another quantity. It can search through its memory to find individual entries having certain prescribed characteristics. (For example, if it has been determined that an accident has been caused by a blond man of about 35 driving a late model Buick, it can print out all people having blond hair, say between 30 and 40, who own Buicks.) The computer cannot perform tasks that a person could not do, given sufficient time, but it can perform them at lightning speeds where a person might require several lifetimes or more. The problem, in order to be solvable by computer, must be reduced to concrete terms, the required inputs must all be given, and the computer must be programmed.

11–4.5 Programming

Nothing occurs in the computer without a program. It does not create anything spontaneously. Rather, it must be given specific directions as to what steps to carry out. These directions must be accurate. If the directions are correctly phrased, the computer carries them out and accomplishes the task set for it. This set of directions is known as the *program,* and the person creating the program is the programmer. The programmer often first draws up a flowchart from which he writes the program. When the program is written, it does not necessarily work properly the first time. If it does not, it must be "debugged"—i.e., any flaws in it must be removed so that the computer operates properly with it. Here the tacit assumption, usually, though not always, justified because of the great reliability of present-day machines, is that malfunction is due to the program and not the computer itself. Once debugged, the program can be used over and over whenever an identical problem is to be solved, but a different program is required for a different task.

11–4.6 Output

Once all required steps have been performed, the information is fed out in a variety of ways. The most common of these are the *printout,* consisting of copy printed on a roll of paper and copy displayed on a CRT (cathode ray tube) terminal. The display looks physically like a television set, but its output, instead of a picture, comprises lines of type or diagrams. Additionally, the computer can produce outputs in many other forms by resorting to digital-to-analog converters, the inverse of the analog-to-digital converters discussed above. These accept the computer's digital outputs and convert them to analog form. Due to the speed with which the modern computer can operate, the limitation on its performance is not the internal speed of computation, but rather the capabilities of the input–output devices. For example, there is some speed above which a mechanical printer cannot operate, and there is a limit on the speed at which a human operator can punch in information. But internally pulses are created and processed at "clock" frequencies in the billion-hertz (gigahertz, or ghz) region and going higher. The clock pulses are precisely timed and control all operations.

QUESTIONS

11–1 Explain the difference in approach between analog and digital systems.

11–2 What is the effect of noise on an analog system?

11–3 What is the reason for the relative noise immunity of the digital system?

11-4 Among different number systems, why is the binary system superior to others?

11-5 What mathematical function is performed by a capacitor?

11-6 What mathematical functions can a potentiometer perform?

11-7 Give and explain three uses for an operational amplifier.

11-8 Explain the operational amplifier's ability to function properly despite great variations in its components.

11-9 Explain the capacitor's role as a timing device.

11-10 What capacitor defect is corrected by the Miller circuit?

11-11 Explain the importance of *address,* and explain how a similar function is carried out in the analog computer.

11-12 Explain the difference between *bit* and *byte.*

11-13 What is the purpose of memories?

11-14 What is the purpose of a program?

11-15 Explain the four basic functions of logic circuits.

11-16 When does an operational amplifier employ negative feedback? Positive feedback?

PROBLEMS

11-1 Devise an analog method for computing $y = x^2$.

11-2 Devise an analog method for dividing x by y. *Hint:* Consider feedback.

11-3 Using graphical methods, estimate the departure from linearity of the curve of Fig. 11-9 at 1/2 time constant, 1/4 time constant, and 1/10 time constant.

11-4 Choose values of resistors in Fig. 11-7 to solve the equation

$$y = x_1 - 2x_2 + 3x_3 - x_4$$

11-5 Convert the decimal number 25 to binary, duodecimal, and hexadecimal equivalents.

11-6 Express the decimal numbers 17, 21, and 27 in the binary system.

11-7 Had man been born with four fingers on each hand instead of five, we might today have a number system based on eight, an octal system. Prepare a column of figures similar to those in Table 11-1 for this system.

11-8 A digital system employs 5 V to represent binary 1, and 0 to represent binary 0. How large would a noise pulse have to be to produce an erroneous result?

Chapter 12

Digital Mathematics, Logic, and Program

12–1 DIGITAL MATHEMATICS

12–1.1 Introduction

Primitive number systems used symbols to represent numbers, much as is done today, but with a difference. Consider the Roman system of numerals:

I, V, X, L, C, D, M, etc.

A defect in this system is that there is no short-cut available for addition and subtraction as is the case with the Arabic numeral system. The reason for this is that the Roman system did not embody the 0. For example, if 65 had to be added to 88, the result

LXV + LXXXVIII = CLIII

had to be known in advance; there was no way of performing the addition by the short-cut method. On the other hand, with the Arabic system of numerals, there is a 0, and the possibility of "carrying" a digit exists, which is the major contribution of this system of numerals. Performing the addition, one can obtain: $5 + 8 = 13$; write 3, carry 1; $1 + 6 + 8 = 15$; write 5, carry 1; giving the result, 153.

12-1.2 Binary Numbers

In the binary system of numerals, there are only two symbols: the 1 and the 0. But the "carry" feature is retained, as in the decimal system. The sequence of numbers, therefore, becomes:

0, 1, 10, 11, 100, 101, 110, 111, 1000, 1001,
1010, 1011, 1100, 1101, etc.

In the binary system, $1 + 1 = 10$. It is seen that there is a carry produced each time a 1 is added to another 1, giving a 0 and the carry, while, of course, there is no carry when adding 0 to 0 or 1 to 0.

The numbers have been generated as follows:

$0 = 0$

$1 = 1$

$2 = 1 + 1 = 10$, making the right digit 0 and producing a carry

$3 = 10 + 1 = 11$, no carry

$4 = 11 + 1 = 100$, with a carry in the right and middle digits

$5 = 100 + 1 = 101$, no carry

etc.

To convert a decimal number to a binary number, the numbers 2, 4, 8, 16, 32, 64, etc., in the decimal system are important. These numbers, known as the powers of 2—i.e., 2^1, 2^2, 2^3, 2^4, 2^5, 2^6, etc.—are used to convert decimal numbers to binary numbers. The rule is: Find the highest power of 2 contained in the original number, write down a 1, and check whether the next higher power of 2 is contained. If so, write a 1; if not, write a 0. Check similarly for all other powers of 2, writing down a 1 if they are contained and a 0 if they are not contained.

EXAMPLE 12-1

Convert the number 113 to binary form:

$2^7 = 128$ is obviously not present, nor are higher powers.

$2^6 = 64$ is present, write a 1. Subtracting the 64 leaves 49.

$2^5 = 32$ is present, write a 1. Subtracting the 32 leaves 17.

$2^4 = 16$ is present, write a 1. Subtracting 16 leaves 1.

$2^3 = 8$ is not present. Write 0.

$2^2 = 4$ is not present, giving another 0.

$2^1 = 2$ is not present, hence another 0.

1 is present; write a 1. ($1 = 2^0$, as is any number raised to a zero power.)
The end result is the binary number 1110001 = 113. This is now checked by reversing the procedure. This reverse process is best done starting from the righthand side; 1 is present, 2 is not, nor 4, nor 8; 16 is, as are 32 and 64.

$$1 + 16 + 32 + 64 = 113$$

EXAMPLE 12-2

Convert the binary number 110101 to decimal. Here, 1 is present, 2 is not, 4 is, 8 is not, 16 is, 32 is. The number is $32 + 16 + 4 + 1 = 53$.

12–1.3 Binary Addition

Binary addition is carried out just like decimal addition, using the 1 and 0, as well as the "carry" property. The results may appear a little strange at first, but on reflection they appear entirely logical and consistent. Some of the formulations are:

$$1 + 1 = 10$$

$$10 + 1 = 11$$

$$11 + 1 = 100$$

$$111 + 1 = 1000$$

EXAMPLE 12-3

Add the binary numbers 1101011 and 1001101.

$$\begin{array}{r} 1101011 \\ \underline{1001101} \\ 10111000 \end{array}$$

The result is obtained as follows (starting from the right):

$1 + 1 = 10$, write 0, carry 1

$1 + 1 = 10$, write 0, carry 1

$1 + 1 = 10$, write 0, carry 1

$1 + 1 = 10 + 1 = 11$, write 1, carry 1

$0 + 1 = 1$

$1 + 0 = 1$

$1 + 1 = 10$, write 0, carry 1

To check this, the number 1101011 can be converted to a decimal number:

$$64 + 32 + 8 + 2 + 1 = 107$$

The number 1001101 is $64 + 8 + 4 + 1 = 77$.

$$107 + 77 = 184$$

Converting the result of the binary addition, 10111000, gives:

$$128 + 32 + 16 + 8 = 184$$

12-1.4 Binary Subtraction

Binary subtraction is carried out like decimal subtraction, with liberal use of the "borrowing" technique, in which a 1 is borrowed from the next most significant digit to enable the subtraction where required. Borrowing is the inverse of carrying.

EXAMPLE 12-4

Subtract the binary number 11011 from the binary number 1011101.

$$\begin{array}{r} 1011101 \\ \underline{110111} \\ 100110 \end{array}$$

Starting at the right:

$1 - 1 = 0$

$10 - 1 = 1$, with a 1 borrowed from the next line.

$10 - 1 = 1$, again a 1 borrowed

$0 - 0 = 0$

$1 - 1 = 0$

$10 - 1 = 1$

As a check, 1011101 is seen to be $1 + 4 + 8 + 16 + 64 = 95$. 110111 is $1 + 2 + 4 + 16 + 32 = 55$. The result of the subtraction, 100110, is $2 + 4 + 32 = 40$. Checking decimally, the result is seen to be correct.

12-1.5 Binary Multiplication

Binary multiplication is carried out in the same manner as decimal multiplication.

EXAMPLE 12-5

Multiply $13 = 1101$ by $15 = 1111$.

$$\begin{array}{r} 1111 \\ \underline{1101} \\ 1111 \\ 11110 \\ \underline{1111} \\ 11000011 \end{array}$$

This equals $1 + 2 + 64 + 128 = 195$.

12–1.6 Binary Division

Binary division is carried out in the manner of long division in the decimal system.

EXAMPLE 12-6

Divide the number 10110100 by the number 1111. Using long division,

```
         1100
1111 ) 10110100
         1111
         01111
          1111
          000000
```

The dividend is 128 + 32 + 16 + 4 = 180. The divisor is 8 + 4 + 2 + 1 = 15. The quotient is 8 + 4 = 12, which checks.

EXAMPLE 12-7

Divide 111000 by 1000.

```
         111
1000 ) 111000
        1000
         1100
         1000
         1000
         1000
         0000
```

Checking:

$$111000 = 32 + 16 + 8 = 56$$

$$1000 = 8$$

$$111 = 4 + 2 + 1 = 7$$

12-1.7 Conclusion

Although the binary system may not be used directly in a computer, and a binary coded decimal system, a hexadecimal system, or some other method of encoding may be employed instead, the basis of all the systems is the use of the 1's and 0's. Hence, the techniques described above have a general applicability.

12-2 LOGIC

12-2.1 Introduction

Logic is defined as the science that deals with principles of reaching conclusions from known facts. Considering the example given in Sec. 11-4.4, the computer has to find a man who is blond AND about 35 AND drives a Buick; i.e., all three characteristics must be present if the man can have caused the accident. On the other hand, if nothing about the man is known except the fact that he drove a Buick, a Ford, or a Dodge, there is a requirement for the computer to pick out anyone driving a Buick OR a Ford OR a Dodge. If, instead, it is known that the car involved was neither a Ford product, NOR a GM product, NOR a Chrysler product, then the task becomes one of eliminating all three and printing out only those cars made by other manufacturers. The third case can be looked at in two ways: one can search for a car that is non-Ford AND non-GM AND non-Chrysler; or one can also consider Ford OR GM OR Chrysler and reject in each case, both viewpoints yielding the same result. The first case above illustrates an AND condition, the second an OR situation, and the third is an example of NAND, or of NOR. These are the basic building blocks of computer logic. By combining these elements into more complex configurations, a great many logic problems can be solved. To analyze complex logic circuits, Boolean algebra is useful, but that is beyond the scope of this discussion. It is, however, useful to define the complementary functions. If the functions to be considered are A, B, and C, then the complementaries are $\overline{A}$, $\overline{B}$, and $\overline{C}$, defined as NOT A, NOT B, and NOT C. Or it can be said that when A is 1, $\overline{A}$ is 0; when A is 0, $\overline{A}$ is 1, etc. An inverter stage can always be used to change A to $\overline{A}$, if desired.

12-2.2 The AND Circuit

For illustrative purposes, three input functions—A, B, and C—are used, as well as an output function D. However, one can employ only two functions, or there can be four, five, or more. What is important is that, in order to activate an AND circuit, all the functions must have the required value; i.e.,

$$\text{if } A = 1, B = 1, C = 1, \text{ then } D = 1$$

if any one of the three = 0, then $D = 0$
if two or three are 0, $D = 0$

The same information can also be presented in a truth table, when there are only two input functions. The values for D are given by the numbers in the table; the values for A and B are along the sides.

		B	
		1	0
A	1	1	0
	0	0	0

Truth table for D

To show that one configuration can be changed to another, one can look at the situation also as $\overline{A}$ *or* $\overline{B}$ giving $\overline{D}$, with the accompanying truth table.

		$\overline{B}$	
		0	1
$\overline{A}$	0	0	1
	1	1	1

Truth table for $\overline{D}$

12-2.3 The OR Circuit

In the OR circuit, any one of the input functions can produce the output, even if the others are absent; i.e.,

if $A = 1, B = 0, C = 0$, then $D = 1$
if $A = B = C = 1, D = 1$
if $A = B = C = 0, D = 0$

The truth table is shown for two functions:

		B	
		1	0
A	1	1	1
	0	1	0

Truth table for D

The OR circuit for the functions can be shown to be the AND circuit for the complementary functions, with the truth table:

		$\bar{B}$	
		0	1
$\bar{A}$	0	0	0
	1	0	1

Truth table for $\bar{D}$

12–2.4 The NOR Circuit

The NOR circuit gives an output of 1 when all inputs are 0; i.e.,

$$\text{if } A = 0, B = 0, C = 0, \text{ then } D = 1$$
$$\text{if } A \text{ or } B \text{ or } C = 1, \text{ then } D = 0$$

with the truth table for two functions:

		B	
		1	0
A	1	0	0
	0	0	1

Truth table for D

The circuit can also be considered as the NAND circuit, using the complementary functions:

		$\bar{B}$	
		0	1
$\bar{A}$	0	0	0
	1	0	1

Truth table for $\bar{D}$

Note that, in this case, both configurations yield D so that they are identical. In the AND and OR circuits, they yielded D in one case and $\bar{D}$ in the other so that an inversion would have been required.

This chapter deals only with digital mathematics. The actual circuits for realizing these functions are given in the following chapter.

12-2.5 Comparison

When making comparisons, the computer answers the question: Is function B greater than A, less than A, or equal to A? This is an important role of the computer since this capability, in combination with the logic, can solve a great many problems. In making comparisons, the computer deals only with those quantities that differ from one another by at least one bit. If they do not differ by a bit, they are considered the same. This is a great advantage over the analog computer where a quantity A applied to a voltage comparator had to differ from another quantity B by some finite amount before the circuits responded. That is, there is a "dead zone," where the difference is too small and quantities that actually differ are judged equal; e.g., because a silicon diode does not pass substantial current until its voltage equals 0.6 volt, a double-sided dead zone of 1.2 volts exists. This condition does not appear in the digital computer so here, as in actual computation, the digitization has an advantage.

Since the computer not only performs mathematics but also handles letters of the alphabet (so that we speak of alphanumerics), it can make comparisons between mathematical quantities as well as between words of the language. Thus, it can answer the question: Is word A the same as word B? Here the concept of "more" or "less" loses significance, so either more or less can be combined into "different," easily handled by a slight change in logic. This enables the computer to search its dictionary to see if a given word exists, the assumption being that, if the word does not exist in its dictionary, then that word has been misspelled, an assumption usually but not always justified. (The word may have been omitted from the dictionary.)

12-3 PROGRAM

12-3.1 Introduction

It is not intended here, of course, to teach computer programming. That is a course of study in itself (computer science), and those who study this discipline devise the programs, analyze and "debug" them, and finish by developing a working program that can then be used over and over whenever a similar problem arises. What is intended here is to supply only the elementary knowledge of programming that is required in order to deal with the electronics of the computer, which is of interest here.

12–3.2 Function of the Program

The program is the heart of the computer in the sense that nothing significant can come out of the computer without it. That is, since the computer cannot "think" in human terms and cannot have original thoughts or inspirations, it must be told what to do at every step of a procedure. The instructions must be very specific.

EXAMPLE 12-8

Devise a program for a simple addition. The program might be something like this:

1. Store input quantity A in register 10001.
2. Store input quantity B in register 10002.
3. Transfer the contents of register 10002 into register 10001.
4. Read out the contents of register 10001.

When computers were first developed, the programmer had to select the addresses (registers) himself, as well as to devise the steps to be taken, which, for this very simple problem, does not seem like much of a task, but can be quite formidable for more complex problems.

A later generation of computers eliminated much of the drudgery of, and also reduced the expertise required for, programming, in that a direct knowledge of computer electronics is no longer required. The programmer must supply proper instructions, but need not know how these instructions are electronically carried out. Instead, a number of computer languages—COBOL, FORTRAN, BASIC, and a great many others—have been, and are still being developed. Whatever language is to be used is first fed to a *compiler,* which translates from that language into the language of the computer. The use of these languages eliminates the need for the programmer to select the addresses; that is done automatically. The directions are given in the form of plain English or something similar, so that they are more easily learned and retained than would be the case if the computer had to be programmed directly. As more and more progress is made in the computer languages, and as compilers improve, the task of programming is further simplified, though perhaps the problems that can be solved become more complex, nullifying that factor.

Whatever the language, all steps must be spelled out accurately and definitely. Often a flow diagram is prepared first, from which the program is developed. Though the program can be extremely complex, the computer, like a baby, takes one little step at a time. If any step is wrong, the entire program is invalid. Consider the following:

EXAMPLE 12-9

What is wrong if a programmer writes:

"If A is greater than B, go to step 8,
"If A is less than B, go to step 9"?

The answer is that, if the programmer does not say what to do if A equals B, then the program cannot function. All logical possibilities must be taken into account; there can be no loose ends.

Furthermore, all directions must be precise, so that, if the correct instruction is

ADD*,

the program does not function if the programmer says just

ADD (without the asterisk)

The attention to such details is, of course, part of learning to program.

Typical instructions that can be given to a computer, then, are the following:

add, subtract, multiply, divide, raise to a power, take root, take logarithm, find sine, cosine, tangent, etc.
transfer from one register to another
make comparisons
decide AND, OR, NOR, NAND questions
search the memory for a specific quantity
print out
reject either programming instructions or inputs as unacceptable

For an input to be acceptable, it must meet the format. For example, if inputs must consist of names of people, and it is decided that names cannot contain more than 22 letters, then a name involving 23 letters is unacceptable. That name must either be abbreviated or the format must be increased. Also, if it is known that the sine of an angle can never exceed 1, then, if someone attempted to insert a sine of 1.1, he would find that the computer cannot accept it.

What is of interest to the student of electronics, however, is not so much the program itself as the electronic circuits that are employed to accomplish the tasks, some of which are given in the next chapter. A computer jargon phrase that occurs widely enough to have entered the general English language is GIGO (garbage in, garbage out), which means that either an input or an instruction is improper, giving a useless result.

A widely publicized recent example of GIGO was the case in which a computer approved a locality experiencing a great boom as qualified for federal aid to depressed areas. The reason: The computer was fed census data that were almost ten years old, representing, in this case, garbage in.

QUESTIONS

12–1 What is the main advantage of the Arabic system of numerals?

12–2 Explain the main advantage and the main disadvantage in using a binary system of numerals.

12–3 What are the advantages and disadvantages of a binary coded decimal system? (Binary numbers are used to encode each digit, but that digit is encoded separately from others.)

12–4 Explain this statement: The "carry" is the heart of binary addition and "borrowing" is the heart of binary subtraction.

12–5 Explain the principles involved in converting decimal numbers to binary and vice versa.

12–6 Explain the principles of binary addition and subtraction. Do they differ fundamentally (other than the use of two characters) from the rules for decimal addition and subtraction?

12–7 Give an example of an AND problem.

12–8 Give an example of an OR problem.

12–9 Give an example of a NOR problem.

12–10 Give an example of a NAND problem.

12–11 Explain why format is important.

12–12 What is a compiler?

PROBLEMS

12–1 Convert the following decimal numbers to binary: 394, 1753, 87, 2543.

12–2 Convert the following binary numbers to decimal: 110101110, 101011001, 11000111, 110010101.

The following problems, which would be trivial when done in the decimal system, are to be done by writing the numbers first in the binary system, performing the operation, and converting back.

12–3 Perform the following additions:

a. $186 + 213 = 399$
b. $285 + 88 = 373$
c. $17 + 44 = 61$

12–4 Perform the following subtractions:

a. $87 - 34 = 53$
b. $101 - 87 = 14$
c. $168 - 42 = 126$

12–5 Perform the following multiplications:

a. $18 \times 17 = 306$
b. $25 \times 29 = 725$
c. $16 \times 28 = 448$

12–6 Perform the following divisions:

a. $56 \div 14 = 4$
b. $168 \div 12 = 14$
c. $272 \div 17 = 16$

Chapter 13

Digital Circuits

13–1 INTRODUCTION

In the chapters devoted to the analysis of semiconductor amplifying devices (Chapters 3–5), the objective has been the relatively undistorted amplification of signals (e.g., sine waves and waves composed of sine waves). This objective was optimally achieved by operating the devices near the center of the operating region. In proceeding from this to the operation of transistors in digital circuits, the emphasis shifts. In digital circuits, the transistors are operated in what is called the *switching mode.*

13–2 THE SWITCHING MODE

13–2 Definition

For analysis, Fig. 3–16 is reproduced as Fig. 13–1, with the addition of points *a* and *b*. Where in the operation of the transistor as a linear amplifier a *Q*-point is established, and the amplification takes place around that point, in the operation of switching devices such as the flip-flop, the linear region is of interest only in a transitional phase as explained below. The points of major interest are points *a* and *b*. Point *a* represents the transistor saturation (full conduction), the

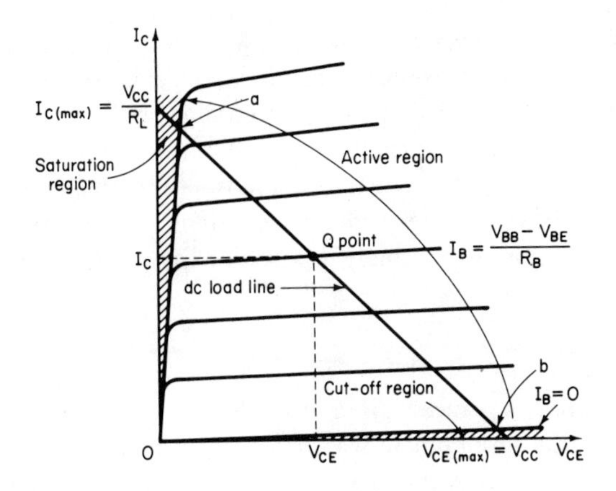

Fig. 13–1 Collector characteristics of a BJT in the CE configuration, showing cutoff and saturation.

transistor ON condition. Point b is the point of transistor cut-off, or the transitor OFF condition. At a, $V_{CE} \approx 0.2$ V. At b, $V_{CE} \approx V_{CC}$.

13–2.2 Power Dissipation

$$\text{Power dissipated} = VI \text{ (product of potential and current)} \qquad (13\text{–}1)$$

Consider Example 3-5. First, for purposes of comparison, it is useful to determine the power dissipated under linear operation (class A).

$I_C = 5$ ma at saturation, and about half of this, or 2.5 ma, at the Q-point. V_{CE} at this point $= 5$ V.

$$\text{Dissipation} = V_{CE} I_C = 5(2.5) = 12.5 \text{ mW}$$

At point a, V_{CE} can be assumed to be equal to 0.2 V and $I_C = 5$ mA.

$$\text{Dissipation} = 0.2(5) = 1 \text{ mW}$$

At point b, I_C can be assumed $= 0$, conservatively assume 10 μA; $V_{CE} = 10$ V.

$$\text{Dissipation} = 10(.01) = 0.1 \text{ mW}$$

It can be seen from this calculation that the power dissipated at both points *a* and *b* is considerably less than it would be under class A operating conditions. If the circuit is redesigned so as to reduce the saturation current (R_C is increased), an even more spectacular reduction can be achieved in the dissipation. It can be seen that a transistor operating as a switch, being most of the time either at point *a* or at point *b*, and subject to intermediate conditions during only a very brief transitional interval, has an extremely low average power dissipation.

13–2.3 Integrated Circuits

As technology advanced from discrete semiconductor components to integrated circuits, the development was greatly facilitated by the above consideration. That is, the smaller the average power dissipation, the smaller the components that can be designed. The smaller the components, the more of them that can be made to occupy a given area; hence the allowable component density is increased. Furthermore, the smaller the component, the shorter, generally, the rise time (the time required for transition between the two points), also decreasing the power consumed during the transition. It can thus be seen that the great progress in integrated circuits, leading to LSI (large-scale integration) was made possible by the fact that those circuits in which the integrated components were to be used were switching circuits, not class A circuits. If class A circuits had had to be used, nowhere near the degree of density accomplished for the switching circuits could have been achieved. Integrated class A circuits such as operational amplifiers occupy much more space.

13–3 FLIP-FLOPS

13-3.1 The Basic Flip-Flop

By far the most widely used circuit configuration in the digital computer is the *flip-flop*, so-called because it has two stable states between which it "flips" and "flops"—i.e., it remains most of the time in one or the other of these states, while being for only very short intervals in the transition between them. This circuit is also known as a *bistable multivibrator.*

If in Fig. 13–2 transistor Q_1 is initially in the OFF condition, the collector is at a potential of V_{CC}. This is point *b* in Fig. 13–1. The high potential causes a high base current to be applied to transistor Q_2, making that element fully conducting—i.e., putting it in the ON condition. This is point *a* of Fig. 13–1. If a pulse going positive is applied to both transistors, it does not affect Q_2 since that element is ON. But the pulse turns on Q_1 to some extent. This lowers the potential at the Q_1 collector, lowering the base current of Q_2 and driving that transistor partially OFF. This raises the collector voltage of Q_2, applying greater base current to Q_1 and thus driving Q_1 further ON. This is seen to be a positive feedback condition, which continues the process until such time as Q_1 has been driven fully into the ON condition, while Q_2 is completely OFF. "Switching" is complete. If another pulse appears later, it can act only on Q_2 but does not affect Q_1. Therefore, the action of the circuit is the reverse of that just described, and Q_1 reverts to the OFF condition, while Q_2 again goes ON.

It is seen that this circuit can be used to represent one digital *bit* (the name is a contraction of "binary digit"). That is, the state Q_1-ON, Q_2-OFF can be designated 1, while Q_1-OFF, Q_2-ON can be called 0, or vice versa (whichever

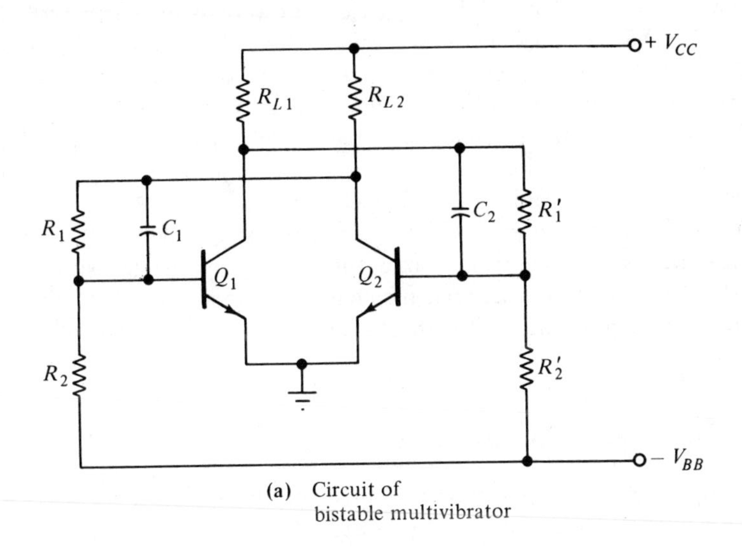

(a) Circuit of bistable multivibrator

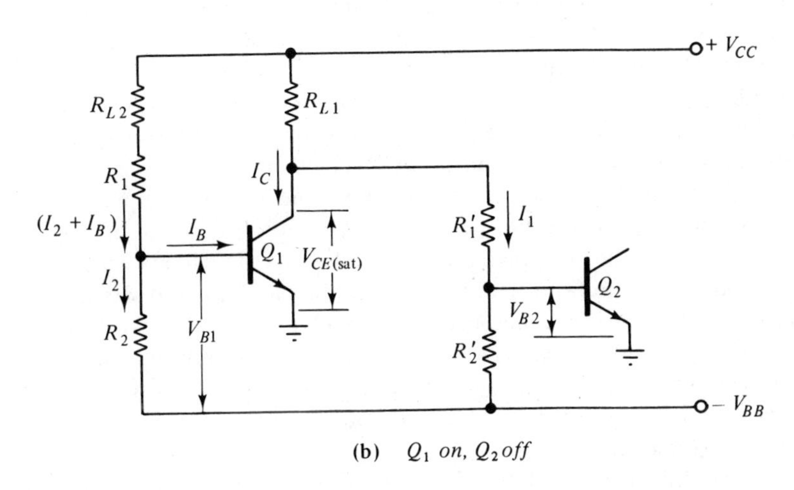

(b) Q_1 *on*, Q_2 *off*

Fig. 13–2 Flip-flop circuit.

is convenient), since the circuit is symmetrical. The circuit can also be used as a divide-by-two element, supplying one output pulse at one of the collectors for every two input pulses supplied. It must be emphasized that the switching process rather lengthily described above actually takes place very rapidly, while the two possible stable conditions remain until the next pulse arrives. The time to switch is very short, limited only by the properties of the components and the amount of positive feedback, while the time the circuit remains in its two stable states can last indefinitely. It can be appreciated that formerly, when these flip-flops comprised relays, vacuum tubes, or transistors, an enormous amount of hardware was necessary if the above elements were required to represent just one bit. But today, with integrated circuits, where thousands of flip-flops can be constructed on a small chip, this is no longer a problem, and a small area can now house what in the past required rooms full of equipment.

In the analysis given above, the role of capacitors C_1 and C_2 has been ignored. The reason for this is that they serve to facilitate coupling between the collector and the base, and thus speed up the switching process. They also acquire different charges while the flip-flop is in one of its stable states. When the circuit is switched, these charges insure that the OFF transistor is triggered ON during the brief instant when both may be temporarily OFF. This process is known as *commutation*. But the capacitors do not otherwise affect the basic analysis.

13–3.2 Types of Flip-Flops

Many variations are possible on the basic flip-flop circuit of Fig. 13–2. Consider the emitter–coupled multivibrator of Fig. 13–3. The only change from Fig. 13–2 is the addition of emitter resistor R_E, which aids in the coupling process in the following manner. If Q_1 is going toward heavier conduction, the voltage across R_E rises. This has a tendency to bias Q_2 more negatively, driving Q_2 more towards cutoff. R_E acts similarly on circuit reversal. However, this aid in coupling is not the main reason for employing this circuit. The main advantage lies in the fact that this circuit can be operated with only one power supply, while the circuit of Fig. 13–2 requires both a positive and a negative supply voltage.

Triggering of the flip-flop can be accomplished either by applying a pulse to the bases or by applying it to the collectors. The fundamental flip-flop operation is not significantly different since the base of one transistor and the collector of the other are connected by the coupling capacitors. The base and the collector are at the same point electrically, except for any dc charge the capacitors may have acquired. Triggering can be applied to each transistor separately (known as asymmetrical triggering), or both transistors can be triggered from one input (symmetrical triggering).

To ensure that all flip-flop triggering occurs at the proper time to function well in a computer, it is possible to provide logic such that no triggering

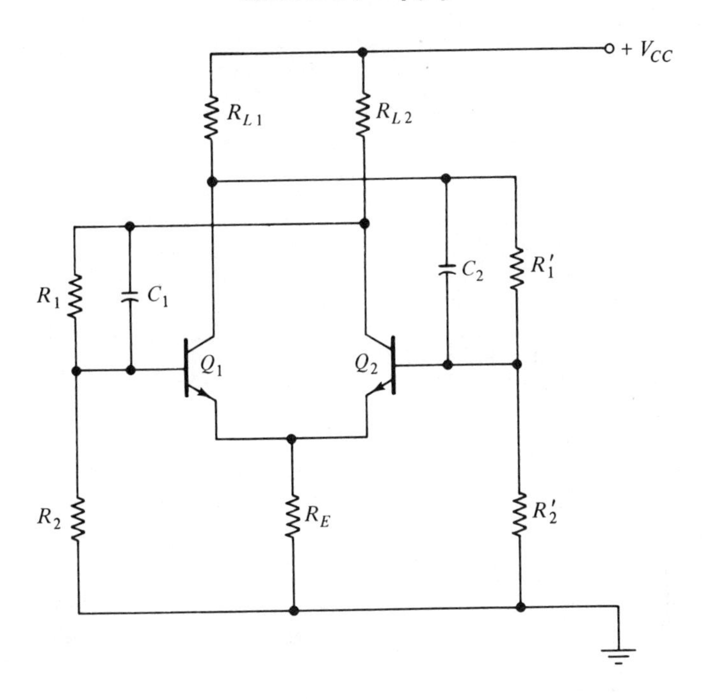

Fig. 13-3 Emitter-coupled flip-flop.

takes place unless a pulse from the clock (computer standard operating frequency source) is present. This gives rise to the various types of flip-flops, available in integrated-circuit form, discussed below.

13-3.3 The T Flip-Flop

The T (for toggle) flip-flop (Fig. 13-4) is symmetrically triggered (only one trigger input applied to both transistors). It has two outputs, Q and $\overline{Q}$. $\overline{Q}$ is the "complement" of Q; i.e., whenever Q is 1, $\overline{Q}$ is 0, and vice versa. The small circle in the block diagram for the circuit indicates that a negative trigger input is required. When the diagram appears without the circle, a positive trigger is needed; i.e., the circle represents a phase inverter stage. The timing diagram indicates what the outputs are for the input shown. Note that the input shown is that at terminal T, not before the inverter.

13-3.4 The SC Flip-Flop

The SC flip-flop, shown in Fig. 13-5, so-called because it sets and clears, is also known as the RS flip-flop, for set and reset. It has two trigger inputs; i.e., it is asymmetrically triggered. The timing diagram shows circuit behavior. The

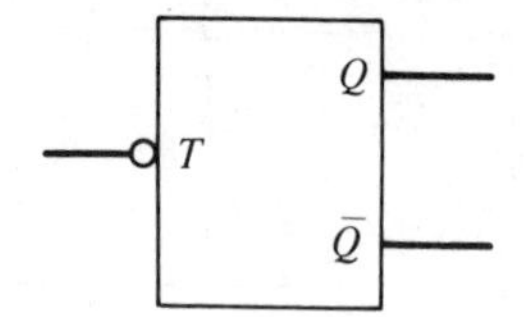

(a) Logic symbol for T flip-flop triggered by a negative-going edge.

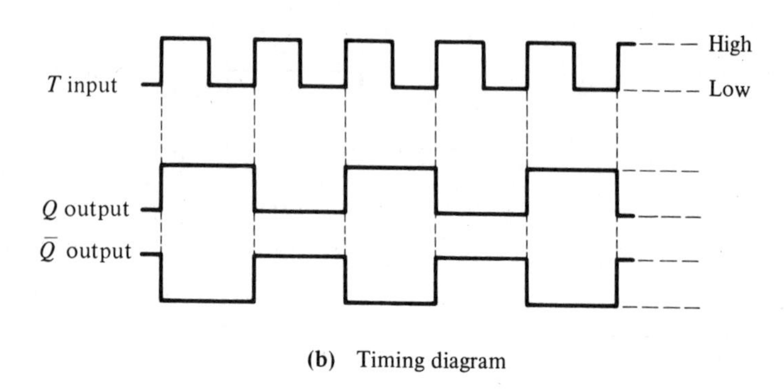

(b) Timing diagram

Fig. 13–4 Symbol and timing diagram of a T flip-flop.

circles again indicate that a negative trigger is required. At t_1 in the timing diagram, S goes negative, and Q goes to 1, while $\overline{Q}$ goes to 0. At t_2 there is no action because the trigger goes positive. At t_3 there is no change because the output is already in the set condition. When C goes negative (t_4), Q goes to 0, and $\overline{Q}$ goes to 1. At t_5 nothing happens because S goes positive. At t_6 again nothing happens because C is going positive. At t_7 nothing happens because the circuit is already in the clear condition. At t_8 S goes negative so the circuit returns to set; i.e., Q is again 1, while $\overline{Q}$ is 0. Figure 13–6 is a circuit diagram of the flip-flop with two phase inverter transistors added. When pulses are applied simultaneously to both S and C, circuit behavior is indeterminate.

13–3.5 The Clocked SC Flip-Flop

The clocked SC flip-flop, shown in Fig. 13-7, has a clock input as well as an S input and a C input. In order for the circuit to switch to S, a positive clock pulse must occur, and the S input must be at 1. In order for the circuit to switch to C, a positive clock pulse must occur, and the C input must be at 1. If there is a 1 at both S and C, the circuit behavior is indeterminate.

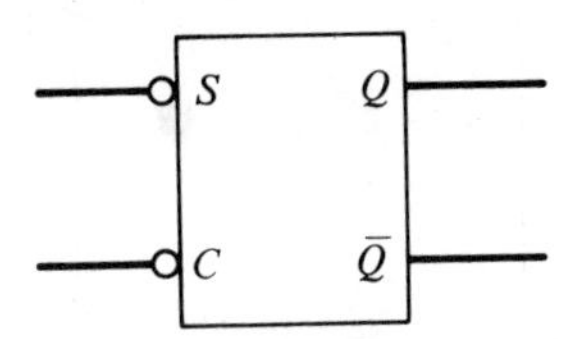

(a) Logic symbol for SC flip-flop triggered by negative-going signals

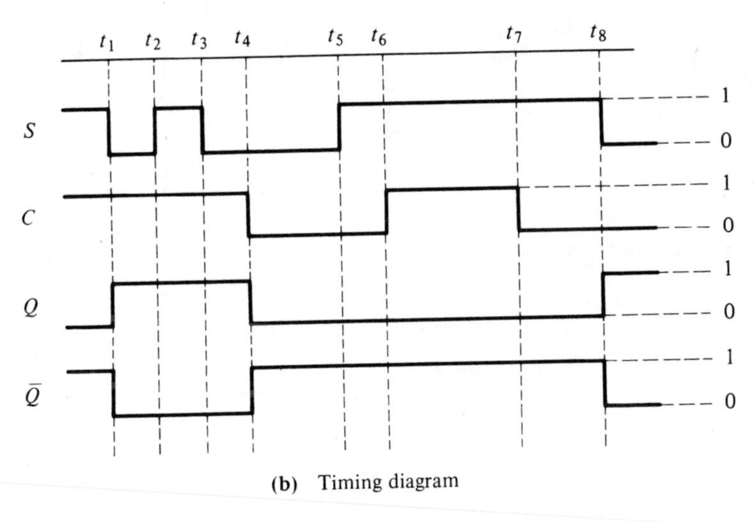

(b) Timing diagram

Fig. 13–5 Symbol and timing diagram of an SC flip-flop.

13–3.6 The D Flip-Flop

When the T flip-flop is arranged to require clock pulses, it becomes a D flip-flop. A positive clock pulse is required, and when it occurs, the circuit switches, where called for, to the state of the input *D*. Figure 13–8 shows the timing diagram.

13–3.7 The JK Flip-Flop

Finally, there is the JK flip-flop (Fig. 13–9), identical to the clocked SC flip-flop, but with the additional feature that the ambiguity mentioned in Sec. 13–3.5 has been resolved by having the circuit switch (or toggle) each time a positive clock pulse occurs, just in case the inputs *J* and *K* are both in the 1 state. When this is not the case, and it seldom is, since one normally would not wish the

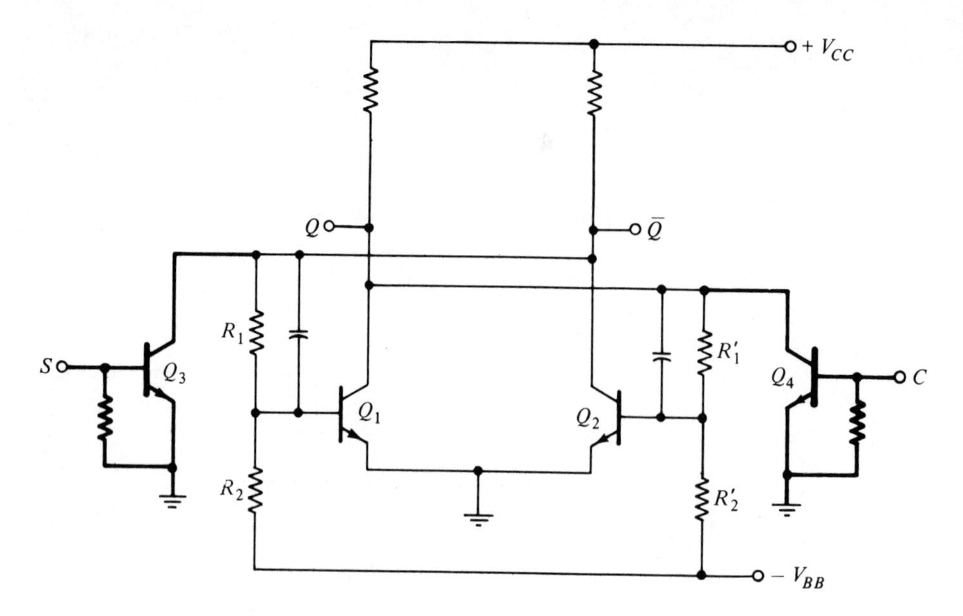

Fig. 13–6 Circuit of SC flip-flop.

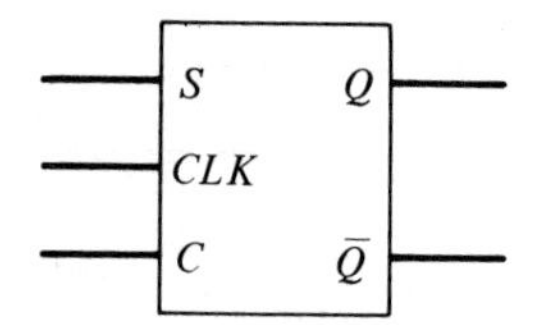

(a) Logic symbol for clocked SC flip-flop

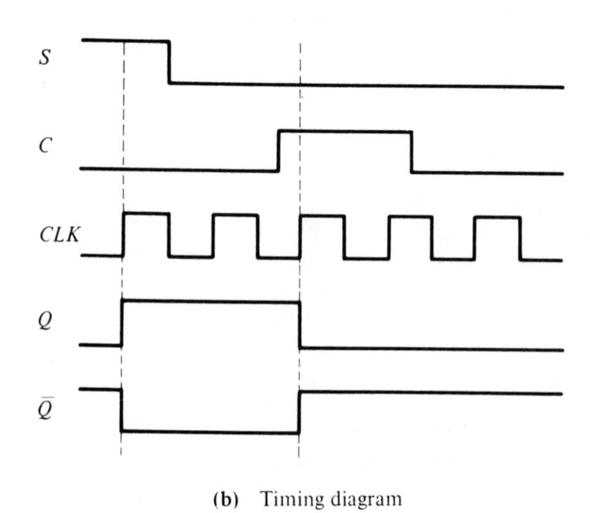

(b) Timing diagram

Fig. 13–7 Symbol and timing diagram of clocked SC flip-flop.

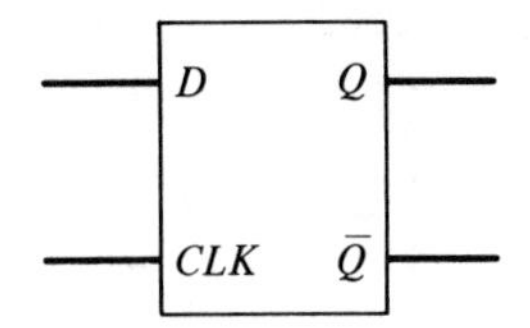

(a) Logic symbol for D flip-flop

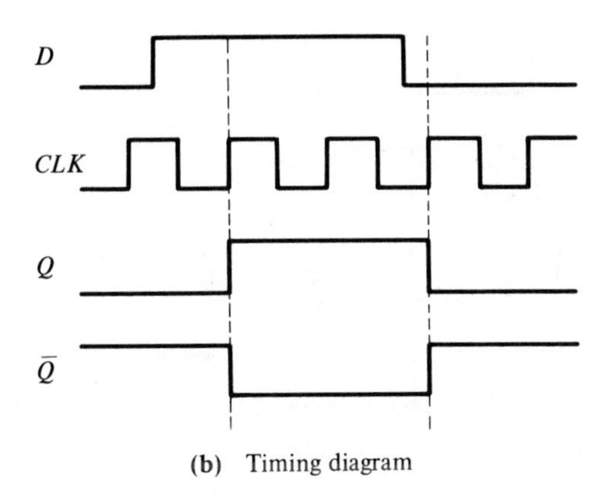

(b) Timing diagram

Fig. 13-8 Symbol and timing diagram of D flip-flop.

circuit to set and clear at the same time, the circuit operates identically to the clocked SC flip-flop.

13-3.8 The Monostable Circuit

The flip-flop is also known as the *bistable multivibrator*. This designation distinguishes it from the monostable and astable circuits. Making some modifications in the circuit of Fig. 13-2, Fig. 13-10 is obtained. The circuit remains a positive-feedback configuration. But the removal of resistor R_1 and its replacement by capacitor C_1 prevents Q_1 from cutting off Q_2 indefinitely. So long as C_1 remains uncharged, the pulse from the Q_1 collector is coupled to the Q_2 base, as was the case with the flip-flop. But as soon as C_1 is charged, the influence of Q_1 and Q_2 ceases. There is, therefore, only one stable state—namely, Q_1-OFF, Q_2-ON. If a positive pulse is applied to the base of Q_1, it is driven ON. Then, as the capacitor charges at a rate determined by the resistors and the value of C_1, Q_2 returns to conduction, and is driven by the feedback to its fully ON state. This circuit is known as a monostable circuit, to distinguish it from the bistable flip-flop. Its application is as a generator of standardized, longer pulses from input pulses that are narrower and of various magnitudes.

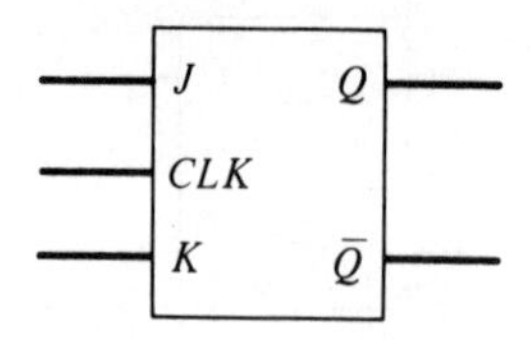

(a) Logic symbol for *JK* flip-flop

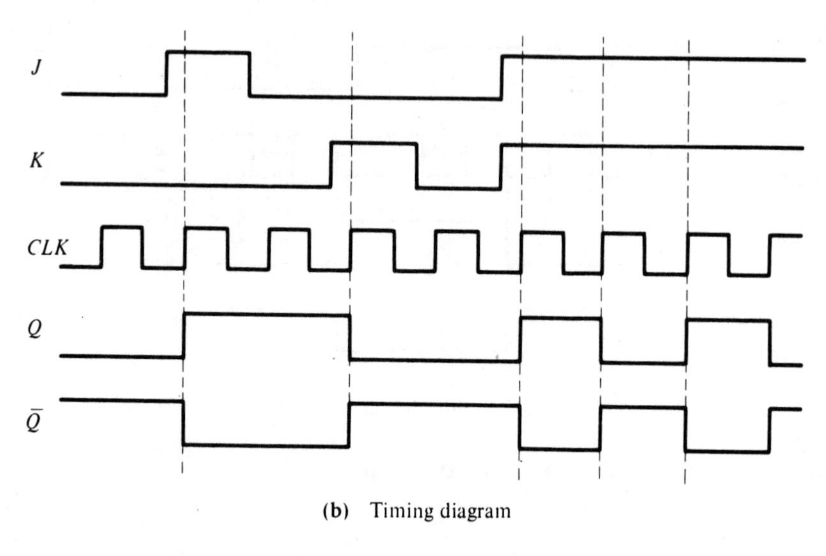

(b) Timing diagram

Fig. 13–9 Symbol and timing diagram of JK flip-flop.

13–3.9 The Astable Circuit

There is a third variation of this type of circuit. It is shown in Fig. 13–11, and has two coupling capacitors, C_1 and C_2. Now there is no longer a stable state, giving the circuit its name *astable*. It is also known as a *multivibrator*. It is used to generate a square wave. The period of the square wave is determined by the RC time constants.

13–3.10 Schmitt Trigger

A type of monostable circuit known as the Schmitt trigger is shown in Fig. 13–12. It has a common emitter resistor R_E. When there is no input at the base of Q_1, that transistor if OFF. A voltage divider consisting of R_{L1}, R_1, and R_2 provides V_{B2} sufficient to turn Q_2 ON, and this is the stable state. The voltage V_E

across the common emitter resistor R_E is determined by the current in Q_2. If a voltage is now applied to the base of Q_1, and the voltage is sufficient to overcome V_E plus the base-emitter voltage of Q_1, that transistor begins to turn ON. This causes the collector voltage of Q_1 to drop, reducing base current to Q_2, turning that transistor partially off. This reduces the current in Q_2 and, hence, reduces V_E, which gives the base input voltage V_{B1} of Q_1 less voltage to overcome; as a result Q_1 conducts more heavily. This is seen to be a positive-feedback configuration, resulting in the eventual state of Q_1 ON, Q_2 OFF. This state is maintained until V_{B1} is dropped below V_E. When that happens, Q_1 begins to turn OFF, the action reverses, and eventually the stable state is regained. Since switching takes place very rapidly, the output is either V_{CC} or V_C (the stable state), depending only on whether or not V_{B1} exceeds the critical value needed to switch. The circuit serves to sense whether or not this condition has occurred.

13–4 COUNTING

13–4.1 Single Flip-Flop

When a flip-flop is used by itself, it has two basic capabilities. It can store a single bit of information, which, for example, it can accept from an input to the

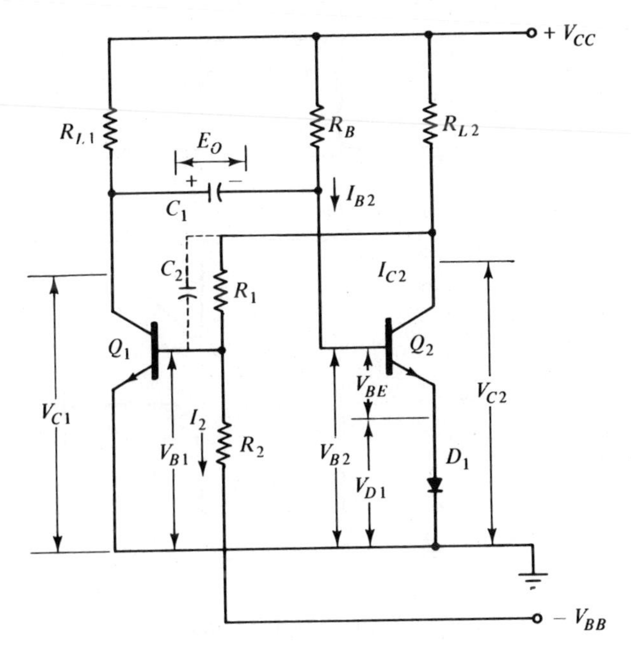

Fig. 13–10 Monostable circuit.

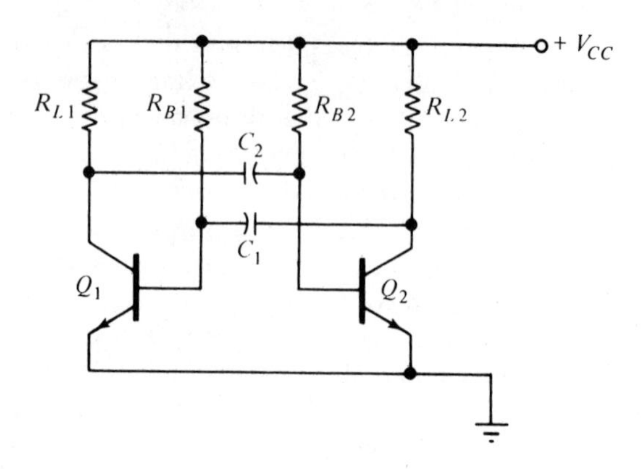

(a) Collector-coupled astable multivibrator circuit

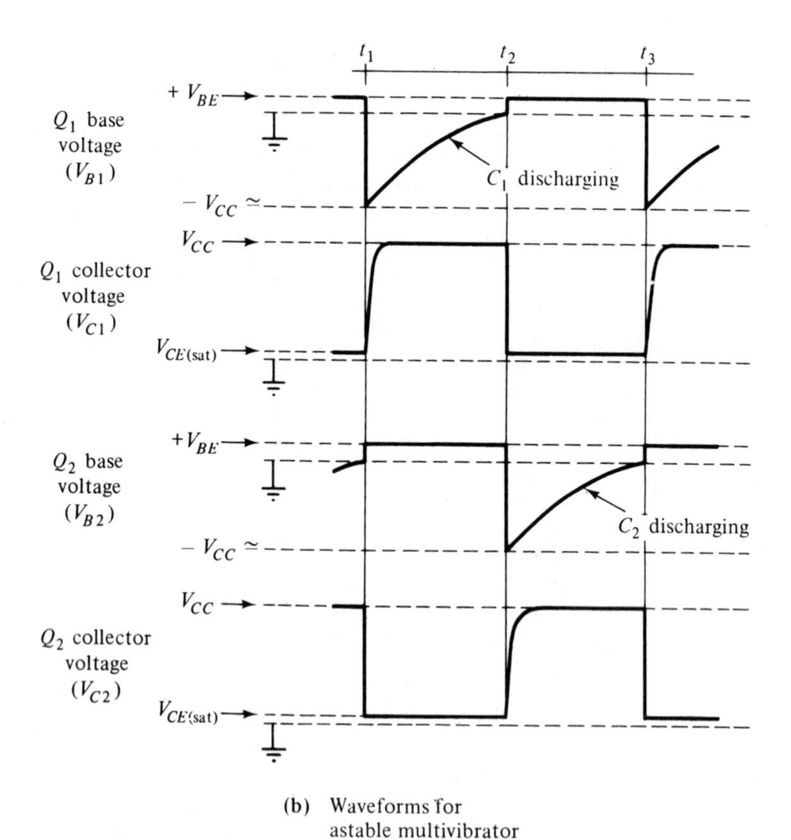

(b) Waveforms for astable multivibrator

Fig. 13–11 Circuit and waveforms of an astable circuit.

computer in order to pass it on. It can also store an output quantity before it is printed. Or it can store the bit in intermediate storage while computing is in progress. Obviously, a single bit is not too useful by itself as a memory; hence, very large numbers of flip-flops have to be combined. The second capability of the flip-flop is as a divide-by-two circuit. Whether triggering is by positive or negative pulse, two triggers are required to produce an output pulse at either the Q or $\overline{Q}$ output. Therefore, for every two pulses at the input, one output pulse occurs. Again, the use of one flip-flop has limited applications; hence, several are

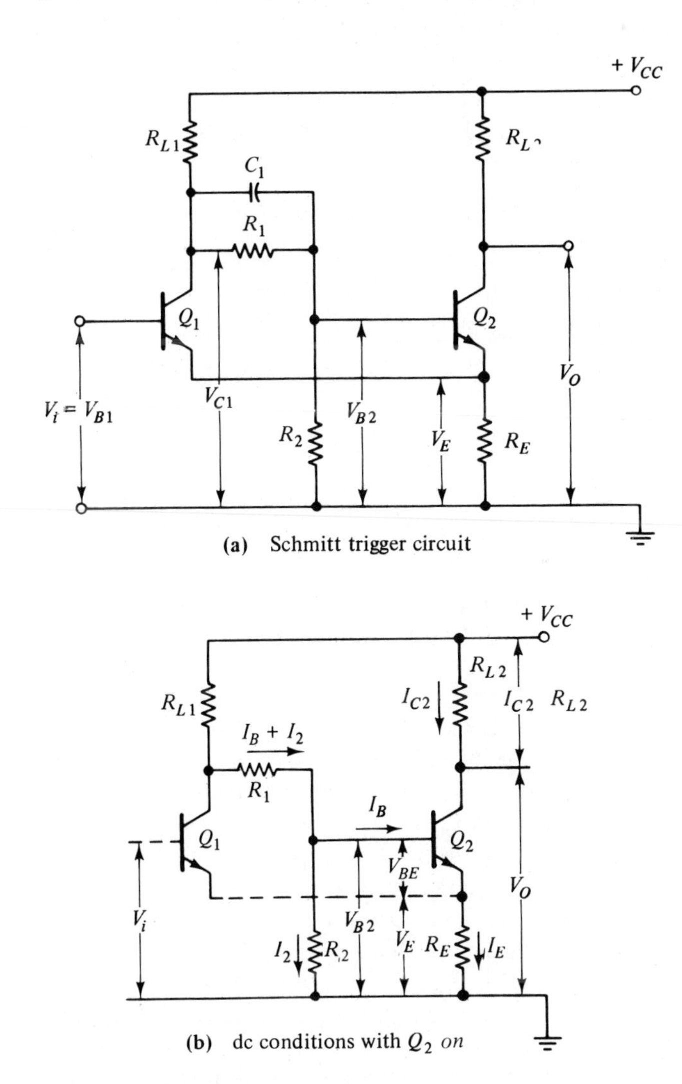

Fig. 13–12 Schmitt trigger circuit.

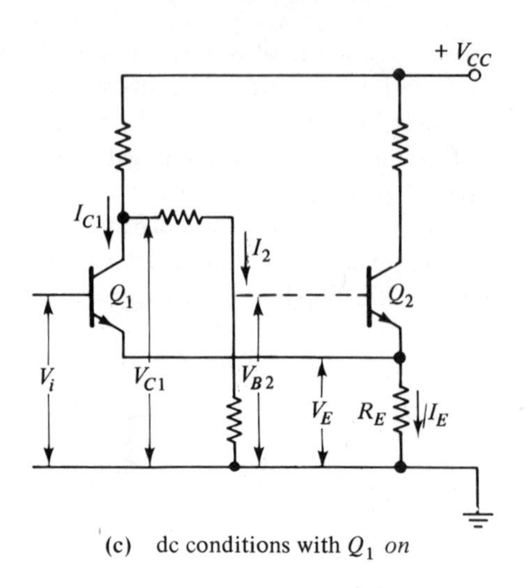

(c) dc conditions with Q_1 *on*

Fig. 13–12 *(continued)*

cascaded. Since one flip-flop divides by 2, two flip-flops divide by 4. Division by all the powers of 2 ($2^2 = 4$, $2^3 = 8$, $2^4 = 16$, $2^5 = 32$, $2^6 = 64$, $2^7 = 128$, etc.) is directly possible.

13–4.2 Four-Stage Counter

The schematic diagram for a four-stage counter is shown in Fig. 13–13, and the block diagram and truth table are shown in Fig. 13–14. Signal travel is shown in a direction from right to left—rather than the usual left to right—because when done this way it agrees with the way numbers are written. As an initial state, assume that transistors Q_2, Q_4, Q_6, and Q_8 are ON, or conducting, making their collector voltages 0. At this time, Q_1, Q_3, Q_5, and Q_7 are OFF, making their collector voltages V_{CC}. This is defined as the zero or 0000 condition of the counter. Also assume that a negative trigger is required. If a trigger pulse is now applied to transistor Q_2, that element is turned OFF, turning Q_2 ON. This is the 0001 condition of the counter. At the collector of Q_2, the voltage goes to V_{CC}, which means that a positive pulse appears at this collector. However, since a negative trigger is required, this has no effect on the second stage. When another trigger pulse is applied at the input, Q_1 goes OFF and Q_2 goes ON, generating a negative trigger pulse. This is coupled to the second stage, turning Q_3 ON, and turning Q_4 OFF. This is the 0010 condition of the counter since the first stage has returned to 0, while the second stage is at 1. Applying further triggers, all the conditions in the truth tables are generated in turn. There are 16 stable states, with the counts of 16 and zero being identical. The states are given in the binary and decimal systems. Waveforms are given in Fig. 13–15.

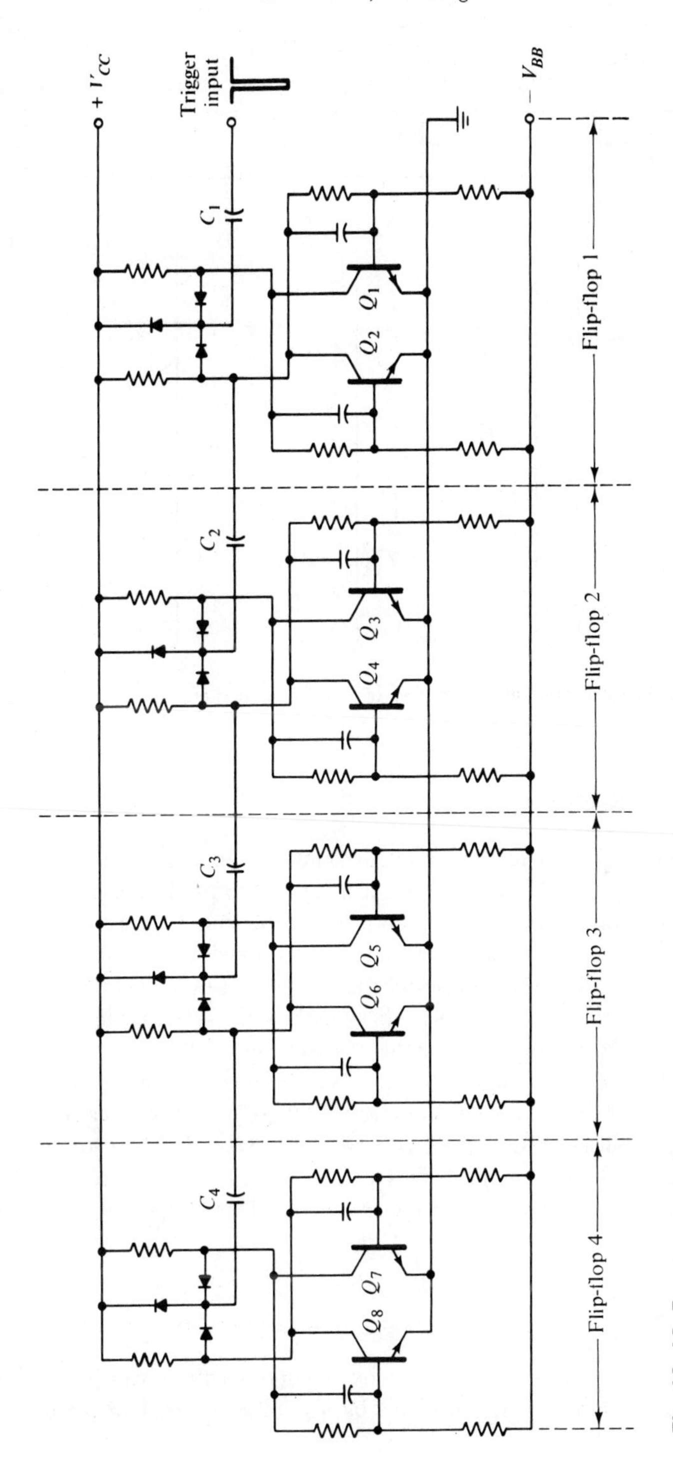

Fig. 13–13 Four-stage counter.

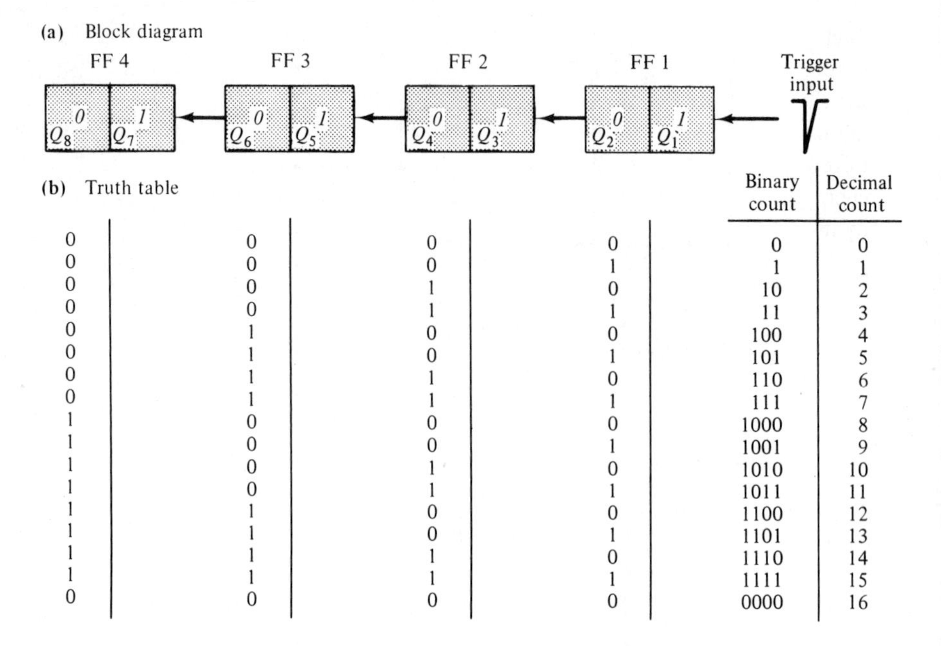

FF 4	FF 3	FF 2	FF 1	Binary count	Decimal count
0	0	0	0	0	0
0	0	0	1	1	1
0	0	1	0	10	2
0	0	1	1	11	3
0	1	0	0	100	4
0	1	0	1	101	5
0	1	1	0	110	6
0	1	1	1	111	7
1	0	0	0	1000	8
1	0	0	1	1001	9
1	0	1	0	1010	10
1	0	1	1	1011	11
1	1	0	0	1100	12
1	1	0	1	1101	13
1	1	1	0	1110	14
1	1	1	1	1111	15
0	0	0	0	0000	16

Fig. 13–14 Block diagram and truth table of four-stage counter.

13–4.3 Decade Counter

The most efficient way to utilize counters is in a system based on 16, the hexadecimal system. However, often it is desired to produce a decade counter; i.e., one that will count to 10, even if it is not as efficient. This is easily done if the four-stage counter of the previous section is equipped with feedback. As can be seen in Fig. 13–16, the output pulse that is produced by stage four when it goes to 0000, the 16 state or 0 state, is fed back to the two previous stages, setting both of them to 1 instead of zero. Thus there is no longer a 0000 state, nor any other low state between 0000 and 0110. In other words, counting begins with 0110. This is now defined as zero. Counting from there to full counter requires only nine trigger pulses; hence, the feedback has converted the 16-counter into a 10-counter or decade counter. This device is very useful since it makes the computer adaptable to our own number system, which happens to be the decimal system.

13–4.4 Other Counts

As can be seen from the previous section, feedback can be employed to preset a counter so that it can divide by a number other than the number by which it would divide without feedback. Extending this theory, feedback can also be applied to a three-stage counter to make it divide by 5, 6, or 7, and to a four-stage counter to make it divide by 9, 10, 11, 12, 13, 14, and 15. Such ratios may not be extensively required in the computer, but they are useful when it is desired to produce precision frequency standards. That is, a voltage can be generated, if that is convenient, at a high frequency and then divided down by any integral ratio to achieve a desired precise lower frequency.

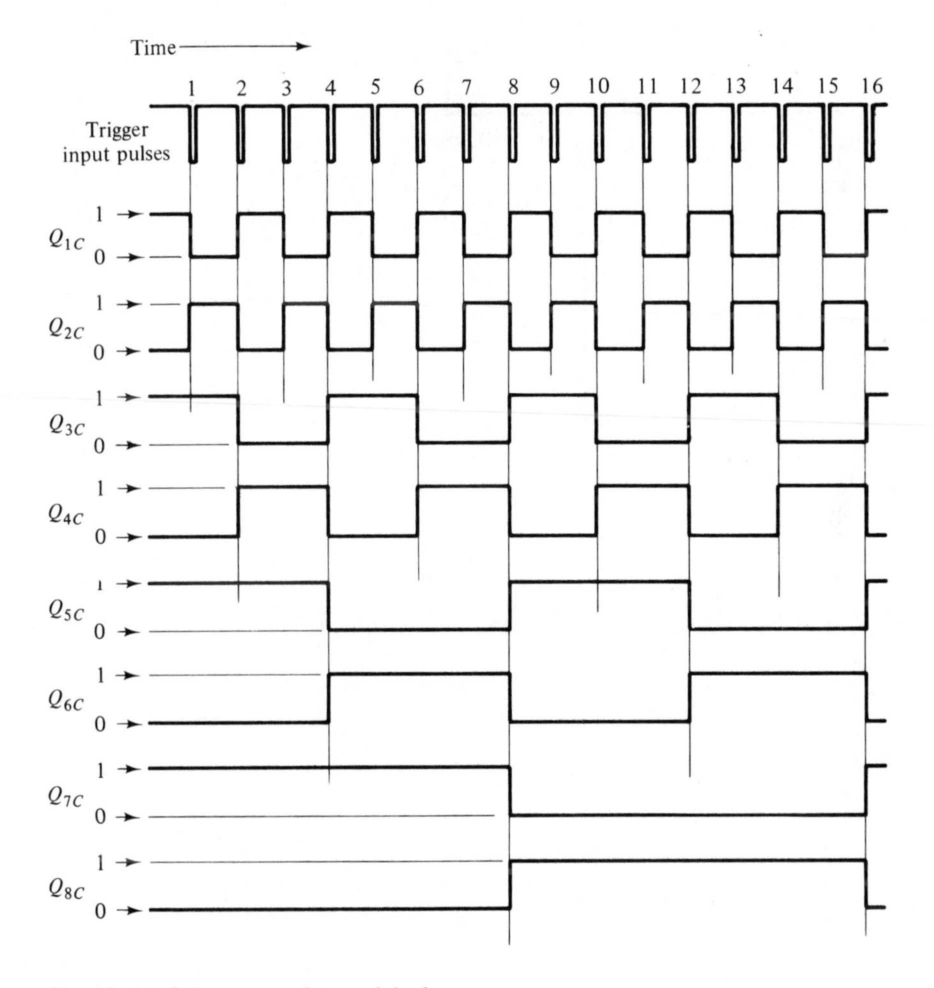

Fig. 13–15 Collector waveforms of the four-stage counter.

13–4.5 Memory

The flip-flop counter, as it stands, can be used as a short-term memory. A number is simply “read in” by feeding a pulse train to the input, and can later be “read out” again as needed. There is only one caution required. The format must be considered. By format is meant the number of digits that are expected in the information to be processed. This must be either known or assumed. Once a

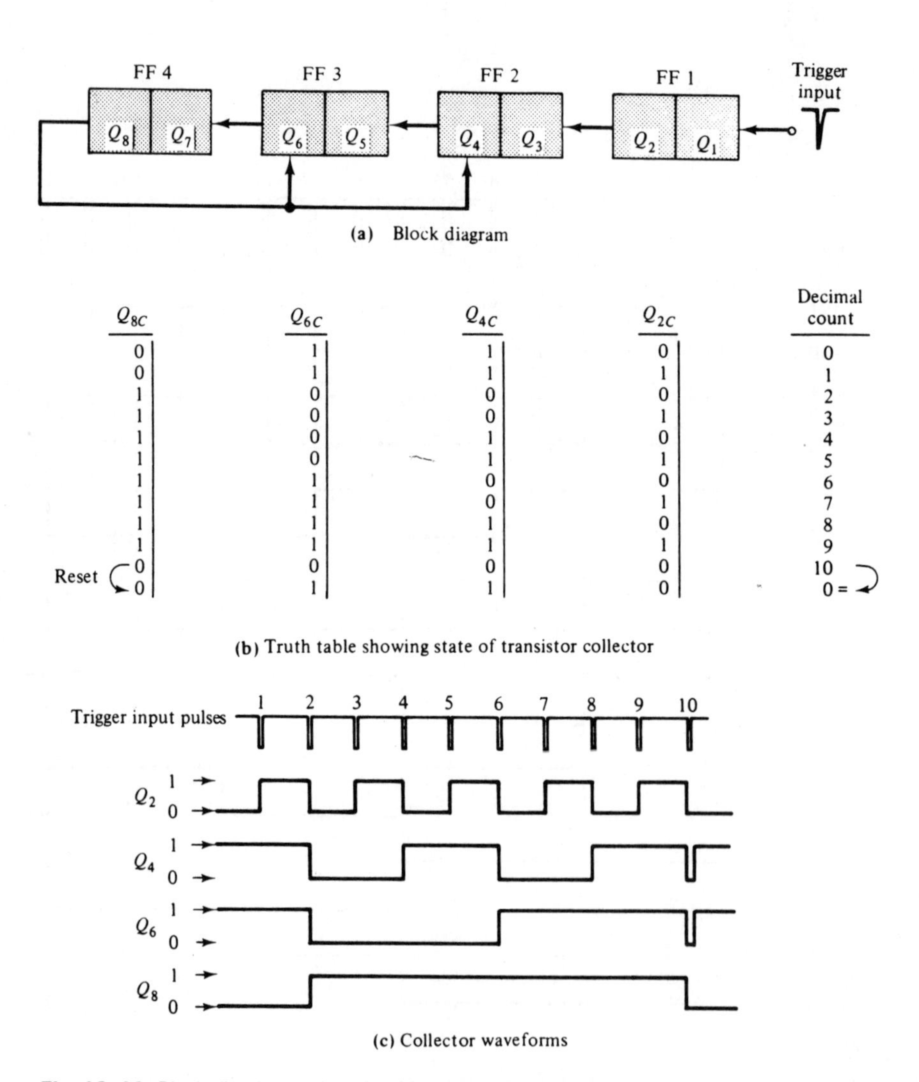

Q_{8C}	Q_{6C}	Q_{4C}	Q_{2C}	Decimal count
0	1	1	0	0
0	1	1	1	1
1	0	0	0	2
1	0	0	1	3
1	0	1	0	4
1	0	1	1	5
1	1	0	0	6
1	1	0	1	7
1	1	1	0	8
1	1	1	1	9
0	0	0	0	10
0	1	1	0	0 =

Fig. 13–16 Block diagram and truth table of decade counter.

format has been established, the counter must have a number of digits adequate to handle it. In an ordinary calculator the format is established once and for all by the number of digits in the calculator read-out display. But in a computer, the format must be established by the programmer, and once that is done, the memories must be capable of handling it, or a GIGO condition appears.

13–5 CALCULATION

13–5.1 Addition

If the number A is to be added to the number B, a flip-flop counter with an adequate format can be used. First, the number A is read into the counter, which in this case is called a *register* or *accumulator.* Once this is completed, the number B is read into the same register. The pulses from B advance the register from its setting for A to the setting for the sum of A and B; hence the process is direct.

13–5.2 Subtraction

Since a register or counter can only go up in pulse count, but cannot go down, pulses cannot be subtracted. But subtraction can be accomplished, nevertheless, indirectly. In Sec. 12–2.1 the concept of complements was introduced. If A is 1, $\overline{A}$ is 0, if A is 0, $\overline{A}$ is 1, etc. Let F represent the number achieved when a register is full, or the register capacity, which is a series of 1's only; i.e., $F = 11111$ for this case. We can write

$$A + \overline{A} = F, \quad \text{regardless of what is meant by } A \tag{13–2}$$

Similarly, of course, $B + \overline{B} = \text{F}$; hence $\overline{B} = F - B$ and $B = F - \overline{B}$. When a subtraction is to be performed,

$$\begin{aligned} C &= A - B \\ &= A - (F - \overline{B}) = A - F + \overline{B} \end{aligned} \tag{13-3}$$

But the quantity F is a string of 1's; therefore, the addition of just one more pulse returns the register to 0, which eliminates F from the equation. Hence, a way of accomplishing subtraction is to add the complement $\overline{B}$ to A and add one more pulse to the result:

$$C = A + \overline{B} + 1 \tag{13–4}$$

EXAMPLE 13-1

Subtract 18 from 25.

SOLUTION

In binary terms, 25 is the number 11001 and 18 is 10010. The complement of 18 is 01101 (each 1 is replaced by 0, each 0 replaced by 1). Adding 11001 and 01101 gives 100110. To this is added one more pulse, giving 100111.

The figure 1 at the extreme left is ignored since it exceeds the register capacity and has no significance except to indicate that the register was filled and then returned to zero. The remainder of the result is 111 in binary terms, which is 7 in decimal terms, as expected. As was pointed out above in discussing flip-flops, each flip-flop has two outputs, Q and $\overline{Q}$. Due to this fact, the process of subtraction is simple since the complement of a number does not have to be separately computed; it is always available at the other flip-flop outputs.

13-5.3 Integration

Integration in the digital computer is accomplished by use of a register or accumulator to generate the required output. Into this register are fed periodic samples of the input function multiplied by the appropriate time interval. Consider Fig. 13-17, showing a function to be integrated. It is assumed that integration is equivalent to finding the area under the curve. First, the time (x-axis) is divided into appropriate intervals. Obviously, the more intervals used, the shorter the time between intervals becomes; hence, the results are more accurate. A sample is taken at the end of each interval. For example, at time t_4 the function has the value y_4. The cruder approach is to multiply the value of y_4 by the time interval and insert the result into the accumulator. At all intervals from t_1 through t_n another sample is inserted into the accumulator. A more accurate way of accomplishing the same thing is to take not y_4, but the average of the value at t_3 and t_4:

$$\frac{y_3 + y_4}{2}$$

multiply that by the time interval, and insert:

$$S_4 = \left(\frac{y_3 + y_4}{2}\right)(t_4 - t_3) \qquad (13\text{–}5)$$

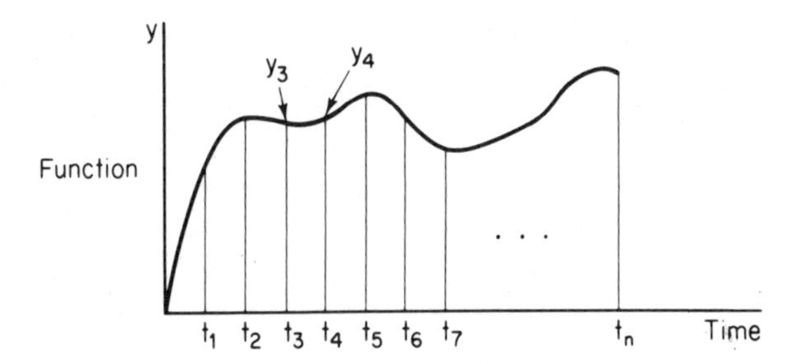

Fig. 13-17 Integration of a function.

where S stands for sample. The difference between the two approaches is based on "chopping up" the function into rectangles in the first case and into trapezoids in the second. As mentioned above, the shorter the time interval, the more accurate the result; however, at the same time, there is less advantage to the trapezoidal approach. An ordinary digital register is adequate as an integrator, though it is necessary to take care that the format be adequate—i.e., that the result of the integration does not overfill the register. At any time in the process, the instantaneous state of the register is the integral of the function.

13-5.4 Differentiation

Differentiation is the opposite of integration; it can be defined for present purposes as the determination of the slope of a function. Figure 13-17 can again be used for analysis. To determine the slope at time t_4, it is not only required to have y_4, the value of the function at t_4, but also the value at the time t_3—namely y_3. The slope, then, is the ratio

$$\frac{y_4 - y_3}{t_4 - t_3} \tag{13-6}$$

This value is not fed to an accumulator or register, but straight to the output, because slope is an instantaneous value, as contrasted to the integral, in the previous section, which is an accumulation.

13–5.5 Comparison

When two quantities A and B are to be compared, it is possible to consider the process a subtraction, which is carried out as in Sec. 12–1.4. If the procedure given there is followed, then three cases can arise. If $A > B$, the subtraction can be carried out, and an overflow pulse is produced. If $A = B$, an overflow pulse is also produced, but the register is 0. If $A < B$, no overflow pulse is produced.

This is one method for comparing digital quantities that is simple because it requires little more than the subtraction process. The concepts of "same" and "different" are clearly defined in a digital system since two quantities, in order to be "different," must vary from each other by at least one digit. But in an analog system, in order to activate the comparator circuits, a certain minimum voltage is required, which leaves open the question of whether or not the two voltages might differ from each other by a quantity less than this minimum.

13–6 LOGIC

13–6.1 Definition

Whatever may be meant by logic in ordinary English usage, in the computer, since all concepts must be concrete, the word *logic* simply describes the realization or embodiment of the logic concepts discussed in Sec. 12-2. These are the AND, OR, NOR, and NAND circuits, each of which can be realized in a variety of ways, as described below. When many of these simple logic circuits are combined, complex logic problems can be solved.

13–6.2 Diode AND Circuit

The diode AND circuit shown in Fig. 13–18 has three diodes, any one of which can be forward biased by an input of 0, since $V_{cc} > 0$. This makes the output voltage V_o equal to the forward voltage of a diode. Actually this is either 0.3 volt for germanium or 0.7 for silicon, but for practical purposes it is assumed to be 0. Hence, if any of the inputs is 0, the output is 0. But if all of the inputs are 1 (i.e., the voltage applied to each diode is approximately V_{CC}), all the diodes are reverse biased, and the output voltage is V_{CC}, defined as 1. It is seen that this circuit functions as an AND configuration.

13–6.3 Diode OR Circuit

In Fig. 13–19, if any of the inputs is 1, the diode connected to that input is forward biased, and the applied voltage less the diode drop appears across R_1,

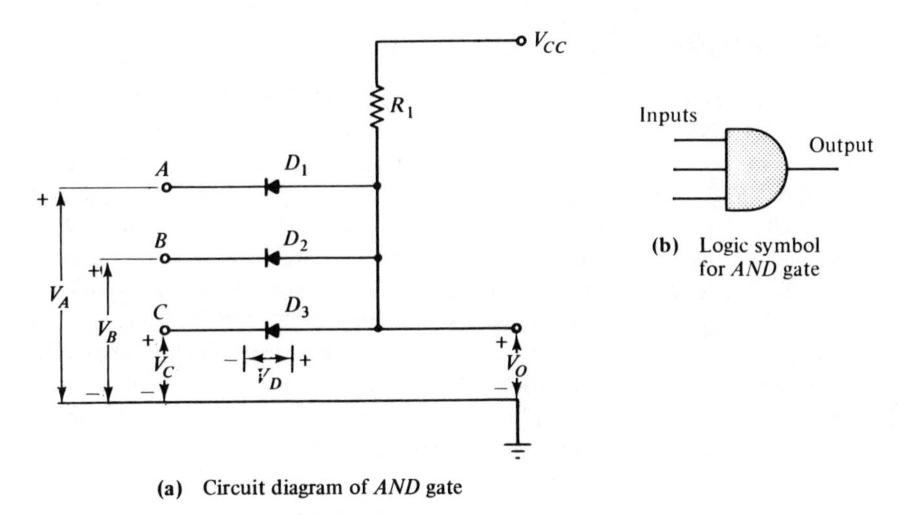

(a) Circuit diagram of *AND* gate

Fig. 13–18 Diagram and symbol of AND gate.

giving a 1 output. This is true regardless of what happens to the other inputs since their diodes are either reverse biased, or they have the same voltage as the first input. Hence, this circuit functions as an OR configuration.

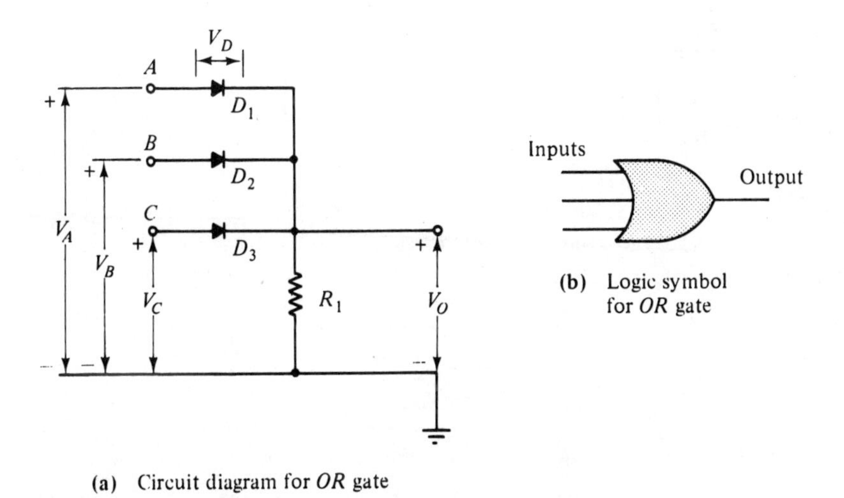

(a) Circuit diagram for *OR* gate

Fig. 13–19 Diagram and symbol of OR gate.

13-6.4 NAND Gate

The circuit of Fig. 13-20 operates as a NAND gate in that the left section is an AND circuit (all inputs must be 1 to bring V_X to 1), and the right section is an inverter which produces a 0 when its input is 1, and vice versa. This circuit, in this particular application, can be called a NOT gate. The combination of an AND gate and a NOT gate gives the NAND configuration. The circuit would operate without diodes D_4 and D_5, were it not for the forward voltage of diodes D_1, D_2, and D_3. But since this voltage is present, it is possible that transistor Q_1 could be ON when it should properly be OFF. Diodes D_4 and D_5 ensure that V_X is at least sufficient to forward bias these diodes; hence, V_X must be greater than two diode forward voltages, requiring a 1. Thus, V_X cannot be a 0 falsely turning Q_1 ON. The symbol for the NAND circuit is the symbol for an AND circuit followed by the small circle to indicate the inverter stage.

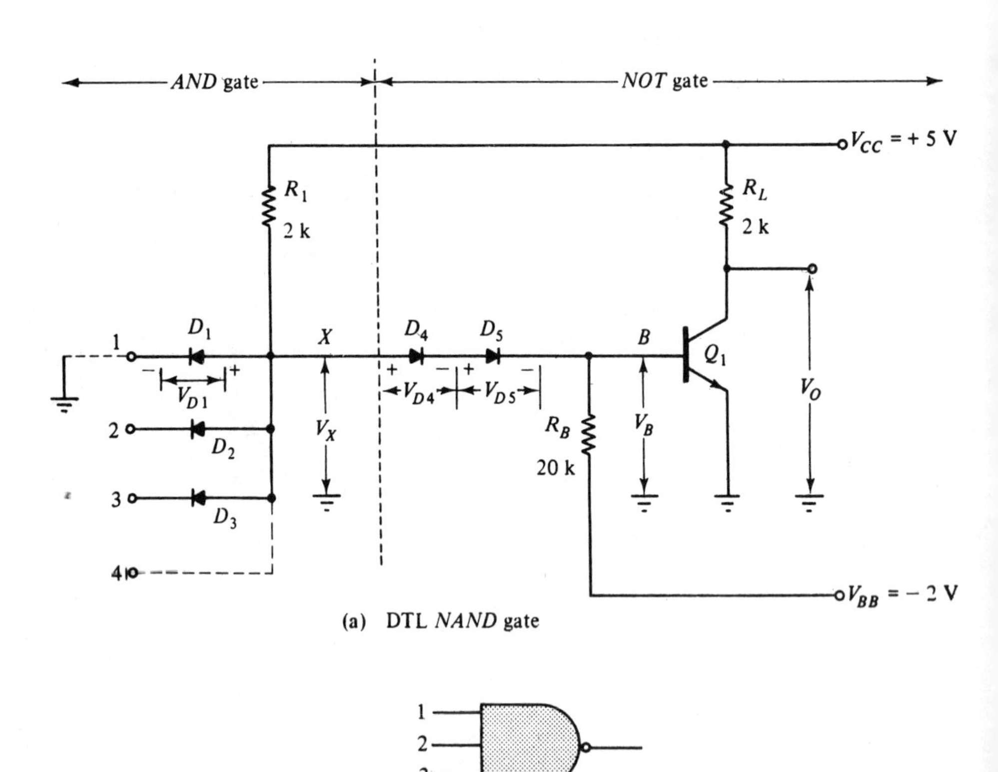

(a) DTL *NAND* gate

(b) Logic symbol for *NAND* gate

Fig. 13-20 Diagram and symbol of NAND gate.

13-6.5 NOR Gate

If an OR circuit is combined with an inverter stage, or NOT gate, as shown in Fig. 13-21, the NOR gate results; i.e., any one of the inputs can produce a 1 across R_1, turning transistor Q_1 ON, regardless of what appears at the other inputs. When Q_1 is ON, the voltage at the collector is 0. The symbol is the one used for the OR circuit followed by the small circle indicating inversion.

13-6.6 TTL Logic

TTL logic (transistor–transistor logic) is obtained by replacing the diodes in the circuits shown above by transistors. Figure 13-22 shows a TTL NAND gate. It is seen that the diodes have been replaced by transistor stages. In Fig.

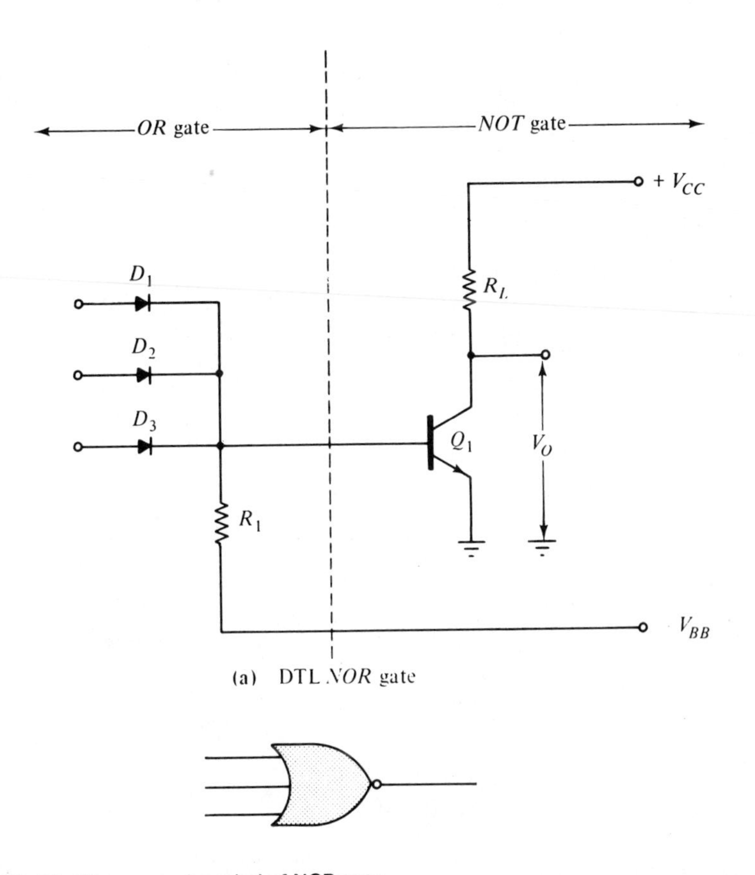

Fig. 13-21 Diagram and symbol of NOR gate.

13–22, Q_1 is a transistor with three emitter terminals, and this can be easily fabricated as an integrated circuit, but three transistors could have been employed. The advantage of the TTLs over diode logic consists of the fact that less loading is caused on whatever circuits are feeding the inputs, and a lower

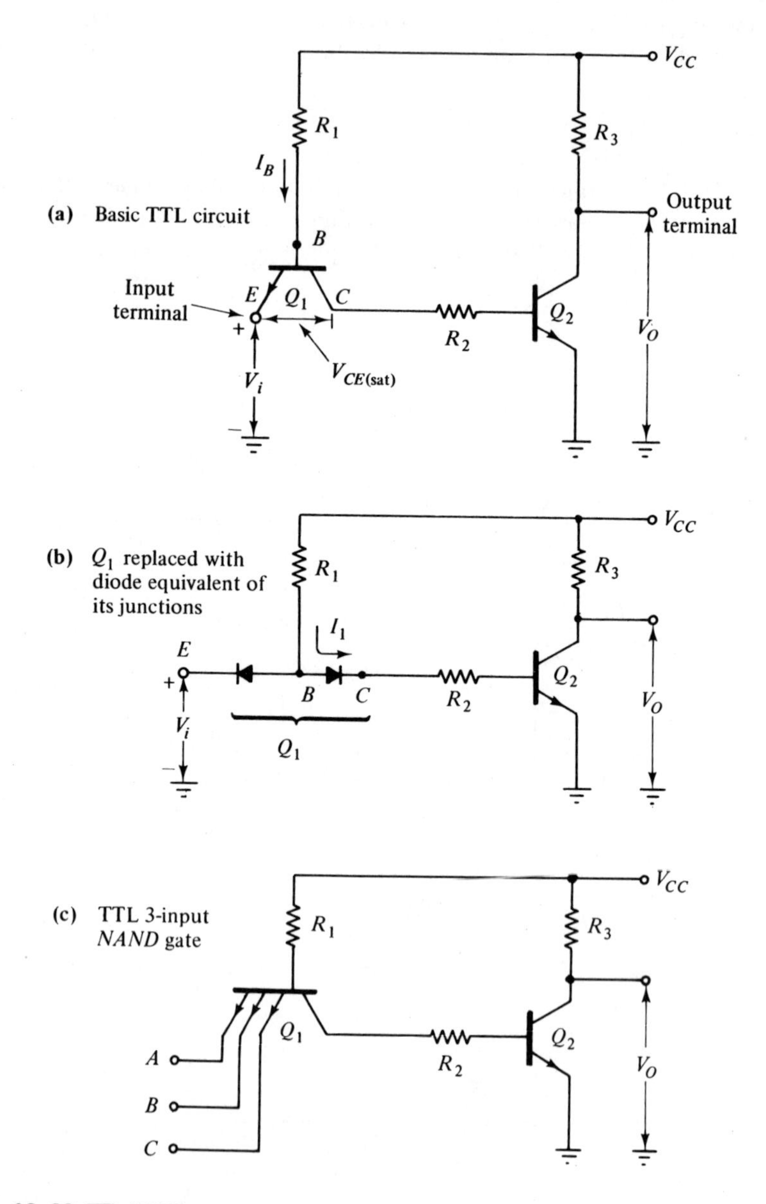

Fig. 13–22 TTL NAND gate.

output impedance is presented to whatever circuits are being fed from the logic output. This influences fan-in and fan-out as discussed below.

13–6.7 Fan-In and Fan-Out

For illustrative purposes, the circuits analyzed above have been shown with three inputs each. However, in actual practice, the number of inputs can vary, and the outputs are not necessarily fed to only one point, but may feed several other circuits. It must be remembered that each point to which the output is fed causes some loading. Therefore, the manufacturer's specification limits the number of outputs that can be connected. This specification is called *fan-out*. The number of inputs which a circuit accepts is known as *fan-in*. The concepts of fan-in and fan-out must be considered when selecting components.

EXAMPLE 13-2

Assume that $V_{CC} = 5$ V and that a circuit supplying logic information has an output impedance of 2 kΩ. The circuit is to supply this information to a number of circuits with input impedance of 3 kΩ each, requiring at least 1.5 V for proper operation. Determine the fan-out.

An equivalent circuit can be drawn as in Fig. 13–23, showing a 5-V generator, an internal impedance of 2 kΩ, and a load, to which at least 1.5 V must be applied. (The figure of 1.5 V is a conservative number which must ensure that the voltage for the 1 state is considerably greater than 0.5 V, the value that might possibly be reached with a 0 state, so that it is certain that a 1, not a 0, is present.)

By Ohm's law,

$$I = \frac{5\text{ V}}{2\text{ k}\Omega + R}$$

$$1.5 = IR$$

$$\frac{1.5}{R} = \frac{5\text{ V}}{2\text{ k}\Omega + R}$$

$$3\text{ k}\Omega + 1.5\,R = 5\,R$$

$$3.5\,R = 3\text{ k}\Omega$$

$$R = \frac{3\text{ k}\Omega}{3.5} = 857\ \Omega$$

Since each of the circuits to be supplied has an input resistance of 3 kΩ, they have a total input impedance of

$$\frac{3}{N}, \text{ where } N \text{ is the number of such circuits}$$

$$\frac{3}{N} = 857\ \Omega$$

$$N = \frac{3\ \text{k}\Omega}{857} = 3.5$$

and since an integral number is required, the fan-out is 3.

13–6.8 Application

Applications of logic circuits are legion, but to illustrate the possibilities one application can be mentioned. This is the binary-to-decimal conversion shown in Fig. 13–24. Here the states of the flip-flops have to be interpreted so as to give decimal numbers. Taking the circuit for number 2, and consulting the table, it is seen that the binary count in this case (since feedback is used) is 1000; i.e., transistors Q_8, Q_5, Q_3, and Q_1 must read 1. Hence, diodes D_9, D_{10}, D_{11}, and D_{12} constitute an AND gate, turning ON transistor Q_{12}. As a further example, the circuit for number 7 requires a binary count of 1101; hence, transistors Q_8, Q_6, Q_3, and Q_2 must be 1, and diodes D_{29}, D_{30}, D_{31}, and D_{32} make up the AND gate, turning ON transistor Q_{17}. A similar AND gate exists for each of the ten numbers.

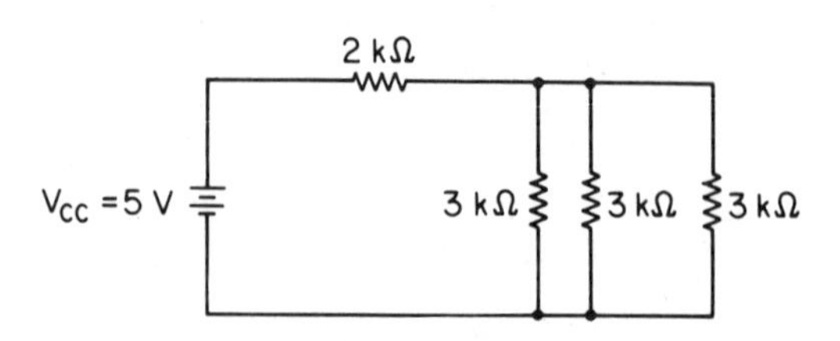

Fig. 13–23 Fan-out.

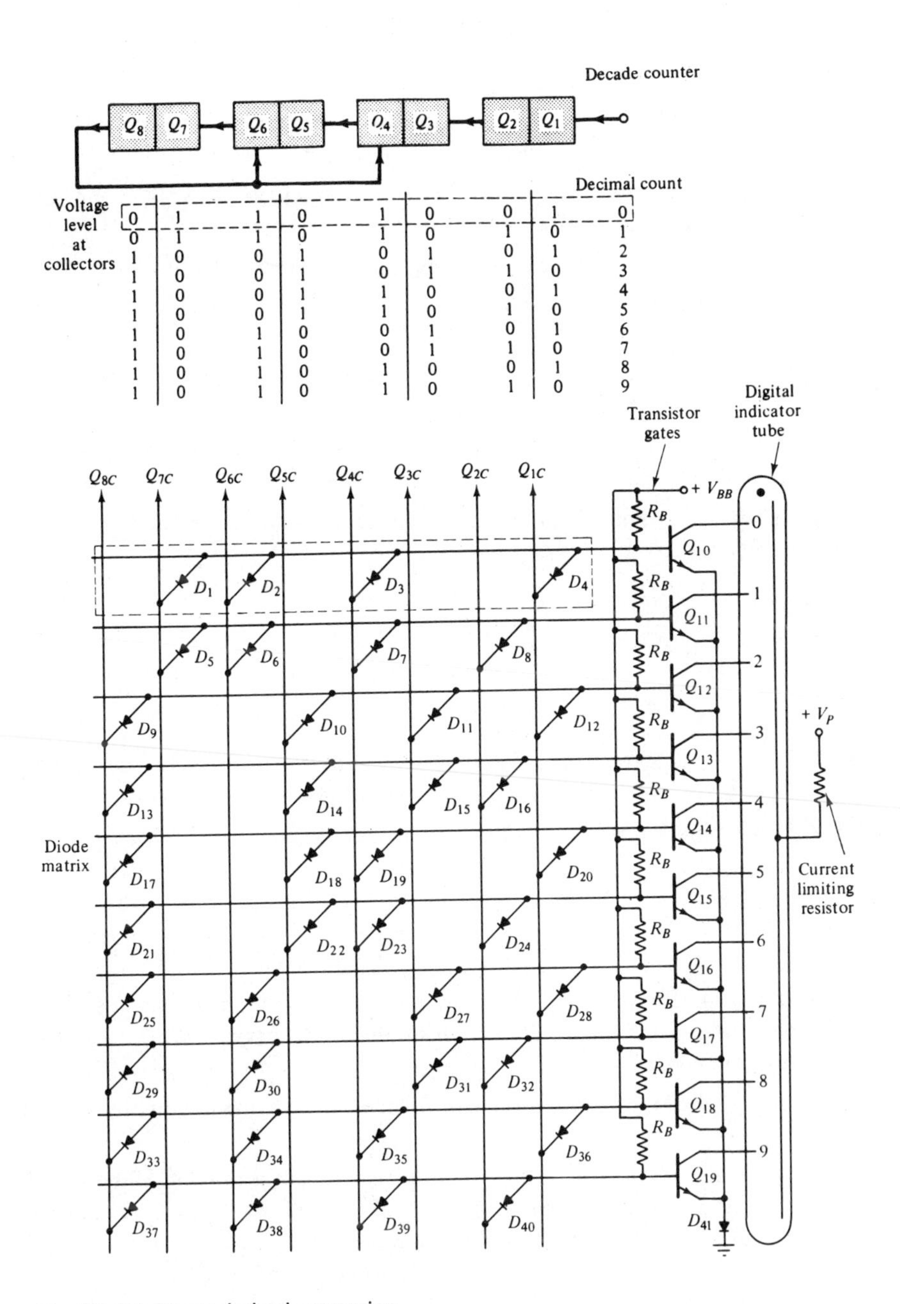

Fig. 13–24 Binary-decimal conversion.

QUESTIONS

13–1 Explain the difference between analog circuits and digital circuits insofar as their operation of active devices is concerned.

13–2 Explain the relationship between nonlinear circuit operation and LSI (large-scale integration).

13–3 Summarize the importance to the digital computer of the flip-flop circuit.

13–4 Explain the difference between analog circuits and digital circuits insofar as their use of feedback is concerned.

13–5 If the letters of the alphabet are to be represented by a binary system, how many bits are required per letter?

13–6 What might be a possible use for a monostable circuit? For an astable circuit?

13–7 What is the complement of a number? What is the complement of a logic function?

PROBLEMS

13–1 In Sec. 13–2.2 reconsider Example 13-5 with the load resistor increased to three times its value. Determine dissipated power.

13–2 Draw a block diagram of a flip-flop divider to divide by
a. 7
b. 12
c. 100
d. 35

13–3 A precise frequency of 25 kHz is required with a crystal oscillator operating at 1 MHz available. Draw the block diagram for the required flip-flop divider.

13–4 In Fig. 13–24, determine the logic involved in deriving the numbers 5 and 9.

13–5 "A picnic will take place if it is not too hot and not too humid; if there is no rain; if at least two people bring food; if either John, Tom, or Bill brings his car; and if that car has gasoline but does not get a flat." Devise a block diagram for logically determining whether the picnic takes place.

13–6 If a calculator is to have the capability of handling six-digit decimal numbers, determine the number of flip-flops required to construct a register.

13-7 A circuit with a supply voltage of 10 V and an internal output impedance of 5 kΩ is to supply circuits with an input impedance of 10 kΩ, and the voltage supplied must be at least 2 V. What is the fan-out?

13-8 Plot the curve $y = x^2$. Graphically determine the area under the curve between $x = 0$ and $x = 4$, and the slope at $x = 2$ and $x = 4$.

13-9 Do Problem 12-4 using complements.

Chapter 14

Integrated Circuits

14–1 INTEGRATED CIRCUITS

14–1.1 General Description

The planar technique provides the possibility of manufacturing a large number of semiconductor devices (such as diodes, BJTs, FETs, or MOSTs) on a single silicon crystal, covering an area of only a few square millimeters. A logical extension of the planar technique is to produce complete circuits, consisting of active devices, resistors, and capacitors, on a single crystal. The required connections between the various circuit elements are made by vapor-depositing metal strips (aluminum) between the appropriate points on the crystal. Deposited connections made right on the crystal have the advantage of being much more reliable than wired connections.

It is entirely possible to produce resistors on the same crystal that contains the planar transistors. The conductivity of silicon can be precisely controlled by doping. For example, a diffused n layer in a p-type substrate can have a typical sheet resistance of 100 Ω/mm^2, so a strip 1 mm long and 10 μm wide has a resistance of 10,000 Ω. Vapor deposition of aluminum contacts to the ends of this small strip then provides a resistor as an integral part of the silicon crystal structure. It should be noted that resistors manufactured in this manner occupy a relatively large area of the crystal; cost plays a role here.

Capacitors can be manufactured by the MOS technique. Here a thin layer of silicon dioxide is deposited onto the crystal and a metal plate is placed on top of the SiO_2 layer in exactly the same manner as is used in the gate structure of a MOST. Alternatively, a reverse-biased diode may be used as a capacitor.

The planar technique of etching, diffusing, and depositing can therefore be used to manufacture both active devices, such as transistors, and passive elements, such as resistors, on a single silicon crystal. A complete circuit, containing transistors, diodes, resistors, and capacitors, all produced on a single crystal chip, with a single process, is called an *integrated circuit* (IC).

In changing from discrete components to integrated circuits, several factors have to be taken into account. For example, in many ordinary or conventional circuits the negative poles of diodes, transistors, or capacitors are not interconnected. In the integrated circuit, therefore, the various elements on the common crystal substrate must be isolated from one another. There are several ways of doing this. One of the commonly used schemes places the complete IC in an n-type epitaxial layer which is grown onto a low-conductivity p-type substrate. The substrate only serves to provide mechanical support and adequate heat transfer (if required). Frames of wedge-shaped bars of p-type material are then diffused into the n-type epitaxial layer, sufficiently deep to penetrate into the p-type substrate. Together with the substrate, these wedge-shaped frames isolate the various elements from one another.

Figure 14–1 shows a cross section of an integrated circuit containing several components. The bipolar transistor in Fig. 14–1 is manufactured by diffusing a p-type base region into the n-type epitaxial layer. A second diffusion of n^+-type material (heavily doped n material) into this base region produces the emitter, while the epitaxial layer itself forms the collector. A connection to the bottom of the epitaxial layer can no longer be made, and the collector connection is now provided by making an additional n^+-type diffusion into the top of the epitaxial layer. The npn transistor so constructed is isolated from its neighboring

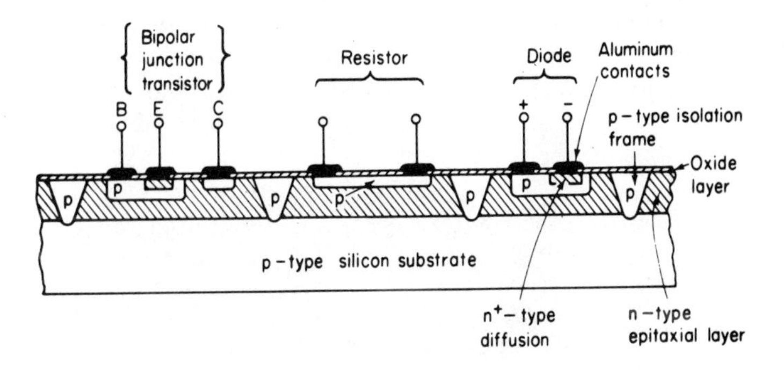

Fig. 14–1 Different elements of the integrated circuit, manufactured on a p-type silicon substrate by the planar technique.

elements by the surrounding wedge-shaped frame, diffused into the epitaxial layer and penetrating into the substrate.

The resistor is made by a single diffusion of *p*-type material into the epitaxial layer. The doping concentration and the dimensions of the diffused area determine the resistance. Contacts are provided by vapor-depositing aluminum to the ends of the resistive strip.

The capacitor can be made by the MOS technique of depositing a very thin layer of SiO_2 on the epitaxial layer and then covering this by vapor deposition with a metal plate. Alternatively, a reverse-biased diode can be used as a capacitor.

The internal connections to the various elements of the IC are made by bonding thin wires to small aluminum contact pads on the IC. These thin wires are then welded to the heavy terminal strips of the IC package, as in Fig. 14–2.

14.1–2 Integrated Circuits with MOSTs

The simple structure and low power dissipation of MOS transistors make them ideally suited for production in large numbers per unit area of the silicon crystal chip. Moreover, since a MOST has a very high input resistance and allows current flow in both directions, some MOS circuits require fewer components than the corresponding bipolar circuits. These features make the MOST eminently suitable for use in integrated circuits.

The voltages applied to a MOST are always arranged so that the drain and the inversion layer (channel) are reverse-biased with respect to the substrate. Drain and channel are therefore isolated from the substrate by a depletion layer (see Fig. 4–22). This natural isolation of drain and channel by the depletion layer presents considerable advantages in the manufacture of ICs made up of MOS devices, since there is now no need for separate isolation diffusions. As a result, the number of photoetching processes is drastically reduced, and this of course is a cost-saving factor.

Since isolation diffusions are not necessary, a very large number of MOS-type components can be formed on a single silicon chip. In addition, since the power dissipation of MOSTs is very low, a very high packing density

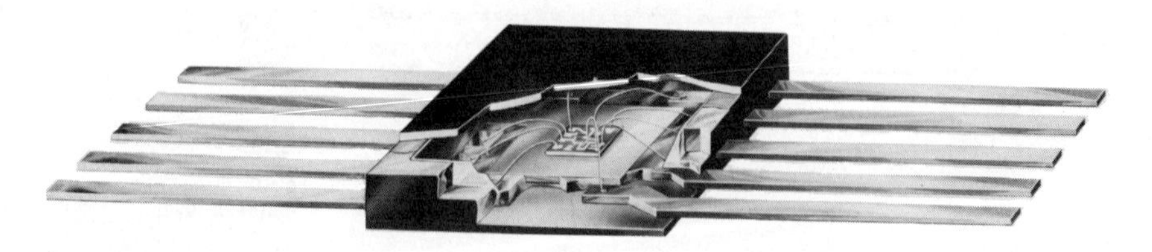

Fig. 14–2 Example of IC packaging (*Courtesy N. V. Philips Gloeilampenfabrieken, Eindhoven, the Netherlands*).

(number of devices per unit surface area) is now a practical proposition. The average packing density for MOS devices is typically 20,000 per square centimeter, while ICs with bipolar devices have a packing density of only a few thousand per square centimeter. A typical monocrystalline silicon chip measuring 1 mm^2 accommodates from 100 to 1,000 complete MOS circuits.

14-1.3 Medium-Scale and Large-Scale Integration

Particularly in large and complex digital systems, such as computers or telephone switching equipment, the integrated circuit plays an increasingly important role. Such systems are generally constructed from a very large number of identical circuit elements, which usually include discrete components such as transistors or diodes, or complete integrated circuits such as gates or flip-flops.

One of the problems encountered in manufacturing these large systems concerns the matter of *interconnections* to be made between the many elements. In an individual IC, for example, each internal connection to the IC element is bonded to aluminum contact strips deposited on the crystal, and these in turn are welded to the terminal strips of the IC package. A large array of ICs, mounted for example on a printed circuit (PC) board, then requires many additional interconnections to produce the desired final circuit. Rather than cutting the crystal up into its individual IC elements, it is extremely attractive, both from the point of view of cost and of circuit reliability, to connect ICs into the desired total circuit configuration right on the crystal by depositing aluminum strips to the appropriate IC terminals in the same operation in which normal contacts are made.

This then leads to a new stage in the development of IC techniques: medium-scale integration and large-scale integration.

In *medium-scale integration,* a number of IC units are manufactured on a single silicon chip measuring approximately 4 mm × 4 mm. The required connections between the IC elements are deposited on the silicon chip by the technique of aluminum deposition. Only a few external lead-in wires are required for final connection into the remaining part of the circuit.

If the number of interconnections to be made on the chip is too large to be accommodated in one plane, a second intermediate layer of silicon dioxide can be applied and appropriate cross connections are made from one plane to the next.

The fault risk due to local imperfections on the crystal (dust particles or dislocation of the masks during the etching or diffusion process) is the same for a large number of circuits on the chip as it is for one single circuit, and the reject ratio is therefore quite small.

In *large-scale integration* a very large number of similar elements to be interconnected are made on one single crystal slice. Circuits like these may be used in computer memories for example, where large arrays of thousands of MOST flip-flops are expected to compete against the conventional magnetic core memories. The problem with large-scale integration is basically one of reject. If, for example, one of several thousand flip-flops on a chip is faulty, the entire chip may be useless. To overcome this problem, a larger number of circuits than is needed (say 20 percent) is manufactured on the silicon slice. During the automatic measuring procedure, a computer marks the faulty elements and stores their location on the chip into its memory. This same computer then controls an automatic drawing machine which draws an interconnection pattern to join the good circuits and bypass the faulty ones. This method is called *discretionary wiring*. Making these connections is no more difficult than any of the steps needed to produce the circuit itself.

14–1.4 Thick-Film and Thin-Film Circuits

Fully integrated circuits are derived from a monocrystalline silicon substrate, into which diodes and transistors are deposited by the planar technique of etching and diffusion. This technique also allows the manufacture of capacitors and resistors on the same silicon chip. The fully integrated circuit is extremely useful and very economical when large numbers of identical circuits are to be produced. For smaller batches of circuits, perhaps as low as 200 or so, *film circuits* are often used to advantage.

In the film technique, the starting point of the circuit is the conductor rather than the semiconductor device. Generally, resistor–conductor or purely conductive (interconnect) patterns are deposited onto a nonconductive substrate by the film method, and discrete components, especially designed for attachment to the film circuit, are then added to form what is called a *hybrid circuit*. Hybrid circuits fill the gap between an assembly of discrete components and the fully integrated circuit chip. This gap exists because the cost of designing and making small numbers of integrated circuits is prohibitive. The use of film circuits to interconnect pretested logic chips, for example, enables a move toward medium-scale integration without the use of discretionary wiring.

Film circuits are made by various processes, such as chemical or electrolytic deposition, vacuum deposition, screen-and-fire techniques, vapor plating, and direct wiring. Film circuits are considered to be *thin-film* if made by vacuum-deposition methods, where the thickness of the deposited film varies between 100 Å and 2 μm.

In the thin-film technique, conductors, resistors, and sometimes capacitors are vapor-deposited onto a suitable substrate such as glass or glazed alumina. Conductors are often made of nickel, resistors of nickel–chromium (nichrome). It is also possible to make thin-film capacitors by depositing alternate layers of conductor and dielectric materials.

The thin-film process involves the deposition of a uniform layer of metal onto the substrate by heating the metal (nickel, if a conductive pattern is required) in a high vacuum to just below its melting point. With the substrate correctly positioned in the vacuum chamber, the nickel evaporates and settles on the substrate in a very thin layer or film. To delineate the desired pattern of conductors, a photoresist is applied to the entire surface area and exposed through a mask of the correct design. The unexposed areas are etched away from the substrate and the conductive pattern remains behind. Resistors can be manufactured in a very similar manner.

A complete circuit containing both resistors and conductors can be produced in a fairly simple process. First, a film of nichrome (resistive material) is deposited onto the substrate, and then a second deposition of nickel (conductive material) is made, so that the substrate is covered with a resistive film and a conductive film. A mask is then produced which delineates the area of both nichrome and nickel to be removed from the substrate. The plate is exposed through this mask and the unexposed areas are etched off, so that a pattern of nichrome plus nickel remains on the substrate.

A second mask, which delineates only the resistors, is then produced. After exposure the nickel is etched off, leaving the resistive areas exposed. Finally, the plate is immersed in a tin bath, where the tin adheres to the nickel but not to the nichrome. The result is a plate with a pattern of resistive elements and conductive interconnections, as in Fig. 14–3. Appropriate contact pads are incorporated in the structure to which semiconductor devices and other discrete elements can be soldered.

Thick-film circuits are made by depositing conductive and resistive pastes in the desired pattern on a substrate by means of a stencil screen. These pastes are a mixture of solvents, binders, glasses, and metals. The circuit is made

Fig. 14–3 Thin-film circuit (*Courtesy N. V. Philips Gloeilampenfabrieken, Eindhoven, the Netherlands*).

permanent by passing it through a drying tunnel where the solvents evaporate, and then through a furnace where the binders are burnt off and the glass phase is melted to bond the metal to the substrate.

A general-purpose conductor paste is palladium–silver; palladium oxide–silver paste is often used for resistors. Pastes with dielectric properties are also available and they can be used to make capacitors, to overglaze resistors, or as a sealant. Print thickness depends on several factors, such as the volume of paste deposited, viscosity, temperature, type of screen used, and so on. Typical print thickness for thick-film circuits is from about 0.01 to 0.03 mm, depending on the type of paste.

Screens are made from silk, nylon, polyester, or stainless steel mesh. A light-sensitive emulsion is attached to the screen and exposed through a suitable mask to ultraviolet light. The unexposed areas on the emulsion are washed off and the remaining parts of the emulsion fill the mesh voids. The screen is then brought into contact with the substrate and the desired paste is applied through the mesh openings.

To obtain high yields, almost all thick-film resistors are adjusted by a microengraving technique, where resistance values are adjusted by eroding resistive areas away with a jet of fine abrasive powder.

When the film circuit has been constructed, normal and subminiature components can be mounted on the substrate by various techniques. IC chips can be attached "face up" to the film circuit simply by bonding the chip to a mounting area. Fine wires between the chip and circuit contact pads then complete the attachment. The most recent method for "face-down" bonding of IC chips is used mainly with thin-film circuits and involves devices of the *beam-lead* type, where thin leads extend from the side of a normal IC chip. The chip is bonded face down to the substrate and the leads are attached to substrate pads by ultrasonics or thermocompression. Other semiconductor devices are available from the manufacturers in special form for mounting on strip lines.

Finally, the whole assembly is pressed or molded in plastic to make it moisture-resistant and to protect it mechanically.

14–1.5 IC Design

Though ICs can be produced to duplicate the functions of most circuits comprising discrete components (especially where some discrete components such as inductors and capacitors of large values are externally added because they do not lend themselves to IC realization), the configurations are often quite different. An equivalent circuit for an IC can always be drawn in terms of discrete components, but that equivalent circuit does not necessarily resemble its discrete-component counterpart. This is due to several considerations. Among discrete components, a resistor is the simplest element, followed in complexity by the capacitor, with the active element the most complicated. In IC technology, on

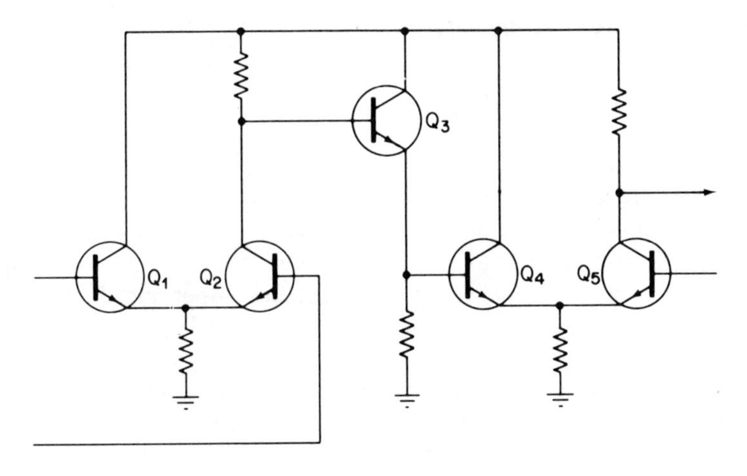

Fig. 14–4 Transistor as coupling element.

the other hand, the active element is the simplest both in terms of its production and in terms of achievable density. Then comes the resistor, with an intermediate desirability, and finally the capacitor is most complex. Hence, it is desirable to eliminate as many capacitors as possible, and to substitute active elements for resistors where possible. Due to the complexity of the internal interconnections, it is also desirable to minimize the number of these. Figure 14–4 shows how an additional transistor Q_3 can take the place of a coupling capacitor as an interstage coupling element. Figure 14–5 shows how a MOSFET can be substituted for a resistor. The MOSFET provides a space saving and permits greater power dissipation.

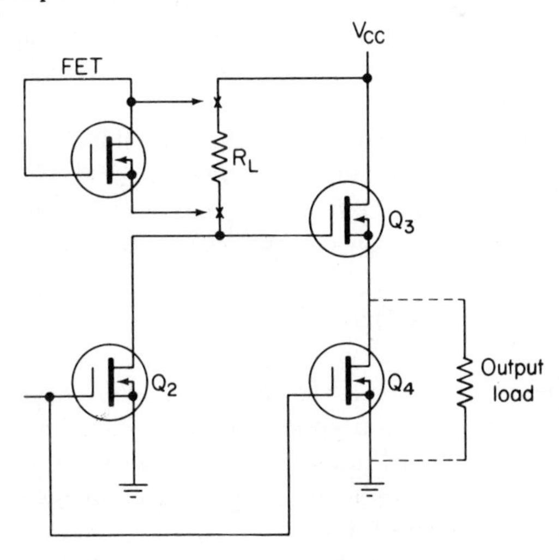

Fig. 14–5 FET replaces resistor.

14–2 OPERATIONAL AMPLIFIERS

14–2.1 Introduction

The operational amplifier, discussed in Sec. 11–3.3, is so versatile that it lends itself particularly well to integrated-circuit realization. Once an IC operational amplifier, often abbreviated as "op amp," is well-designed and tested, it can be applied to a great many uses. Other than power supply connections, the only terminals of concern are the output and the two input terminals, marked + for the noninverting terminal and − for the inverting terminal. It should also be remembered that any feedback to the + terminal is positive, tending towards oscillation, while feedback to the − terminal is negative and is the type of feedback employed for applications where the objective is amplification. Figures 14–6(a) and 14–6(b) show two methods of applying negative feedback, inverting and noninverting, together with their respective characteristics.

14–2.2 Monostable Circuit

The circuit shown in Fig. 14–7 is a monostable multivibrator employing an IC op amp. The + terminal is connected to a positive voltage determined by $+V_{CC}$ and resistors R_1 and R_2, constituting a voltage divider. If a positive pulse is applied to the − terminal, that pulse is amplified and inverted. It is also coupled via C_1 to the + terminal and, because of positive feedback, reinforces the action until the amplifier is driven almost all the way down to the negative supply voltage $-V_{EE}$. Before the input pulse was applied, C_1 had acquired a charge approximately equal to the difference between $+V_{CC}$ and V_{R2}. As long as the charge remains on the capacitor, the − terminal is no longer at potential V_{R2} but is, instead, at approximately $-V_{EE}$ minus the charge, a highly negative voltage. The capacitor begins to discharge, and, once it has done so, V_{R2} returns to its original value. The circuit thus returns to its original stable state. It can be seen that in this configuration the IC op amp performs just like the monostable circuit comprising discrete components discussed in Sec. 13–3.8.

14–2.3 Astable Circuit

An astable multivibrator is shown in Fig. 14–8. Assume C_1 originally uncharged. Assume the output at approximately $+V_{CC}$. This voltage, divided down by R_2 and R_3, is applied to the + terminal where it only reinforces the existing condition. Then C_1 begins to charge to a positive voltage, driving V_o down. This reduces the voltage fed to the − terminal, causing a further drop in V_o, a positive-feedback effect which continues until V_o has gone almost all the way to $-V_{EE}$. Once this has occurred, C_1 begins to charge to a negative voltage.

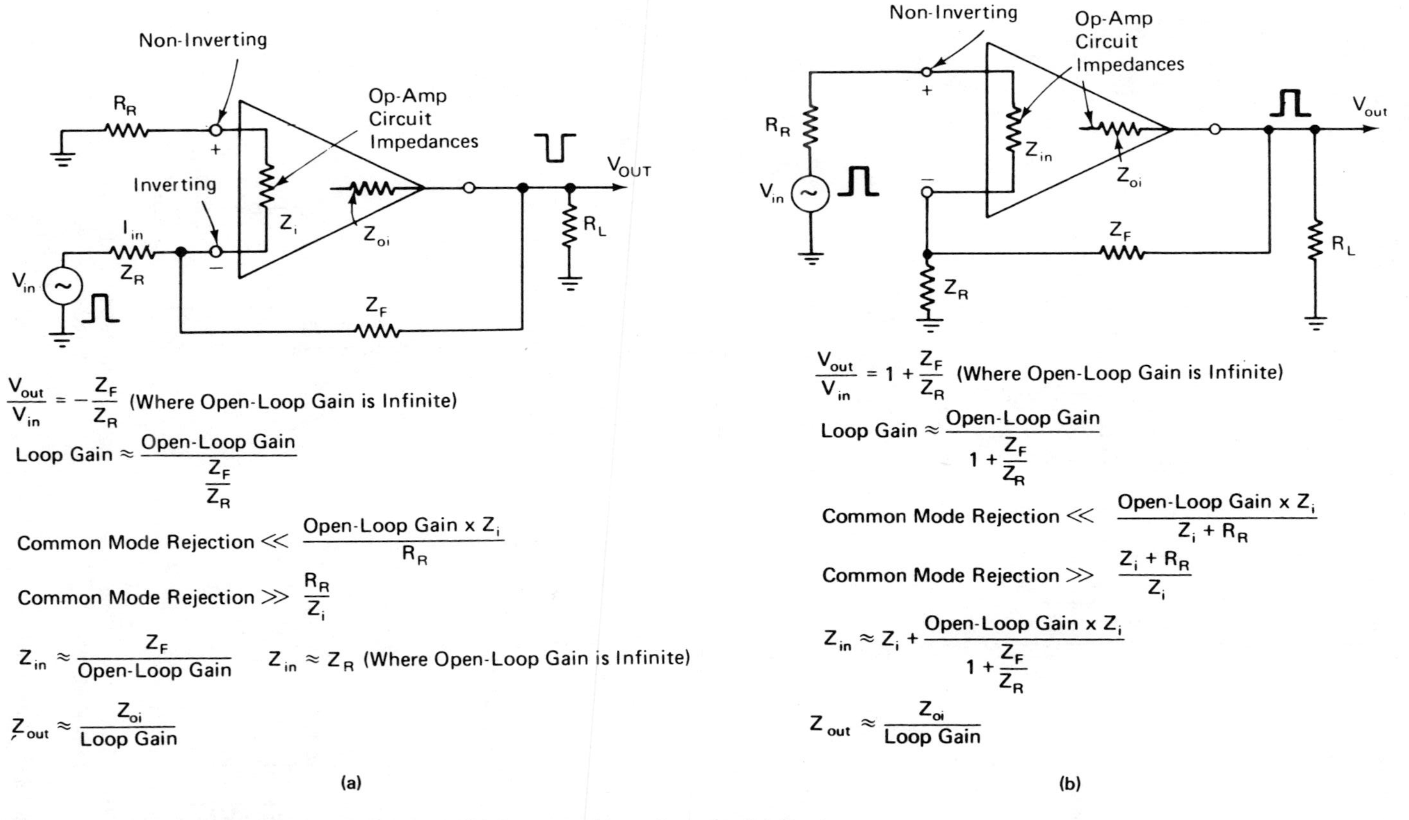

Fig. 14–6 (a) Op amp with inverting feedback. (b) Op amp with noninverting feedback.

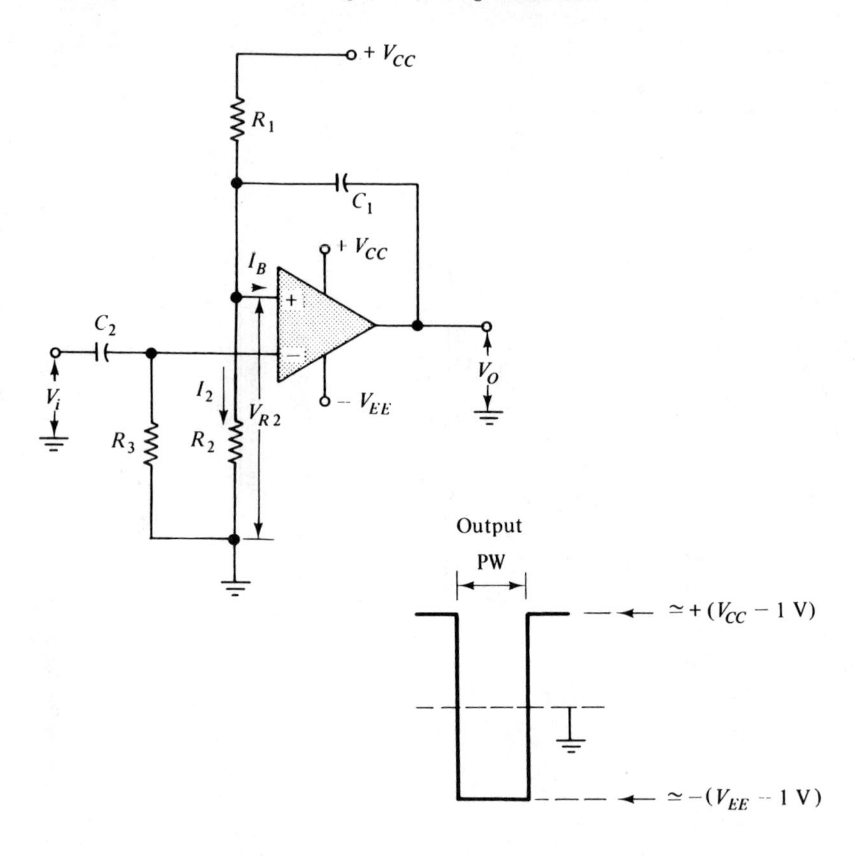

(a) Circuit and output of monostable multivibrator using IC operational amplifier

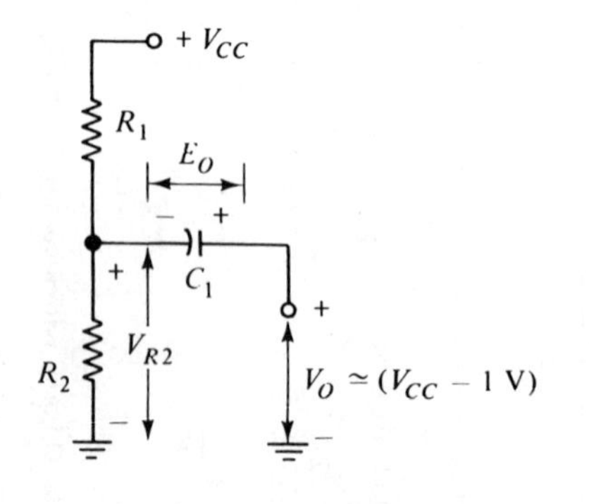

(b) Charge on C_1 when output is positive

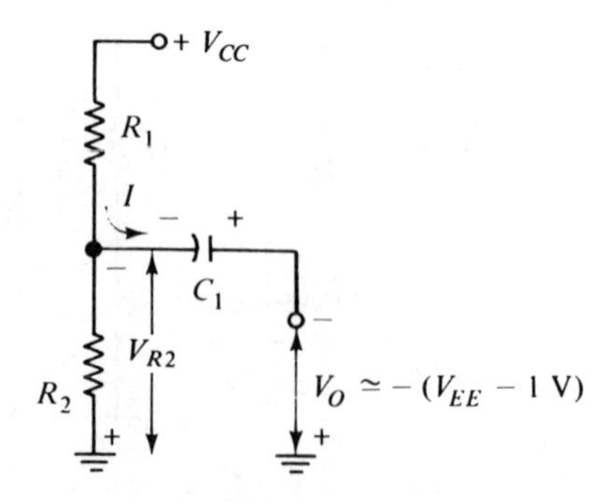

(c) Initial conditions when the output goes negative

Fig. 14–7 IC monostable circuit.

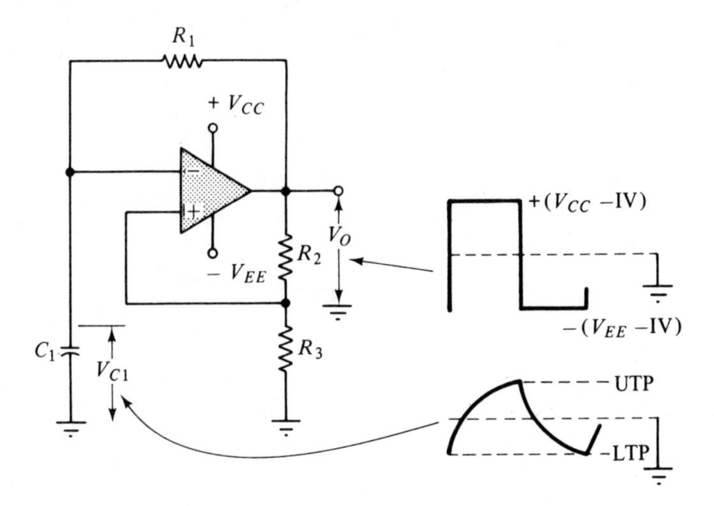

Fig. 14–8 IC astable circuit.

When that voltage is sufficiently large, V_o is driven up, and positive feedback takes it all the way almost to $+V_{CC}$ again. The time constant R_1C_1 determines the period of oscillation, and a square wave is generated at the output. The circuit performs similarly to the astable circuit using discrete components described in Sec. 13–3.9. An advantage, other than the use of the IC, is the fact that only one RC combination is needed, where the discrete version required two, eliminating the necessity of perfect component equality when a symmetrical square wave is to be generated. However, instead, it is necessary to ensure that V_{CC} is numerically very close to V_{EE}.

14–2.4 Schmitt Trigger

A comparison between Figs. 14–9 and 14–8 shows that the R_1C_1 circuit has been removed. Instead, an input voltage V_i is applied to the – terminal. When this voltage is sufficiently small or zero, the R_1R_2 divider keeps the output voltage V_o at about $+V_{CC}$. As the input voltage reaches a sufficiently high positive voltage, V_o drops, and the positive feedback drives the circuit almost all the way down to $-V_{EE}$, where it remains until the input voltage drops below a critical value when $+V_{CC}$ is restored, the stable state. It is seen that this is a Schmitt trigger functioning similarly to the one discussed in Sec. 13–3.10. There is a difference in that this circuit uses a negative supply and the other does not. This affects primarily the point at which switchover occurs. Each Schmitt trigger circuit is characterized by an upper switchover point and a lower switchover

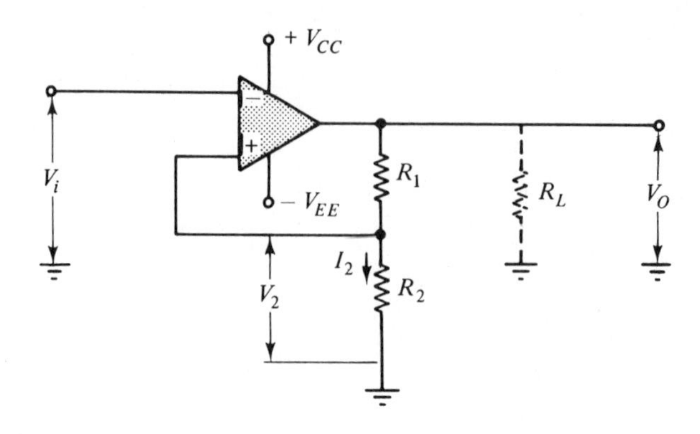

Fig. 14–9 IC Schmitt trigger.

point. In some designs, these points have the same value, or approximately so; in other cases there is a considerable difference between them. In that case the term *hysteresis* is used to describe the difference.

14–2.5 IC Voltage Comparator

As the name indicates, the purpose of this circuit is to compare two input voltages to each other. Each time one input voltage exceeds the other, the circuit switches to its upper output voltage. When the one input drops below the other, the circuit switches back to its lower output voltage. This circuit is seen to be a version of the Schmitt trigger. But where in the Schmitt a fixed reference voltage is used, against which an input is compared, in the comparator two input voltages are employed. Also, where in the Schmitt the upper and lower switching points are not necessarily identical and may be characterized by hysteresis, in the comparator it is attempted to bring the two switching points as close to another as possible to enable detection of the smallest possible differences between the two voltages. ICs designed for comparator applications are also characterized by short switching times.

14–2.6 Voltage Follower

A voltage follower is a device that has the same output voltage as input voltage (a gain of 1), but a low output impedance with a high input impedance. The op amp is connected as in Fig. 14–10, with the output fully fed back to the – input. We have

$$V_o = AV_{\text{in}} \tag{14–1}$$

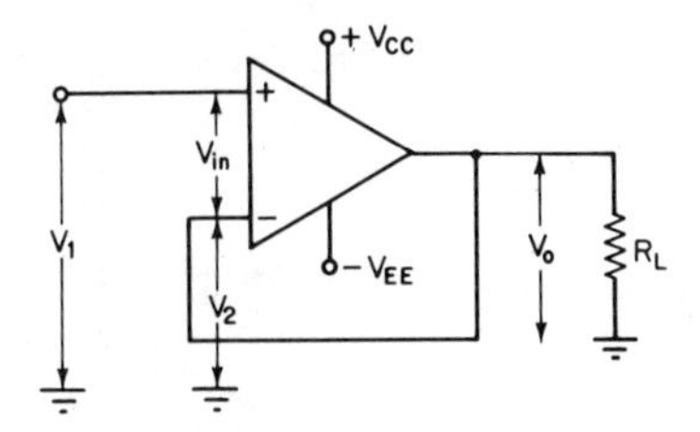

Fig. 14–10 Voltage follower.

where A is the open-loop gain of the amplifier and V_{in} is the input voltage.

$$V_{in} = V_1 - V_2 \qquad (14\text{-}2)$$

where V_1 is the voltage applied to the noninverting (+) input and V_2 is the voltage at the inverting (−) input. But in this case

$$V_2 = V_o = A\,V_{in} \qquad (14\text{-}3)$$

$$V_{in} = V_1 - A\,V_{in} \qquad (14\text{-}4)$$

$$V_{in}(1 + A) = V_1 \qquad (14\text{-}5)$$

$$V_o = A\,V_{in} = \frac{A\,V_1}{(1 + A)} \qquad (14\text{-}6)$$

If $A \gg 1$, $A \approx 1 + A$; hence, $V_o = V_1$.

The circuit has the advantages over an emitter follower of coming closer to a gain of 1 and lower output impedance.

14–2.7 IC Bootstrap Ramp Generator

An application of the voltage follower of the previous section is as an IC ramp generator. Comparing Fig. 14–11 with Fig. 11–12, it is seen that the voltage follower has been substituted for the emitter follower. Due to the large gain of the op amp, the gain can come far closer to 1 than is possible with the emitter follower, and greater linearity can be achieved. Other than these considerations, the two circuits function identically.

14–2.8 The IC OTA

The IC OTA (Operational Transconductance Amplifier) is a variation of op amp that is characterized by two features: It has a very high output

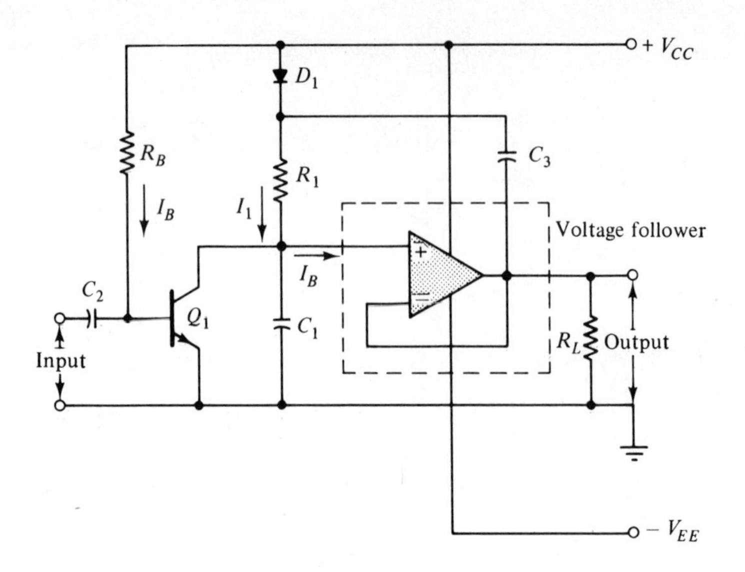

Fig. 14–11 IC bootstrap ramp generator.

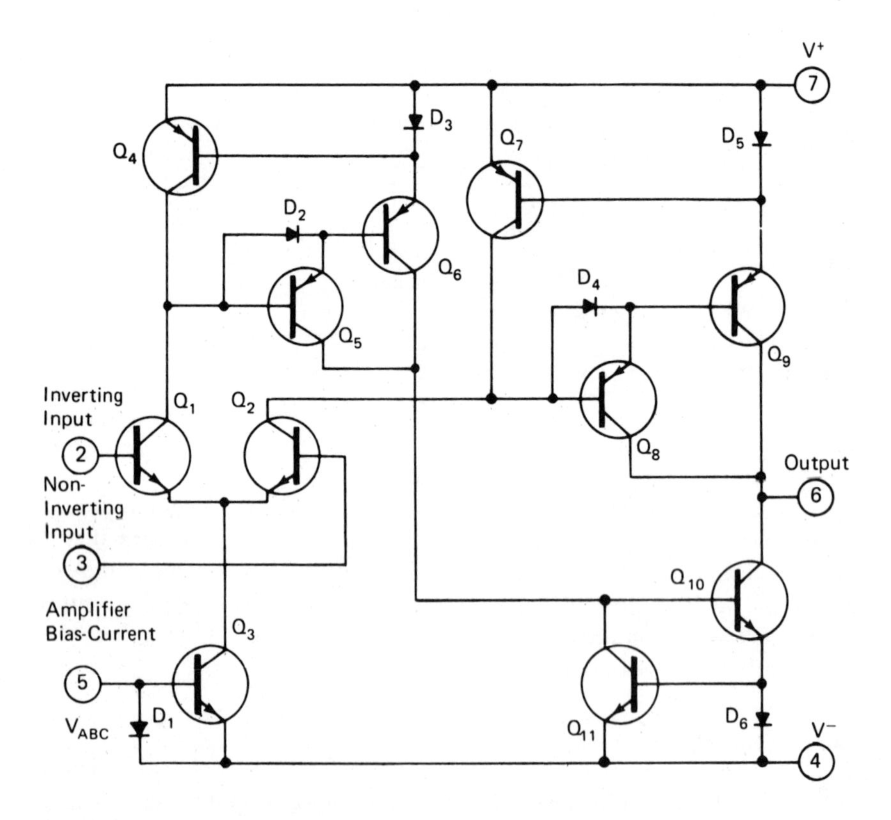

Fig. 14–12 Operational transconductance amplifier.

impedance and therefore functions more as a current source than a voltage source (hence the use of the term *transconductance*), and the value of the transconductance is variable. The transconductance, defined as the ratio of output current to input voltages, is variable. This is accomplished by using, in addition to the two inputs ordinarily applied to the op amp, a third input. This is a control current designated as I_{ABC}. Figure 14–12 shows one type of OTA, the RCA CA3080 and CA3080A. At terminal 5, a voltage designated V_{ABC} is applied, and controls the amount of current in Q_3—i.e., the emitter current of the differential amplifier comprising Q_1 and Q_2. As V_{ABC} is varied, the transconductance is varied.

14–3 TIMER CIRCUITS

14–3.1 Introduction

In the applications of the operational amplifier described above, the op amp was not primarily used in its linear region but, instead, was switched between its two end points, with the linear region only involved during the short switching interval. This suggests that for those applications an op amp is not really required, but a flip-flop might serve as well. Embodying this idea, timers—for example, the 555 IC Timer—have been developed. There are, of course, many others, but for illustrative purposes the 555 is discussed here. It can be obtained in two different forms: the flat dip type and the transistor can. As shown in Figure 14–13, the 555 embodies two voltage comparators, a flip-flop of the RS (set-reset) or SC (set-clear) variety, and an output stage. The voltage divider network consisting of R_1, R_2, and R_3 provides reference inputs to both comparator circuits. If a voltage applied at terminal 6 (threshold) exceeds the voltage from the voltage divider, comparator 1 changes state and resets the flip-flop. If the voltage at terminal 2 (trigger) goes below the voltage provided by the voltage divider, comparator 2 changes state and sets the flip-flop. Flip-flop output is fed through a low-impedance output stage, which can be used to control a large current of up to 200 ma.

Since the RS flip-flop requires a positive pulse at either the R or the S terminal for operation, the negative pulses that are created as the comparators return to their other states have no effect. Thus, in order for the flip-flop to change states, an input at either the threshold or trigger terminals is always required. There is also a second reset for the flip-flop. Q_2, a *pnp* transistor with its emitter at V_{REF}, can be turned ON by connecting terminal 4 to ground. This resets the flip-flop.

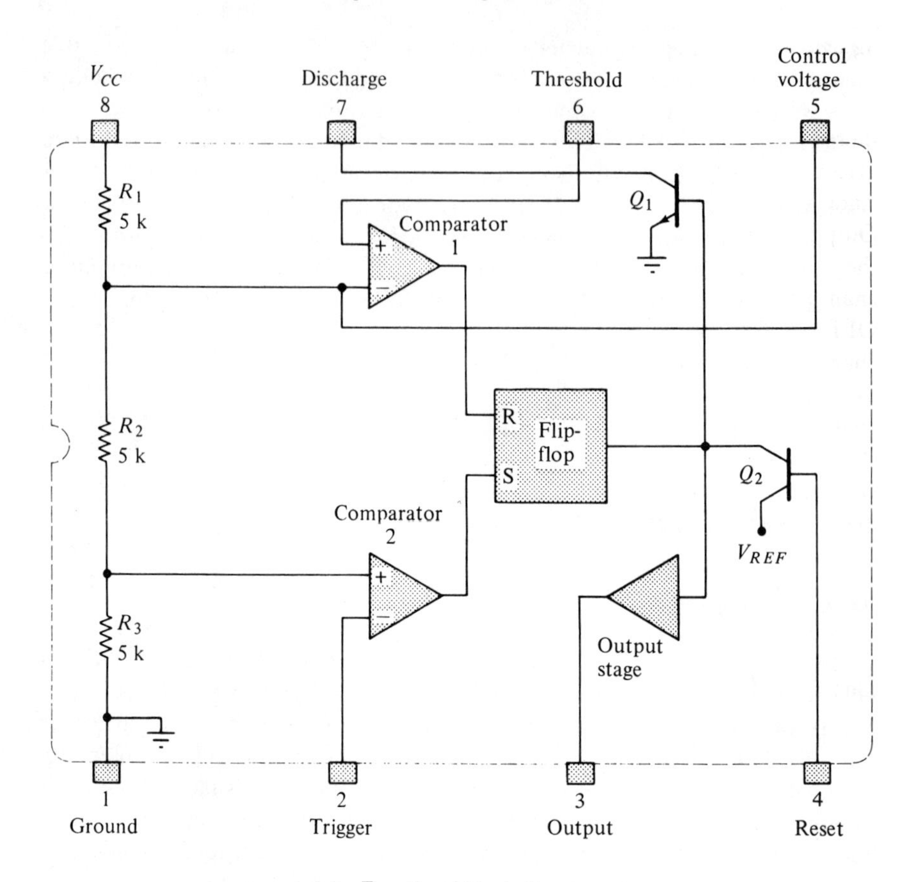

(a) Functional block diagram

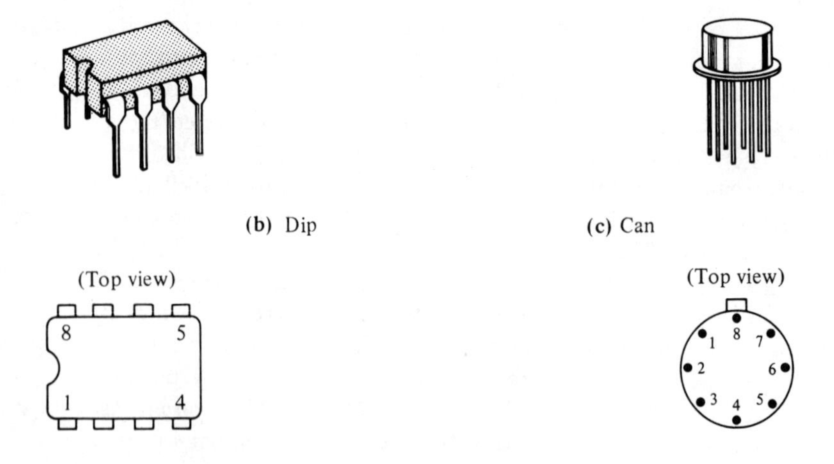

(b) Dip

(c) Can

Fig. 14–13 The 555 timer.

14–3.2 555 as a Monostable Circuit

Figure 14–14 shows the 555 connected as a monostable circuit. Initially, there is no trigger, and terminal 2 is at V_{CC}. Comparator 2 is in its 0 state. Capacitor C_A is discharged; hence, terminal 6 is 0, and comparator 1 is in its 0 state. The flip-flop is in its reset condition; i.e., the collector is at V_{CC}. Q_1 is ON. Output terminal 3 is at 0. As soon as the negative trigger is applied at terminal 2, the – terminal of comparator 2 is driven down, causing the comparator to change states and causing the flip-flop to be set. It is now at 0, switching Q_1 OFF. This removes Q_1 as a short-circuit across C_A, allowing the capacitor to charge through R_A from supply voltage V_{CC}. The charging process continues until the capacitor voltage exceeds the reference voltage at the – terminal of comparator 1. When that happens, the comparator changes state and triggers the flip-flop back into its reset condition. This turns Q_1 ON and discharges the capacitor. The action repeats when the next trigger occurs. PW, the pulse width, depends on the product of R_A and C_A.

14–3.3 555 as an Astable Circuit

The circuit of Fig. 14–15 shows the 555 connected as an astable circuit. Comparing it to Fig. 14–14, the difference is that R_A has been replaced by the voltage divider comprising R_A and R_B, with the voltage divider center connected to terminal 7, while the lower end is connected to terminal 6. Also, terminal 2

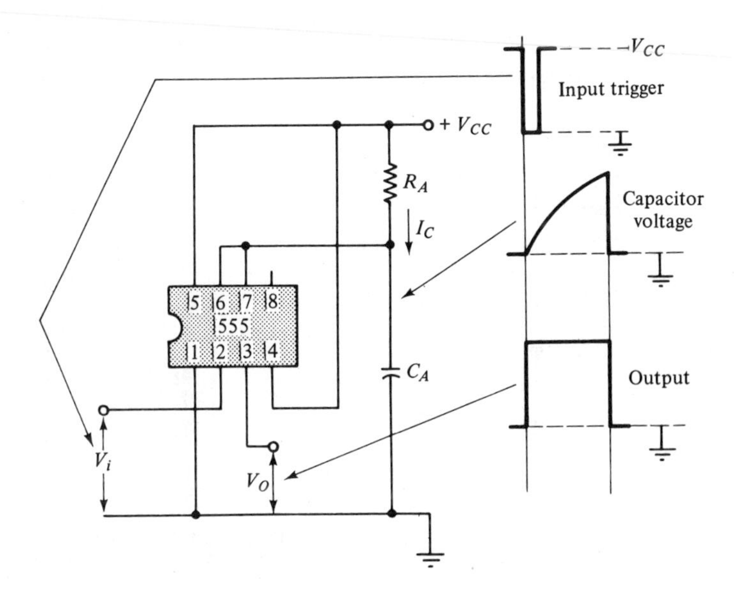

Fig. 14–14 555 as monostable circuit.

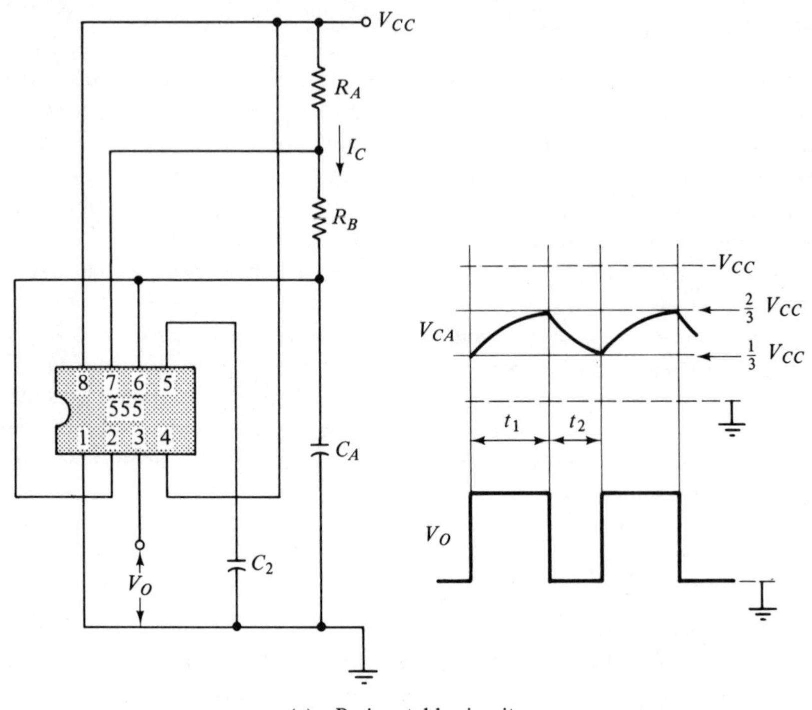

(a) Basic astable circuit

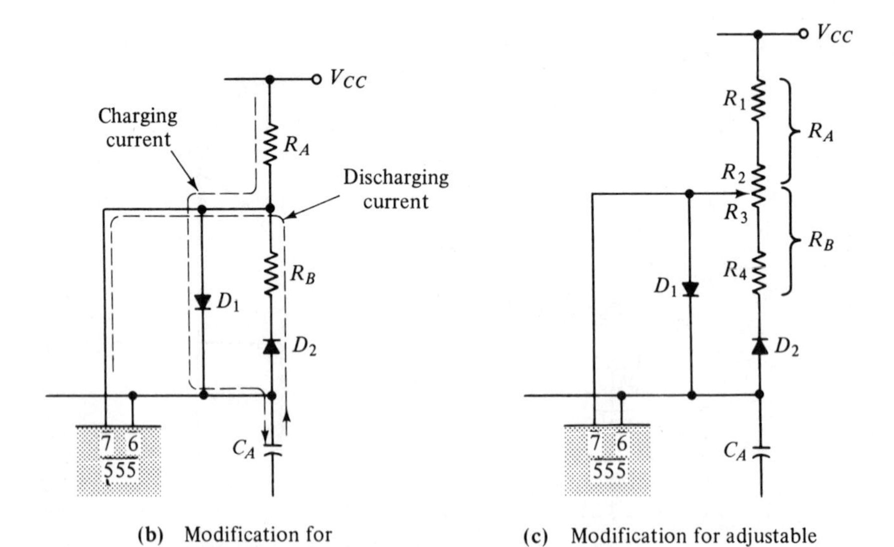

(b) Modification for 50% duty cycle

(c) Modification for adjustable duty cycle with constant PRF

Fig. 14–15 555 as astable circuit.

(trigger) has been connected to terminal 6 (threshold). When the voltage of capacitor C_A goes below the reference voltage for comparator 2, the comparator goes to 1 and sets the flip-flop. This turns Q_1 OFF, and C_A can charge through R_A and R_B. It charges until it reaches the reference voltage for comparator 1, when that comparator goes to 1 and resets the flip-flop. This turns Q_1 ON and discharges capacitor C_A through resistor R_B. Once the C_A voltage has dropped to the reference value for comparator 2, that comparator changes state, and the cycle recommences. The two time intervals involved are

$$t_1 = 0.693\, C_A(R_A + R_B)$$
$$t_2 = 0.693\, C_A R_B \tag{14-7}$$

[The number 0.693 = In 2, resulting from substitution in Eq. (11-15).]

Under these circumstances, t_1 cannot equal t_2; i.e., a 50% duty cycle cannot be obtained. If this is desired, however, all that is required is that diodes D_1 and D_2 be added [Fig. 14-15(b)]. If a further modification is made [Fig. 14-15(c)], the duty cycle can be varied.

14-3.4 Sequential 555 Timer

By cascading three monostable 555 circuits, for example, as shown in Fig. 14-16, the negative-going pulse from the first monostable 555 triggers the second, and the pulse from the second 555 triggers the third. This gives sequential outputs as shown. More than three can, of course, be cascaded as well.

14-4 INTEGRATED-CIRCUIT LOGIC

14-4.1 Introduction

In Sec. 13-6 the basic logic circuits and their embodiments in terms of discrete components were discussed. Logic circuits lend themselves particularly well to IC realization. This is true because the logical functioning of a computer can be extremely complex (analyzable only with the aid of Boolean algebra and logic maps), but despite their complexity, they are built up of the basic four logic functions: AND, OR, NAND, and NOR. Hence, a great number of identical circuits are called for, which constitutes ideal conditions for IC application. Due to this great need for logic circuits in computers, many different types of logic have been developed. Here a number of different performance criteria enter the picture. Switching time must be kept as low as possible. Noise immunity must be

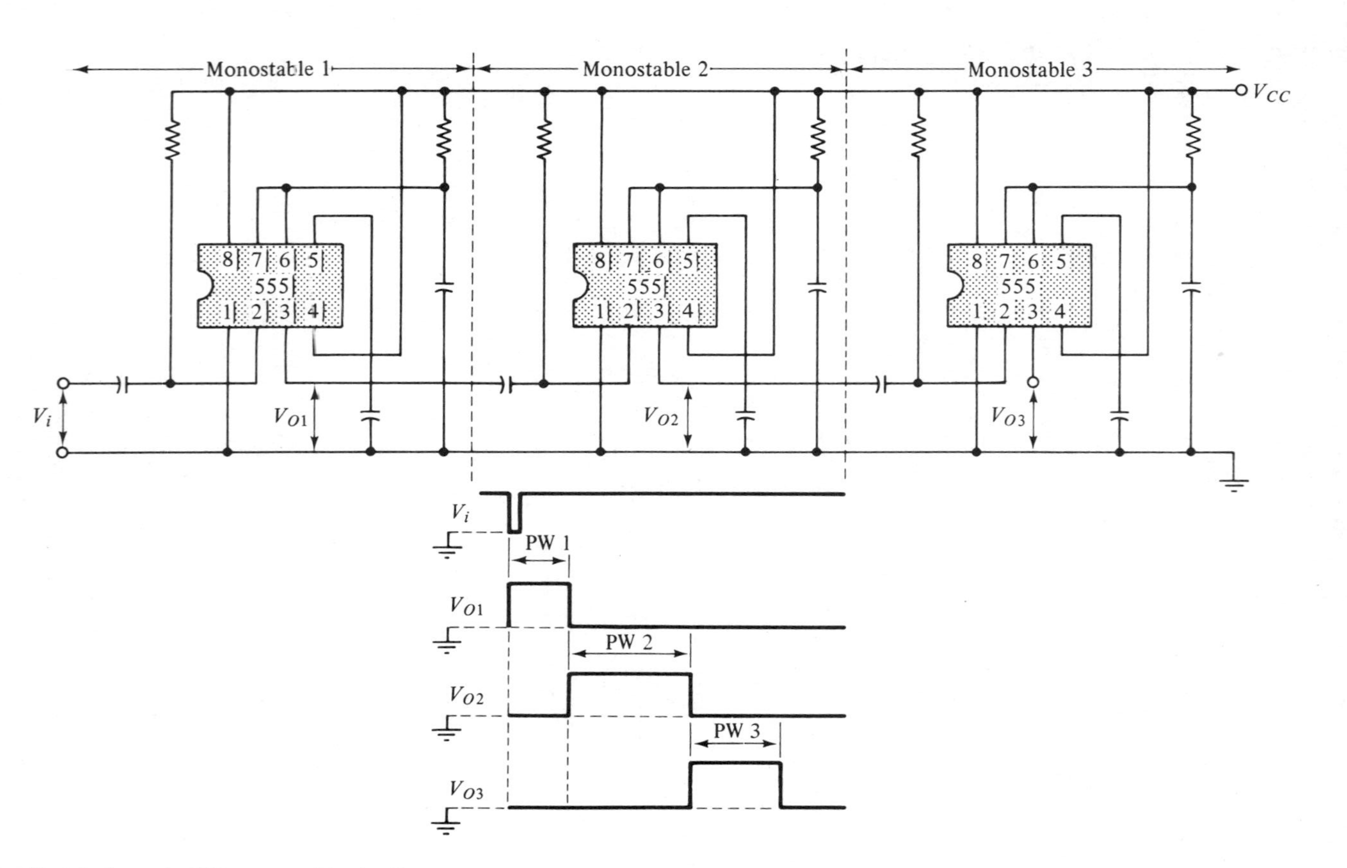

Fig. 14–16 Cascade 555 monostable circuits.

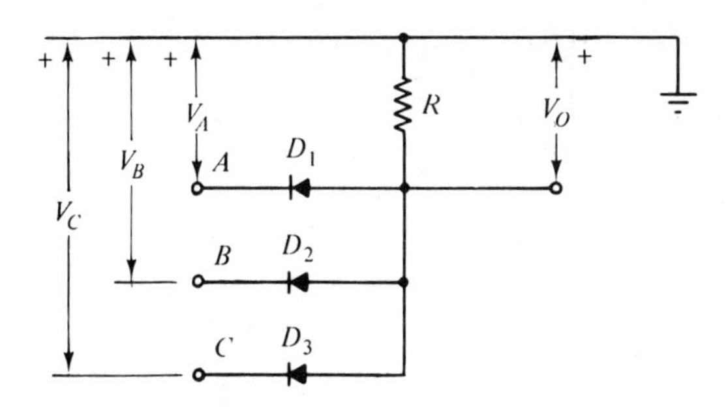

Fig. 14–17 Negative logic OR gate.

provided to the maximum possible extent. Power consumption must be kept as low as possible in order to permit maximum component density. Fan-out has to be considered. The supply voltage or voltages are a consideration in order that the logic circuits fit into the rest of the system. There is also a choice between positive and negative logic. Using positive logic, a positive input voltage is provided and a positive output is produced. With negative logic, a negative input is applied and a negative output is obtained. Figures 14–17 and 14–18 show how the negative logic circuits look in comparison with the positive logic circuits shown in Sec. 13–6. It is possible to construct the logic circuits discussed in Sec. 13–6 in IC form so that they duplicate their discrete-component counterparts as diode logic and DTL (diode transistor logic), and these have been made. But for better performance criteria, other configurations have been developed which are more advantageous from the IC viewpoint.

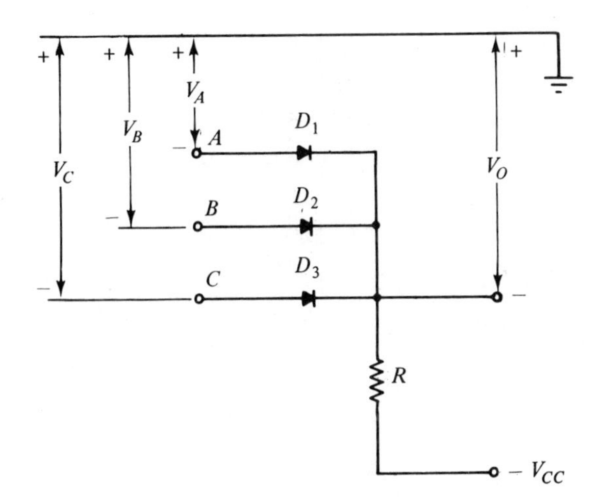

Fig. 14–18 Negative logic AND gate.

14–4.2 High Threshold Logic

High threshold logic (HTL) was developed to improve noise immunity. In the circuit of Fig. 14–19, a Zener diode has been included. The voltage of this diode must be overcome in order to turn on transistor Q_1. Hence, any voltage not sufficient to do this, such as a noise voltage, has no effect. To achieve this benefit, however, the supply voltage has to be increased sufficiently to overcome the Zener voltage.

14–4.3 Emitter Coupled Logic

Emitter coupled logic (ECL) is shown in Fig. 14–20. This circuit has been designed so that the transistors are not saturated. This eliminates a factor called *storage time*. It involves the accumulation of carriers in a transistor while it is heavily conducting. Eliminating these carriers when the transistor is switched OFF takes time. The phenomenon is largely eliminated by not saturating the transistors. Additionally, this circuit provides two outputs which are the complements of each other: OR and NOR.

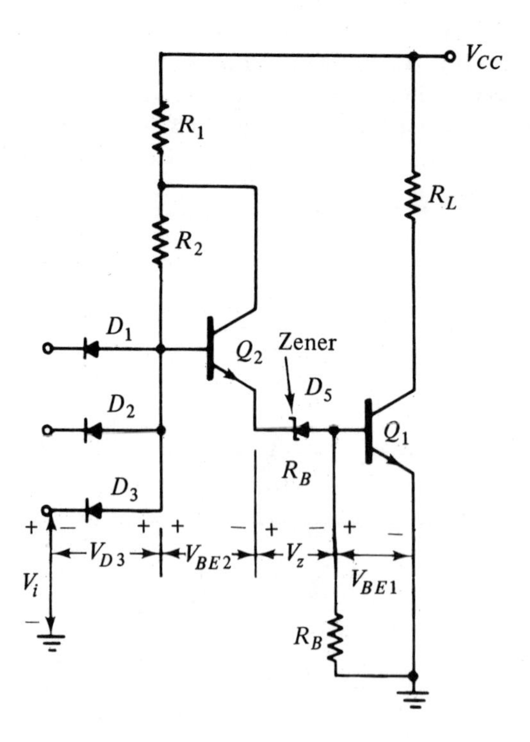

Fig. 14–19 High threshold logic.

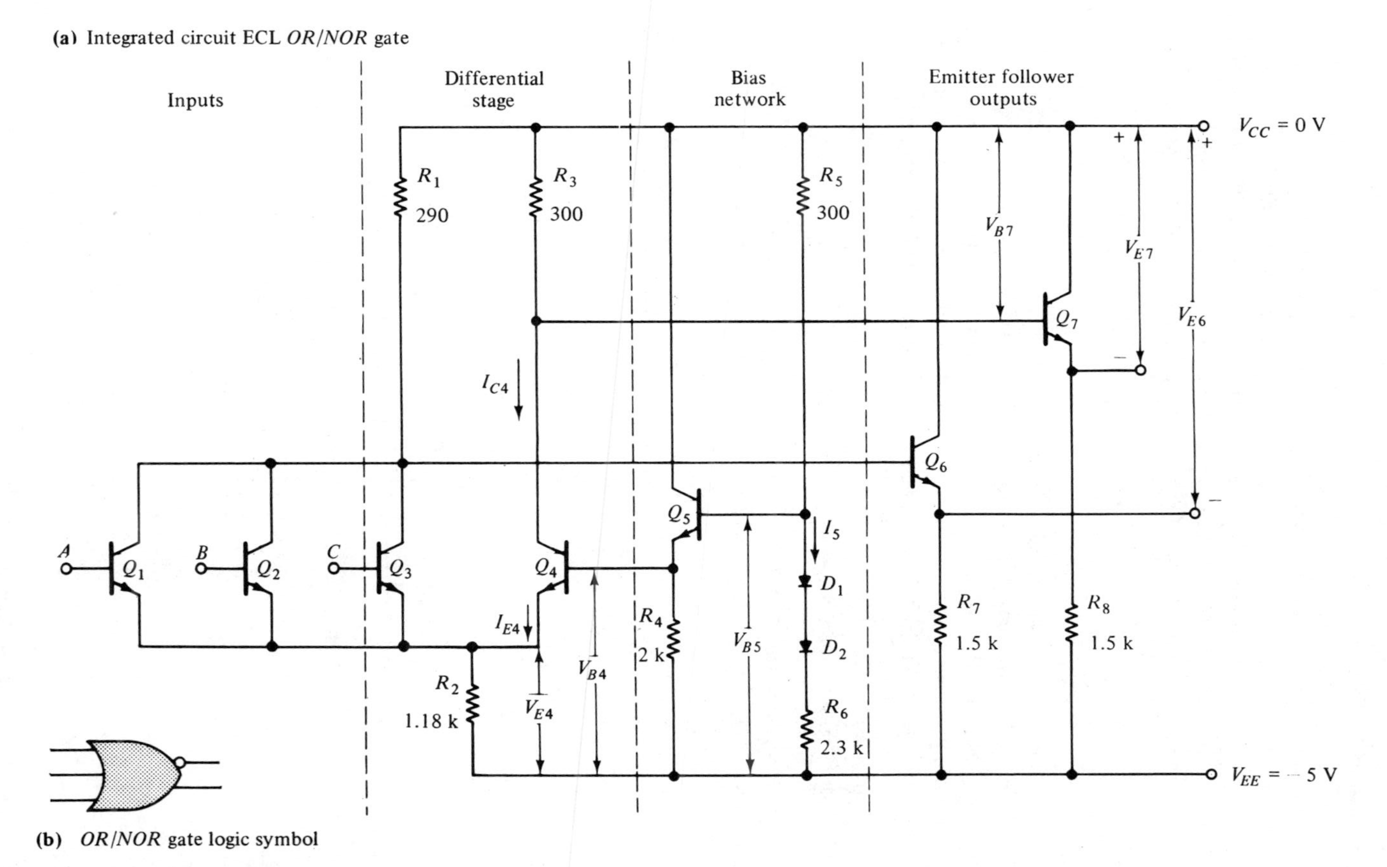

Fig. 14-20 Emitter coupled logic.

14–4.4 MOSFET Logic

Due to the fact that MOSFET circuits have high resistances and dissipate very little power, they permit a great density and, hence, lend themselves very well to use in IC logic circuits. Figure 14–21 shows a P-MOS NAND gate, and Fig. 14–22 shows an N-MOS NOR gate. It is seen that there are no components other than MOSFETs, a desirable fact for IC production.

14–4.5 CMOS Logic

The devices discussed above employed either *p*-type or *n*-type MOSFETs. It is also possible to use complementary devices, combining *n*-types and *p*-types. With this construction, power consumption is reduced even further, noise immunity is excellent, and good fan-out is obtained. Figures 14–23 and 14–24 show the CMOS NAND gate and the CMOS NOR gate, respectively.

14–4.6 Integrated Injection Logic

Integrated injection logic, also known as IIL or as I^2L, is based on the principle that, for IC purposes, it is desirable to substitute transistors for

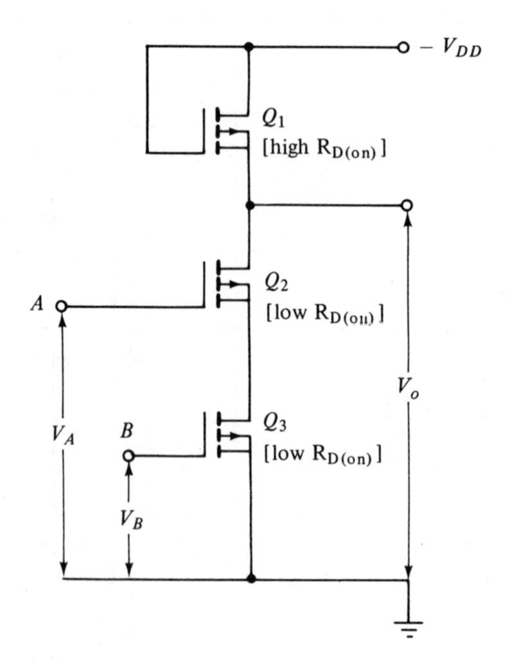

Fig. 14–21 P-MOS NAND gate.

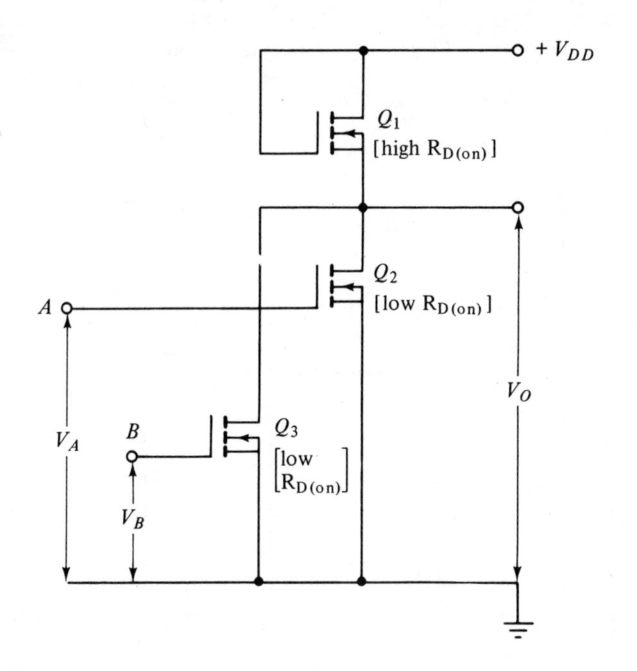

Fig. 14–22 N-MOS NOR gate.

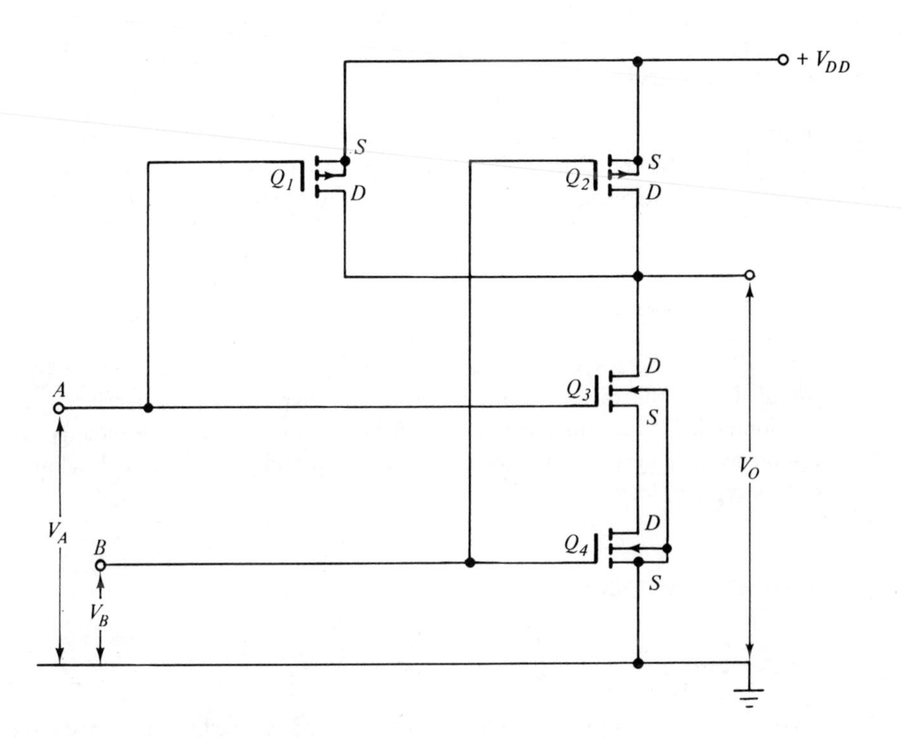

Fig. 14–23 CMOS NAND gate.

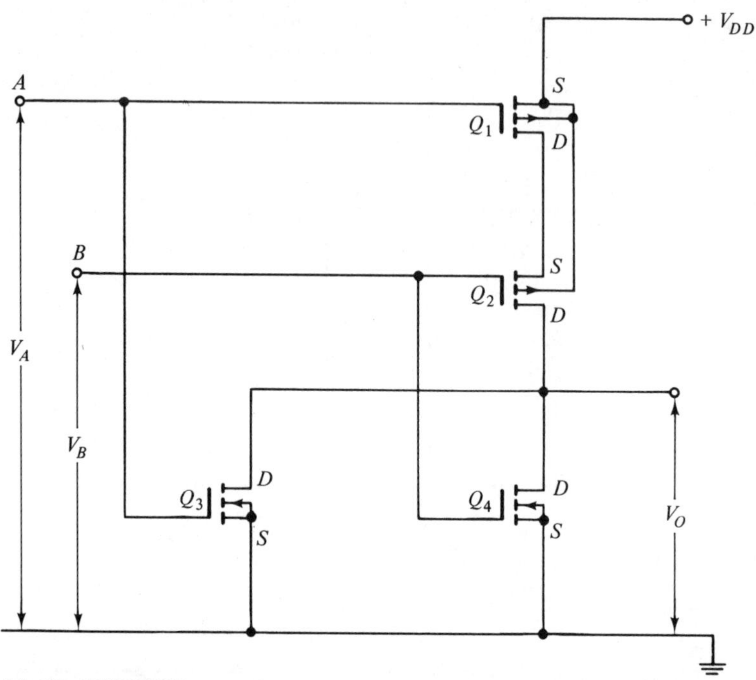

Fig. 14–24 CMOS NOR gate.

resistors. Figure 14–25 shows such a circuit. First, it is shown how two I^2L inverters can be constructed. Then, in Fig. 14–26, the I^2L NAND and NOR gates are shown. Finally, Fig. 14–27 shows the construction of an I^2L logic circuit.

14–4.7 Comparison

To choose a system of logic for a particular application, Table 14-1 is useful. It lists the various available types and gives their characteristics. Since development is constantly continuing, such a table is subject to change, both insofar as the characteristics given are concerned and also as to the development of new types of logic.

14–5 MICROPROCESSORS

14–5.1 Introduction

As the techniques of producing integrated circuits advanced from printed circuits with discrete components through medium- and large-scale integration, it became possible to produce the equivalents of resistors, capacitors, diodes, transistors, and interconnections between them. As progress was made in fabricating techniques, and in the reduction of dissipated power (see Sec. 13-2.2), more and more complete circuits could be produced on a single chip. Where at previous stages of development, say, a dozen flip-flops could be included in an integrated circuit, now the trend was toward producing complete counters or flip-flop memories. This made possible tremendous economies of labor, time, and cost since the individual circuits did not have to be interconnected externally, but had the interconnections within the integrated circuit. It can also be appreciated that, as the circuit components become ever smaller in their dimensions, the limiting factor is no longer the size of the components but rather the necessity of making electrical connections to the components. That is,

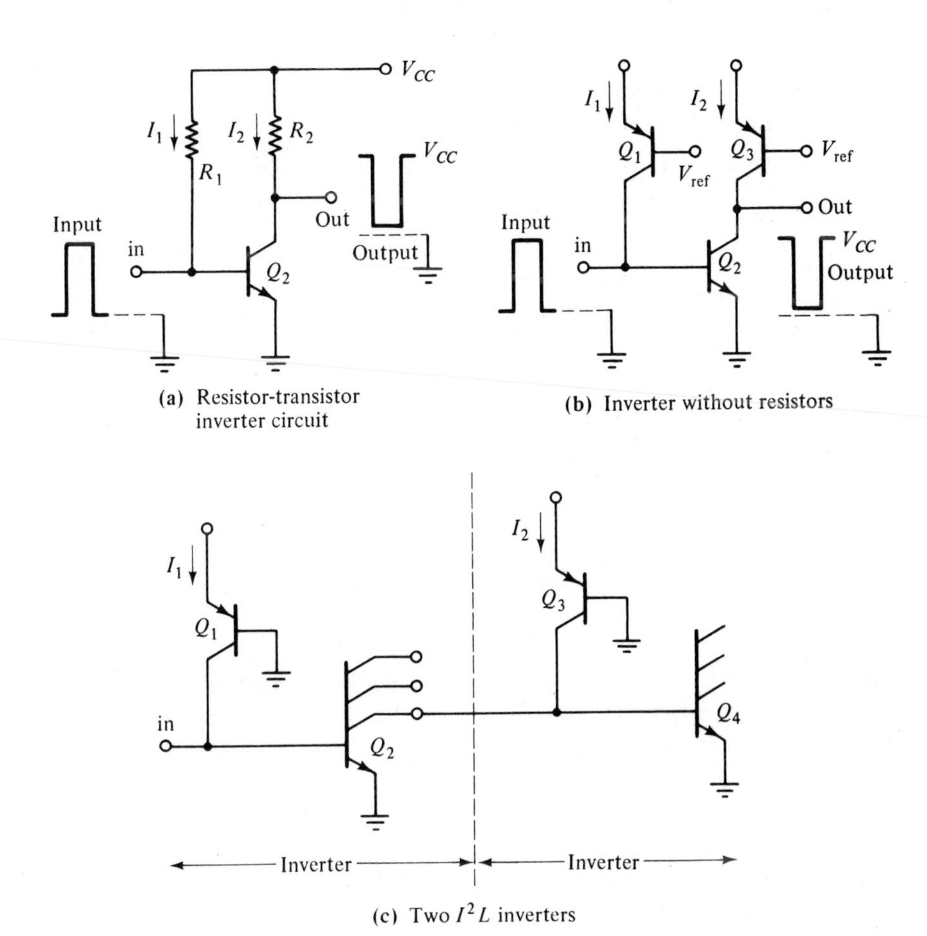

Fig. 14-25 Integrated injection logic.

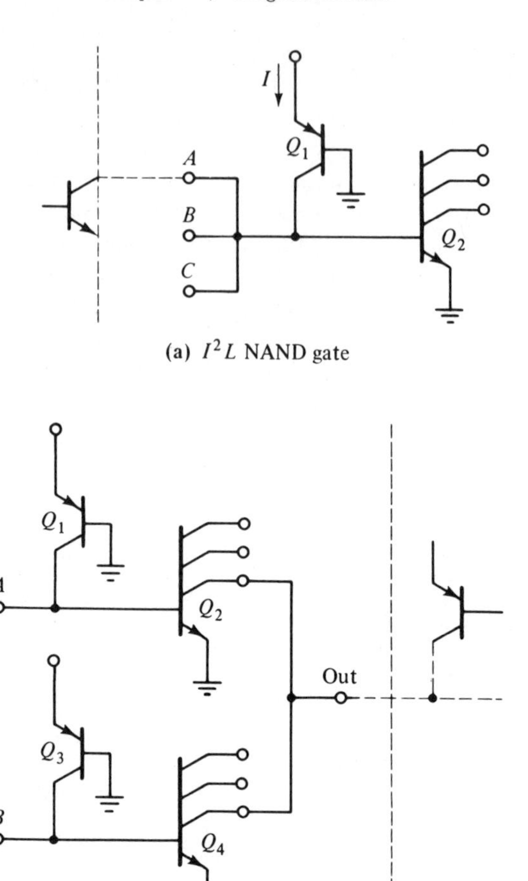

Fig. 14–26 Integrated injection logic NAND and NOR gates.

even though the components themselves might be microscopic in size, the external connections, if any, must be of such size as to be handled by personnel doing the interconnecting.

14–5.2 Interconnections

When, in view of the above considerations, attempts were made at further miniaturization, the obvious way to proceed was to produce all interconnections internally, except for those that were indispensable: connections to the power supply such as $+V_{CC}$, $-V_{CC}$, and ground; and connections for supplying inputs, for supplying the output, and for accepting programming instructions. All other interconnections were done internally. The result of this development was the microprocessor, which can best be described briefly as a computer on a chip. In other words, one chip is sufficient to do all that a computer needs to do in the particular application for which it is designed.

14–5.3 General Purpose Microprocessors

Perhaps the term *general purpose microprocessor* is a bit exaggerated, and the term *multipurpose microprocessor* might be more appropriate, but, in

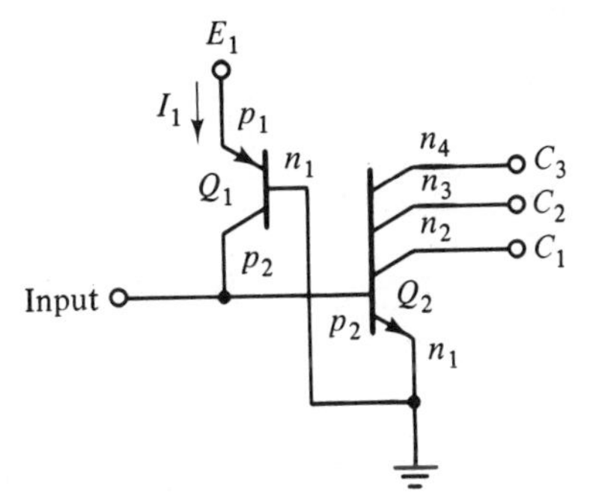

(a) p and n regions of inverter transistors

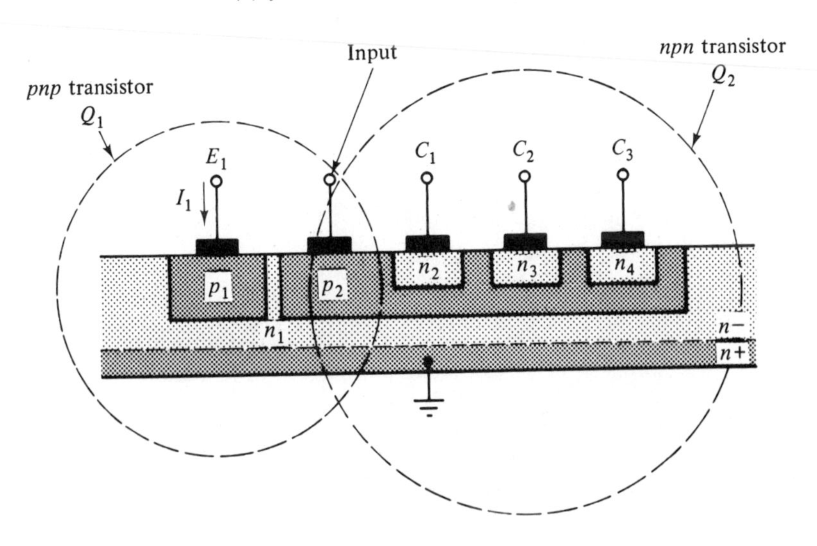

(b) Cross-section of inverter transistors

Fig. 14–27 Construction of I^2L logic.

TABLE 14–1 COMPARISON OF IC LOGIC TYPES

				TTL										
	DTL	RTL	HTL	74	74H	74L	74S	74AS	74LS	74ALS	ECL	P-MOS	CMOS	I^2L
Propagation delay time (ns)	30	12	119	10	6	33	3	1.5	9	4	2	100	25	*10 to 250
Power dissipation per gate (mW)	15	15	50	10	22.5	1	20	20	2	1	25	1	10 nW	*6 nW to 70 μW
Noise margin (V)	1.4	0.7	7.5	0.4	0.4	0.4	0.4	0.4	0.4	0.4	0.25	2	0.3 V_{DD}	0.25
Noise immunity rating	good	poor	excellent	good	good	good	good	good	good	good	fair	excellent	excellent	fair
Fan-out	8	5	10	10	10	10	10	10	10	10	25	> 50	> 50	*depends upon injection current

any case, the term is meant to designate a microprocessor chip that can perform all the functions of a computer except programming. That is, it has all the computational facilities, the memories, the clock, the logic, and the ability to accept inputs and to feed out outputs. All that is then required to operate the chip is an external program. This does not give the microprocessor the flexibility of a large computer because it can accept only certain types of programs especially designed for it, while a larger computer can be reprogrammed at will. However, it does have the flexibility of having many different programs run through it.

14-5.4 Special Purpose Microprocessors

The other category of microprocessors, the special purpose microprocessors, has gone a step further in miniaturization in that these microprocessors contain, as well as the elements mentioned in the previous section, the program itself. That is, they have been programmed as part of the design procedure so that the program is fixed. This eliminates completely the need for an externally supplied program but, of course, at a sacrifice in flexibility. This is not always a very limiting factor, as can be appreciated by considering the ordinary calculator. Though it is inflexible in the sense that it can only calculate (it cannot, say, turn your heating system on and off), it can perform a tremendous variety of mathematical calculations. Even a very complicated program or a series of many different programs can be produced on a chip so that, going far beyond the elementary procedures of adding, subtracting, multiplying, and dividing, the program can be made to perform dozens of other calculations (roots, powers, logs, etc.). In contrast, an example of a multipurpose microprocessor might be a chip that enables the playing of games on television. This chip requires the input of a different tape program for each game to be played, but it is capable, with this proviso, of playing a great variety of games.

14-5.5 Design

To design a microprocessor, initially the same procedure is followed as for a large computer. A flow diagram is prepared, from which a program is evolved. Then the circuit is designed on paper so as to embody either the program itself or the facilities for feeding in the program from an external source. The necessary computational facility and the necessary memories and logic circuits are provided. Then the paper design is converted to printable form in macrodimensions, or dimensions easily manipulable by a draftsman. When the macrodiagram representing the entire chip is complete, it is photographically reduced to microdimensions for production.

14–5.6 Production

Starting with the microdimensional photograph of the circuit, the production techniques described above, and considerable refinements of those techniques, are applied to produce a single chip that is essentially an entire computer. All that is required to operate this computer is a power supply and (possibly) a program input. It is then ready to accept inputs and produce outputs with no interconnections other than those incorporated within the chip itself.

14–5.7 Applications

The microprocessor is characterized by an extremely low unit cost, considering the work it performs. However, it is also characterized by a high initial cost since the design, preparation, and debugging involve considerable labor. These facts make it attractive for applications involving thousands of identical units such as calculators or games. But for other applications such as one-of-a-kind equipment or equipment involving the production of only a few units, integrated circuits and even discrete components continue to find applications. Hence, they are not likely to be completely displaced by the microprocessor.

Furthermore, there are two types of electrical components that have not to date lent themselves to integrated-circuit procedures. The coil or inductor is one; the other is the large capacitor. The nature of the production process of ICs does not permit making large capacitors, although it is not useful to specify an upper limit at this point since it is possible that that limit can be extended by further development. In most cases, it is possible to design circuits that do not require such components. Where this is not possible, these components must be connected to the IC externally.

QUESTIONS

14–1 What is an integrated circuit (IC)? What is meant by large-scale integration?

14–2 Name some of the advantages of using MOS devices in integrated circuits.

14–3 What is the difference between a thick-film circuit and a thin-film circuit? Why are film circuits sometimes more advantageous than integrated circuits?

14–4 Explain why it is not necessarily desirable to produce an IC that resembles its discrete-component counterpart.

14–5 Explain why it is desirable to make the maximum number of interconnections internally and the minimum externally.

14–6 Explain why the op amp, though versatile, cannot achieve the miniaturization possible with digital circuits.

14–7 Explain why large capacitors present a problem with IC technology.

14–8 Why cannot a coil be produced with present IC technology?

14–9 What is the advantage of a voltage follower over an emitter-follower?

14–10 How do microprocessors carry miniaturization a step beyond LSI?

Appendix

Ratings and Characteristics

AC126 GERMANIUM ALLOY TRANSISTOR

pnp transistor in a TO-1 metal envelope intended for use in pre-amplifier or driver stages.

QUICK REFERENCE DATA

Collector-base voltage (open emitter)	$-V_{CBO}$	max.	32	V
Collector-emitter voltage (open base)	$-V_{CEO}$	max.	12	V
Collector current (dc)	$-I_C$	max.	100	mA
Total power dissipation up to T_{amb} = 45°C with cooling fin No. 56227 on a heatsink of at least 12.5 cm^2	P_{tot}	max.	500	mW
Junction temperature	T_j	max.	90	°C
dc current gain at T_{amb} = 25°C				
$-I_C = 2$ mA; $-V_{CE} = 5$ V	h_{FE}	>	65	
		typ.	140	

Small signal current gain at $T_{amb} = 25°C$		
$I_E = 2$ mA; $-V_{CB} = 5$ V; $f = 1$ kHz	h_{fe}	typ. 180 130 to 300
Transition frequency		
$-I_C = 10$ mA; $-V_{CE} = 2$ V	f_T	typ. 2.3 MHz

RATINGS

Voltages				
Collector-base voltage (open emitter)	$-V_{CBO}$	max.	32	V
Collector-emitter voltage (open base)	$-V_{CEO}$	max.	12	V
Collector-emitter voltage with $R_{BE} < 1$ kΩ	$-V_{CER}$	max.	32	V
Emitter-base voltage (open collector)	$-V_{EBO}$	max.	10	V
Currents				
Collector current (dc)	$-I_C$	max.	100	mA
Emitter current (peak value)	I_{EM}	max.	200	mA
Power dissipation				
Total power dissipation up to T_{amb} = 45°C with cooling fin No. 56227 mounted on a heatsink of at least 12.5 cm²	P_{tot}	max.	500	mW
Temperatures				
Storage temperature	T_{stg}		−55 to +90	°C
Junction temperature	T_j	max.	90	°C

THERMAL RESISTANCE

From junction to ambient in free air	$R_{th\ j\text{-}a}$	=	0.3 °C/mW
From junction to ambient with cooling fin No. 56227 mounted on a heatsink of at least 12.5 cm²	$R_{th\ j\text{-}a}$	=	0.09 °C/mW

CHARACTERISTICS $T_{amb} = 25°C$ unless otherwise specified

Collector cut-off current				
$I_E = 0$; $-V_{CB} = 10$ V	$-I_{CBO}$	<	10	μA
$I_E = 0$; $-V_{CB} = 32$ V; $T_j = 75°C$	$-I_{CBO}$	<	800	μA
Emitter cut-off current				
$I_C = 0$; $-V_{EB} = 5$ V; $T_j = 75°C$	$-I_{EBO}$	<	550	μA
Emitter-base voltage				
$I_E = 2$ mA; $-V_{CB} = 5$ V	V_{EB}	typ.	105	mV
$I_E = 100$ mA; $V_{CB} = 0$	V_{EB}	<	400	mV
dc current gain				
		>	65	
$-I_C = 2$ mA; $-V_{CE} = 5$ V	h_{FE}	typ.	140	
$-I_C = 50$ mA; $V_{CB} = 0$	h_{FE}	typ.	135	
$-I_C = 100$ mA; $V_{CB} = 0$	h_{FE}	typ.	105	

Collector capacitance at $f = 0.45$ MHz				
$I_E = I_e = 0; -V_{CB} = 5$ V	C_c	typ.	40	pF
		<	50	pF
Feedback impedance at $f = 0.45$ MHz				
$-I_C = 1$ mA; $-V_{CE} = 5$ V	$\lvert z_{rb} \rvert$	typ.	90	Ω
Transition frequency				
$-I_C = 10$ mA; $-V_{CE} = 2$ V	f_T	>	1.7	MHz
		typ.	2.3	MHz
Cut-off frequency				
$-I_C = 10$ mA; $-V_{CE} = 2$ V	f_{hfe}	>	10	kHz
		typ.	17	kHz
Noise figure at $f = 1$ kHz				
$-I_C = 0.5$ mA; $-V_{CE} = 5$ V; $R_S = 500\ \Omega$; Bandwidth = 200 Hz	F	typ.	4	dB
		<	10	dB
h parameters at $f = 1$ kHz				
$-I_C = 2$ mA; $-V_{CE} = 5$ V				
Input impedance	h_{ie}	typ.	2.4	kΩ
		1.7 to	3.8	kΩ
Reverse voltage transfer	h_{re}	typ.	8.0	10^{-4}
		<	13.0	10^{-4}
Small signal current gain	h_{fe}	typ.	180	
		130 to	300	
Output admittance	h_{oe}	typ.	100	$\mu\Omega^{-1}$
		<	170	$\mu\Omega^{-1}$

2N929 SILICON PLANAR TRANSISTOR

npn transistor in a TO-18 metal envelope with the collector connected to the case.

This device is primarily intended for use in high performance, low level, low noise amplifier applications both for direct current and for frequencies of up to 100 MHz

QUICK REFERENCE DATA

Collector-base voltage (open emitter)	V_{CBO}	max.	45	V
Collector-emitter voltage (open base)	V_{CEO}	max.	45	V
Collector current (peak value)	I_{CM}	max.	60	mA
Total dissipation up to $T_{amb} = 25°C$	P_{tot}	max.	300	mW
Junction temperature	T_j	max.	175	°C
dc current gain at $T_j = 25°C$				
$I_C = 10\ \mu A$; $V_{CE} = 5$ V	h_{FE}	40 to 120		
$I_C = 10$ mA; $V_{CE} = 5$ V	h_{FE}	100 to 350		

Transition frequency				
$I_C = 0.5$ mA; $V_{CE} = 5$ V	f_T	typ.	80	MHz
Noise figure ($f =$ 10 Hz to 15 kHz)				
$I_C = 10\ \mu$A; $V_{CE} = 5$ V; $R_S = 10$ kΩ	F	typ.	2.5	dB
		<	4	dB

RATINGS

Voltages				
Collector-base voltage (open emitter)	V_{CBO}	max.	45	V
Collector-emitter voltage (open base)	V_{CEO}	max.	45	V
Collector-emitter voltage at $V_{EB} = 0$	V_{CES}	max.	45	V
Emitter-base voltage (open collector)	V_{EBO}	max.	5	V
Currents				
Collector current (dc or average over any 50 ms period)	I_C	max.	30	mA
Collector current (peak value)	I_{CM}	max.	60	mA
Emitter current (dc or average over any 50 ms period)	$-I_E$	max.	35	mA
Emitter current (peak value)	$-I_{EM}$	max.	70	mA
Power dissipation				
Total power dissipation up to $T_{amb} =$ 25°C	P_{tot}	max.	300	mW
Temperatures				
Storage temperature	T_{stg}	−65 to +175		°C
Junction temperature	T_j	max.	175	°C

THERMAL RESISTANCE

From junction to ambient in free air	$R_{th\ j\text{-}a}$	=	0.5	°C/mW
From junction to case	$R_{th\ j\text{-}c}$	=	0.25	°C/mW

CHARACTERISTICS $T_j = 25$°C unless otherwise specified

Collector cut-off current				
$I_E = 0$; $V_{CB} = 45$ V	I_{CBO}	<	10	nA
$I_B = 0$; $V_{CE} = 5$ V	I_{CEO}	<	2	nA
$V_{EB} = 0$; $V_{CB} = 45$ V	I_{CES}	<	10	nA
Emitter cut-off current				
$I_C = 0$; $V_{EB} = 5$ V	I_{EBO}	<	10	nA
Emitter-base voltage				
$-I_E = 0.5$ mA; $V_{CB} = 5$ V	$-V_{EB}$	0.6 to	0.8	V
Saturation voltages				
$I_C = 10$ mA; $I_B = 0.5$ mA	$V_{CE\,sat}$	<	1	V
	$V_{BE\,sat}$	0.6 to	1	V

dc current gain			
$I_C = 10\ \mu A$; $V_{CE} = 5$ V	h_{FE}	40 to 120	
$I_C = 10\ \mu A$; $V_{CE} = 5$ V; $T_j = -55°C$	h_{FE}	> 10	
$I_C = 500\ \mu A$; $V_{CE} = 5$ V	h_{FE}	> 60	
$I_C = 10$ mA; $V_{CE} = 5$ V	h_{FE}	100 to 350	
Collector capacitance at $f = 1$ MHz			
$I_E = I_e = 0$; $V_{CB} = 5$ V	C_c	< 8	pF
Transition frequency			
$I_C = 0.5$ mA; $V_{CE} = 5$ V	f_T	> 50	MHz
Cut-off frequency			
$I_C = 0.5$ mA; $V_{CE} = 5$ V	f_{hfe}	> 200	kHz
Noise figure ($f = 10$ Hz to 15 kHz)			
$I_C = 10\ \mu A$; $V_{CE} = 5$ V; $R_S = 10$ kΩ	F	typ. 2.5	dB
		< 4	dB
h parameters at $f = 1$ kHz			
$I_C = 1$ mA; $V_{CE} = 5$ V			
Input impedance	h_{ie}	typ. 5.0	kΩ
Reverse voltage transfer	h_{re}	typ. 2.5	10^{-4}
Small signal current gain	h_{fe}	typ. 200 60 to 350	
Output admittance	h_{oe}	typ. 14	$\mu\Omega^{-1}$

1N3138 TUNNEL DIODE

gallium-arsenide high-speed tunnel diode
(Switching service)

MAXIMUM RATINGS:

Instantaneous forward current	i_F	100	max.	mA
Instantaneous reverse current	i_R	200	max.	mA
Dissipation at $T = 25°C$	P_{tot}	75	max.	mW
Ambient temperature	T_{amb}	−65 to +150		°C
Storage temperature	T_{stg}	−65 to +175		°C

ELECTRICAL CHARACTERISTICS ($T_{amb} = 25°C$):

STATIC:		*Min*	*Typ*	*Max*	
Peak point current	I_P	47.5	50	52.5	mA
Valley point current	I_V	—	2.5	3.5	mA
Peak point voltage	V_P	120	—	260	mV
Valley point voltage	V_V	510	—	620	mV
Positive voltage at peak point I_P = 52.5 mA	$V_{F'}$	1100	—	1400	mV

DYNAMIC:

Terminal valley point capacitance	C_V	—	10	30	$\mu\mu$F
Total series inductance	L_S	—	—	0.6	mμH
Total series resistance	R_S	—	—	2.6	Ω
Negative resistance	$-R_D$	—	2.6	—	Ω
Rise time	t_r	—	—	2	mμs
Figure of merit	$I_P/2C_{max}$	—	0.9	—	mA/$\mu\mu$F

2N3228 SILICON CONTROLLED RECTIFIER

Power control/switching service

MAXIMUM RATINGS:

Non-repetitive peak reverse voltage	V_{RSM}	300	V
Repetitive peak reverse voltage	V_{RRM}	200	V
Repetitive peak OFF-state voltage	V_{DRM}	600	V
ON-state current (dc)	I_T	3.2	A
Non-repetitive peak ON-state current	I_{TSM}	60	A
Rate of rise of ON-state current	dI_T/dt	200	A/μs
Average gate power dissipation	$P_{G(AV)}$	0.5	W
Peak gate power dissipation	P_{GM}	13	W
Storage temperature	T_{stg}	−40 to 125	°C
Junction temperature	T_j	−40 to 125	°C

ELECTRICAL CHARACTERISTICS (T_{amb} = 25°C):

		Min	*Typ*	*Max*	
Forward breakover voltage	$V_{(BO)}$	200	—	—	V
Peak OFF-state current	I_{DM}	—	0.1	1.5	mA
Repetitive peak reverse current	I_{RRM}	—	0.05	0.75	mA
Continuous ON-state voltage	V_T	—	2.15	2.8	V
Gate current	I_{GT}	—	8	15	mA
Gate-cathode trigger voltage	V_{GT}	—	1.2	2	V
Holding current	I_H	—	10	20	mA
Rate-of-rise of OFF-state voltage	dV_D/dt	10	200	—	V/μS
Turn-on time	t_{on}	0.75	1.5	—	μs
Turn-off time	t_q	—	15	50	μs
Thermal resistance from junction to case	θ_{Jc}	—	—	4	°C/W

References

1. Angelo, Jr., E. James, *Electronics: BJTs, FETs, and Microcircuits,* New York: McGraw-Hill Book Company, Inc., 1969.
2. Comer, Donald T., *Large-signal Transistor Circuits,* Englewood Cliffs, N.J.: Prentice-Hall, Inc., 1967.
3. Corning, John J., *Transistor Circuit Analysis and Design,* Englewood Cliffs, N.J.: Prentice-Hall, Inc., 1965.
4. Cowles, Laurence G., *Transistor Circuits and Applications,* Englewood Cliffs, N.J.: Prentice-Hall, Inc., 1968.
5. Cutler, Philip, *Semiconductor Circuit Analysis,* New York: McGraw-Hill Book Company, Inc., 1964.
6. *Transistor Manual, 7th ed.,* Syracuse, N.Y.: General Electric Company, Semiconductor Products Department, 1969.
7. Lenert, Louis H., *Semiconductor Physics, Devices and Circuits,* Columbus, Ohio: Charles E. Merrill Publishing Company, 1968.
8. Malvino, Albert Paul, *Transistor Circuit Approximations,* New York: McGraw-Hill Book Company, Inc., 1968.
9. Motorola, Inc., *Integrated Circuits,* New York: McGraw-Hill Book Company, Inc., 1965.
10. Pullen, Keats A., *Handbook of Transistor Circuit Design,* Englewood Cliffs, N.J.: Prentice-Hall, Inc., 1961.

11. *Transistor, Thyristor & Diode Manual,* Somerville, N.J.: RCA Corporation, Solid-state Division, 1971.
12. Ristenbatt, Marlin P., and Robert L. Riddle, *Transistor Physics and Circuits, 2nd ed.,* Englewood Cliffs, N.J.: Prentice-Hall, Inc., 1966.
13. Thomas, Harry E., *Handbook of Transistors, Semiconductors, Instruments, and Microelectronics,* Englewood Cliffs, N.J.: Prentice-Hall, Inc., 1968.

Index

C

D

E

F

N

O

P

Q

R

S

T

U

V